**Nonlinear Dynamics
of Nanosystems**

*Edited by
Günter Radons, Benno Rumpf,
and Heinz Georg Schuster*

Related Titles

H.G. Schuster (Ed.)

Reviews of Nonlinear Dynamics and Complexity
Volume 1

2008
ISBN 978-3-527-40729-3

H.G. Schuster (Ed.)

Reviews of Nonlinear Dynamics and Complexity
Volume 2

2009
ISBN 978-3-527-40850-9

E. Vedmedenko

Competing Interactions and Patterns in Nanoworld

2007
ISBN 978-3-527-40484-1

R. Kelsall, I.W. Hamley, M. Geoghegan (Eds.)

Nanoscale Science and Technology

2005
ISBN 978-0-470-85086-2

R. Waser (Ed.)

Nanoelectronics and Information Technology

2005
ISBN 978-3-527-40542-8

Nonlinear Dynamics of Nanosystems

Edited by
Günter Radons, Benno Rumpf,
and Heinz Georg Schuster

WILEY-VCH Verlag GmbH & Co. KGaA

The Editors

Günter Radons
Technische Universität Chemnitz
Institut für Physik
Reichenhainer Str. 70
09126 Chemnitz
Germany

Benno Rumpf
Technische Universität Chemnitz
Institut für Physik
Reichenhainer Str. 70
09126 Chemnitz
Germany

Heinz Georg Schuster
Illweg 10
66113 Saarbrücken
Germany

■ All books published by Wiley-VCH are carefully produced. Nevertheless, authors, editors, and publisher do not warrant the information contained in these books, including this book, to be free of errors. Readers are advised to keep in mind that statements, data, illustrations, procedural details or other items may inadvertently be inaccurate.

Library of Congress Card No.: applied for

British Library Cataloguing-in-Publication Data:
A catalogue record for this book is available from the British Library.

Bibliographic information published by the Deutsche Nationalbibliothek
The Deutsche Nationalbibliothek lists this publication in the Deutsche Nationalbibliografie; detailed bibliographic data are available on the Internet at <http://dnb.d-nb.de>.

© 2010 WILEY-VCH Verlag GmbH & Co. KGaA, Weinheim

All rights reserved (including those of translation into other languages). No part of this book may be reproduced in any form – by photoprinting, microfilm, or any other means – nor transmitted or translated into a machine language without written permission from the publishers. Registered names, trademarks, etc. used in this book, even when not specifically marked as such, are not to be considered unprotected by law.

Printed in the Federal Republic of Germany
Printed on acid-free paper

Cover Design   Schulz Grafik-Design, Fußgönheim
Typesetting   le-tex publishing services GmbH, Leipzig
Printing and Binding   betz-druck GmbH, Darmstadt

ISBN 978-3-527-40791-0

List of Contributors

Martin Aeschlimann
Christian-Albrechts-Universität
Institut für Experimentelle und
Angewandte Physik
24118 Kiel
Germany
ma@physik.uni-kiel.de

Sequoyah Aldridge
California Institute of Technology
Department of Physics
Pasadena, CA 91125
USA
sequoyah@caltech.edu

Atef Fadl Amin
Jacobs University Bremen
School of Engineering and Science
Campus Ring 1
28759 Bremen
Germany
a.amin@jacobs-university.de

Michael Bauer
Christian-Albrechts-Universität
Institut für Experimentelle und
Angewandte Physik
24118 Kiel
Germany
bauer@physik.uni-kiel.de

Daniela Bayer
University Kaiserslautern
Department of Physics and Research
Center OPTIMAS
67663 Kaiserslautern
Germany

Michael C. Cross
California Institute of Technology
Department of Physics
Pasadena, CA 91125
USA
mcc@caltech.edu

Cornelia Denz
Westfälische Wilhelms-Universität
Münster
Institute for Applied Physics and
Center for Nonlinear Science (CeNoS)
48149 Münster
Germany
denz@uni-muenster.de

Thorsten Emig
Universität zu Köln
Institut für Theoretische Physik
Zülpicher Strasse 77
50937 Köln
Germany
emig@lptms.u-psud.fr

Nonlinear Dynamics of Nanosystems. Edited by Günter Radons, Benno Rumpf, and Heinz Georg Schuster
Copyright © 2010 WILEY-VCH Verlag GmbH & Co. KGaA, Weinheim
ISBN: 978-3-527-40791-0

List of Contributors

Denis J. Evans
Australian National University
Research School of Chemistry
Canberra, ACT 0200
Australia
evans@rsc.anu.edu.au

Kerstin Falk
Universität Erlangen–Nürnberg
Institut für Theoretische Physik
Staudtstrasse 7
91058 Erlangen–Nürnberg
Germany
kerstin.falk@physik.uni-erlangen.de

Pierre Gaspard
Université Libre de Bruxelles
Center for Nonlinear Phenomena
and Complex Systems
Code Postal 231, Campus Plaine
1050 Brussels
Belgium
gaspard@ulb.ac.be

Dennis Göries
Westfälische Wilhelms-Universität
Münster
Institute for Applied Physics and
Center for Nonlinear Science (CeNoS)
48149 Münster
Germany
Dennis.Goeries@uni-muenster.de

Milena Grifoni
Universität Regensburg
Institut für Theoretische Physik
93040 Regensburg
Germany
milena.grifoni@physik.uni-regensburg.de

Takashi Hikihara
Kyoto University
Department of Electrical Engineering
Katsura, Nishikyo
615-8510 Kyoto
Japan
hikihara@kuee.kyoto-u.ac.jp

Jörg Imbrock
Westfälische Wilhelms-Universität
Münster
Institute for Applied Physics and
Center for Nonlinear Science (CeNoS)
48149 Münster
Germany
imbrock@uni-muenster.de

Franz J. Kaiser
Universität Augsburg
Institut für Physik
Universitätsstrasse 1
86135 Augsburg
Germany
franz.josef.kaiser@physik.uni-augsburg.de

Ulrich Kleinekathöfer
Jacobs University Bremen
School of Engineering and Science
Campus Ring 1
28759 Bremen
Germany
u.kleinekathoefer@jacobs-university.de

Sigmund Kohler
Instituto de Ciencia de
Materiales de Madrid (CSIS)
28049 Madrid
Spain
sigmund.kohler@physik.icmm.cics.es

List of Contributors

Joachim Krug
Universität zu Köln
Institut für Theoretische Physik
Zülpicher Strasse 77
50937 Köln
Germany
krug@thp.uni-koeln.de

Ron Lifshitz
Tel Aviv University
Raymond and Bervely Sackler School of Physics & Astronomy
Tel Aviv 69978
Israel
ronlif@tau.ac.il

Klaus Mecke
Universität Erlangen–Nürnberg
Institut für Theoretische Physik
Staudtstrasse 7
91058 Erlangen–Nürnberg
Germany
klaus.mecke@physik.uni-erlangen.de

Ciprian Padurariu
Jacobs University Bremen
School of Engineering and Science
Campus Ring 1
28759 Bremen
Germany

Markus Rauscher
Universität Stuttgart
ITAP
Pfaffenwaldring 57
70569 Stuttgart
Germany
rauscher@mf.mpg.de

Klaus Richter
Universität Regensburg
Institut für Theoretische Physik
93040 Regensburg
Germany
Klaus.Richter@physik.uni-regensburg.de

Patrick Rose
Westfälische Wilhelms-Universität Münster
Institute for Applied Physics and Center for Nonlinear Science (CeNoS)
48149 Münster
Germany
patrick.rose@uni-muenster.de

Debra J. Searles
Griffith University
Queesland Micro- and Nanotechnology Centre and School of Biomolecular and Physical Sciences
Brisbane, Qld 4111
Australia
D.Bernhardt@griffith.edu.au

Eckehard Schöll
Technische Universität Berlin
Institut für Theoretische Physik
Hardenbergstrasse 36
10623 Berlin
Germany
schoell@physik.tu-berlin.de

Bernd Terhalle
Westfälische Wilhelms-Universität Münster
Institute for Applied Physics and Center for Nonlinear Science (CeNoS)
48149 Münster
Germany
bernd.terhalle@uni-muenster.de

Daniel Waltner
Universität Regensburg
Institut für Theoretische Physik
93040 Regensburg
Germany
Daniel.Waltner@physik.uni-regensburg.de

Carsten Wiemann
Institut für Festkörperforschung (IFF),
Elektronische Eigenschaften
Forschungszentrum Jülich
52425 Jülich
Germany

Stephen R. Williams
Australian National University
Research School of Chemistry
Canberra, ACT 0200
Australia
swilliams@rsc.anu.edu.au

Kohei Yamasue
Kyoto University
Department of Electronic Science and
Engineering
Katsura, Nishikyo
615-8510 Kyoto
Japan
k-yamasue@kuee.kyoto-u.ac.jp

Contents

List of Contributors V

Preface XVII

Part I Fluctuations

1 Nonequilibrium Nanosystems 1
Pierre Gaspard
1.1 Introduction 1
1.2 Statistical Thermodynamics of Nonequilibrium Nanosystems 4
1.2.1 From Newton's Equations to Stochastic Processes 4
1.2.2 Entropy and the Second Law of Thermodynamics 10
1.2.3 Identifying the Nonequilibrium Constraints and the Currents with Graph Analysis 11
1.2.4 Fluctuation Theorem for the Currents 13
1.2.5 Consequences for Linear and Nonlinear Response Coefficients 15
1.2.6 Temporal Disorder 16
1.2.7 Nanosystems Driven by Time-Dependent Forces 18
1.3 Mechanical Nanosystems 21
1.3.1 Friction in Double-Walled Carbon Nanotubes 21
1.3.1.1 Translational Friction 23
1.3.1.2 Rotational Friction 28
1.3.2 Electromagnetic Heating of Microplasmas 30
1.3.2.1 The Undriven System and Its Hamiltonian 30
1.3.2.2 The Driven System and the Fluctuation Theorem 31
1.4 Mechanochemical Nanosystems 32
1.4.1 F_1-ATPase Motor 32
1.4.2 Continuous-State Description 35
1.4.3 Discrete-State Description 41
1.5 Chemical Nanosystems 45
1.5.1 Chemical Transistor 46
1.5.2 Chemical Multistability 50
1.5.3 Chemical Clocks 53

1.5.4	Chemical Clocks Observed in Field Emission Microscopy	56
1.5.5	Single-Copolymer Processes	60
1.5.5.1	Copolymerization without a Template	61
1.5.5.2	Copolymerization with a Template	63
1.5.5.3	DNA Replication	64
1.6	Conclusions and Perspectives	65
	References	71

2 Thermodynamics of Small Systems 75
Denis J. Evans, Stephen R. Williams, and Debra J. Searles

2.1	Introduction	75
2.2	Thermostated Dynamical Systems	76
2.3	The Transient Fluctuation Theorem	79
2.4	Thermodynamic Interpretation of the Dissipation Function	82
2.5	The Dissipation Theorem	84
2.6	Nonequilibrium Work Relations	86
2.7	Nonequilibrium Work Relations for Thermal Processes	91
2.8	Corollaries of the Fluctuation Theorem and Nonequilibrium Work Relations	94
2.8.1	Generalized Fluctuation Theorem	94
2.8.2	Integrated Fluctuation Theorem	94
2.8.3	Second Law Inequality	95
2.8.4	Nonequilibrium Partition Identity	96
2.8.5	The Steady State Fluctuation Theorem	97
2.8.6	Minimum Average Work Principle	100
2.9	Experiments	100
2.10	Conclusion	102
	References	107

3 Quantum Dissipative Ratchets 111
Milena Grifoni

3.1	Introduction to Microscopic Ratchets	111
3.2	The Feynman Ratchet	113
3.3	Tunneling Ratchets: Temperature Driven Current Reversal	114
3.4	Rocked Ratchets in the Deep Quantum Regime	116
3.5	Rocked Shallow Ratchets	118
3.6	Spin Ratchets	119
	References	120

Part II Surface Effects

4 Dynamics of Nanoscopic Capillary Waves 121
Klaus Mecke, Kerstin Falk, and Markus Rauscher

4.1	Stochastic Hydrodynamics	122
4.1.1	Stochastic Interfaces	122
4.1.2	Acoustic Waves	124

4.1.3	Capillary Waves	*125*
4.1.4	Linearized Stochastic Hydrodynamics	*126*
4.2	Surface Tension at Nanometer Length Scales: Effect of Long Range Forces and Bending Energies	*129*
4.3	Thermal Noise Influences Fluid Flow in Nanoscopic Films	*132*
4.3.1	Dynamics of the Film Thickness	*133*
4.3.2	Comparison with Experiments	*135*
4.3.3	Linearized Stochastic Thin Film Equation	*136*
	References	*141*
5	**Nonlinear Dynamics of Surface Steps**	*143*
	Joachim Krug	
5.1	Introduction	*143*
5.2	Electromigration-Driven Islands and Voids	*143*
5.2.1	Electromigration of Single Layer Islands	*144*
5.2.2	Continuum vs. Discrete Modeling	*147*
5.2.3	Nonlocal Shape Evolution: Two-Dimensional Voids	*150*
5.2.4	Nonlocal Shape Evolution: Vacancy Islands with Terrace Diffusion	*151*
5.3	Step Bunching on Vicinal Surfaces	*152*
5.3.1	Stability of Step Trains	*153*
5.3.2	Strongly and Weakly Conserved Step Dynamics	*154*
5.3.3	Continuum Limit, Traveling Waves and Scaling Laws	*155*
5.3.4	A Dynamic Phase Transition	*157*
5.3.5	Coarsening	*159*
5.3.6	Nonconserved Dynamics	*160*
5.3.7	Beyond the Quasistatic Approximation	*161*
5.4	Conclusions	*162*
	References	*162*
6	**Casimir Forces and Geometry in Nanosystems**	*165*
	Thorsten Emig	
6.1	Casimir Effect	*166*
6.2	Dependence on Shape and Geometry	*168*
6.2.1	Deformed Surfaces	*169*
6.2.2	Lateral Forces	*176*
6.2.3	Cylinders	*180*
6.2.4	Spheres	*186*
6.3	Dependence on Material Properties	*187*
6.3.1	Lifshitz Formula	*188*
6.3.2	Nanoparticles: Quantum Size Effects	*189*
6.4	Casimir Force Driven Nanosystems	*192*
6.5	Conclusion	*199*
	References	*199*

Part III Nanoelectromechanics

7 The Duffing Oscillator for Nanoelectromechanical Systems 203
Sequoyah Aldridge
- 7.1 Basics of the Duffing Oscillator 203
- 7.2 NEMS Resonators and Their Nonlinear Properties 205
- 7.3 Transition Dynamics of the Duffing Resonator 208
- 7.4 Energy for "Uphill" Type Transitions 210
- 7.5 Energy Calculation Using a Variational Technique 214
- 7.6 Frequency Tuning 216
- 7.7 Bifurcation Amplifier 217
- 7.8 Conclusion 218
  - References 218

8 Nonlinear Dynamics of Nanomechanical Resonators 221
Ron Lifshitz and M.C. Cross
- 8.1 Nonlinearities in NEMS and MEMS Resonators 221
  - 8.1.1 Why Study Nonlinear NEMS and MEMS? 222
  - 8.1.2 Origin of Nonlinearity in NEMS and MEMS Resonators 222
  - 8.1.3 Nonlinearities Arising from External Potentials 223
  - 8.1.4 Nonlinearities Due to Geometry 224
- 8.2 The Directly-Driven Damped Duffing Resonator 227
  - 8.2.1 The Scaled Duffing Equation of Motion 227
  - 8.2.2 A Solution Using Secular Perturbation Theory 228
  - 8.2.3 Addition of Other Nonlinear Terms 235
- 8.3 Parametric Excitation of a Damped Duffing Resonator 236
  - 8.3.1 Driving Below Threshold: Amplification and Noise Squeezing 239
  - 8.3.2 Linear Instability 241
  - 8.3.3 Nonlinear Behavior Near Threshold 242
  - 8.3.4 Nonlinear Saturation above Threshold 245
  - 8.3.5 Parametric Excitation at the Second Instability Tongue 247
- 8.4 Parametric Excitation of Arrays of Coupled Duffing Resonators 250
  - 8.4.1 Modeling an Array of Coupled Duffing Resonators 250
  - 8.4.2 Calculating the Response of an Array 252
  - 8.4.3 The Response of Very Small Arrays and Comparison of Analytics and Numerics 255
  - 8.4.4 Response of Large Arrays and Numerical Simulation 257
- 8.5 Amplitude Equation Description for Large Arrays 258
  - 8.5.1 Amplitude Equations for Counter Propagating Waves 259
  - 8.5.2 Reduction to a Single Amplitude Equation 260
  - 8.5.3 Single Mode Oscillations 261
  - References 263

9	**Nonlinear Dynamics in Atomic Force Microscopy and Its Control for Nanoparticle Manipulation** *267*	
	Kohei Yamasue and Takashi Hikihara	
9.1	Introduction *267*	
9.2	Operation of Dynamic Mode Atomic Force Microscopy *269*	
9.3	Nonlinear Dynamics and Control of Cantilevers *270*	
9.3.1	Nonlinear Oscillation and Its Influence on Imaging *270*	
9.3.2	Model of a Cantilever under Tip–Sample Interaction *272*	
9.3.3	Application of Time-Delayed Feedback Control *273*	
9.3.4	Experimental Setup for Control of Nonlinear Cantilever Dynamics *274*	
9.3.4.1	Circuit Implement of Time-Delayed Feedback Control *274*	
9.3.4.2	Frequency Response of Magnetic Actuators and Deflection Sensors *275*	
9.3.5	Experimental Demonstration of the Stabilization of Cantilever Oscillations *275*	
9.4	Manipulation of Single Atoms at Material Surfaces *277*	
9.4.1	Model of Single Atoms and Molecules *277*	
9.4.2	Analysis Based on an Action-Angle Formulation *279*	
9.4.3	Dynamics of Single Atoms Induced by Probes *281*	
9.4.4	Control of Manipulation *283*	
9.5	Concluding Remarks *283*	
	References *284*	
Part IV	**Nanoelectronics**	
10	**Classical Correlations and Quantum Interference in Ballistic Conductors** *287*	
	Daniel Waltner and Klaus Richter	
10.1	Introduction: Quantum Transport through Chaotic Conductors *287*	
10.2	Semiclassical Limit of the Landauer Transport Approach *289*	
10.3	Quantum Transmission: Configuration Space Approach *291*	
10.3.1	Diagonal Contribution *292*	
10.3.2	Nondiagonal Contribution *293*	
10.3.3	Magnetic Field Dependence of the Nondiagonal Contribution *297*	
10.3.4	Ehrenfest Time Dependence of the Nondiagonal Contribution *298*	
10.4	Quantum Transmission: Phase Space Approach *299*	
10.4.1	Phase Space Approach *299*	
10.4.2	Calculation of the Full Transmission *301*	
10.5	Semiclassical Research Paths: Present and Future *303*	
	References *304*	
11	**Nonlinear Response of Driven Mesoscopic Conductors** *307*	
	Franz J. Kaiser and Sigmund Kohler	
11.1	Introduction *307*	
11.2	Wire-Lead Model and Current Noise *308*	
11.2.1	Charge, Current, and Current Fluctuations *310*	

11.2.2	Full Counting Statistics 311
11.3	Master Equation Approach 312
11.3.1	Perturbation Theory and Reduced Density Operator 312
11.3.2	Computation of Moments and Cumulants 313
11.3.3	Floquet Decomposition 315
11.3.3.1	Fermionic Floquet Operators 315
11.3.3.2	Master Equation and Current Formula 316
11.3.4	Spinless Electrons 318
11.4	Transport under Multi-Photon Emission and Absorption 318
11.4.1	Electron Pumping 319
11.4.2	Coherent Current Suppression 320
11.5	Conclusions 322
	References 323

12	**Pattern Formation and Time Delayed Feedback Control at the Nanoscale** 325
	Eckehard Schöll
12.1	Introduction 325
12.2	Control of Chaotic Domain and Front Patterns in Superlattices 329
12.3	Control of Noise-Induced Oscillations in Superlattices 333
12.4	Control of Chaotic Spatiotemporal Oscillations in Resonant Tunneling Diodes 341
12.5	Noise-Induced Spatiotemporal Patterns in Resonant Tunneling Diodes 350
12.6	Conclusion 361
	References 363

Part V Optic-Electronic Coupling

13	**Laser-Assisted Electron Transport in Nanoscale Devices** 369
	Ciprian Padurariu, Atef Fadl Amin, and Ulrich Kleinekathöfer
13.1	Open Quantum Systems 370
13.1.1	Quantum Master Equation Approach 371
13.1.2	Time-Local and Time-Nonlocal Master Equations 373
13.1.3	Full Counting Statistics 377
13.2	Model System Describing Molecular Wires and Quantum Dots 385
13.3	The Single Resonant Level Model 391
13.4	Influence of Laser Pulses 398
13.5	Summary and Outlook 403
	References 403

14	**Two-Photon Photoemission of Plasmonic Nanostructures with High Temporal and Lateral Resolution** 407
	Michael Bauer, Daniela Bayer, Carsten Wiemann, and Martin Aeschlimann
14.1	Introduction 407

14.2	Experimental *410*	
14.3	Results and Discussion *414*	
14.3.1	Localized Surface Plasmons Probed by TR-2PPE *414*	
14.3.2	Single Particle Plasmon Spectroscopy by Means of Time-Resolved Photoemission Microscopy *417*	
14.4	Conclusion *423*	
	References *424*	
15	**Dynamics and Nonlinear Light Propagation in Complex Photonic Lattices** *427*	
	Bernd Terhalle, Patrick Rose, Dennis Göries, Jörg Imbrock, and Cornelia Denz	
15.1	Introduction *427*	
15.2	Wave Propagation in Periodic Photonic Structures *428*	
15.2.1	Linear Propagation *429*	
15.2.2	Nonlinear Propagation *430*	
15.3	Optically-Induced Photonic Lattices in Photorefractive Media *431*	
15.3.1	Mathematical Description of Photorefractive Photonic Lattices *431*	
15.3.2	Experimental Configuration for Photorefractive Lattice Creation *432*	
15.4	Complex Optically-Induced Lattices in Two Transverse Dimensions *433*	
15.4.1	Triangular Lattices *434*	
15.4.2	Multiperiodic Lattices *437*	
15.5	Vortex Clusters *440*	
15.5.1	Necessary Stability Criterion *441*	
15.5.2	Compensation of Anisotropy in Hexagonal Photonic Lattices *441*	
15.5.3	Ring-Shaped Vortex Clusters *442*	
15.5.4	Multivortex Clusters *446*	
15.6	Summary and Outlook *447*	
	References *448*	

Index *451*

Preface

Nanotechnology has rapidly developed in recent years, making it possible to engineer mechanical, optical and electronic devices that are the size of only a few hundred atomic diameters. The equations of motion of such nanoscopic systems are generically nonlinear and frequently operate in a regime where a linear approximation is not justified. The comprehension of nonlinear dynamical processes in nanosystems is a new field of research that is certainly of considerable technological importance.

Miniaturization leads to new effects that radically change the dynamical properties. Thus, nanoscopic systems do not operate in the same way as their macroscopic counterparts. In particular, *scaling effects, stochasticity*, and *quantum effects* distinguish nanosystems from macroscopic systems. *Scaling* the size of a physical system changes the dominant forces, for example, macroscopic systems are frequently dominated by bulk effects that are proportional to mass. In contrast, surface effects such as adhesion and surface tension add new sources of nonlinearity in small systems. *Stochasticity* of nanoscopic devices is caused by thermal motion at the atomic level. Nanosystems and their environment form a high-dimensional chaotic system by the strong nonlinearity of interatomic forces. By their nature as averages, thermodynamic laws cannot be directly applied to systems that are so small that the motion of single atoms is relevant. *Quantum mechanics* enters macroscopic physics via large ensemble averages, while nanoscopic devices can directly achieve quantum states.

These effects are both an impediment and an opportunity for applications. On one hand, devices may not work when they are simply scaled down, for example, a miniaturized electric motor may be locked by adhesive forces and can be randomly forced by thermal motion. On the other hand, nanosystems can perform functions which cannot be achieved with larger devices. Current research is increasingly concerned with two fundamental objectives The first is the *analysis of nonlinear dynamical effects in nanosystems*. As it turns out, nonlinear effects cannot be suppressed in many nanoscopic applications, and indeed, they offer new opportunities in engineering. *Control of nonlinear nanosystems* is therefore the fundamental task for applications.

This book introduces the crucial fields of nonlinear dynamics of nanosystems. The topics cover a wide range of current research in this field. It includes 15 reviews organized in the five parts: *Fluctuations, Surface effects, Nanoelectromechanics, Nanoelectronics, and Optic-electronic coupling*.

Part 1: Fluctuations

The chapters by Gaspard and by Evans, Williams, and Searles survey the extremely important recent generalizations of the second law of thermodynamics to a group of theorems that are significant for such small systems. Gaspard treats the statistical thermodynamics of nonequilibrium nanosystems with mechanical, mechanochemical, and chemical applications. Evans, Williams and Searles present the fluctuation theorem, the nonequilibrium work relation and the dissipation theorem. Regarding applications, randomness seems to impede the directional motion of nanosystems. The chapter by Grifoni introduces the theoretical framework of Brownian motors, that is, engines that make use of thermal noise and quantum tunneling in order to achieve directed motion.

Part 2: Surface Effects

Surface effects are fundamentally important in nanoscopic systems. Mecke, Falk and Rauscher show that the effect of stochasticity is particularly important for interfaces and films in contrast to bulk fluid mechanics in hydrodynamics. They discuss the new phenomena in films at the nanometer scale resulting from thermal fluctuations. Surfaces of solids show a rich dynamic behavior. Krug investigates the intrinsic nonlinear dynamics of surface steps, and in particular of single layer islands under external forces. Casimir and van der Waals forces are fundamentally important quantum mechanical surface effects in small systems, as Emig shows.

Part 3: Nanoelectromechanics

Oscillating micro- and nanoscopic beams are exploited in atomic force spectroscopy. Nanomechanical oscillators such as miniaturized cantilevers or carbon nanotubes can easily be excited into a strongly nonlinear regime with an amplitude-dependent frequency. As the chapters by Aldridge and by Lifshitz and Cross show, nanoelectromechanical systems are experimental realizations of the Duffing oscillator. The frequency of cantilever oscillations increases for a decreasing system size. Therefore, in the near future mechanical oscillators may be manufactured for which energy quantization is relevant. The control of the nonlinear dynamics of nanomechanical devices is a crucial task for future applications of force microscopy, as Yamasue and Hikihara show. In particular, this implies the possibility of catching and releasing single atoms from a surface.

Part 4: Nanoelectronics

Richter and Waltner study the fundamental connection of classical chaotic dynamics of charge carriers with quantum wave interference and hence, discuss the transition of electronic transport from micro- to nanoscales. Kaiser and Kohler give a detailed investigation of the emergence of nonlinearity in electric conductivity, giving a quantum mechanical analysis of the Coulomb blockade. Schöll investigates nonlinearities in the conductance and the resulting patterns. Time delayed feedback is demonstrated to be a powerful tool for the control of these structures.

Part 5: Optic-Electronic Coupling

Optic-electronic coupling is an important method of controlling electronic devices. Padurariu, Amin and Kleinekathöfer analyze the electron flux through molecular junctions and quantum dots. Their means of controlling the dynamics at the nanoscale is the use of ultra-short laser pulses. Bauer, Bayer, Wiemann and Aeschlimann employ localized surface plasmons as a tool to probe nanoscopic devices. Resonance of plasma oscillations gives information about the size and shape of the device or particle. The interaction of optical waves and complex photonic lattices is discussed by Terhalle, Rose, Göries, Imbrock and Denz.

This book reflects an ongoing interdisciplinary discussion initiated by a Volkswagen-Symposium on the same topic that was held in Chemnitz in 2006. The editors thank the Volkswagen Foundation for financial support.

Chemnitz and Kiel, December 2009

Günter Radons
Benno Rumpf
Heinz Georg Schuster

Part I Fluctuations

1
Nonequilibrium Nanosystems

Pierre Gaspard

1.1
Introduction

The nanometer is the length scale just above that of atoms. Accordingly, the nanoscale is the basis of higher order structures made of atoms: molecules, macromolecules, polymers, fullerenes and nanotubes, atomic or molecular clusters, supramolecular assemblies, molecular machines, and even viruses, organelles or cells in the organic world, and gases, liquids, or solids in the inorganic world. It should be emphasized that the importance of these atomic systems lies not only in their 3D spatial structure, but also in the 4D spatiotemporal paths they can execute, as is the case for catalysts or molecular motors.

In principle, their motions are ruled by Newton's equations based on molecular forces which typically have nonlinear dependences on the interatomic distances. A key feature of atomic motions is their randomness that results from the incessant collisions among the atoms or molecules composing the nanosystem. This randomness manifests itself in the thermal and molecular fluctuations affecting, to some extent, every observable at the nanoscale. Accordingly, nanosystems are often described in terms of stochastic processes, as is the case for the Brownian motion of micrometric particles suspended in a liquid. In this example, the forces between the Brownian particle and the molecules of the surrounding liquid are random on long time scales, the heavy Brownian particle being much slower than the light molecules of the liquid. Accordingly, the Newtonian equation for the Brownian particle contains a Langevin fluctuating force because of the interaction with its surroundings. Since Brownian motion is stochastic, its description is based on a probability distribution which obeys a time-evolution equation called the master equation. As already pointed out by Einstein [1], Brownian particles are examples of mesoscopic systems which are larger than the molecules obeying microscopic Newtonian dynamics, but smaller than the macroscopic systems where the molecular fluctuations are so small with respect to their size that a deterministic description should be considered. In this regard, the stochastic description developed for Brownian motion is expected to apply at the nanoscale as well. A stochastic ap-

Nonlinear Dynamics of Nanosystems. Edited by Günter Radons, Benno Rumpf, and Heinz Georg Schuster
Copyright © 2010 WILEY-VCH Verlag GmbH & Co. KGaA, Weinheim
ISBN: 978-3-527-40791-0

proach is valid if one or more degrees of freedom of the nanosystem is intrinsically weighted heavier than the others. This results in a separation of time scales in the system that occurs between the slow degrees of freedom and the others responsible for the fast thermal fluctuations.

Because of the interaction between the slow and fast degrees of freedom, energy is exchanged. The asymmetry of the coupling between a few slow and many fast degrees of freedom leads to an energy flux from the former to the latter, which is the phenomenon of energy dissipation. Dissipation happens, for instance, if an excess of energy is initially deposited on the slow degrees of freedom and progressively dissipated over the fast degrees of freedom during a relaxation which is determined by the interaction. If the subsystem is interacting with the many degrees of freedom of a thermal bath at a given temperature, its probability distribution undergoes a relaxation towards the equilibrium Boltzmann–Maxwell distribution at the temperature of the bath. During this relaxation, the subsystem is transiently out of equilibrium, though ultimately reaches thermal equilibrium after enough time. These relaxation processes occur in isolated nanosystems such as atomic or molecular clusters where the slow degrees of freedom are associated with the spherical or nonspherical shape of the cluster, and the fast degrees of freedom describe the motion of the individual atoms relative to the global shape. Statistical ensembles of clusters can be described in terms of probability distribution for all of the internal degrees of freedom of the cluster, allowing several possible distributions for the total energy depending on the experimental technique producing the beam of clusters [2]. Such nanosystems remain out of equilibrium during some relaxation time, though finally reach the thermodynamic equilibrium state after a long enough time [3]. Strictly speaking, the concepts of equilibrium state or relaxation times are associated with the probability distribution and its time evolution. The probability distribution describes a statistical ensemble of copies of the nanosystem, each launched from different initial conditions statistically distributed according to the initial probability density. Consequently, the concepts of equilibrium states or relaxation times do not apply to individual nanosystems, but instead ensembles composed of infinitely many copies of the nanosystem with statistically distributed degrees of freedom.

In addition to the aforementioned nanosystems which relax towards an equilibrium state, nanosystems exist which are in contact with at least two heat or particle reservoirs at different temperatures or chemical potentials. These nanosystems present the remarkable feature of reaching a *nonequilibrium steady state* after some transient behavior. Contrary to the previous cases, such nanosystems sustain currents or fluxes of heat or particles and remain out of equilibrium due to a supply of energy from the external reservoirs. Although the instantaneous currents fluctuate in time, they are described by a probability distribution which remains stationary in the nonequilibrium steady state. The mean values of the fluctuating currents are not vanishing and controlled by the differences of temperatures or chemical potentials between the external reservoirs. These mean currents are sustained at the expense of energy dissipation. Therefore, such nonequilibrium nanosystems are characterized by a positive entropy production according to the second law of thermodynamics. In contrast, the entropy production vanishes in the equilibrium stationary state reached by nanosystems in contact with a single heat or particle reservoir. Examples of nonequilibrium nanosystems include the electronic circuits considered in semiconductor or molecular electronics, the chemical nanoreactors in heterogeneous catalysis, or the molecular motors in biology. These examples show the variety of nonequilibrium nanosystems and their importance for nanoscience and nanotechnology.

Further nonequilibrium nanosystems are those which are driven by a time-dependent external force. Examples of such nanosystems are macromolecules such as RNA undergoing repeated unfolding and folding processes by optical tweezers [4] or nanosystems with electric charges driven by electromagnetic fields. In these cases as well, the energy supplied by the external forces is dissipated during the process, leading to thermodynamic entropy production.

The nonequilibrium nanosystems are also of fundamental importance in biology [5]. Indeed, one of the key features of biological systems is their metabolism, meaning that biological systems are functioning out of equilibrium as open thermodynamic systems with an internal dissipation of the chemical energy from the nutrients supplied by the environment. The thermodynamic aspects of metabolism are traditionally envisaged at the macroscale. However, the biological systems are hierarchically structured from the nanoscale up to the macroscale. Indeed, molecules, such as lipids, form cellular membranes while copolymers, such as proteins, RNA, and DNA, combine into supramolecular assemblies functioning as machines: polymerases, ribosomes, flagellar motors, linear motors for cargo transport or muscle contraction. Many of these molecular structures exist only because of their ability to perform a specific motion powered by some energy source as provided by transmembrane pH differences or the hydrolysis of adenosine triphosphate (ATP). In this regard, energy transduction plays a fundamental role at the molecular level in all the biochemical processes of metabolism [6]. The directionality so essential to biological functions is acquired at the nanoscale when the molecular structures are driven out of equilibrium by metabolism. In this respect, the time scale over which a correlated motion can be maintained in some 3D molecular structure is here an essential property characterizing a biological function [7–10]. Thus, the nonequilibrium nanosystems find their importance not only for technological applications, but also for our fundamental understanding of biological systems from the viewpoint of the physico-chemical laws of nature.

The purpose of the present contribution is to give an overview of nonequilibrium nanosystems and to outline their statistical thermodynamics.

In Section 1.2, the statistical thermodynamics of nanosystems is presented starting from the problem of their multi-scale description with, on the one hand, Newton's equations ruling the microscopic dynamics of their constituent atoms over the scales of picometers and femtoseconds and, on the other hand, stochastic processes describing the motion of some of their degrees of freedom on the spatial scales of nanometers or larger and over the time scales of picoseconds or longer. Recent advances in statistical thermodynamics are reviewed, such as the fluctuation theorems that are large-deviation relations for the fluctuations of nonequilibrium work, currents, or other quantities [11–27], as well as a new relationship established between the thermodynamic entropy production and the breaking of time reversal in the property of temporal disorder [28–31]. The latter is at the basis of a new understanding of information processing at the molecular level [32, 33].

Section 1.3 is devoted to mechanical nanosystems and, in particular, the study of friction in double-walled carbon nanotubes. This is an example of an isolated nanosystem evolving towards an equilibrium stationary state because it is not powered by a continuous energy supply.

In Section 1.4, the case of mechanochemical nanosystems is considered. These nanosystems – such as the F_1-ATPase nanomotor – are continuously powered by chemical energy and can thus be driven into nonequilibrium stationary states. Due to their large molecular architecture, the mechanics of these molecular motors can be tightly coupled to their chemistry, allowing sustained rotary or linear motions under nonequilibrium conditions.

Section 1.5 touches on the existence of chemical nanosystems such as chemical clocks in far-from-equilibrium oscillatory regimes. In such systems, the directionality is maintained in a noisy limit cycle of the populations of small molecules involved in a network of coupled chemical reactions.

Conclusions and perspectives are drawn in Section 1.6.

1.2
Statistical Thermodynamics of Nonequilibrium Nanosystems

1.2.1
From Newton's Equations to Stochastic Processes

The same nanosystem may be described in several different ways depending upon the spatial and temporal scales at which its motion is observed.

At room temperature, the dynamics of atoms can be supposed to be classical in many circumstances. Under these conditions, the microscopic dynamics are ruled by Newton's equations for all the atoms of masses $\{m_a\}_{a=1}^N$ and positions $\{r_a\}_{a=1}^N$ composing the system. These atoms are coupled by interatomic forces $F(r_a - r_b) = -\nabla U(r_a - r_b)$, deriving from the Born–Oppenheimer potential energy $U(r_a - r_b)$ of the interaction between the atoms in the electronic quantum state of the molecular system. Besides, an external force can be applied to the system, $F_{\text{ext}}(r_a) = -\nabla U_{\text{ext}}(r_a)$. The sum over all the forces acting on the atom is thus equal to its acceleration multiplied by its mass

$$m_a \frac{d^2 r_a}{dt^2} = F_{\text{ext}}(r_a) + \sum_{b(\neq a)} F(r_a - r_b), \qquad (1.1)$$

where $a, b = 1, 2, \ldots, N$ are the labels of all the atoms composing not only the nanosystem, but also the reservoirs which are in contact with it if the latter is not isolated. Equivalently, Hamilton's equations

$$\begin{cases} \dfrac{d\mathbf{r}_a}{dt} = +\dfrac{\partial H}{\partial \mathbf{p}_a} \\ \dfrac{d\mathbf{p}_a}{dt} = -\dfrac{\partial H}{\partial \mathbf{r}_a} \end{cases} \tag{1.2}$$

govern the time evolution of the positions $\{\mathbf{r}_a\}_{a=1}^{N}$ and momenta $\{\mathbf{p}_a = m_a d\mathbf{r}_a/dt\}_{a=1}^{N}$ of all the atoms in the system. This formulation is expressed in terms of the Hamiltonian function which represents the total energy of the whole system

$$H = \sum_{1 \leq a \leq N} \left[\frac{\mathbf{p}_a^2}{2 m_a} + U_{\mathrm{ext}}(\mathbf{r}_a) \right] + \sum_{1 \leq a < b \leq N} U(\mathbf{r}_a - \mathbf{r}_b) \,. \tag{1.3}$$

The Hamiltonian dynamics are deterministic in the sense that, according to Cauchy's theorem, a unique trajectory is issued from initial conditions taken as a point in the phase space of the positions and momenta

$$\boldsymbol{\Gamma} = (\mathbf{r}_1, \mathbf{r}_2, \ldots, \mathbf{r}_N, \mathbf{p}_1, \mathbf{p}_2, \ldots, \mathbf{p}_N) \,. \tag{1.4}$$

Therefore, the time evolution of a time-independent system is given by a flow, for example, a one-parameter continuous group defined in the phase space:

$$\boldsymbol{\Gamma}(t) = \boldsymbol{\Phi}^t \left[\boldsymbol{\Gamma}(0) \right]. \tag{1.5}$$

Moreover, the Hamiltonian system (1.3) is symmetric under the time reversal defined by the operation

$$\boldsymbol{\Theta}(\mathbf{r}_1, \mathbf{r}_2, \ldots, \mathbf{r}_N, \mathbf{p}_1, \mathbf{p}_2, \ldots, \mathbf{p}_N) = (\mathbf{r}_1, \mathbf{r}_2, \ldots, \mathbf{r}_N, -\mathbf{p}_1, -\mathbf{p}_2, \ldots, -\mathbf{p}_N) \tag{1.6}$$

because the Hamiltonian (1.3) is an even function of the momenta. Accordingly, the time reversal of every solution of Hamilton's equations (1.2) is also a solution, a property called *microreversibility*. It is fundamental to notice that microreversibility does not necessarily imply the coincidence of a trajectory with its time reversal so that the selection of initial conditions can break the time reversal symmetry of the actual history followed by the system [34–37].

Since the phase space is a continuum, the real numbers (1.4) defining the initial conditions are practically known by their few first digits so that the effective knowledge of the initial conditions is always limited. Therefore, an error always affects the preparation of initial conditions launching a trajectory. This inherent limitation of the knowledge of initial conditions taking their values in a continuum justifies the introduction of a probability distribution for the initial positions and momenta compatible with the precision with which they are prepared, $p_0(\boldsymbol{\Gamma})$. This probability distribution evolves in time according to the Liouville equation

$$\partial_t p = \{H, p\}_{\mathrm{Poisson}} \equiv \hat{L} p \,, \tag{1.7}$$

where the Poisson bracket with the Hamiltonian defines the Liouvillian operator $\hat{L}$. According to Liouville's theorem, the probability density at time t is given in terms of the initial probability density by

$$p(\pmb{\Gamma}, t) = p_0\left[\pmb{\Phi}^{-t}(\pmb{\Gamma})\right], \tag{1.8}$$

which defines the so-called Perron–Frobenius operator [38]. If the total system is isolated, the probability distribution may converge in the weak sense towards a stationary probability distribution defining an invariant probability measure, which should correspond to the thermodynamic equilibrium state, $p_{\text{eq}}(\pmb{\Gamma})$. The condition for this weak convergence is the property of mixing [39, 40]. We notice that nonequilibrium states can also be defined as conditionally invariant measures by suitably renormalizing the transient probability distribution evolving in time under given nonequilibrium constraints [38]. In this way, conditionally invariant measures have been constructed in the escape-rate theory or in the hydrodynamic-mode theory [38].

The determinism of Hamiltonian systems does not preclude the possibility of dynamical randomness, for example, temporal disorder in the long-time evolution of such systems. This dynamical randomness finds its origin in the sensitivity to initial conditions. We notice that this property manifests itself in the so-called chaotic systems, but does not appear in integrable systems having as many constants of motion as degrees of freedom. The sensitivity to initial conditions is characterized by the positivity of at least one Lyapunov exponent [41]. These latter quantities are the rates of exponential separation

$$\lambda_i = \lim_{t \to \infty} \frac{1}{t} \ln \frac{\|\delta\pmb{\Gamma}_i(t)\|}{\|\delta\pmb{\Gamma}_i(0)\|} \tag{1.9}$$

between a reference trajectory (1.5) and perturbed trajectories issued from infinitesimally close initial conditions $\pmb{\Gamma}(0) + \delta\pmb{\Gamma}_i(0)$ taken in any possible direction i in the $6N$-dimensional phase space (1.4). In molecular dynamics, typical Lyapunov exponents are of the order of the inverse of the intercollisional time which corresponds to the time scale of the thermal fluctuations [42–45]. The dynamical instability characterized by positive Lyapunov exponents implies that trajectories issued from nearby initial conditions may have very different histories which are thus unpredictable beyond the time scale given by the inverse of the maximal Lyapunov exponent. Over time scales longer than this Lyapunov horizon of predictability, the trajectory appears random, listing in time the digits of the real numbers defining its initial conditions. Therefore, dynamical randomness can be characterized as temporal disorder in terms of the so-called Kolmogorov–Sinai entropy per unit time which is equal to the sum of positive Lyapunov exponents according to Pesin's theorem [41]

$$h_{\text{KS}} = \sum_{\lambda_i > 0} \lambda_i . \tag{1.10}$$

This property of temporal disorder manifests itself in the stochasticity of the random processes describing the slow degrees of freedom of the system where it is characterized by the (ε, τ)-entropy per unit time [46]. Accordingly, the motion of atoms in condensed phases at room temperature is highly random as observed by their thermal and molecular fluctuations.

Methods have been developed in chaos theory to construct solutions of Liouville's equation, in particular, conditionally invariant measures, by using the Lyapunov exponents associated with each trajectory, as it is the case in the periodic-orbit theory [38, 47]. The idea is that the larger the positive Lyapunov exponents of a trajectory, the higher its instability and the lower its probability weight. Methods based on this idea allow us to construct exact solutions of Liouville's equation on fine scales in the phase space of the system. In this regard, the methods of chaos theory fundamentally justify the existence of relaxation times which are intrinsic to the dynamics.

In many systems, these relaxation times can be obtained with excellent approximations thanks to the coarse-grained descriptions established by the pioneering work of Boltzmann [48], Einstein [1], Langevin [49], Fokker [50], Planck [51], Pauli [52], and others. Such coarser descriptions focus on a few relevant observables among all the degrees of freedom of the total system. Examples are given by the indicator functions of subsets ω taken inside the phase space: $I_\omega(\Gamma) = 1$ if $\Gamma \in \omega$ and zero otherwise. The probability that the system visits this subset at the time t is given by the mean value of this observable taken over the phase-space probability distribution (1.8) at the time

$$P(\omega, t) \equiv \int I_\omega(\Gamma) p(\Gamma, t) d\Gamma . \tag{1.11}$$

If these probabilities evolve slower than the other observables, the memory of the fast degrees of freedom may be lost over the time scale of variation of these probabilities, which may justify that their time evolution is ruled by a Markovian master equation such as

$$\frac{dP(\omega, t)}{dt} = \sum_{\rho, \omega'} \left[P(\omega', t) W_{+\rho}(\omega'|\omega) - P(\omega, t) W_{-\rho}(\omega|\omega') \right] , \tag{1.12}$$

where $W_\rho(\omega'|\omega)$ is the rate of the transition $\omega' \xrightarrow{\rho} \omega$ induced by some elementary mechanism ρ [53–56]. The relaxation times of the stochastic process ruled by this master equation can be obtained in terms of the eigenvalues of this equation. It is interesting to note that stochastic processes have a dual description either in terms of the probabilities ruled by the master equation or in terms of individual random realizations of the time evolution as simulated, for instance, by Gillespie's algorithm [57, 58]. Such random realizations are paths in the space of the coarse-grained states $\{\omega\}$

$$\boldsymbol{\omega} = \omega_0 \xrightarrow{\rho_1} \omega_1 \xrightarrow{\rho_2} \omega_2 \xrightarrow{\rho_3} \cdots \xrightarrow{\rho_n} \omega_n , \tag{1.13}$$

with random jumps $\omega_{j-1} \xrightarrow{\rho_j} \omega_j$ between dwelling time intervals $t_j < t < t_{j+1}$, during which the system stays in the state ω_j. By construction, such a random path should be statistically equivalent to the path that would be obtained from the deterministic trajectory starting from some compatible initial conditions: $\omega\left[\boldsymbol{\Gamma}(t)\right] = \omega\left\{\boldsymbol{\Phi}^t\left[\boldsymbol{\Gamma}(0)\right]\right\}$.

In the case of Brownian motion in an external force field $\boldsymbol{F}_{\text{ext}}(\boldsymbol{r})$, random paths can be simulated by integrating Langevin's equation

$$m\frac{d^2\boldsymbol{r}}{dt^2} = \boldsymbol{F}_{\text{ext}}(\boldsymbol{r}) - \zeta\frac{d\boldsymbol{r}}{dt} + \boldsymbol{F}_{\text{L}}(t)\,, \tag{1.14}$$

which is a Newtonian equation where the friction force $\boldsymbol{F}_{\text{frict}} = -\zeta d\boldsymbol{r}/dt$ and the associated fluctuating Langevin force $\boldsymbol{F}_{\text{L}}(t)$ represent the contributions of the forces between the Brownian particle and the molecules of the surrounding fluid [59]. The Langevin forces can be modeled as Gaussian white noises

$$\langle F_{\text{L},i}(t)\rangle = 0 \tag{1.15}$$

$$\langle F_{\text{L},i}(t) F_{\text{L},j}(t')\rangle = 2\zeta k_{\text{B}} T \delta(t-t')\delta_{ij}\,, \tag{1.16}$$

where $i,j = x,y,z$ denote the Cartesian components of the force, ζ the friction coefficient, T the temperature of the surrounding fluid, and $k_{\text{B}} = 1.38 \times 10^{-23}$ J/K Boltzmann's constant. The Langevin equation (1.14) is a stochastic differential equation. Its solutions $\boldsymbol{r}(t)$ mimic the motion of the Brownian particle as given by typical solutions of Hamilton's equations (1.2), $\boldsymbol{r}(t) = \boldsymbol{r}_1\left[\boldsymbol{\Gamma}(t)\right] = \boldsymbol{r}_1\left\{\boldsymbol{\Phi}^t\left[\boldsymbol{\Gamma}(0)\right]\right\}$, supposing that the Brownian particle has the label $a = 1$ among the N particles of the system.

In the stochastic model by Langevin, the time correlation functions of the fluctuating force coming from the fluid are delta-correlated, meaning that the time over which the correlation functions decay to zero is much shorter than the time scale of the described process. This correlation time is of the order of the intercollisional time of the Brownian particle with surrounding molecules. We should notice that the friction coefficient can generally be calculated in terms of the integral of the time correlation function of the fluctuating force according to the Kirkwood formula [60]

$$\zeta = \frac{1}{2k_{\text{B}} T}\int_{-\mathcal{T}}^{+\mathcal{T}} \langle F_{\text{L},i}(t) F_{\text{L},i}(0)\rangle\, dt\,, \tag{1.17}$$

where $\mathcal{T}$ is a time scale longer than the correlation time but shorter than the time over which the conservation of the total linear momentum of the total system might manifest itself. If the system is infinite, the limit $\mathcal{T} \to \infty$ may be taken. The Kirkwood formula for the friction coefficient has been extended to the famous Green–Kubo formulas for the coefficients of transport properties such as the viscosities, the conductivities, as well as the diffusivities [61–63].

The master equation corresponding to the Langevin equation (1.14) is the Fokker–Planck equation

$$\frac{\partial \mathcal{P}}{\partial t} + \boldsymbol{v}\cdot\frac{\partial \mathcal{P}}{\partial \boldsymbol{r}} + \frac{\boldsymbol{F}_{\text{ext}}}{m}\cdot\frac{\partial \mathcal{P}}{\partial \boldsymbol{v}} = \frac{\zeta}{m}\frac{\partial}{\partial \boldsymbol{v}}\cdot(\boldsymbol{v}\mathcal{P}) + \frac{\zeta k_{\text{B}} T}{m^2}\frac{\partial^2 \mathcal{P}}{\partial \boldsymbol{v}^2}\,, \tag{1.18}$$

where $\mathcal{P}$ denotes the probability density in order to find the Brownian particle with the position r and the velocity v at the time t [59]. This probability density corresponds in principle to the probability distribution obeying Liouville's equation (1.7) according to

$$\mathcal{P}(r, v, t) \equiv \int \delta(r - r_1) \, \delta(v - p_1/m) \, p(\Gamma, t) \, d\Gamma \, . \tag{1.19}$$

In this regard, the delta distributions play a similar role as the indicator function in (1.11).

In the case where the external force is time independent and derives from a potential $F_{\text{ext}}(r) = -\nabla U_{\text{ext}}(r)$, the solution of the Fokker–Planck equation undergoes a relaxation towards an asymptotic equilibrium state given by the Boltzmann–Maxwell distribution

$$\lim_{t \to \infty} \mathcal{P}(r, v, t) = \mathcal{P}_{\text{eq}}(r, v) \equiv \mathcal{N} \exp\left[-\frac{mv^2}{2k_B T} - \frac{U_{\text{ext}}(r)}{k_B T}\right] \tag{1.20}$$

with a normalization constant $\mathcal{N}$ such that $\int \mathcal{P}_{\text{eq}}(r, v) \, drdv = 1$.

In contrast, if the external force is time dependent, the system remains out of equilibrium. If the Brownian particle is dragged by an optical trap moving at the velocity u, the external potential is given by $U_{\text{ext}} = (\kappa/2)(r - ut)^2$ and the probability density can reach a nonequilibrium stationary solution in the frame of the optical trap. In this nonequilibrium state, the energy supplied by the optical trap is dissipated by the friction of the Brownian particle on the surrounding fluid, leading to a positive thermodynamic entropy production.

Brownian motion is the paradigm of physico-chemical stochastic processes. This paradigm can be extended down to the nanoscale and applied to mechanical systems such as multiwalled carbon nanotubes as well as to molecular motors where mechanics is coupled to chemistry. Multiwalled carbon nanotubes have slow and fast degrees of freedom and thus qualify for a description in terms of stochastic processes (see Section 1.3). In molecular motors, the stochastic process is a combination of diffusive mechanical motions interrupted by random jumps between discrete chemical states due to reactive events. Such diffusion-reaction stochastic processes are governed by coupled Fokker–Planck equations (see Section 1.4).

The paradigm also extends to mesoscopic chemical systems where reactions transform populations of molecules. In mesoscopic chemical systems, such as nanoreactors or nanoelectrodes, the numbers of molecules are random variables jumping at each reactive event. Therefore, the molecular numbers obey a stochastic process compatible with the mass-action law of chemical kinetics [53–56]. At the macroscale, the molecular fluctuations disappear and the chemical concentrations follow deterministic differential equations of chemical kinetics. At the mesoscale, chemical systems can be described as continuous-time jump processes ruled by a master equation for the probability $P(\{N_i\}, t)$ of finding N_i molecules of species $i = 1, 2, \ldots, c$ in the system, or as diffusive processes ruled by a Fokker–Planck equation for the probability density $\mathcal{P}(\{x_i\}, t)$ defined in the space of chemical concentrations $x_i = N_i/N$ where $N = \sum_{i=1}^{c} N_i$ is the total number of molecules in the system (see Section 1.5).

Master equations have also been deduced for quantum systems at the nanoscale [64].

1.2.2
Entropy and the Second Law of Thermodynamics

Since its historical origin in the pioneering work by Sadi Carnot on the efficiency of steam engines, the concept of entropy is associated with the idea of partitioning the system into microscopic degrees of freedom having their own dynamics and macroscopic ones which can be manipulated at will. In a steam engine, the former are the degrees of freedom of the water molecules and the latter the piston and the valves of the engine. In this regard, the thermodynamic entropy appears as a property of the system of microscopic degrees of freedom with respect to their manipulation by a coarser device which is external to the described system. A priori, the thermodynamic entropy is thus a property of the system with respect to a coarse graining superimposed by some external apparatus.

Accordingly, the concept of entropy applies to nanosystems described in terms of the probabilities (1.11) to visit some coarse-grained states ω. The thermodynamic entropy associated at time t can be defined as

$$S(t) = \sum_{\omega} S(\omega) P(\omega, t) - k_B \sum_{\omega} P(\omega, t) \ln P(\omega, t) . \tag{1.21}$$

The first term is the mean contribution of the entropy $S(\omega)$ due to the statistical distribution of all the degrees of freedom which are not specified by the coarse-grained state ω [20]. For instance, if the coarse-grained state ω only specifies the numbers of the molecules of the different chemical species, $S(\omega)$ is the entropy of the statistical distribution of the positions and momenta of the particles enumerated by ω. The second term characterizes the disorder in statistical distribution $P(\omega, t)$ over the different coarse-grained states $\{\omega\}$.

Since the probability distribution $\{P(\omega, t)\}$ evolves in time according to the master equation (1.12), the entropy (1.21) varies accordingly. It is well known that the time variation of the entropy can be decomposed as [65, 66]

$$\frac{dS}{dt} = \frac{d_e S}{dt} + \frac{d_i S}{dt} \tag{1.22}$$

into the entropy flow or entropy exchange between the system and its environment and the entropy production which is internal to the system. The *entropy flow* is given by

$$\begin{aligned}\frac{d_e S}{dt} = \sum_{\rho,\omega,\omega'} & \left[P(\omega', t) W_{+\rho}(\omega'|\omega) - P(\omega, t) W_{-\rho}(\omega|\omega') \right] \\ & \times \left[S(\omega) - \frac{k_B}{2} \ln \frac{W_{+\rho}(\omega'|\omega)}{W_{-\rho}(\omega|\omega')} \right],\end{aligned} \tag{1.23}$$

which can be either positive or negative. On the other hand, the *entropy production*

$$\frac{d_i S}{dt} = \frac{k_B}{2} \sum_{\rho,\omega,\omega'} \left[P(\omega', t) W_{+\rho}(\omega'|\omega) - P(\omega, t) W_{-\rho}(\omega|\omega') \right]$$
$$\times \ln \frac{P(\omega', t) W_{+\rho}(\omega'|\omega)}{P(\omega, t) W_{-\rho}(\omega|\omega')} \geq 0$$
(1.24)

is always nonnegative, in agreement with the second law of thermodynamics [65, 66]. In a system without external nonequilibrium constraint, the probability distribution $P(\omega, t)$ undergoes a relaxation towards the equilibrium state $P_{eq}(\omega)$ for which the entropy production vanishes because of the detailed balancing conditions

$$P_{eq}(\omega') W_{+\rho}(\omega'|\omega) = P_{eq}(\omega) W_{-\rho}(\omega|\omega'),$$
(1.25)

which should hold for all the possible transitions $\omega' \xrightarrow{\rho} \omega$ [67, 68]. During relaxation, the system is transiently out of equilibrium so that the entropy production is positive. The entropy production vanishes asymptotically as the time goes to infinity and the thermodynamic equilibrium is reached.

If external nonequilibrium constraints are imposed on the system, the relaxation proceeds towards a nonequilibrium steady state, $(d/dt) P_{neq}(\omega) = 0$, in which the detailed balancing conditions (1.25) do not hold and the entropy production remains positive. Therefore, the thermodynamic entropy production allows us to distinguish between nonequilibrium and equilibrium steady states among all the stationary solutions of the master equation (1.12) such that $(d/dt) P_{st}(\omega) = 0$.

1.2.3
Identifying the Nonequilibrium Constraints and the Currents with Graph Analysis

The nonequilibrium constraints are the control parameters driving the nanosystem out of equilibrium. These control parameters are the differences of temperatures or chemical potentials between the heat or particle reservoirs in contact with the nanosystem. In the case of chemical reactions, the difference of chemical potentials is taken between the reactants and the products of each reaction and are controlled by chemical concentrations. These control parameters are hidden in the transition rates $W_\rho(\omega'|\omega)$ of the stochastic process and it is of great important to identify them.

A systematic method is provided with graph theory, as developed by Hill and Schnakenberg [6, 65]. A graph is associated with the stochastic process as follows. Each state ω of the system defines a vertex or node of the graph while each allowed transition $\omega \underset{-\rho}{\overset{+\rho}{\rightleftharpoons}} \omega'$ corresponds to an edge. In this respect, two states can be connected by several edges if several elementary processes ρ allow transitions between them.

An orientation is assigned to each edge of the graph G. The directed edges are thus defined by

$$e \equiv \omega \xrightarrow{\rho} \omega'. \tag{1.26}$$

Let F be a directed subgraph of G. The orientation of the subgraph F with respect to its edges $\{e\}$ is described by introducing the quantity

$$\varsigma_e(F) \equiv \begin{cases} +1 & \text{if } e \text{ and } F \text{ are parallel} \\ -1 & \text{if } e \text{ and } F \text{ are antiparallel} \\ 0 & \text{if } e \text{ is not in } F \end{cases} \tag{1.27}$$

where e and F are said to be parallel (respectively antiparallel) if F contains the edge e in its reference (respectively opposite) orientation.

In order to identify all the cycles of a graph, a concept of maximal tree is introduced [65]. Every maximal tree $T(G)$ of the graph G should satisfy the following properties:

1. $T(G)$ is a covering subgraph of G, that is, $T(G)$ contains all the vertices of G and all the edges of $T(G)$ are edges of G;
2. $T(G)$ is connected;
3. $T(G)$ contains no cycle (i.e., no cyclic sequence of edges).

The edges l of the graph G which do not belong to the maximal tree $T(G)$ are called the chords of $T(G)$. If we add a chord l to $T(G)$, the resulting subgraph $T(G) + l$ contains exactly one cycle C_l, which is obtained from $T(G) + l$ by removing all the edges which are not part of the cycle. The orientation is taken such that $\varsigma_l(C_l) = 1$, that is, the cycles are oriented as the chords l. A maximal tree $T(G)$ together with its associated fundamental set of cycles $\{C_1, C_2, \ldots, C_l, \ldots\}$ provides a decomposition of the graph G.

We notice that a given graph G has several maximal trees $T(G)$ and that all the maximal trees of a graph can be obtained by linear combinations of a given maximal tree $T(G)$ with its associated cycles, as described in [65].

A remarkable property is that the ratio of the products of the transition rates $W_\rho(\omega'|\omega)$ along the two possible directions of any cycle C_l of the graph is independent of the states composing the cycle and will thus only depend on the external nonequilibrium constraints imposed to the system [65]. Thanks to this property, the thermodynamic forces, also called the affinities [69, 70], can be introduced according to [65]

$$\prod_{e \in C_l} \frac{W_{+\rho}(\omega|\omega')}{W_{-\rho}(\omega'|\omega)} = \exp A(C_l), \tag{1.28}$$

where $e \in C_l$ denotes the edges (1.26) in the cycle C_l. In the equilibrium state, the affinities vanish and we recover the conditions of detailed balancing between every forward and backward transition. An important observation is that many of these affinities are equal, $A(C_l) = A_\alpha$ for all $C_l \in \alpha$, which defines the macroscopic affinities A_α imposed by the external reservoirs.

The instantaneous current on the chord l is defined by [22]

$$j_l(t) \equiv \sum_{n=-\infty}^{+\infty} \varsigma_l(e_n)\delta(t-t_n),\qquad(1.29)$$

where t_n is the time of the random transition e_n during a path of the stochastic process. The convention is used that j_l is oriented as the graph G since $\varsigma_l(e_n)$ is equal to $(-)1$ if the transition e_n is (anti)parallel to the chord l. The current (1.29) is a fluctuating random variable. The different microscopic currents corresponding to a given macroscopic process α can now be regrouped as [22]

$$j_\alpha(t) \equiv \sum_{l\in\alpha} j_l(t) = \sum_{l\in\alpha}\sum_{n=-\infty}^{+\infty} \varsigma_l(e_n)\delta(t-t_n).\qquad(1.30)$$

Examples of nonequilibrium stochastic processes described by the Markovian master equation (1.12) and their graph analysis are provided in the following sections.

1.2.4
Fluctuation Theorem for the Currents

In the previous framework, a fundamental result can be obtained for the full counting statistics of the fluctuating currents (1.30) which are flowing across the nanosystem in some nonequilibrium steady state. The generating function of all the statistical cumulants of the fluctuating currents is defined as

$$Q(\lambda,A) \equiv \lim_{t\to\infty} -\frac{1}{t}\ln\left\langle\exp\left[-\lambda\cdot\int_0^t dt'\, j(t')\right]\right\rangle\qquad(1.31)$$

with $\lambda = \{\lambda_\alpha\}$, $A = \{A_\alpha\}$, and $j(t) = \{j_\alpha(t)\}$. We notice that the generating function depends on the affinities because the statistical average $\langle\cdot\rangle$ is carried out in the steady state corresponding to the values A of the affinities. The mean value of a current is given by differentiating the generating function with respect to the parameter λ_α and afterward setting all these parameters to zero:

$$J_\alpha(A) \equiv \left.\frac{\partial Q}{\partial \lambda_\alpha}\right|_{\lambda=0} = \lim_{t\to\infty}\frac{1}{t}\int_0^t \langle j_\alpha(t')\rangle\, dt'.\qquad(1.32)$$

The diffusivities or second cumulants of the fluctuating currents can be defined as

$$D_{\alpha\beta}(A) \equiv -\frac{1}{2}\left.\frac{\partial^2 Q}{\partial \lambda_\alpha \partial \lambda_\beta}\right|_{\lambda=0} = \frac{1}{2}\int_{-\infty}^{+\infty}\langle[j_\alpha(t)-\langle j_\alpha\rangle][j_\beta(0)-\langle j_\beta\rangle]\rangle\, dt.$$

$$(1.33)$$

Higher-order statistical cumulants can be defined similarly such as the third and fourth cumulants

$$C_{\alpha\beta\gamma}(\mathbf{A}) \equiv \left.\frac{\partial^3 Q}{\partial\lambda_\alpha \partial\lambda_\beta \partial\lambda_\gamma}\right|_{\lambda=0} \quad (1.34)$$

$$B_{\alpha\beta\gamma\delta}(\mathbf{A}) \equiv -\frac{1}{2}\left.\frac{\partial^4 Q}{\partial\lambda_\alpha \partial\lambda_\beta \partial\lambda_\gamma \partial\lambda_\delta}\right|_{\lambda=0} \quad (1.35)$$

which all characterize the full counting statistics of the coupled fluctuating currents $\{j_\alpha(t)\}$.

We have the fluctuation theorem for the currents:

Theorem 1.1: Fluctuation Theorem for Currents

The generating function (1.31) obeys the symmetry relation

$$Q(\lambda, \mathbf{A}) = Q(\mathbf{A} - \lambda, \mathbf{A}) . \quad (1.36)$$

This theorem has been proved in the framework of graph theory [21–23] and can also be proved for open quantum systems as a consequence of microreversibility [26, 27].

The Legendre transform of the generating function (1.31)

$$H(\xi, \mathbf{A}) = \text{Max}_\lambda \left[Q(\lambda, \mathbf{A}) - \lambda \cdot \xi\right] \quad (1.37)$$

is the decay rate of the probability that the fluctuating currents averaged over the finite time interval t have their values in the range $(\xi, \xi + d\xi)$, yielding

$$H(\xi, \mathbf{A}) \equiv \lim_{t\to\infty} -\frac{1}{t} \ln P\left[\frac{1}{t}\int_0^t j(t')\,dt' \simeq \xi\right], \quad (1.38)$$

where P denotes the probability distribution of the nonequilibrium steady state corresponding to the affinities $\mathbf{A}$. In terms of these decay rates, the fluctuation theorem (1.36) can be written as

$$H(-\xi, \mathbf{A}) - H(\xi, \mathbf{A}) = \mathbf{A} \cdot \xi , \quad (1.39)$$

which means that the ratio of the probabilities that the fluctuating currents take oppositive values behaves exponentially in time with a rate equal to the affinities $\mathbf{A}$ multiplied by the supposed values ξ for the currents:

$$\frac{P\left[\frac{1}{t}\int_0^t j(t')\,dt' \simeq +\xi\right]}{P\left[\frac{1}{t}\int_0^t j(t')\,dt' \simeq -\xi\right]} \simeq \exp(\mathbf{A} \cdot \xi\, t) \quad \text{for} \quad t \to \infty . \quad (1.40)$$

If the fluctuating currents take their mean values (1.32), $\xi = J$, the decay rate vanishes by the law of large numbers, $H(J, \mathbf{A}) = 0$, so that (1.40) shows that the probability of the opposite values $-J$ decays at a rate equal to

$$\frac{1}{k_B}\frac{d_i S}{dt} = \mathbf{A} \cdot J \geq 0 , \quad (1.41)$$

which is the well-known expression of the entropy production in nonequilibrium thermodynamics [70]. This relation shows that the fluctuation theorem provides an extension of the second law of thermodynamics to small systems. At equilibrium, the affinities vanish with the currents and the thermodynamic entropy production, as expected. When not at equilibrium, the fluctuation theorem (1.40) shows that an asymmetry appears between the probabilities of opposite fluctuations: the farther from equilibrium, the lower the probability of reversed fluctuations. Since the ratio of probabilities depends exponentially on the time and the affinities, the reversed fluctuations rapidly become negligible as the system is driven far from equilibrium. Ultimately, the probability of the reversed fluctuations vanishes in fully irreversible regimes where the entropy production is infinite.

1.2.5
Consequences for Linear and Nonlinear Response Coefficients

Typically, the currents flowing across the nanosystem have a nonlinear dependence on the affinities. It is only if the nonequilibrium constraints are weak and the system remains close to equilibrium that the currents may have a linear dependence on the affinities. This is the case for transport properties such as heat conductivity, viscosity, or diffusion in macroscopic fluids. However, nonlinearities tend to manifest themselves in nanosystems because of their inherent heterogeneities. These nonlinearities are well known in chemical reactions which are completed after the breaking of chemical bonds over subnanometric distances [54, 70]. Accordingly, we should expect that nanosystems might present highly nonlinear properties.

The affinities are the thermodynamic forces driving the system out of equilibrium. In this regard, they represent the control parameters probing the responses of the system to external perturbations. If the perturbations are weak, the system remains in the linear regime around its state of thermodynamic equilibrium. If the perturbations are stronger, the effects of the nonlinear responses become observable. Therefore, the response properties of the system with respect to the nonequilibrium constraints can be defined by expanding the currents in powers of the affinities as

$$J_\alpha = \sum_\beta L_{\alpha,\beta} A_\beta + \frac{1}{2} \sum_{\beta,\gamma} M_{\alpha,\beta\gamma} A_\beta A_\gamma + \frac{1}{6} \sum_{\beta,\gamma,\delta} N_{\alpha,\beta\gamma\delta} A_\beta A_\gamma A_\delta + \cdots \quad (1.42)$$

The linear response of the currents J_α with respect to a small perturbation in the affinities A_β is characterized by the Onsager coefficients $L_{\alpha,\beta}$ and the nonlinear response by the higher-order coefficients $M_{\alpha,\beta\gamma}$, $N_{\alpha,\beta\gamma\delta}$, ...

Since the currents can be deduced from the generating function (1.31) according to (1.32), any symmetry of the generating function will imply special relations among the linear and nonlinear response coefficients in the expansion (1.42). This is the case for the symmetry relation given by the fluctuation theorem (1.36). Indeed, the response coefficients can be found by differentiating the relation (1.36) with respect to the parameters λ and the affinities A.

The linear response coefficients $L_{\alpha,\beta}$ are given by differentiating twice the generating function with respect to λ_α and A_β. By using the fluctuation theorem (1.36), the linear response coefficients can be shown to be equal to the diffusivities (1.33) taken at equilibrium, $L_{\alpha,\beta} = D_{\alpha\beta}(0)$. We notice that this is the content of the fluctuation-dissipation theorem and the Green–Kubo formulas [61–63]. Since the diffusivities are symmetric under the exchange of the indices α and β, we recover Onsager's reciprocity relations [71]

$$L_{\alpha,\beta} = L_{\beta,\alpha}. \tag{1.43}$$

The remarkable result is that this method can proceed to higher orders, leading to new relations between the nonlinear response coefficients and quantities characterizing the nonequilibrium fluctuations of the currents [21, 72]. The second-order response coefficients can be related to the diffusivities according to

$$M_{\alpha,\beta\gamma} = \left(\frac{\partial D_{\alpha\beta}}{\partial A_\gamma} + \frac{\partial D_{\alpha\gamma}}{\partial A_\beta} \right)_{A=0}. \tag{1.44}$$

Similarly, the third-order response coefficients turn out to be related to the fourth and second cumulants by

$$N_{\alpha,\beta\gamma\delta} = \left(\frac{\partial^2 D_{\alpha\beta}}{\partial A_\gamma \partial A_\delta} + \frac{\partial^2 D_{\alpha\gamma}}{\partial A_\beta \partial A_\delta} + \frac{\partial^2 D_{\alpha\delta}}{\partial A_\beta \partial A_\gamma} - \frac{1}{2} B_{\alpha\beta\gamma\delta} \right)_{A=0}, \tag{1.45}$$

while the third and fourth cumulants are linked by

$$B_{\alpha\beta\gamma\delta}(0) = \left(\frac{\partial C_{\alpha\beta\gamma}}{\partial A_\delta} \right)_{A=0}. \tag{1.46}$$

Such relations exist at arbitrary orders as consequences of the fluctuation theorem [72]. Similar relations can be deduced in the presence of an external magnetic field [26]. They characterize the nonlinear response properties of nonequilibrium nanosystems.

1.2.6
Temporal Disorder

At the nanoscale, the currents are fluctuating either at equilibrium or out of equilibrium. These fluctuations are the manifestation of dynamical randomness due to the incessant collisions among the particles composing the system. This dynamical randomness can be characterized as a property of disorder in the successive pictures of the system in movies of the stochastic process. The time series of the fluctuating currents can be analyzed and its temporal disorder characterized by an entropy per unit time. Such a quantity is defined as an (ε, τ)-entropy per unit time for the fluctuating signal sampled with a resolution ε and a sampling time τ [46]. In deterministic dynamical systems, the (ε, τ)-entropy per unit time would converge to the Kolmogorov–Sinai entropy per unit time in the limit where ε goes to zero.

Such dynamical entropies are the decay rates of the probabilities

$$P(\boldsymbol{\omega}) = P(\omega_0 \omega_1 \omega_2 \ldots \omega_{n-1}) \tag{1.47}$$

that the system follows given paths or histories $\boldsymbol{\omega} = \omega_0 \omega_1 \omega_2 \ldots \omega_{n-1}$ where the symbols ω_j are the coarse-grained states observed with the resolution ε at the successive times $t = j\tau$ with $j = 0, 1, 2, \ldots, n-1$. In (1.47), P denotes the stationary probability distribution of the process. Because of the temporal disorder, the probability of a typical path is known to decay exponentially at a rate defining the (ε, τ)-entropy per unit time [46]

$$h(\varepsilon, \tau) \equiv \lim_{n \to \infty} -\frac{1}{n\tau} \sum_{\boldsymbol{\omega}} P(\boldsymbol{\omega}) \ln P(\boldsymbol{\omega}) . \tag{1.48}$$

In nonequilibrium steady states, we expect that the time-reversal symmetry is broken at the level of the invariant probability distribution so that a path $\boldsymbol{\omega} = \omega_0 \omega_1 \omega_2 \ldots \omega_{n-1}$ and its time reversal $\boldsymbol{\omega}^R = \omega_{n-1} \ldots \omega_2 \omega_1 \omega_0$ should have different probabilities [34–38]

$$\text{out of equilibrium:} \quad P(\boldsymbol{\omega}) \neq P(\boldsymbol{\omega}^R) . \tag{1.49}$$

Accordingly, the probability of a time-reversed path should decay at a rate which is different from the entropy per unit time (1.48). This observation motivates the introduction of the time-reversed (ε, τ)-entropy per unit time [28]

$$h^R(\varepsilon, \tau) \equiv \lim_{n \to \infty} -\frac{1}{n\tau} \sum_{\boldsymbol{\omega}} P(\boldsymbol{\omega}) \ln P(\boldsymbol{\omega}^R) , \tag{1.50}$$

where the average is still carried out with the path probabilities themselves. If the average was performed with the time-reversed path probabilities in (1.50), we would recover the quantity (1.48) because the sum over the paths $\boldsymbol{\omega}$ is equivalent to the sum over their reversals $\boldsymbol{\omega}^R$. In this regard, the time-reversed entropy per unit time characterizes the temporal disorder of the time-reversed paths among the set of the typical paths of the forward process.

The remarkable result is that the difference between the time-reserved and the standard entropies per unit time is equal to the thermodynamic entropy production

$$\frac{1}{k_B} \frac{d_i S}{dt} = \lim_{\varepsilon, \tau \to 0} \left[h^R(\varepsilon, \tau) - h(\varepsilon, \tau) \right] \geq 0 , \tag{1.51}$$

as can be shown for several classes of nonequilibrium stochastic processes as well as in other frameworks [28, 36]. Furthermore, this fundamental connection has been verified experimentally for driven Brownian motion and RC electric circuits, providing evidence for the breaking of time-reversal symmetry in nonequilibrium fluctuations down to the nanoscale [30, 31].

The difference of entropies, $h^R - h$, is always nonnegative in agreement with the second law of thermodynamics. At equilibrium, both entropies are equal. Therefore, the equilibrium temporal disorder looks the same for the typical paths and their time reversals, which is an expression of the principle of detailed balancing. In contrast, the time-reversed entropy per unit time is larger than the standard one if the system is driven out of equilibrium because the nonequilibrium constraints perform a selection of typical paths, whereupon the time-reversal symmetry is broken. The probabilities of the time-reversed paths decay faster than for the corresponding typical paths so that the time-reversed paths appear more random in this regard. We have the theorem of nonequilibrium temporal ordering:

> **Theorem 1.2: Theorem of Nonequilibrium Temporal Ordering**
>
> In nonequilibrium steady states, the typical paths are more ordered in time than their time reversals in the sense that their temporal disorder characterized by h is smaller than the temporal disorder of the corresponding time-reversed paths characterized by h^R [29].

This theorem mathematically expresses the fact that nonequilibrium systems manifest a directionality. For instance, the mean current flowing across a resistance goes downhill in the chemical potential landscape in spite of its upward or downward fluctuations. The farther away from equilibrium, the more regular the flow will look. In the limit of ballistic transport, the current is perfectly regular. This result applies to nonequilibrium nanosystems, showing the potentialities of evolving out of equilibrium to generate or process information at the nanoscale. In particular, Landauer's principle according to which the erasure of information generates thermodynamic entropy production can be deduced from the relationship (1.51) [73]. The consequences of these results for nanosystems will be discussed below.

1.2.7
Nanosystems Driven by Time-Dependent Forces

Fundamental results have also been obtained for systems driven by some time-dependent control parameter $\lambda(t)$ [11, 12]. Let us suppose that the dynamics are described by the Hamiltonian function $H(\boldsymbol{\Gamma}, \lambda)$. The work performed on the system while the control parameter varies from λ_A to λ_B is given by

$$W \equiv H(\boldsymbol{\Gamma}_B, \lambda_B) - H(\boldsymbol{\Gamma}_A, \lambda_A) \tag{1.52}$$

if $\boldsymbol{\Gamma}_A$ is the initial condition of the trajectory followed by the system. Therefore, the work is a random variable depending on the probability distribution of the initial conditions. Following Jarzynski [11], this initial probability distribution is taken as the canonical ensemble

$$p_A(\boldsymbol{\Gamma}_A) = \frac{1}{Z_A} e^{-\beta H(\boldsymbol{\Gamma}_A, \lambda_A)} \tag{1.53}$$

with the inverse temperature $\beta = (k_B T)^{-1}$. The free energy of the system in this initial canonical state is equal to $F_A = -k_B T \ln Z_A$. The probability density of the work performed on the system driven by the *forward protocol* while the control parameter $\lambda(t)$ varies as $\lambda_A \to \lambda_B$ is defined by

$$p_F(W) \equiv \langle \delta[W - (H_B - H_A)] \rangle_A \tag{1.54}$$

with $H_A = H(\mathbf{\Gamma}_A, \lambda_A)$ and $H_B = H(\mathbf{\Gamma}_B, \lambda_B)$.

A *reversed protocol* can be similarly defined by the reversed driving $\lambda_B \to \lambda_A$ with $\lambda(\mathcal{T} - t)$ from the other initial state

$$p_B(\mathbf{\Gamma}_B) = \frac{1}{Z_B} e^{-\beta H(\mathbf{\Gamma}_B, \lambda_B)} \tag{1.55}$$

at the same inverse temperature as in the canonical distribution (1.53). The work performed on the system submitted to this *reversed protocol* has the following probability distribution:

$$p_R(W) \equiv \langle \delta[W - (H_A - H_B)] \rangle_B. \tag{1.56}$$

In this framework, the following fluctuation theorem of Crooks can be proved on the basis of Hamiltonian dynamics by using the Liouville theorem and microreversibility [35]:

Theorem 1.3: **Crook's Fluctuation Theorem**

The probability densities of the work W performed on the system during the forward and reversed protocols have the universal ratio

$$\frac{p_F(W)}{p_R(-W)} = e^{\beta(W - \Delta F)} \tag{1.57}$$

which only depends on the inverse temperature β, the work W itself, and the free energy difference $\Delta F = F_B - F_A$ between the thermodynamic equilibria at λ_B and λ_A [12].

This result has been verified in experiments on the unfolding of single RNA molecules [4].

A similar relation as (1.57) holds if the work is measured on n successive intermediate times $t_A = t_0 < t_1 < t_2 < \cdots < t_{n-1} < t_n = t_B$. The multivariate probability density that the work takes given successive values is defined for the *forward protocol* as

$$p_F(W_1, W_2, \ldots, W_n) \equiv \left\langle \prod_{j=1}^{n} \delta[W_j - (H_j - H_{j-1})] \right\rangle_A \tag{1.58}$$

and for the *reversed protocol* as

$$p_R(W_1, W_2, \ldots, W_n) \equiv \left\langle \prod_{j=1}^{n} \delta[W_j - (H_{j-1} - H_j)] \right\rangle_B \tag{1.59}$$

with $H_j = H\left[\boldsymbol{\Gamma}(t_j), \lambda(t_j)\right]$. For these protocols, the fluctuation theorem reads

$$\frac{p_F(W_1, W_2, \ldots, W_n)}{p_R(-W_1, -W_2, \ldots, -W_n)} = e^{\beta(W_1 + W_2 + \cdots + W_n - \Delta F)} . \tag{1.60}$$

A consequence of Crooks' fluctuation theorem (1.57) is

Theorem 1.4: Jarzynski's Nonequilibrium Work Theorem

The free energy difference $\Delta F = F_B - F_A$ between the thermodynamic equilibria at λ_B and λ_A can be evaluated in terms of the nonequilibrium work W by

$$\left\langle e^{-\beta W}\right\rangle = e^{-\beta \Delta F} , \tag{1.61}$$

where $\langle \cdot \rangle$ denotes the statistical average over an ensemble of random realizations of the forward protocol [11].

This theorem allows the measurement of free energy landscapes with single-molecule force spectroscopy [74]. Extensions to quantum systems have also been obtained [26, 27, 75, 76]. Moreover, Jarzynski's theorem implies Clausius' thermodynamic inequality:

$$\langle W \rangle \geq \Delta F . \tag{1.62}$$

More recently, Kawai, Parrondo and Van den Broeck have shown [77] that the average value of the nonequilibrium work can be expressed as

$$\langle W \rangle = \Delta F + k_B T \int d\boldsymbol{\Gamma}\, p_F(\boldsymbol{\Gamma}, t) \ln \frac{p_F(\boldsymbol{\Gamma}, t)}{p_R(\boldsymbol{\Theta \Gamma}, t)} \tag{1.63}$$

in terms of the phase-space probability distributions of the positions and momenta of the particles at some intermediate time t during the aforementioned protocol. We notice that the equality (1.63) of statistical mechanics completes Clausius' thermodynamic inequality (1.62).

The difference between the work W performed on the system and the free energy ΔF gained by the system is the work dissipated in the process: $W_{\text{diss}} \equiv W - \Delta F$. In this regard, Clausius' inequality (1.62) means that the average dissipated work is always nonnegative, $\langle W_{\text{diss}} \rangle \geq 0$, which is a statement of the second law of thermodynamics. The last term of (1.63) thus provides an exact expression for the work dissipated in the process. If the time-dependent driving is such that a coupling is switched-on between the nanosystem and reservoirs at different temperatures or chemical potentials over a long enough time interval $\mathcal{T}$ for reaching a steady state, (1.63) can be used to obtain the entropy production in nonequilibrium steady states. The driving can be chosen to be time-reversal symmetric, $\lambda(t) = \lambda(\mathcal{T} - t)$, in order for the forward and reversed protocols to be identical. In such circumstances, the thermodynamic entropy production is given by

$$\frac{d_i S}{dt} = \frac{1}{T} \lim_{\mathcal{T} \to \infty} \frac{1}{\mathcal{T}} \langle W_{\text{diss}} \rangle \geq 0 . \tag{1.64}$$

On the other hand, the phase-space integral in (1.63) can be partitioned into the cells $C_{\omega_0 \omega_1 \ldots \omega_{n-1}} = C_{\omega_0} \cap \boldsymbol{\Phi}^{-\tau} C_{\omega_1} \cap \cdots \cap \boldsymbol{\Phi}^{-(n-1)\tau} C_{\omega_{n-1}}$ obtained by sampling the dynamics at the successive times $t = j\tau$ with $j = 0, 1, 2, \ldots, n-1$. If these phase-space cells are supposed of volume $\Delta\Gamma$, the probability densities in (1.63) give approximations for the stationary probabilities $P(\omega) \simeq p_F(\boldsymbol{\Gamma}, t) \Delta\Gamma$ and $P(\omega^R) \simeq p_R(\boldsymbol{\Theta\Gamma}, t) \Delta\Gamma$ of the paths ω and their reversals ω^R. In this way, the thermodynamic entropy production can be expressed as a relative entropy between path probabilities [78] and the relationship (1.51) is recovered, confirming that the thermodynamic entropy production finds its origin in the breaking of the time-reversal symmetry at the level of the probability distribution describing the nonequilibrium steady state.

In the following section, selected studies of nonequilibrium nanosystems will be reviewed.

1.3
Mechanical Nanosystems

The mechanical systems considered in this section are Hamiltonian systems which conserve their total energy in the absence of external driving force.

1.3.1
Friction in Double-Walled Carbon Nanotubes

Carbon nanotubes have remarkable properties which have been systematically investigated since their discovery in 1991 [79]. They appear in the form of nested coaxial tubes called multiwalled carbon nanotubes (MWCNT), which can move relative to one another, presenting the possibility of fabricating mechanical devices at the nanoscale. The relative sliding motion of nested carbon nanotubes was demonstrated in the experiment of Cumings and Zettl [80]. More recently, Fennimore et al. [81] and Bourbon et al. [82] used multiwalled carbon nanotubes as the shaft of rotary motors or actuators. In multiwalled carbon nanotubes, the different coaxial tubes interact with each other by the same van der Waals interactions as between graphene sheets in graphite. Whether the relative motion of nested nanotubes is translational or rotational, the mutual interaction between the nanotubes is the cause of friction and energy dissipation. This friction is a fundamental preoccupation in nanotribology, which requires the use of nonequilibrium statistical mechanics at the nanoscale, as explained below.

We now turn towards double-walled carbon nanotubes (DWCNT) [83–86]. Carbon nanotubes can have different geometries depending on the way the graphene sheet is rolled onto itself in order to form the nanotube. The different geometries are specified by the integers (n, m) with $0 \leq |m| \leq n$, which define the chiral vector $n\boldsymbol{a}_1 + m\boldsymbol{a}_2$ giving the equator of the nanotube in terms of the lattice vectors $\boldsymbol{a}_1$ and $\boldsymbol{a}_2$ of the hexagonal lattice of graphene. The diameter of the nanotube can be

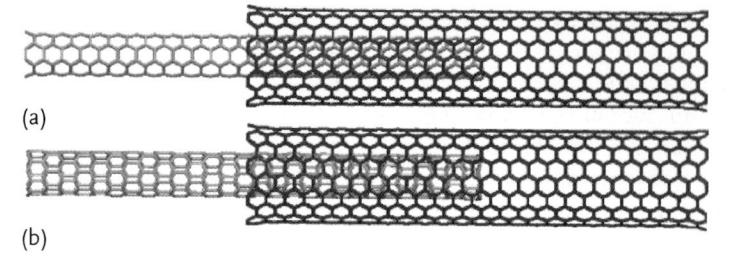

Figure 1.1 (a) The armchair–armchair DWCNT (4,4)@(9,9) with respectively $N_1 = 400$ and $N_2 = 900$ carbon atoms. (b) The zigzag-armchair DWCNT (7,0)@(9,9) with $N_1 = 406$ and $N_2 = 900$. Both DWCNTs have outer diameter 13.2 Å and length 61.5 Å [84].

evaluated by

$$d = \frac{a}{\pi}\sqrt{n^2 + m^2 + nm} \tag{1.65}$$

with $a \simeq 2.5$ Å [87]. The so-called armchair nanotubes correspond to the integers (n, n), the zigzag ones to $(n, 0)$, and the chiral nanotubes to (n, m) with $n \neq m$ [87]. Double-walled carbon nanotubes are denoted as $(n_1, m_1)@(n_2, m_2)$ (see two examples in Figure 1.1).

The carbon atoms within the inner or outer nanotube interact by Tersoff–Brenner potentials, $V_{TB}^{(1)}$ or $V_{TB}^{(2)}$ respectively [88]. The intertube potential is commonly modeled by the 6–12 Lennard-Jones potential

$$V_{LJ}(r) = 4\epsilon\left[\left(\frac{\sigma}{r}\right)^{12} - \left(\frac{\sigma}{r}\right)^{6}\right] \tag{1.66}$$

with $\epsilon = 2.964$ meV and $\sigma = 3.407$ Å, which was successfully used to study C_{60} solids [89] and the sliding of nanotubes on a graphite surface [90]. Accordingly, the total Hamiltonian describing a DWCNT can be written as

$$H = T^{(1)} + T^{(2)} + V_{TB}^{(1)} + V_{TB}^{(2)} + \sum_{i=1}^{N_1}\sum_{j=1}^{N_2} V_{LJ}\left(\left\|\mathbf{r}_i^{(1)} - \mathbf{r}_j^{(2)}\right\|\right), \tag{1.67}$$

where $T^{(1)}$ and $T^{(2)}$ are respectively the kinetic energies of the inner and outer nanotubes. The positions and momenta of the carbon atoms of both nanotubes are denoted by $\{\mathbf{r}_i^{(a)}\}_{i=1}^{N_a}$ and $\{\mathbf{p}_i^{(a)}\}_{i=1}^{N_a}$ with $a = 1$ (resp. $a = 2$) for the inner (resp. outer) tube. The kinetic energies are given by $T^{(a)} = \sum_{i=1}^{N_a}\left(\mathbf{p}_i^{(a)}\right)^2/(2m)$, where $m = 12$ amu is the mass of a carbon atom.

The molecular dynamics of the DWCNT system can be simulated by Hamilton's equations (1.2). The molecular dynamics conserves the total energy $E = H$, the total linear momentum, as well as the total angular momentum. The phase-space volumes are preserved according to Liouville's theorem. The molecular dynamics are carried out with a velocity Verlet algorithm with a time step of 2 fs. The total energy corresponds to a room temperature of about $T = 300$ K.

Within each nanotube, the carbon atoms undergo thermal fluctuations around their equilibrium position. Moreover, large amplitude motions are possible between the two nanotubes which interact with each other by an attractive van der Waals potential and constitute a mechanical oscillator if they form an isolated system. Molecular dynamic simulations reveal three different time scales characteristic of these motions [83, 84]:

- The time scale of the vibration of the carbon atoms around their equilibrium position is determined by the inverse of the Debye frequency of graphene: $t_C \simeq 50$ fs.
- The translational relative motion of the two nanotubes presents inertial oscillations with a period of about $t_P \simeq 10$ ps if the inner nanotube is initially extracted from the outer nanotube by a fraction of their common length.
- The relative sliding motion of the two nanotubes is damped by the dissipation of the energy contained in the relative motion. This dissipation is caused by dynamic friction between the nanotubes, resulting into a rise in temperature. The relaxation time of the inertial oscillations is of the order of $t_R \simeq 1000$ ps.

We notice that these time scales are separated from each other by several orders of magnitudes: $t_C \ll t_P \ll t_R$.

The sliding motion of the nanotubes can be translational or rotational. Although both types of motion can manifest themselves during a single simulation, their friction properties can be separately investigated.

1.3.1.1 Translational Friction

The translational sliding motion of two nanotubes concerns the relative position r and velocity $v = \dot{r}$ between the nanotubes. The relative position can be defined as [84]

$$r(t) \equiv \mathbf{e}_\parallel(t) \cdot \left[\mathbf{R}^{(2)}(t) - \mathbf{R}^{(1)}(t) \right] \tag{1.68}$$

in terms of the centers of mass $\mathbf{R}^{(1)}$ and $\mathbf{R}^{(2)}$ of both nanotubes. The unit vector $\mathbf{e}_\parallel$ points in the direction of the axis of the DWCNT and can be obtained by diagonalizing the inertia tensor of the total system and selecting the eigenvector associated with its smallest eigenvalue. This eigenvector slightly fluctuates around its initial orientation during the time evolution, which justifies its use in order to define the relative position by (1.68).

The relative position of the nanotubes admits a reduced description in terms of a Newtonian equation of Langevin type:

$$\mu \frac{d^2 r}{dt^2} = -\frac{d V_{LJ}(r)}{dr} + F_{\text{frict}} + F_{\text{fluct}}(t) \tag{1.69}$$

where $\mu = m \left(N_1^{-1} + N_2^{-1} \right)^{-1}$ is the relative mass of the DWCNT system. The force in the right-hand side of (1.69) has three contributions.

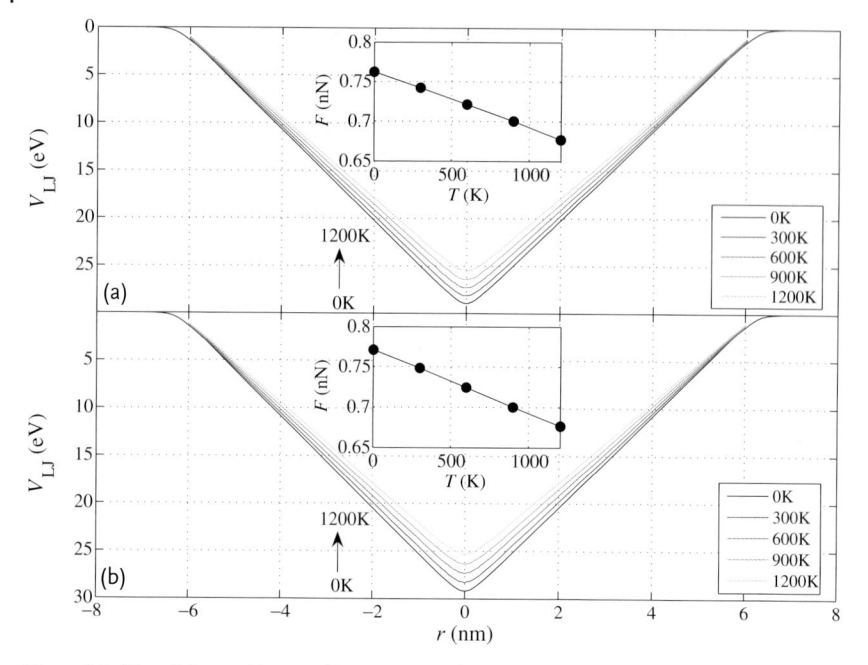

Figure 1.2 Plot of the total Lennard-Jones potential V_{LJ} versus the distance r between the centers of mass of the nanotubes for (a) the armchair–armchair (4,4)@(9,9) and (b) the zigzag-armchair (7,0)@(9,9) DWCNTs [84]. The insets show that the force decreases with the temperature because of dilation.

The first contribution is the force of the V-shaped potential

$$V(r) = F\sqrt{r^2 + \ell^2} - C \simeq F|r| - C \tag{1.70}$$

due to the van der Waals interaction between the two nanotubes (see Figure 1.2). This potential is obtained by averaging the interaction at a fixed relative position r between the nanotubes. We notice the absence of corrugation because of the averaging. The V shape finds its origin in the proportionality of the interaction potential with the number of van der Waals bonds between the nanotubes. The potential is parabolic around its minimum because of thermal fluctuations around the configurations with the maximum number of bonds. If the energy of the relative motion is not too high, the potential forms a well in which the motion presents oscillations which would persist if dissipation could be neglected [91]. This inertial oscillator is anharmonic with a period of about $t_P \simeq 10$ ps.

The second contribution to the total force in (1.69) is the dynamic friction force:

$$F_{\text{frict}} = -\zeta \frac{dr}{dt} + O\left(\frac{dr}{dt}\right)^2 \tag{1.71}$$

with the friction coefficient ζ given by Kirkwood formula (1.17). The force–force correlation function decreases to zero over the time scale of vibration of the car-

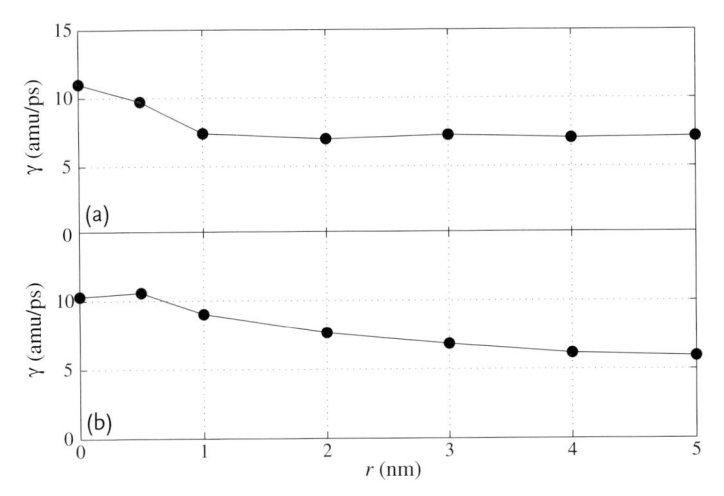

Figure 1.3 Friction coefficient calculated by the Kirkwood formula (1.17) versus the relative position r between the nanotubes for (a) the armchair–armchair (4,4)@(9,9) and (b) the zigzag-armchair (7,0)@(9,9) DWCNTs [84].

bon atoms around their equilibrium position in graphene [83]. This time scale is determined by the inverse of the Debye frequency ω_D of graphene, that is, $t_C \simeq 2\pi/\omega_D \simeq 50\,\text{fs}$. According to the Kirkwood formula (1.17), the friction coefficient can be estimated to be $\zeta \simeq t_C \Delta F^2/(2k_B T) = \pi \Delta F^2/(\omega_D k_B T)$ where ΔF is the standard deviation of the fluctuating intertube force. The latter increases with the temperature T approximately as $\Delta F \sim T$ so that the friction coefficient also increases as $\zeta \sim T$ [83].

The force–force correlation function should be evaluated by fixing the relative position r between the nanotubes. This is carried out by constrained molecular dynamic simulations [83, 84]. The constraint is enforced by modifying the force on all of the atoms of each nanotube according to $F_i^{(a)} \rightarrow F_i^{(a)} - (1/N_a)\sum_{j=1}^{N_a} F_j^{(a)}$. This modification has the required effect of canceling the acceleration of the centers of mass of each nanotube, while conserving the total energy E. This method gives the dependence of the friction coefficient on position as seen in Figure 1.3. There we observe that the friction coefficient has a slow dependence on position, justifying that the friction coefficient is taken to be a constant in (1.71) at the approximate value $\zeta \simeq 6\,\text{amu/ps}$ for the present case [84]. We notice that the friction coefficient also depends on properties affecting the intertube interaction, such as the distance between the nanotubes, the ends of the nanotubes, deformations, defects, or possible impurities composed of atomic species other than carbon.

Although the dynamic friction force is proportional to the velocity at moderate sliding velocities, nonlinear dependences on the sliding velocity become important at larger velocities. These nonlinear effects appear in the form of resonances at specific values of the sliding velocity where dynamic friction is enhanced because of the excitation of radial breathing modes of the outer nanotube. This phenomenon was first observed in the oscillatory system described above [84], as well as in other systems where a finite inner nanotube moves at given sliding velocity inside an infinitely long outer nanotube [92]. This latter configuration allows a precise determination of the resonant velocities. If the nanotubes move at the relative velocity v, the spatial period $a = 2.5$ Å of the corrugation of the intertube potential results in the periodic driving at the washboard frequency $\omega_{wb} = 2\pi v/a$. Thus, resonances are possible when the washboard frequency ω_{wb} coincides with some vibration frequency. The vibration modes of the nanotubes and, in particular, of the outer tube form a spectrum of dispersion relations $\omega = \omega_i(k)$ characterizing the acoustic and optical phonons of each nanotube. Phonons of type i and wave number k are excited if the resonance condition $\omega_{wb} = \omega_i(k)$ is satisfied together with a further condition selecting the resonant values of the sliding velocity. This further condition can be shown to be given by the equality of the sliding velocity with the group velocity of the excited phonons: $v = d\omega_i(k)/dk$ [92]. More recently, a related phenomenon of chiral symmetry breaking has been discovered in the sliding dynamics of DWCNTs made of perfectly left–right symmetric and nonchiral nanotubes [93]. These phenomena of dynamic friction enhancement find their origin in the nonlinear dynamics of DWCNTs.

We notice that the DWCNT system is a weakly under-damped oscillator [83–85]. Therefore, molecular dynamics simulations show many oscillations which are slowly damped because the friction coefficient is relatively small. Figure 1.4 depicts the amplitudes $R(t)$ of the successive oscillations during the relaxation. The results of molecular dynamics are compared with the prediction of the model (1.69) that these amplitudes are exponentially damped as $R(t) = R(0)\exp(-\Gamma_R t)$ with the

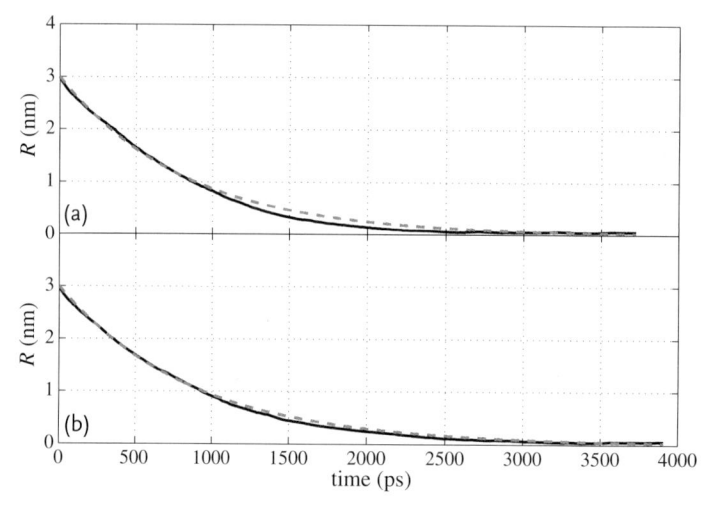

Figure 1.4 Evolution of the amplitude of the damped oscillations for (a) the armchair–armchair (4,4)@(9,9) and (b) the zigzag-armchair (7,0)@(9,9) DWCNTs [84]. The numerical results of molecular dynamics (solid lines) are compared with the theoretical expectation of the model (1.69) (dashed lines).

damping rate $\Gamma_R = 2\zeta/(3\mu)$. The damping of the oscillations means that the energy of the one-dimensional sliding motion of the two nanotubes is dissipated in the many vibrational degrees of freedom of each nanotubes, which indeed undergo a rise in temperature from 300 K at the beginning of the simulation up to 338 K after relaxation.

The third contribution to the total force in (1.69) is the fluctuating Langevin force which is present as a corollary of dynamic friction force by the fluctuation-dissipation theorem. Accordingly, the Langevin force is taken as a Gaussian white noise satisfying

$$\langle F_{\text{fluct}}(t) \rangle = 0 \tag{1.72}$$

$$\langle F_{\text{fluct}}(t) F_{\text{fluct}}(t') \rangle = 2\zeta k_B T \delta(t - t') \tag{1.73}$$

for $|t-t'| \gg t_C$, in consistency with the Kirkwood formula (1.17). As a consequence of the smallness of friction, the fluctuating force is also small and plays a significant role only after the large amplitude oscillations have been damped and no longer overwhelm thermal fluctuations in the relative motion between the nanotubes. The Langevin fluctuating force thus describes a state of thermodynamic equilibrium in the sliding motion. For the total system, this equilibrium state is microcanonical at the energy of the initial conditions of each molecular dynamics simulation. Since the total system has many degrees of freedom, $f = 3(N_1 + N_2) \simeq 3900$, the equilibrium statistical distribution of each degree of freedom is practically canonical at the temperature corresponding to the initial total energy. The equilibrium fluctuations of the relative position between the nanotubes remain very small, on the order of a fraction of a nanometer as seen in Figure 1.5.

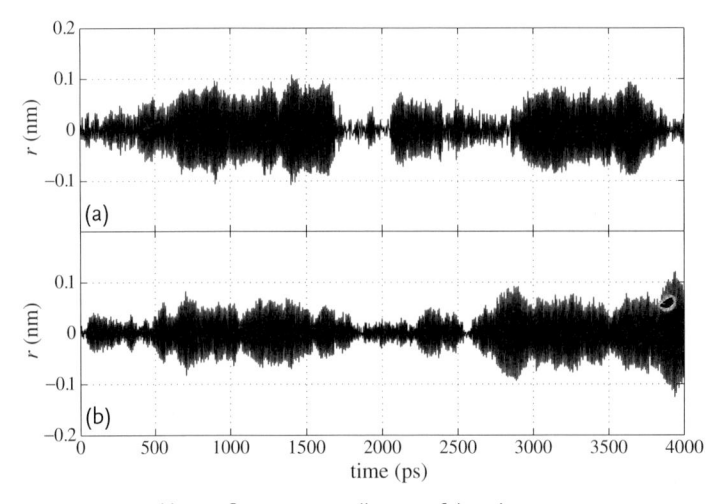

Figure 1.5 Equilibrium fluctuating oscillations of the relative position between the nanotube mass centers for (a) the armchair–armchair (4,4)@(9,9) and (b) the zigzag-armchair (7,0)@(9,9) DWCNTs [84].

As explained in Section 1.2.1, the stochastic process described by the Langevin equation (1.69) admits an equivalent description in terms of a Fokker–Planck equation such as (1.18) for the probability density of the relative position and velocity between the two nanotubes:

$$\mathcal{P}(r,v,t) \equiv \int \delta\left[r - r(\Gamma)\right] \delta\left[v - v(\Gamma)\right] p(\Gamma,t) \, d\Gamma \, . \tag{1.74}$$

This Fokker–Planck equation describes, in particular, the relaxation towards a state of equilibrium such as (1.20) [84]. This relaxation is characteristic of isolated systems with a few slow degrees of freedom coupled to baths of many fast degrees of freedom. The Poincaré recurrences back close to initial conditions are extremely long in an individual system, but they never occur in the statistical ensemble composed of infinitely many copies of the system described by the probability distribution $p(\Gamma, t)$, if the dynamics has the ergodic property of mixing. Molecular dynamics shows that this property is practically fulfilled at a temperature of 300 K.

1.3.1.2 Rotational Friction

Besides translational sliding motion, the two nanotubes may rotate relative to one another (see Figure 1.6a). A friction property can be associated with this rotational motion [85], which is of importance in shafts of nanomachinery made of MWCNTs [81, 82]. For an isolated DWCNT, the total angular momentum is conserved. For sufficiently long DWCNTs, the inner and outer nanotubes essentially rotate around a common axis with their respective angular velocities $\{\omega_1, \omega_2\}$ and angular momenta $\{L_1, L_2\}$. Under these circumstances, the rotational motion can be supposed to be one-dimensional and ruled by the coupled equations

$$\begin{cases} \dfrac{dL_1}{dt} = I_1 \dfrac{d\omega_1}{dt} = N_1 \\ \dfrac{dL_2}{dt} = I_2 \dfrac{d\omega_2}{dt} = N_2 = -N_1 \end{cases} \tag{1.75}$$

where I_a denote the moments of inertia around the common axis and N_a the torques acting on each nanotube. These torques are opposite by the conservation of the total angular momentum $L_1 + L_2$. The moment of inertia of a nanotube of radius R_a and length l is given by $I_a = 2\pi \sigma l R_a^3$ in terms of the surface mass density $\sigma = 4m/(3\sqrt{3} a_{CC}^2) \simeq 4.55 \, \text{amu/Å}^2$ where $m = 12$ amu is the atomic mass of carbon and $a_{CC} = 1.42$ Å the carbon–carbon bond length in the hexagonal lattice of graphene.

When the relative angular velocity and the angular velocity of the center of inertia are written as

$$\omega \equiv \omega_1 - \omega_2 \tag{1.76}$$

$$\Omega \equiv \frac{I_1 \omega_1 + I_2 \omega_2}{I_1 + I_2} \tag{1.77}$$

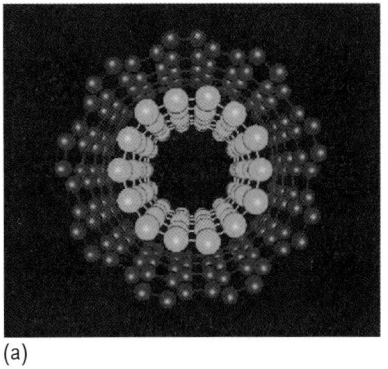

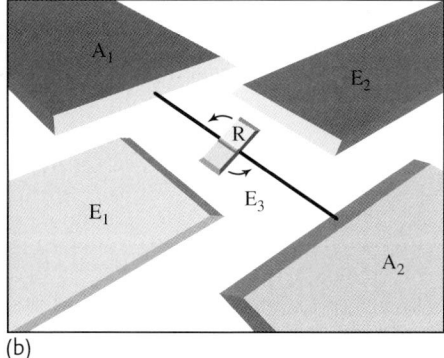

Figure 1.6 (a) Coaxial view of a DWCNT showing that the outer nanotube can rotate around the inner nanotube as in the shaft of a rotary motor [83]. (b) Schematic picture of the rotary motor built in Berkeley [81]. The metal plate rotor R has a size of about 300 nm. It rotates around a shaft made of a MWCNT attached to its anchor pads (A_1, A_2). The rotor is driven by an oscillating electric field between the electrodes E_1, E_2 and E_3 (adapted from [81]).

the equations (1.75) become

$$\begin{cases} I \dfrac{d\omega}{dt} = N_1 \\ \dfrac{d\Omega}{dt} = 0 \end{cases} \tag{1.78}$$

with the relative moment of inertia $I \equiv (I_1^{-1} + I_2^{-1})^{-1}$.

The torque $N_1 = -N_2$ between the nanotubes is determined by the intertube van der Waals interaction. Since the rotation of one nanotube with respect to the other does not change the number of van der Waals bonds, the average potential is essentially flat with negligible corrugation as it is the case for translational motion. However, a dynamic friction torque proportional to the angular velocity and the corresponding Langevin fluctuating torque should be taken into account, as in (1.69). Accordingly, the relative sliding rotation between the two nanotubes is described by the Langevin equation

$$I \frac{d\omega}{dt} = -\chi \omega + N_{\text{fluct}}(t), \tag{1.79}$$

where χ is the rotational friction coefficient and $N_{\text{fluct}}(t)$ is the Gaussian white noise

$$\langle N_{\text{fluct}}(t) \rangle = 0 \tag{1.80}$$

$$\langle N_{\text{fluct}}(t) N_{\text{fluct}}(t') \rangle = 2\chi k_B T \delta(t-t') \tag{1.81}$$

for $|t - t'| \gg t_C$ [85].

This stochastic model accurately describes the molecular dynamics simulations [85]. If the relative angular velocity has a non-vanishing initial value, the sliding rotation is damped exponentially with the relaxation time $\tau = I/\chi$ until an equilibrium state is reached where the angle between the nanotubes undergoes a random walk of diffusion coefficient $D = k_B T/\chi$. The relaxation time is observed to behave as $\tau = \tau_\infty l/(l + l_0)$ in terms of the length l of the DWCNT and a constant l_0 of the order of the nanometer [85]. Hence, the relaxation time becomes independent of the length if the DWCNT is long enough. On the other hand, the relative moment of inertia is proportional to the length $I = 2\pi \sigma l(R_1^{-3} + R_2^{-3})^{-1}$ so that the friction coefficient is also proportional to the length of the DWCNT. This dependence has been shown to be consistent with the proportionality between the friction force and the intertube contact area [85]. Additionally, the rotational friction coefficient is observed to increase with temperature as $\chi \sim T^\nu$ with the exponent $\nu = 1.53 \pm 0.04$ [85]. As for translational sliding motion, the isolated DWCNT is an undriven nonequilibrium system reaching a state of equilibrium after relaxation since friction dissipates kinetic rotational energy.

The rotational friction is the cause of energy dissipation in rotary motors using a DWCNT or MWCNT shaft (see Figure 1.6b). Such nanomotors have been fabricated by attaching a metal rotor plate to a single MWCNT suspended between two anchor pads, as carried out by a group at Berkeley [81]. The motor is controlled by voltages between the rotor plate and three surrounding electrodes. All of these components are integrated on a silicon chip, forming an electromechanical system with a rotor of about 300 nm and an angular frequency of several Hertz. The fabrication of a similar system has been carried out by a collaboration between Paris and Lausanne [82]. Such nanoelectromechanical devices are driven nonequilibrium nanosystems where energy dissipation due to rotational friction is compensated by the electric energy supply.

1.3.2
Electromagnetic Heating of Microplasmas

1.3.2.1 The Undriven System and Its Hamiltonian

Microplasmas are small mechanical systems composed of atomic ions moving in a Penning trap [94–96]. Their spatial extension is in the range of micrometers [95]. These systems can be considered as isolated Hamiltonian systems in which energy is conserved, as long as the system is not subjected to a time-dependent driving. As for isolated DWCNTs, these systems undergo a relaxation towards a microcanonical equilibrium state if their initial conditions correspond to a given total energy. Remarkable crystalline-like configurations of the ions have been observed at low mean kinetic energy [94, 95]. These ordered configurations melt as their kinetic energy is increased (see Figure 1.7). The dynamics are known to be chaotic with a spectrum of positive Lyapunov exponents [97].

In a frame rotating at the Larmor frequency associated with the magnetic field of the Penning trap, the Hamiltonian of the microplasma is given by

$$H_0 = \sum_a \left[\frac{1}{2} \mathbf{p}_a^2 + \left(\frac{1}{8} - \frac{\gamma^2}{4} \right)(x_a^2 + y_a^2) + \frac{\gamma^2}{2} z_a^2 \right] + \sum_{a<b} \frac{1}{r_{ab}}, \qquad (1.82)$$

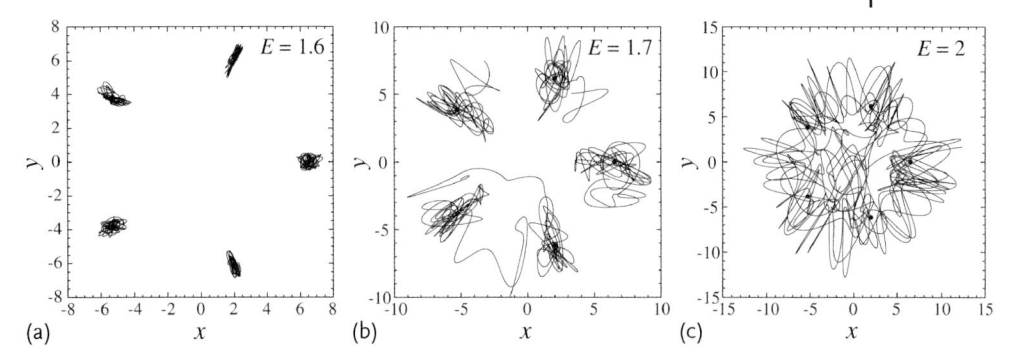

Figure 1.7 Simulation of Hamiltonian trajectories of five atomic ions in an oblate Penning trap with $\gamma = 0.7$. The total angular momentum in the z-direction is vanishing. The total energy (1.82) is (a) $E = 1.6$, (b) $E = 1.7$, and (c) $E = 2$.

in terms of the momenta $\mathbf{p}_a = (p_{xa}, p_{ya}, p_{za})$ and the distances $r_{ab} = \sqrt{(x_a - x_b)^2 + (y_a - y_b)^2 + (z_a - z_b)^2}$ between the ions ($a, b = 1, 2, \ldots, N$). The parameter γ controls the geometry of the trap. The trap is elongated or prolate if $0 < |\gamma| < (1/\sqrt{6})$, spherical if $|\gamma| = (1/\sqrt{6})$, and flat or oblate if $(1/\sqrt{6}) < |\gamma| < (1/\sqrt{2})$.

1.3.2.2 The Driven System and the Fluctuation Theorem

The microplasma can be heated if it interacts with an electromagnetic wave. In this case, the Hamiltonian becomes time dependent:

$$H = H_0 - A \sum_{a=1}^{N} z_a \sin \omega t \tag{1.83}$$

and Crooks fluctuation theorem (1.57) applies. In order for the forward and reverse protocols to be identical, the driving is considered over a time interval with an odd number of half periods, for example, $T = 3\pi/\omega$. In this case, the Hamiltonian (1.83) is the same at the beginning and the end of the driving so that the forward and reversed protocols have the same probability distribution of nonequilibrium work, $p_F = p_R \equiv p$, and the difference of free energy is vanishing, $\Delta F = 0$. In this case, Crooks fluctuation theorem (1.57) can be expressed as

$$\int_{-\infty}^{W} p(W') \, dW' = \int_{-\infty}^{W} e^{\beta W'} p(-W') \, dW' . \tag{1.84}$$

The numerical verification of this result is shown in Figure 1.8 for a heated microplasma of five ions. The effect of heating is seen by the shift of the cumulative functions away from the mid-point at $W = 0$.

We notice that the quantum versions of the fluctuation theorem can also be applied to atoms, molecules, or ions trapped in quantum states, which might be of great interest for the control of quantum information devices and other ultracold systems.

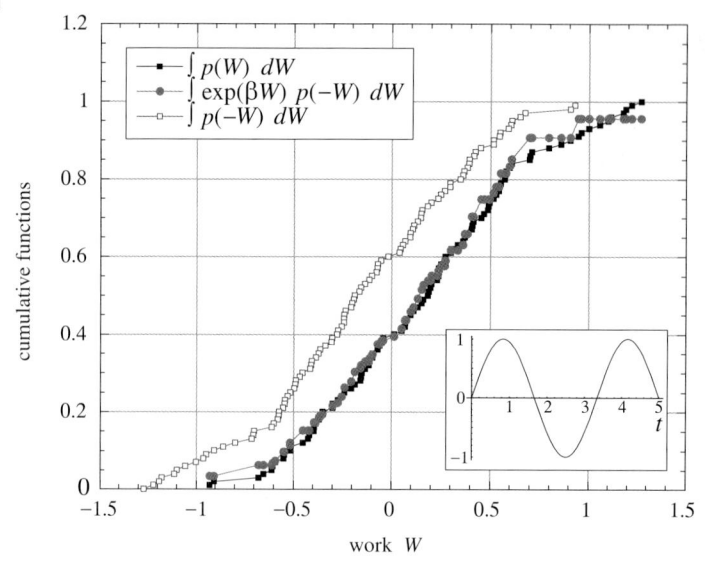

Figure 1.8 Numerical verification of Crooks fluctuation theorem for a microplasma of five atomic ions in an oblate Penning trap with $\gamma = 0.7$ and heated by the time-dependent external field shown in inset over the time interval $\mathcal{T} = 3\pi/\omega = 5$. The verification of the Crooks fluctuation theorem is the coincidence of the cumulative functions in the left-hand side (filled squares) and right-hand side (filled circles) of (1.84). The initial distribution is canonical with temperature $T = 1$. The final distribution is no longer canonical. The cumulative function of the negative values of the work (open circles) is shifted with respect to the others because of heating.

1.4
Mechanochemical Nanosystems

1.4.1
F_1-ATPase Motor

F_1-ATPase is the hydrophilic part of the F_oF_1-ATPase also known as ATP synthase, which is an adenosine triphosphate (ATP) producing protein common to most living organisms [5]. *In vivo*, the two parts of ATP synthase, F_o and F_1, are attached to each other and mechanically coupled by the central γ-subunit. The F_o part is embedded in the inner membrane of mitochondria and is rotating as a turbine when a proton current flows across the membrane. This turbine drives the rotation of the γ-subunit inside the hydrophylic F_1 part. The latter is composed of three α- and three β-subunits spatially alternated as a hexamer $(\alpha\beta)_3$ and forming a barrel for the rotation of the shaft made of the γ-subunit [98, 99]. Upon rotation, the γ-shaft induces conformational changes in the hexamer, leading to the synthesis of ATP in catalytic sites located in each β-subunit.

In their experimental work [100, 101], Kinosita and coworkers have succeeded in building a nanomotor by separating the F_1 part and attaching an actin filament or a colloidal bead to its γ-shaft (see Figure 1.9). *In vitro*, ATP hydrolysis drives the

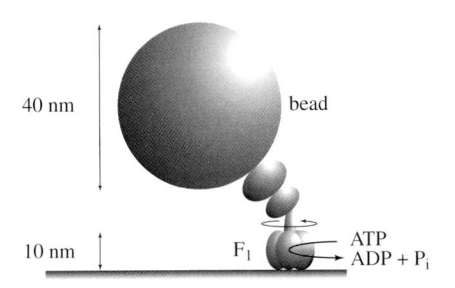

Figure 1.9 Schematic representation of the F_1-ATPase fixed on a surface and with a bead attached to its γ-shaft [101].

rotation of this nanomotor, transforming chemical free energy from ATP into the mechanical motion of the γ-shaft. This motion proceeds in steps of $120°$, revealing the three-fold symmetry of F_1-ATPase [98–100]. The diameter of the F_1-ATPase is 10 nm, which makes it one of the smallest motors in nature, with a power of only about 10^{-18} W.

The rotation of this nanomotor is powered by the chemical energy supplied by the hydrolysis of adenosine triphosphate (ATP) into adenosine diphosphate (ADP) and inorganic phosphate (P_i):

$$\text{ATP} \rightleftharpoons \text{ADP} + P_i \,. \tag{1.85}$$

The thermodynamic force or affinity of this reaction is given by the difference of chemical potential $\Delta\mu$ between the three species:

$$\Delta\mu = \mu_{\text{ATP}} - \mu_{\text{ADP}} - \mu_{P_i} = \Delta\mu^0 + k_B T \ln \frac{[\text{ATP}]}{[\text{ADP}][P_i]} \,, \tag{1.86}$$

where the concentrations are counted in mole per liter (M), T is the temperature, and k_B the Boltzmann's constant. The Gibbs free energy of ATP hydrolysis takes the value $\Delta G^0 = -\Delta\mu^0 = -30.5\,\text{kJ/mol} = -7.3\,\text{kcal/mol} = -50\,\text{pN nm}$ at the temperature of $23\,°\text{C}$, the external pressure of 1 atm, and pH 7 [102]. We notice that ATP hydrolysis provides a significant amount of free energy of $\Delta\mu^0 = -\Delta G^0 = 12.2 k_B T$ above the thermal energy $k_B T = 4.1\,\text{pN nm}$. At equilibrium, where the chemical potential difference (1.86) vanishes, the concentrations of ATP, ADP, and P_i satisfy

$$\left. \frac{[\text{ATP}]}{[\text{ADP}][P_i]} \right|_{\text{eq}} = \exp \frac{\Delta G^0}{k_B T} \simeq 4.9 \times 10^{-6}\,\text{M}^{-1} \,, \tag{1.87}$$

showing that ATP tends to hydrolyze into its products. The motor is in a nonequilibrium state if the concentrations do not satisfy (1.87), whereupon its self-sustained rotation becomes possible thanks to the chemical free energy (1.86) supplied by the reaction.

The motor is functioning along a cycle based on the following kinetic scheme. As reported [101], the first substep, the $90°$ rotation of the γ-shaft, is induced by the binding of ATP to an empty catalytic site. The second substep, the 30° rotation of the γ-shaft, is induced by the release of ADP and P_i. The process can be summarized by the following chemical scheme

$$\text{ATP} + \underbrace{[\emptyset, \gamma(\theta)]}_{\text{state 1}} \underset{W_{-1}}{\overset{W_{+1}}{\rightleftharpoons}} \underbrace{[\text{ATP}^\ddagger, \gamma(\theta + 90°)]}_{\text{state 2}} \underset{W_{-2}}{\overset{W_{+2}}{\rightleftharpoons}} \underbrace{[\emptyset, \gamma(\theta + 120°)]}_{\text{state 1}} + \text{ADP} + P_i \,. \tag{1.88}$$

In state 1, ATP can bind to an empty β-catalytic site $\emptyset$ of F_1 with the γ-shaft at angular position θ. The binding of ATP fills this catalytic site and induces the 90° rotation of the γ-shaft from $\gamma(\theta)$ to $\gamma(\theta + 90°)$. ATP‡ stands for any transition state of ATP between the initial triphosphate molecule to the products of hydrolysis ADP and P_i before the evacuation of the β-catalytic site. State 2 is thus denoted by $[\text{ATP}^\ddagger, \gamma(\theta + 90°)]$. If the F_1-ATPase proceeds to hydrolysis, the products ADP and P_i are released together, which induces the secondary 30° rotation and empties a β-subunit.

The nanomotor can be subjected to an external torque, for instance, coming from the proton turbine F_o. In a nonequilibrium steady state, the nanomotor has the mean rotation rate

$$V \equiv \frac{1}{2\pi} \left\langle \frac{d\theta}{dt} \right\rangle \tag{1.89}$$

in revolution per second and the mean ATP consumption rate

$$R \equiv \left\langle \frac{dN_{\text{ATP}}}{dt} \right\rangle = -\left\langle \frac{dN_{\text{ADP}}}{dt} \right\rangle = -\left\langle \frac{dN_{P_i}}{dt} \right\rangle. \tag{1.90}$$

In this steady state, the thermodynamic entropy production is given by

$$\frac{d_i S}{dt} = \frac{2\pi\tau}{T} V + \frac{\Delta\mu}{T} R \geq 0 \tag{1.91}$$

in terms of the so-called thermodynamics forces or affinities, $2\pi\tau/T$ and $\Delta\mu/T$, and the corresponding fluxes or currents, V and R [103]. The two terms in the entropy production correspond to the possibility of the coupling between the mechanical motion and the chemical reaction. It is thanks to this mechanochemical coupling that ATP is synthesized *in vivo* from the torque induced by the proton turbine F_o and the rotation of the nanomotor F_1 is powered *in vitro* by ATP hydrolysis. We notice that the mechanochemical coupling can be tight or loose depending on the regime of functioning of the nanomotor [104]. In order to further investigate the properties of the nanomotor, stochastic models have been proposed [105–108].

The modeling can be carried out at different levels of coarse graining. The finest level is certainly obtained by molecular dynamics following the phase-space trajectories of all the atoms of the motor and its environment with ATP, ADP and

P_i molecules in water. A stochastic description is obtained by considering the reactive events of the kinetic scheme (1.88) as random events happening upon the random arrival and exit of ATP, ADP or P_i molecules into or out of the catalytic sites of the nanomotor. These reactive events correspond to transitions between the different chemical states of the motor. Since the three catalytic sites can be either occupied or unoccupied, there is a minimum of six states for the motor, which corresponds to the two states of (1.88) for a single catalytic site. On the other hand, the γ-shaft takes an angle θ with respect to the barrel. In each chemical state, this angle moves in a free-energy potential. Since the motor is nanometric, this motion is affected by the thermal fluctuations and is thus similar to a rotational Brownian motion driven by the torque induced by the conformational changes of the barrel. This suggests a continuous-state description in terms of Fokker–Planck equations for the Brownian motion of the angle θ in a free-energy potential corresponding to each chemical state of the motor. These Fokker–Planck equations should be coupled together by the transitions due to the chemical reactions [105, 106]. If the motor has six chemical states, the continuous-state model is thus defined by six diffusion-reaction type, coupled Fokker–Planck equations [107].

However, if the free-energy potentials present important wells and the time intervals to reach these wells are short compared to the dwell times, the angle θ can be supposed to jump between discrete values corresponding to the minima of the potential wells, neglecting the thermal fluctuations of the angle θ around these minima. Under these circumstances, a discrete-state description is appropriate, which further simplifies the modeling [108]. Nevertheless, this simplification carries an assumption of tight coupling between the mechanics and the chemistry of the motor. Indeed, a discrete-state model in which the different angles of the shaft uniquely correspond to the different chemical states of the motor supposes a tight coupling. This is not the case in the continuous-state description where the angle can take several possible values notwithstanding the chemical state. Accordingly, the comparison between both descriptions is necessary in order to determine the regimes of loose or tight coupling [104]. This interesting distinction is important because the chemical and mechanical efficiencies depend on the quality of the mechanochemical coupling.

1.4.2
Continuous-State Description

In the continuous-state model [107], the system is found at a given time t in one out of six chemical states $\sigma = 1, 2, \ldots, 6$ and the γ-shaft at an angle $0 \leq \theta < 2\pi$. There are six chemical states because the three β-subunits can be either empty or occupied by a molecule of ATP or by the products ADP and P_i of hydrolysis. Consequently, the system is described by six probability densities $p_\sigma(\theta, t)$, normalized according to $\sum_{\sigma=1}^{6} \int_0^{2\pi} p_\sigma(\theta, t)\, d\theta = 1$. The time evolution of the probability densities is ruled by a set of six Fokker–Planck equations coupled together by the terms describing the random jumps between the chemical states σ due to the chemical reactions of ATP binding and of the release of the products ADP and P_i with their corresponding reversed reactions [107]:

$$\partial_t p_\sigma(\theta,t) + \partial_\theta J_\sigma(\theta,t)$$
$$= \sum_{\rho=1,2} \sum_{\sigma'(\neq\sigma)} \left[p_{\sigma'}(\theta,t) w_{\rho,\sigma'\to\sigma}(\theta) - p_\sigma(\theta,t) w_{-\rho,\sigma\to\sigma'}(\theta) \right], \tag{1.92}$$

where the probability current densities are given by

$$J_\sigma(\theta,t) = -D\partial_\theta p_\sigma(\theta,t) + \frac{1}{\zeta} \left[-\partial_\theta U_\sigma(\theta) + \tau \right] p_\sigma(\theta,t). \tag{1.93}$$

The diffusion coefficient D is expressed in terms of the friction coefficient ζ according to Einstein's relation $D = k_B T/\zeta$. The friction coefficient ζ can be evaluated for a bead of radius r attached off-axis at a distance $x = r\sin\alpha$ from the rotation axis according to

$$\zeta = 2\pi\eta r^3 \left(4 + 3\sin^2\alpha\right), \tag{1.94}$$

with the water viscosity $\eta = 10^{-9}$ pN s nm^{-2} and $\alpha = \pi/6$ [102, 107].

When the motor is in the chemical state σ, the γ-shaft is subjected to the external torque τ and the internal torque $-\partial_\theta U_\sigma$ due to the free-energy potential $U_\sigma(\theta)$ of the motor with its γ-shaft at the angle θ. Applying an external torque to the motor has the effect of tilting the potentials into $U_\sigma(\theta) - \tau\theta$, which eases the rotation or makes it harder, depending on the sign of τ. These free-energy potentials have been fitted to experimental data and are depicted in Figure 1.10 together with the potentials associated with the transition states of the reactions. We notice that these potentials generate power strokes if their variations are large with respect to the thermal energy $k_B T$, which is the case here, except at the bottom of the potential wells where thermal fluctuations dominate.

The transition rates $w_{\rho,\sigma'\to\sigma}(\theta)$ of the reactions are given by [107]

$$w_+(\theta) = k_0[\text{ATP}] \exp\left\{-\beta \left[U^\ddagger(\theta) - U(\theta) - G^0_{\text{ATP}}\right]\right\} \tag{1.95}$$

$$w_-(\theta) = k_0 \exp\left\{-\beta \left[U^\ddagger(\theta) - \tilde{U}(\theta)\right]\right\} \tag{1.96}$$

$$\tilde{w}_+(\theta) = \tilde{k}_0 \exp\left\{-\beta \left[\tilde{U}^\ddagger(\theta) - \tilde{U}\left(\theta + \frac{2\pi}{3}\right)\right]\right\} \tag{1.97}$$

$$\tilde{w}_-(\theta) = \tilde{k}_0[\text{ADP}][\text{P}_i] \exp\left\{-\beta \left[\tilde{U}^\ddagger(\theta) - U(\theta) - G^0_{\text{ADP}} - G^0_{\text{P}_i}\right]\right\} \tag{1.98}$$

in terms of the concentrations of ATP, ADP, and P_i molecules in the solution surrounding the nanomotor and the free-energy potentials $U(\theta)$ and $\tilde{U}(\theta)$ for the wells and the potentials $U^\ddagger(\theta)$ and $\tilde{U}^\ddagger(\theta)$ for the transition states. Equations 1.95–1.98 represent, respectively, the transition rates of binding and unbinding of ATP, and of unbinding and binding of ADP and P_i to the first β-subunit. The other transitions rates are obtained by 120° rotations of the rates (1.95)–(1.98) in order to

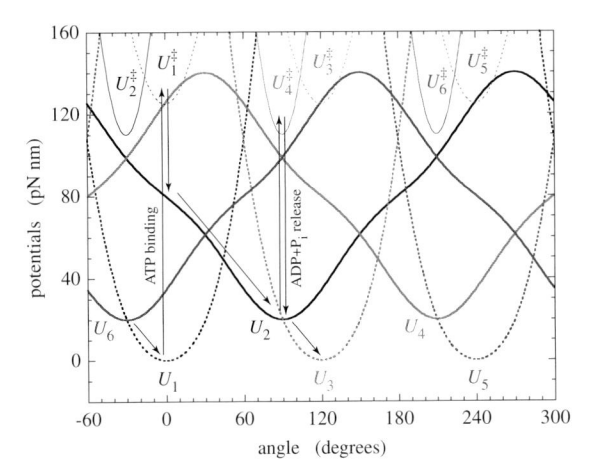

Figure 1.10 The potentials of the chemical states $U_\sigma(\theta)$ and of the transition states $U_\sigma^\ddagger(\theta)$ with a schematic representation of the transitions between them during the motor cycle [107]. Because of the threefold symmetry of the F_1 motor, the different potentials are given by $U_1(\theta) = U(\theta)$, $U_3(\theta) = U(\theta - 2\pi/3)$, $U_5(\theta) = U(\theta - 4\pi/3)$, $U_2(\theta) = \tilde{U}(\theta)$, $U_4(\theta) = \tilde{U}(\theta - 2\pi/3)$, and $U_6(\theta) = \tilde{U}(\theta - 4\pi/3)$ in terms of only the two potentials $U(\theta)$ and $\tilde{U}(\theta)$ corresponding to the empty and occupied catalytic sites. A similar symmetry reduction holds for the transition states.

reproduce the threefold symmetry of F_1-ATPase. We notice that the system has a threefold rotational symmetry but no reflection symmetry, which is attributed to the chirality of the supramolecular architecture of the F_1 molecular motor and is essential for its unidirectional rotation in the presence of its chemical fuel.

The stochastic process ruled by the Fokker–Planck equations (1.92) can be simulated by Gillespie's algorithm [57, 58], which provides realistic random trajectories as shown in Figure 1.11. The rotation proceeds by rapid jumps due to the power strokes generated after each reactive event by the free-energy potentials of Figure 1.10. Between two successive jumps, the angle undergoes thermal fluctuations around the minima of the potential wells of Figure 1.10. In this respect, the shaft performs a random motion with a mean rotation rate fixed by the chemical concentrations of ATP, ADP, and P_i. For vanishing concentrations of the products of ATP hydrolysis, the mean rotation rate depends on ATP concentration in a way characteristic of typical Michaelis–Menten kinetics:

$$V \simeq \frac{V_{\max}[\text{ATP}]}{[\text{ATP}] + K_M}, \tag{1.99}$$

with the constant $K_M \simeq 16\,\mu\text{M}$, as depicted in Figure 1.12. At low ATP concentration, the rotation rate increases with ATP concentration. However, the rate saturates at the maximum value $V_{\max} \simeq 130\,\text{rev/s}$ at high ATP concentration where the speed of the motor is limited by the time scale of the release of the products, ADP and P_i.

In order to determine the regimes of loose and tight couplings between the chemistry and the mechanics of the F_1 motor, both the rotation rate V and the

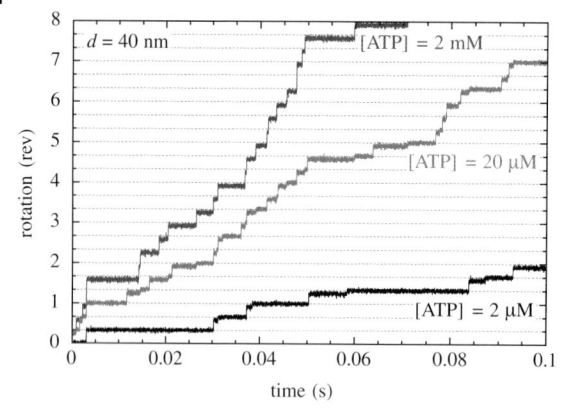

Figure 1.11 Stochastic trajectories of the rotation of the γ-shaft of the F_1 motor [107]. The number of revolutions $\theta(t)/2\pi$ is plotted versus time t in seconds for [ATP] = 2 µM, 20 µM, 2 mM, and [ADP][P_i] = 0. The diameter of the bead is d = 40 nm. The temperature is of 23 °C. The external torque is zero. This figure is to be compared with Figure 4 [101].

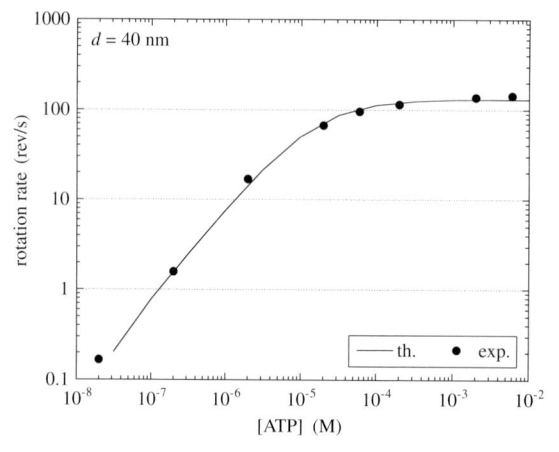

Figure 1.12 Mean rotation rate of the γ-shaft of the F_1 motor in revolutions per second, versus ATP concentration in mole per liter for [ADP][P_i] = 0 [107]. The diameter of the bead is d = 40 nm. The temperature is of 23 °C. The external torque is zero. The circles are the experimental data [101]. The solid line is the result of the present model.

ATP consumption rate R have been simulated with the continuous-angle model (1.92) for different values of the external torque τ and chemical potential difference $\Delta\mu$, which are the corresponding affinities. Figure 1.13 depicts the plane $(\tau, \Delta\mu)$ with the curves where either the rotation stops, $V = 0$, or the ATP consumption rate vanishes, $R = 0$. The value of the external torque where the rotation stops is called the stalling torque. The two curves $V = 0$ and $R = 0$ intersect at

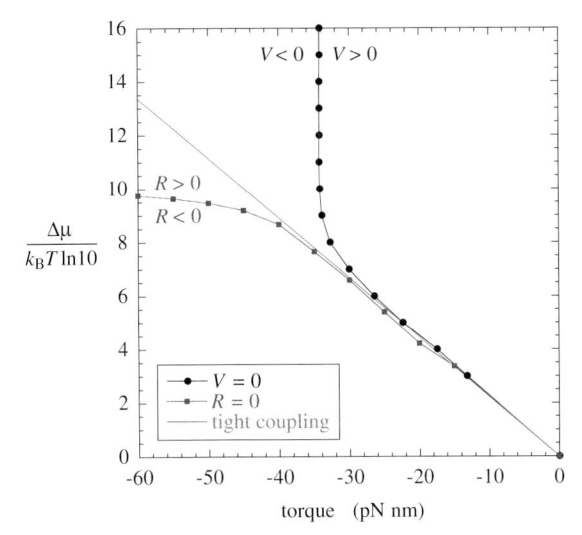

Figure 1.13 Chemical potential difference $\Delta\mu$ in units of $k_B T \ln 10$ versus the external torque τ for the situations where the rotation rate V (circles) and the ATP consumption rate R (squares) vanish in the continuous model (1.92) [108]. The straight line $\Delta\mu = -2\pi\tau/3$ where the chemomechanical affinity (1.101) vanishes, $A = 0$, is drawn for comparison. The concentrations are fixed according to $[\text{ATP}] = 4.9 \times 10^{0.8a-11}$ M and $[\text{ADP}][P_i] = 10^{-0.2a-5}$ M, in terms of the quantity $a = \Delta\mu/(k_B T \ln 10)$. The bead attached to the γ-shaft has the diameter $d = 2r = 80$ nm and the temperature is of 23 °C. The torque where $V = 0$ is called the stall torque. Curves $V = 0$ and $R = 0$ are difficult to determine close to the thermodynamic equilibrium point ($\tau = 0, \Delta\mu = 0$) because both the rotation rate V and the ATP consumption rate R are very small in this region, which explains the absence of dots close to the origin.

the origin ($\tau = 0, \Delta\mu = 0$), which is the thermodynamic equilibrium point. We notice that the curve $V = 0$ is above the curve $R = 0$ in the plane of the chemical potential difference $\Delta\mu$ versus the torque, as it should be in order to satisfy the second law of thermodynamics (1.91).

We observe that the two curves $V = 0$ and $R = 0$ are very close to each other if the external torque is larger than about -30 pN nm. In this regime, the following condition is satisfied:

$$\text{tight coupling:} \quad V = \frac{1}{3} R \qquad (1.100)$$

for which one revolution is driven by the hydrolysis of three ATP molecules. In the tight-coupling regime, there remains a single independent current and its associated affinity defined as [108]

$$A \equiv \underbrace{\frac{2\pi}{3} \frac{\tau}{k_B T}}_{\text{mechanics}} + \underbrace{\frac{\Delta\mu}{k_B T}}_{\text{chemistry}} . \qquad (1.101)$$

Consequently, the thermodynamic entropy production (1.91) becomes

$$\text{tight coupling:} \quad \frac{1}{k_B}\frac{d_i S}{dt} = AR \geq 0, \qquad (1.102)$$

which vanishes under the condition $\Delta\mu = -2\pi\tau/3$, as observed in Figure 1.13 for $-30\,\text{pN nm} < \tau < 0$.

Beyond this regime, the free-energy potentials $U_\sigma(\theta) - \tau\theta$ are so tilted by the external torque τ that the rotation can proceed independent of the reaction, and thus the coupling becomes loose [107]. Therefore, we recover two independent currents and affinities in the loose-coupling regime.

Chemical and mechanical efficiencies can be introduced for such molecular motors [103]. In the regime of ATP synthesis under a negative external torque, the ATP consumption rate as well as the rotation rate are negative, $R < 0$ and $V < 0$. In this regime, a chemical efficiency can be defined as the ratio of the free energy stored in the synthesized ATP over the mechanical power due to the external torque [103]

$$\eta_c \equiv -\frac{\Delta\mu R}{2\pi\tau V}, \qquad (1.103)$$

such that $0 \leq \eta_c \leq 1$. In the regime where the rotation is powered by ATP, a mechanical efficiency can be defined as the inverse of the chemical efficiency [103]

$$\eta_m \equiv -\frac{2\pi\tau V}{\Delta\mu R}. \qquad (1.104)$$

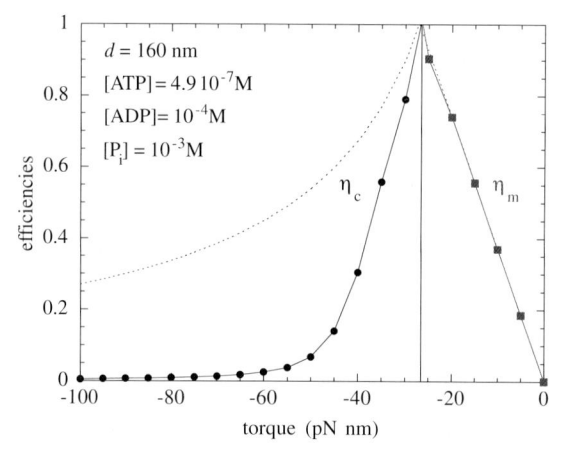

Figure 1.14 Chemical efficiency (1.103) and mechanical efficiency (1.104) versus the external torque τ in the continuous-state model (respectively circles and squares joined by a solid line) and compared with the prediction (1.105) of tight coupling (dashed lines) [108]. The vertical solid line indicates the stalling torque at $\tau = \tau_{\text{stall}} = -27.0\,\text{pN nm}$. The concentrations are [ATP] $= 4.9 \times 10^{-7}$ M, [ADP] $= 10^{-4}$ M, and [P$_i$] $= 10^{-3}$ M. The diameter of the bead is $d = 2r = 160$ nm with a temperature of 23 °C. The predictions of tight coupling (dashed lines) are respectively $\eta_c = \tau_{\text{stall}}/\tau$ for $\tau < \tau_{\text{stall}}$, and $\eta_m = \tau/\tau_{\text{stall}}$ for $\tau_{\text{stall}} < \tau < 0$, with $\tau_{\text{stall}} = -27.0\,\text{pN nm}$.

The mechanical efficiency satisfies $0 \leq \eta_m \leq 1$ in the regime where the external torque is negative, while both the rotation rate and the ATP consumption rates are positive, $V > 0$ and $R > 0$. Both efficiencies are depicted in Figure 1.14 and compared with the values

$$\text{tight coupling:} \quad \eta_c = \frac{1}{\eta_m} = -\frac{3\Delta\mu}{2\pi\tau} \quad (1.105)$$

expected from the tight-coupling conditions (1.100). This plot confirms that the nanomotor is functioning above the stalling torque in the tight-coupling regime and below in a loose-coupling regime. The efficiencies can nearly reach unit values around the stalling torque where the rotational motion of the motor is very slow and nearly adiabatic.

1.4.3
Discrete-State Description

In the tight-coupling regime, the rotation of the shaft is directly driven by each reactive event, which justifies the modeling of the stochastic process by a master equation for the probabilities in order to find the motor in each one of its different chemical states [108]

$$\frac{dP_\sigma(t)}{dt} = \sum_{\rho,\sigma'} \left[P_{\sigma'}(t) W_{+\rho}(\sigma'|\sigma) - P_\sigma(t) W_{-\rho}(\sigma|\sigma') \right], \quad (1.106)$$

with a sum over the reactions ρ and the chemical states σ' before the transition $\sigma' \xrightarrow{\rho} \sigma$ or after the reverse transition $\sigma \xrightarrow{-\rho} \sigma'$. The master equation conserves the total probability $\sum_\sigma P_\sigma(t) = 1$ for all times t.

The discrete-state model (1.106) can, in principle, be obtained by coarse graining the continuous-state model (1.92). Since the discrete states correspond to the angular intervals $\theta_\sigma < \theta < \theta_\sigma + 2\pi/3$ where the γ-shaft spends most of its time while in the chemical state σ, the probabilities ruled by the master equation (1.106) are related to the probability densities of the continuous-state description (1.92) according to

$$P_\sigma(t) = \int_{\theta_\sigma}^{\theta_\sigma + 2\pi/3} p_\sigma(\theta, t) \, d\theta \, . \quad (1.107)$$

In general, this method would lead to a non-Markovian master equation. In the case where there is a net separation of time scales between the dwell times and the jump times, the non-Markovian effects may be negligible and a description in terms of a Markovian equation such as the master equation (1.106) may be obtained. This is the situation we now consider.

The quantities $W_\rho(\sigma'|\sigma)$ are the transition rates per unit time from the state σ' to the state σ due to the reaction ρ. According to the mass-action law of chemical kinetics, the reaction rates W_ρ in (1.88) depend on the molecular concentrations in the solution surrounding the motor as follows [108]

$$W_{+1} = k_{+1}[\text{ATP}] \tag{1.108}$$

$$W_{-1} = k_{-1} \tag{1.109}$$

$$W_{+2} = k_{+2} \tag{1.110}$$

$$W_{-2} = k_{-2}[\text{ADP}][\text{P}_i] \tag{1.111}$$

where the quantities k_ρ ($\rho = \pm 1, \pm 2$) are the constants of the forward and backward reactions of binding and unbinding of ATP or ADP with P_i while [ATP], [ADP], and [P_i] represent the concentrations of these species. k_{+1} is the constant of ATP binding, k_{-1} the ATP unbinding constant, k_{+2} the constant of ATP synthesis, and k_{-2} the constant of product release. These constants can be fit to data from experiments or numerical simulations of the continuous-state model. In the latter case, we notice that the reaction constants are effective constants which depend on the external torque and are not identical with those entering the definition of the continuous-state model [108].

An advantage of the discrete-state model is that its solutions can be obtained analytically [108]. In a stationary state, the mean rotation and ATP consumption rates are given by [108]

$$V = \frac{V_{\text{max}}([\text{ATP}] - K_{\text{eq}}[\text{ADP}][\text{P}_i])}{[\text{ATP}] + K_M + K_P[\text{ADP}][\text{P}_i]} = \frac{1}{3}R \tag{1.112}$$

in terms of the constants

$$V_{\text{max}} \equiv \frac{1}{3} k_{+2} \tag{1.113}$$

$$K_M \equiv \frac{k_{-1} + k_{+2}}{k_{+1}} \tag{1.114}$$

$$K_P \equiv \frac{k_{-2}}{k_{+1}} \tag{1.115}$$

$$K_{\text{eq}} \equiv \frac{k_{-1}k_{-2}}{k_{+1}k_{+2}} = \exp\frac{1}{k_B T}\left(\Delta G^0 - \frac{2\pi}{3}\tau\right) \tag{1.116}$$
$$\simeq 4.9 \times 10^{-6}\,\text{M}^{-1} \exp\left(-\frac{2\pi}{3}\frac{\tau}{k_B T}\right).$$

We recover the Michaelis–Menten kinetics (1.99) for vanishing concentrations of ADP or P_i. An important observation is that the mean rotation and ATP consumption rates, which are the nonequilibrium fluxes of the nanomotor, both have a highly nonlinear dependence on the thermodynamic force or affinity (1.101) driving the motor out of equilibrium. This mechanochemical affinity allows us to express the ATP concentration as

$$[\text{ATP}] = K_{\text{eq}}[\text{ADP}][\text{P}_i]\,e^A. \tag{1.117}$$

The state of thermodynamic equilibrium thus corresponds to the vanishing of the affinity (1.101), as it should. Substituting (1.117) into (1.112), we obtain the following expression for the mean rotation rate [108]

$$V = \frac{V_{max}(e^A - 1)}{e^A - 1 + \frac{3V_{max}}{L}}, \quad (1.118)$$

where the coefficient L depends on the concentrations of ADP and P_i as well as the constants (1.113)–(1.116) and controls the linear response of the molecular motor because

$$V \simeq \begin{cases} \frac{1}{3}LA & \text{for} \quad A \ll 1 \\ V_{max} & \text{for} \quad A \gg 1 \end{cases}. \quad (1.119)$$

The analytic form (1.118) shows that the rotation rate depends on the thermodynamic force A in a highly nonlinear way, in contrast to what is often supposed. The nonlinear dependence is very important as observed in Figure 1.15. The linear regime extends around the thermodynamic equilibrium point at $\Delta\mu = 0$ where the function $V(A)$ is essentially flat because the linear-response coefficient assumes the very small value $L \simeq 10^{-5}\,\text{s}^{-1}$. Since the affinity is about $A \simeq 21.4$ under the physiological conditions [ATP] $\simeq 10^{-3}$ M, [ADP] $\simeq 10^{-4}$ M, and $[P_i] \simeq 10^{-3}$ M [102], the rotation rate would take the extremely low value $V \simeq LA/3 \simeq 6.5$ rev/day if the motor was functioning in the linear regime. Remarkably, the nonlinear depen-

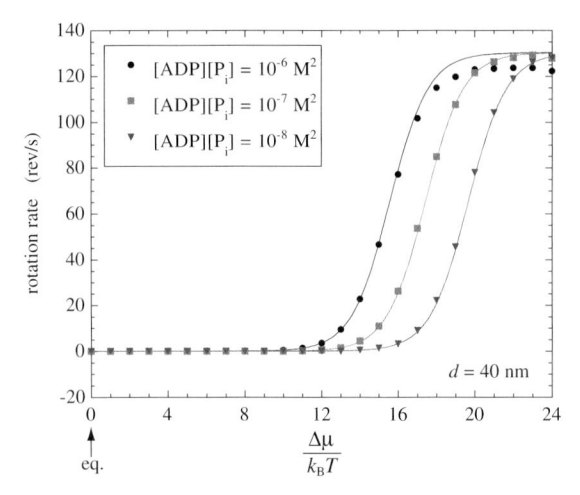

Figure 1.15 Mean rotation rate versus the affinity (1.101) for a zero external torque, in which case the affinity is equal to the chemical potential difference $\Delta\mu$ in units of the thermal energy $k_B T$ [108]. The thermodynamic equilibrium corresponds to $\Delta\mu = 0$. The ATP concentration is given in terms of the chemical potential difference by [ATP] = [ADP][P_i] exp[$(\Delta\mu - \Delta\mu^0)/(k_B T)$] $\simeq$ 4.9 × 10^{-6} M^{-1}[ADP][P_i] exp[$\Delta\mu/(k_B T)$] since $\Delta\mu^0 = -\Delta G^0 = 50$ pN nm. The results of the discrete model (solid lines) are compared with the continuous model (dots) for three different values of [ADP][P_i]. The diameter of the bead is $d = 2r = 40$ nm with a temperature of 23 °C.

dence of (1.118) on the affinity A allows the rotation rate to reach the maximum value $V_{\max} \simeq 130$ rev/s under physiological conditions.

The fluctuation theorem can be verified from the statistics of the random forward and backward substeps undergone by the γ-shaft of the F_1 motor, a full revolution corresponding to six substeps [111]. The graph associated with the stochastic process has six vertices simply connected as the edges of a hexagon. Therefore, there is a single independent current or flux because of the tight coupling between the mechanical rotation and the chemistry. According to (1.40), we should thus expect the following fluctuation relation

$$\frac{P(S_t = +s)}{P(S_t = -s)} = e^{As/2} \tag{1.120}$$

for the probability $P(S_t = s)$ that the nanomotor performs $s = S_t$ substeps over the time interval t. The quantity A is the affinity (1.101) for a zero external torque $\tau = 0$.

Figure 1.16 shows that the fluctuation relation (1.120) is indeed satisfied [111]. As seen in Figure 1.16, the probability distribution of the displacements takes a specific form where the odd displacements are almost never occurring. Indeed, for the values of the chemical concentrations considered in Figure 1.16, the probability to be on odd sites is about four orders of magnitude lower than the probability to be on even sites. Therefore, the system almost never stays on an odd site and immediately jumps to the next or previous site. We notice that the backward substeps of the motor are possible here because the concentrations are close to chemical equilibrium. Under physiological conditions, the motor is already far enough from equilibrium that the backward rotations become very improbable.

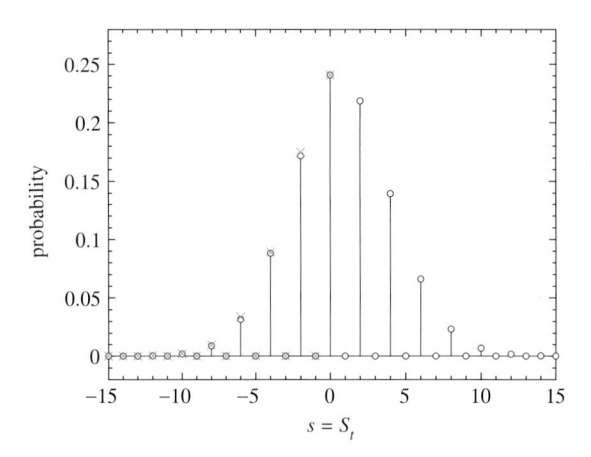

Figure 1.16 Probability $P(S_t = s)$ (open circles) that the F_1 motor performs $s = S_t$ substeps during the time interval $t = 10^4$ s compared with the expression $P(S_t = -s) e^{sA/2}$ (crosses) expected from the fluctuation theorem for [ATP] $= 6 \times 10^{-8}$ M and [ADP][P_i] $= 10^{-2}$ M^2 [111].

The theorem 1.2 of nonequilibrium temporal ordering also applies to the molecular motor, showing that their motion is more ordered out of equilibrium than at equilibrium. If the motion of the shaft of the F_1 rotary motor was recorded with the integers $\omega \in \{1, 2, 3\}$ corresponding to the three main steps, a stochastic trajectory as depicted in Figure 1.11 would correspond to a path $\ldots \omega_{n-1} \omega_n \omega_{n+1} \ldots$ At equilibrium where the principle of detailed balancing holds, the forward and backward motions are equiprobable and a typical path $\ldots 212132131223132\ldots$ would contain short sequences as well as their time reversals, for instance 132 and 231. In contrast, the time reversals of typical sequences are less probable out of equilibrium by the theorem of nonequilibrium temporal ordering. This remarkable property leads to the emergence of directionality in the rotation of the shaft as observed in Figure 1.11 where the paths are now restricted to $\ldots 123123123123123\ldots$ For this nonequilibrium trajectory, the probability of the time reversal 321 of the observed short sequence 123 is essentially vanishing. In the regime of Figure 1.11, the time-reversed temporal disorder (1.50) is thus very large while the temporal disorder (1.48) is very small. According to (1.51), the thermodynamic entropy production is thus large and positive, confirming that the motor is functioning away from equilibrium. This example shows that the directionality of the motion of molecular machines finds its origin in the nonequilibrium driving of these systems. The theorem of nonequilibrium temporal ordering is thus establishing a fundamental relationship between the second law of thermodynamics and the dynamical order that is observed, in particular, in biology. Indeed, the metabolism of biological systems is functioning out of equilibrium thanks to the energy supplied by the environment. This nonequilibrium driving allows the directionality of the various internal machines. This directionality means that the motion is dynamically ordered, a concept often intuitively quoted in biology. Remarkably, this dynamical order finds its fundamental understanding with the theorem of nonequilibrium temporal ordering.

In conclusion, the highly nonlinear dependence of the mean rotation rate of the γ-shaft (1.118) on the chemomechanical affinity (1.101) shows that, typically, the F_1 motor does not function in the linear-response regime defined by Onsager's linear-response coefficients, but instead runs in a nonlinear-response regime which is more the feature of far-from-equilibrium systems than of close-to-equilibrium ones. This remarkable property is attributed to the molecular architecture of the F_1 motor at the nanoscale, which allows for tight coupling between the mechanical motion and the chemical reactions powering the motor.

1.5
Chemical Nanosystems

Besides the aforementioned mechanical and mechanochemical nanosystems, there also exist chemical systems where populations of molecules evolve in time by reactions. These reactions can take place in a small recipient playing the role of a reactor, such as catalytic or electrochemical reactions at the surface of a nanoparticle or nanoelectrode. Other examples concern the biochemical reactions occurring in the nucleus or the cytosol of biological cells. Such reactions may form networks, as in the case for the metabolic networks or the genetic regulatory networks inside cells. Since the number of molecules is limited in such small systems, their time evolution is stochastic. These numbers are jumping at the random times corresponding to the random reactive events. This stochastic process is ruled by a chemical master equation for the probability that the system contains certain numbers of molecules of the species involved in the reactions. Such systems are out of equilibrium as long as the concentrations of these species have not reached their equilibrium ratios. The systems can be maintained out of equilibrium if the reactants are continuously supplied to the reactor from some reservoirs and the products are evacuated. This is the case for heterogeneous catalytic reactions on a solid surface in contact with a mixture of gases at fixed partial pressures. Since the reactions only happen at the surface thanks to its catalytic properties, the gaseous mixture acts as a reservoir containing large amounts of reactants. The reactions proceed out of equilibrium if the ratios of partial pressures do not take their equilibrium values. Since the numbers of molecules at the surface are small with respect to the numbers in the gaseous mixture, the nonequilibrium constraints can be maintained for arbitrarily long time intervals. Such nonequilibrium conditions are also satisfied if the reactants and products are supplied in larger quantities than the intermediate species. The ultimate situation is a reactive process taking place on a single molecule such as a molecular motor or a copolymer in the processes of DNA replication or protein synthesis. The importance of stochasticity has been emphasized in the context of genetic regulatory networks inside the cellular nucleus where the number of DNA molecules is necessarily limited [112, 113].

1.5.1
Chemical Transistor

An example of purely chemical systems is provided by the "chemical transistor" defined by the network of the following three chemical reactions:

$$R_1 \underset{k_{-1}}{\overset{k_{+1}}{\rightleftharpoons}} X, \quad R_2 \underset{k_{-2}}{\overset{k_{+2}}{\rightleftharpoons}} X, \quad R_3 \underset{k_{-3}}{\overset{k_{+3}}{\rightleftharpoons}} X, \tag{1.121}$$

in which the molecules of the species X can be produced from three different reactant or product species coming from different reservoirs. The parameters k_ρ are the reaction constants. The number X of molecules of the intermediate species X is a random variable that is incremented by one every time a molecule coming from a reservoir is converted into X and decreased by one if a reversed reaction occurs.

For this stochastic process, the probability $P(X, t)$ that the system contains X molecules at time t is ruled by the master equation [20, 21, 53, 54, 109, 110]

$$\frac{d}{dt} P(X, t) = \sum_\rho \left[P(X - \nu_\rho, t) W_{+\rho}(X - \nu_\rho | X) - P(X, t) W_{-\rho}(X | X - \nu_\rho) \right],$$

$$\tag{1.122}$$

where ν_ρ denote the stoichiometric coefficients of the reactions $\rho = \pm 1, \pm 2, \pm 3$. The transition rates are given by

$$W_{+\rho}(X|X+1) = k_{+\rho}\langle R_\rho \rangle \qquad \nu_{+\rho} = +1 \tag{1.123}$$

$$W_{-\rho}(X|X-1) = k_{-\rho}X \qquad \nu_{-\rho} = -1 \tag{1.124}$$

with $\rho = 1, 2, 3$ [114]. The concentrations of the species in the reservoir determine the mean numbers $\langle R_\rho \rangle = \Omega [R_\rho]$ where Ω is the volume of the reservoir.

The concentration of the species X is defined in terms of the mean number of molecules as

$$[X] = \frac{\langle X \rangle}{\Omega} = \frac{1}{\Omega} \sum_{X=0}^{\infty} X P(X,t). \tag{1.125}$$

This concentration evolves in time according to the following rate equation of macroscopic chemical kinetics:

$$\frac{d[X]}{dt} = \sum_\rho k_{+\rho}[R_\rho] - \sum_\rho k_{-\rho}[X]. \tag{1.126}$$

We notice that the macroscopic kinetic equation is exactly recovered because the chemical reaction network (1.121) is linear in the sense that the transition rates (1.123)–(1.124) are at most linear in X. For such linear reactions, the stationary solution of the master equation is given by the Poisson probability distribution:

$$P_{st}(X) = e^{-\langle X \rangle} \frac{\langle X \rangle^X}{X!} \tag{1.127}$$

with the mean value

$$\langle X \rangle = \frac{\sum_\rho k_{+\rho} \langle R_\rho \rangle}{\sum_\rho k_{-\rho}}. \tag{1.128}$$

The graph associated with the stochastic process has an infinite number of vertices for $X = 0, 1, 2, \ldots$ Each pair $(X, X+1)$ of vertices is connected by three non-directed edges, one for each of the three reactions (1.121). Figure 1.17 depicts the associated graph as well as alternative choices of maximal tree together with examples of possible chords defining cycles. Every time a transition occurs on one of these chords, the corresponding current (1.30) presents a delta peak. The analogy of this reaction network with a transistor is made by associating the three reservoirs with the source, the drain, and the gate of the transistor. Therefore, two independent affinities that control two independent currents can be defined in such systems.

By using the cycle which starts from the state X, goes to the state $X+1$ by the chord $\rho = \alpha$, and returns to the state X by the edge $\rho = -3$, the corresponding

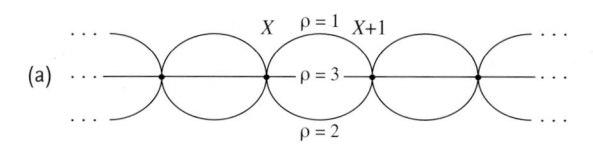

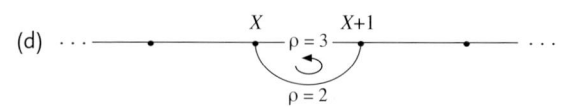

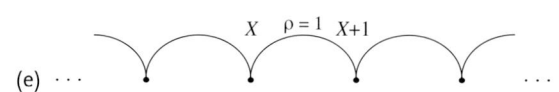

Figure 1.17 (a) Graph G associated with the reaction network (1.121) of the "chemical transistor" [21]. (b) Maximal tree $T(G)$ obtained after removing all the edges corresponding to the reactions $\rho = 1$ and $\rho = 2$. (c) Subgraph $T(G) + l$ composed of the maximal tree $T(G)$ and the chord $l = X \overset{\rho=1}{\to} X+1$, forming a cycle contributing to the current j_1 of affinity A_1. (d) Subgraph $T(G) + l$ composed of the maximal tree $T(G)$ and the chord $l = X \overset{\rho=2}{\to} X+1$, forming a cycle contributing to the current j_2 of affinity A_2. (e) Alternative maximal tree $T'(G)$ obtained after removing all the edges corresponding to the reactions $\rho = 2$ and $\rho = 3$.

macroscopic affinity is defined by (1.28) as

$$A_\alpha \equiv \ln \frac{W_{+\alpha}(X|X+1)\,W_{-3}(X+1|X)}{W_{-\alpha}(X+1|X)\,W_{+3}(X|X+1)} = \ln \frac{k_{-3}k_{+\alpha}\langle R_\alpha\rangle}{k_{-\alpha}k_{+3}\langle R_3\rangle} \qquad (1.129)$$

for $\alpha = 1$ or 2, as shown respectively in Figure 1.17c or 1.17d. Therefore, there are only two independent affinities in this chemical reaction network. These affinities only depend on the concentrations of the external reservoirs. The state of thermodynamic equilibrium is reached if both affinities vanish, that is, if the following detailed balancing conditions are satisfied:

$$\frac{k_{+\rho}}{k_{-\rho}}\langle R_\rho\rangle = \langle X\rangle_{\mathrm{eq}} \quad \text{with} \quad \rho = 1, 2, 3. \qquad (1.130)$$

These conditions fix the concentrations of two reservoirs in terms of the third reservoir R_3. The equilibrium states thus depend on the third concentration $[R_3] = \langle R_3\rangle/\Omega$ and form a hyperplane of codimension-one in the three-dimensional space of the concentrations. The distance with respect to this equilibrium hyperplane is controlled by the two affinities (1.129).

The independent random fluxes or currents corresponding to the two affinities (1.129) can be defined by (1.30). The generating function of the statistical cumulants of these currents can be precisely calculated and is given by [114]

$$Q(\lambda_1, \lambda_2) = \frac{k_{+3}\langle R_3 \rangle}{k_{-3}} \left[k_{-1} e^{A_1} + k_{-2} e^{A_2} + k_{-3} \right.$$

$$\left. - \frac{k_{-1} e^{\lambda_1} + k_{-2} e^{\lambda_2} + k_{-3}}{k_{-1} + k_{-2} + k_{-3}} \left(k_{-1} e^{A_1 - \lambda_1} + k_{-2} e^{A_2 - \lambda_2} + k_{-3} \right) \right], \tag{1.131}$$

which satisfies the fluctuation theorem (1.36)

$$Q(\lambda_1, \lambda_2) = Q(A_1 - \lambda_1, A_2 - \lambda_2) \tag{1.132}$$

in terms of the macroscopic affinities (1.129). The independent macroscopic fluxes (1.32) are given by [21]

$$J_1 = \frac{k_{-1} k_{+3} \langle R_3 \rangle}{k_{-1} + k_{-2} + k_{-3}} \left[e^{A_1} - 1 + \frac{k_{-2}}{k_{-3}} \left(e^{A_1} - e^{A_2} \right) \right] \tag{1.133}$$

$$J_2 = \frac{k_{-2} k_{+3} \langle R_3 \rangle}{k_{-1} + k_{-2} + k_{-3}} \left[e^{A_2} - 1 + \frac{k_{-1}}{k_{-3}} \left(e^{A_2} - e^{A_1} \right) \right], \tag{1.134}$$

which can be expanded in powers of the affinities according to (1.42). The Onsager reciprocity relations (1.43) can be verified as well as their generalizations (1.44) and (1.45) relating the higher-order response coefficients to the cumulants (1.33)–(1.35) [114]. The macroscopic expression of the thermodynamic entropy production (1.41) is thus recovered. In this reaction network, the fluxes have a strong nonlinear dependence on the thermodynamic forces or affinities, A_1 and A_2, in spite of their linear dependence on the concentrations.

This is also the case for the chemical diode which is the special case where the second reservoir is decoupled from the system by setting $k_{\pm 2} = 0$. There remains a single flux between the first and the third reservoir which is given by [21]

$$J_1 = \frac{k_{-1} k_{+3} \langle R_3 \rangle}{k_{-1} + k_{-3}} \left(e^{A_1} - 1 \right). \tag{1.135}$$

The flux can become arbitrarily large for positive values of the affinity, but saturates for negative values. This is the behavior of an electric diode or rectifier.

We notice that the nonlinear dependences of the fluxes on the affinities have the same origin as those found in (1.118) and Figure 1.15 for mechanochemical systems.

1.5.2
Chemical Multistability

An example of a bistable chemical system is given by Schlögl's trimolecular reaction network [115, 116]

$$A \underset{k_{-1}}{\overset{k_1}{\rightleftharpoons}} X \tag{1.136}$$

$$3X \underset{k_{-2}}{\overset{k_2}{\rightleftharpoons}} 2X + B . \tag{1.137}$$

On mesoscopic scales, the reaction is described as a stochastic process ruled by the master equation (1.122) with the transition rates:

$$W_{+1}(X|X+1) = k_{+1}[A]\Omega \qquad \nu_{+1} = +1 \tag{1.138}$$

$$W_{-1}(X|X-1) = k_{-1}X \qquad \nu_{-1} = -1 \tag{1.139}$$

$$W_{+2}(X|X-1) = k_{+2}X\frac{X-1}{\Omega}\frac{X-2}{\Omega} \qquad \nu_{+2} = -1 \tag{1.140}$$

$$W_{-2}(X|X+1) = k_{-2}[B]X\frac{X-1}{\Omega} \qquad \nu_{-2} = +1 . \tag{1.141}$$

The macroscopic kinetic equation for the concentration (1.125) of the intermediate species X is given by

$$\frac{d}{dt}[X] = k_{+1}[A] - k_{-1}[X] - k_{+2}[X]^3 + k_{-2}[B][X]^2 , \tag{1.142}$$

which is obtained from the master equation by neglecting the effects of fluctuations at $O(1/\Omega)$ in the limit $\Omega \to \infty$ [20]. This kinetic equation is nonlinear in the concentration, which leads to a phenomenon of bistability far from thermodynamic equilibrium as observed in Figure 1.18a. Figure 1.18b depicts the entropy production, showing that the regime of bistability exists far from the thermodynamic equilibrium state where the entropy production vanishes. Bistability is a particular case of multistability, which plays an important role in many nonlinear dissipative systems, especially in genetic regulatory networks controlling cell differentiation and its maintenance [54].

A stochastic trajectory simulated by Gillespie's algorithm [57, 58] is depicted in Figure 1.18c in the regime of bistability. Because of the fluctuations, the concentration does not remain forever in one of the two macroscopic steady states but randomly jumps between the upper and lower states. In this nonequilibrium regime, the stationary probability distribution is not Poissonian [20]. As suggested by the reaction network (1.136) and (1.137), a current is flowing between the reservoirs of molecules A and B. The graph associated with the stochastic process is shown in Figure 1.19 confirming the existence of a single independent current associated

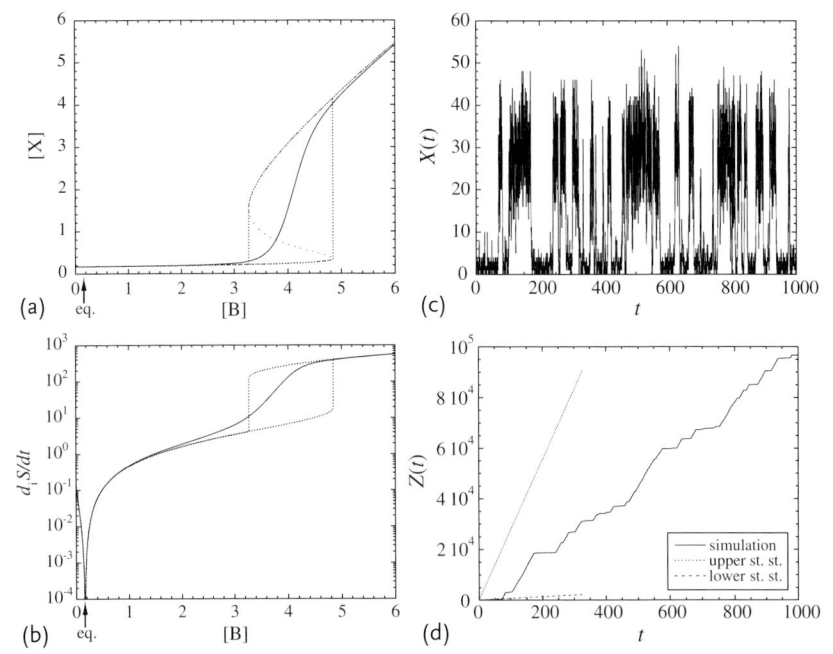

Figure 1.18 Schlögl trimolecular model (1.136)–(1.137) with the control parameters $k_{+1}[A] = 0.5$, $k_{-1} = 3$, and $k_{+2} = k_{-2} = 1$: (a) Bifurcation diagram of the concentration [X], obtained from the macroscopic equation (1.142) (dashed lines) and the stochastic description for $\Omega = 10$ (solid line). (b) Entropy production versus the control concentration [B] given by the macroscopic theory (dashed lines) and by the stochastic description for $\Omega = 10$ (solid line). The thermodynamic equilibrium is located at $[B]_{eq} = \frac{1}{6}$. (c) Stochastic time evolution of the number X of molecules of the intermediate species X, simulated by Gillespie's algorithm for $[B] = 4$ and $\Omega = 10$. (d) Stochastic time evolution of the quantity $Z(t)$ for the trajectory (c) for $[B] = 4$ and $\Omega = 10$. The increase of $Z(t)$ fluctuates between the entropy production rate of the lower (long-dashed line) and upper (dashed line) macroscopic stationary concentrations, in correlation with the jumps seen in (c). (Adapted from [20]).

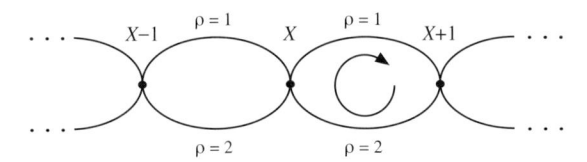

Figure 1.19 Graph associated with the stochastic process of Schlögl's trimolecular model.

with the affinity

$$A \equiv \ln \frac{W_{+1}(X|X+1)\,W_{+2}(X+1|X)}{W_{-1}(X+1|X)\,W_{-2}(X|X+1)} = \ln \frac{k_{+1}k_{+2}[A]}{k_{-1}k_{-2}[B]}. \tag{1.143}$$

The generating function (1.31) of the unique current $j(t)$ obeys the fluctuation theorem $Q(\lambda) = Q(A - \lambda)$.

An alternative fluctuating quantity has been defined by Lebowitz and Spohn [18] as the following ratio

$$Z(t) \equiv \ln \frac{W_{+\rho_1}(X_0|X_1)\, W_{+\rho_2}(X_1|X_2) \cdots W_{+\rho_n}(X_{n-1}|X_n)}{W_{-\rho_1}(X_1|X_0)\, W_{-\rho_2}(X_2|X_1) \cdots W_{-\rho_n}(X_n|X_{n-1})} . \tag{1.144}$$

Over long time intervals, this quantity is proportional to the fluctuating current, $Z(t) \simeq A \int_0^t dt'\, j(t')$. Accordingly, its statistical average in a stationary state gives the entropy production

$$\frac{1}{k_B} \frac{d_i S}{dt} = \lim_{t \to \infty} \frac{1}{t} \langle Z(t) \rangle = AJ \geq 0 . \tag{1.145}$$

The generating function of the statistical cumulants of the quantity (1.144) is defined as

$$q(\eta) \equiv \lim_{t \to \infty} -\frac{1}{t} \ln \langle e^{-\eta Z(t)} \rangle \tag{1.146}$$

and obeys the fluctuation theorem

$$q(\eta) = q(1 - \eta) . \tag{1.147}$$

The behavior of the quantity (1.144) in the bistable regime is shown in Figure 1.18d for the same stochastic trajectory as in Figure 1.18c. Since the entropy production rate is larger in the upper state than in the lower, the quantity (1.144) increases faster during the time intervals when the system is in the upper state. The generating function of this quantity can be calculated numerically and is depicted in Figure 1.20 for different values of the reservoir concentration [B] from the monostable to the bistable regime. In all cases, the generating function is symmetric around $\eta = 1/2$, as predicted by the fluctuation theorem (1.147), which is thus verified in this far-from-equilibrium bistable chemical system.

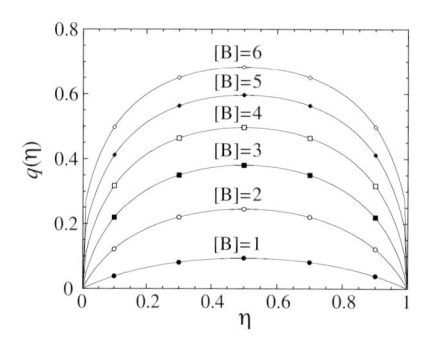

Figure 1.20 Generating function (1.146) of the fluctuating quantity $Z(t)$ versus η in the Schlögl model (1.136)–(1.137) for $k_{+1}[A] = 0.5$, $k_{-1} = 3$, $k_{+2} = k_{-2} = 1$, $[B] = 1, 2, \ldots, 6$, and $\Omega = 10$. We notice that $q(\eta) = 0$ at the equilibrium $[B]_{eq} = 1/6$ (Adapted from [20]).

We point out that the verification of the fluctuation theorem requires the coexistence of direct and reversed reactions in the network (1.136)–(1.137). If the rate constants k_ρ of some reactions were vanishing, the quantity (1.144) could not be defined and the thermodynamic entropy production (1.145) would be infinite. In this case, the reaction network is said to be fully irreversible.

1.5.3
Chemical Clocks

If the chemical reaction network involves two intermediate species X and Y, a self-sustained cyclic process becomes possible if the system is maintained far from equilibrium [70]. In such regimes, the chemical concentrations oscillate in time along a so-called limit cycle which is a periodic solution of the macroscopic kinetic equations [54]. Such rhythmic phenomena have been called chemical clocks [54] and observed not only in the famous Belousov-Zhabotinsky chemical reaction [117, 118], but also in biochemical reactions and in the regulatory networks at the basis of circadian rhythms [119]. On mesoscopic scales, the oscillations are affected by molecular fluctuations and the limit cycle is noisy. The description of such stochastic processes can be carried out in terms of the chemical master equation (1.122) now extended to the time evolution of the multivariate probability in order to find the system with given numbers of the different molecular species [54].

A model of oscillatory chemical reactions is provided by the so-called Brusselator model, which is defined by the following reaction network [120]:

$$A \underset{k_{-1}}{\overset{k_{+1}}{\rightleftharpoons}} X \tag{1.148}$$

$$B + X \underset{k_{-2}}{\overset{k_{+2}}{\rightleftharpoons}} Y + C \tag{1.149}$$

$$2X + Y \underset{k_{-3}}{\overset{k_{+3}}{\rightleftharpoons}} 3X \tag{1.150}$$

involving two intermediate species X and Y. The species A, B, and C are supposed to enter the system with the constant concentrations [A], [B], and [C]. We notice that the trimolecular reaction network (1.148)–(1.150) can be conceived as the reduction of a larger bimolecular reaction network [121]. Because of the autocatalytic reaction (1.150), the macroscopic kinetic equations are nonlinear and this nonlinearity is at the origin of the oscillations. These oscillations persist in the fully irreversible Brusselator where the constants of the reversed reactions are vanishing, $k_{-1} = k_{-2} = k_{-3} = 0$, and the thermodynamic entropy production is infinite. In order to keep the entropy production finite, all of the reaction constants should take non-vanishing values, which is herein supposed.

At the mesoscopic level, the random reactive events are described as a birth-and-death stochastic process for the numbers, $X(t)$ and $Y(t)$, of molecules of the intermediate species. This stochastic process is ruled by a Markovian master equation for the probability $P(X, Y, t)$ that the system contains the numbers X and Y of molecules at time t [20, 21, 53, 54, 109, 110]. For the Brusselator, the transitions rates of the master equation are given by [122]

$$W_{+1}(X, Y | X + 1, Y) = k_{+1} [A] \Omega \qquad (1.151)$$

$$W_{-1}(X, Y | X - 1, Y) = k_{-1} X \qquad (1.152)$$

$$W_{+2}(X, Y | X - 1, Y + 1) = k_{+2} [B] X \qquad (1.153)$$

$$W_{-2}(X, Y | X + 1, Y - 1) = k_{-2} [C] Y \qquad (1.154)$$

$$W_{+3}(X, Y | X + 1, Y - 1) = k_{+3} \frac{X(X-1)Y}{\Omega^2} \qquad (1.155)$$

$$W_{-3}(X, Y | X - 1, Y + 1) = k_{-3} \frac{X(X-1)(X-2)}{\Omega^2} , \qquad (1.156)$$

where Ω is the extensivity parameter characterizing the volume of the system.

Figure 1.21 shows examples of stochastic trajectories numerically simulated by Gillespie's algorithm [57, 58] for different values of the extensivity parameter Ω [122]. The reaction constants and the reservoir concentrations correspond to the same regime of oscillations. Since the numbers of molecules in the system is proportional to the extensivity parameter Ω, the size of the system increases with the parameter Ω. In the small system of Figure 1.21a, the molecular fluctuations are so important that regular oscillations are not visible. Indeed, the time autocorrelation function depicted in the third column rapidly decays to zero before the completion of a single cycle. Regular oscillations emerge if the system contains a few hundred molecules at larger values of Ω, as seen in Figure 1.21b. In this case, the time autocorrelation function presents several oscillations before decaying to zero. The oscillations become more regular as the size further increases in Figure 1.21c and d. For any finite size, the time autocorrelation function presents exponentially damped oscillations

$$C_{XX}(t) = \frac{\langle X(t) X(0) \rangle}{\langle X \rangle^2} - 1 \sim e^{-\gamma t} \cos(\omega t + \phi) \qquad (1.157)$$

with a correlation time proportional to the extensivity parameter $\gamma^{-1} \sim \Omega$. The constant of proportionality can be calculated by the Hamilton–Jacobi method in the weak-noise limit [123] and determines how the nonlinearities of the reaction network controls the robustness of the oscillations with respect to the molecular fluctuations.

Remarkably, the fluctuation theorem (1.147) is satisfied in the far-from-equilibrium oscillatory regime for both the quantity (1.144) and the fluctuating currents between the reservoirs [122]. In particular, the generating function of the quantity (1.144) is symmetric for a reflection around $\eta = 1/2$, as verified in Figure 1.22. For [B] = 7, the system evolves in the oscillatory regime of Figure 1.21.

We notice the analogy between the cyclic processes of molecular motors and chemical clocks. Both types of cyclic processes can be self-sustained if the system

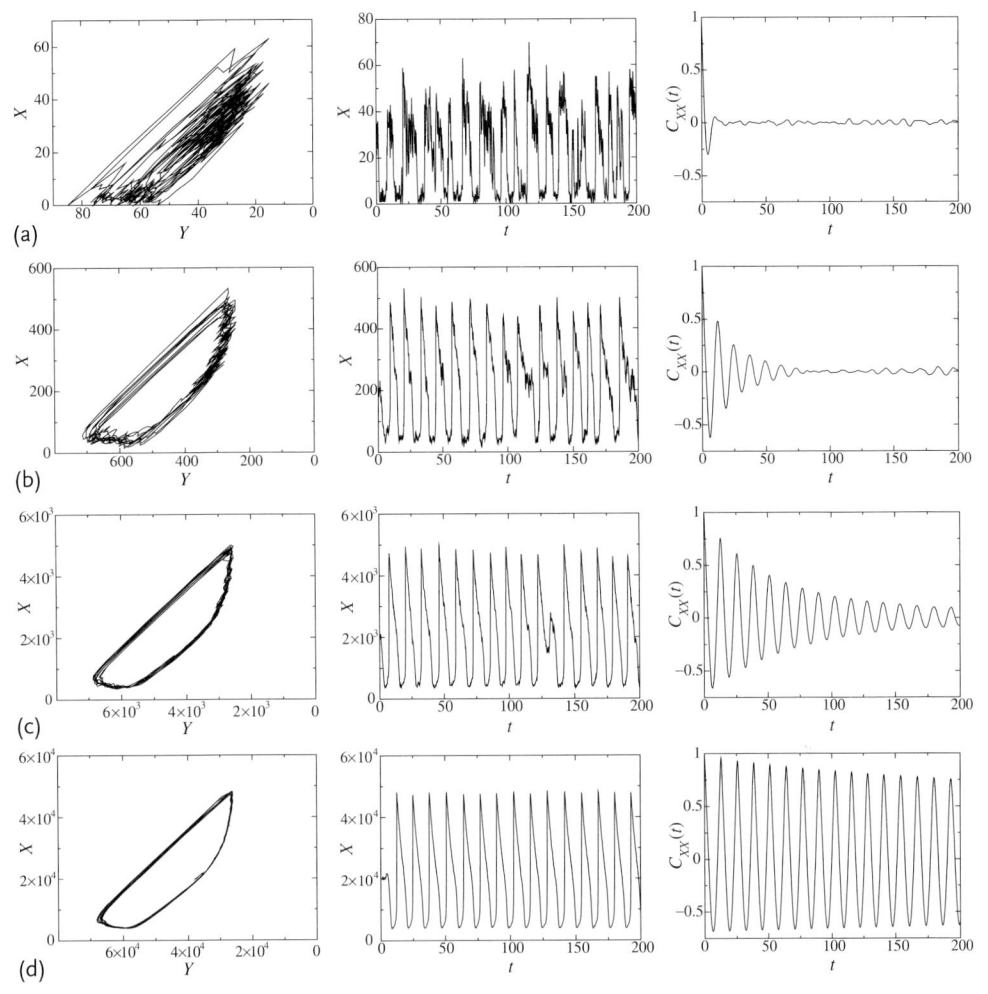

Figure 1.21 Simulation by Gillespie's algorithm of the oscillatory regime for the reversible Brusselator (1.148)–(1.150). The values of the concentrations are [B] = 7, [A] = [C] = 1, and the reaction constants $k_{+1} = 0.5$, $k_{+2} = k_{+3} = 1$, $k_{-1} = k_{-2} = k_{-3} = 0.25$. From (a) to (b), the extensivity parameter takes the values: (a) $\Omega = 10$, (b) $\Omega = 100$, (c) $\Omega = 1000$, and (d) $\Omega = 10\,000$. The first column depicts the phase portrait in the plane of the numbers X and Y of molecules. The second column shows the number X as a function of time. The third one depicts the autocorrelation function (1.157) of the number X, which is normalized to unity. (Adapted from [122]).

is driven out of equilibrium with appropriate thermodynamic forces or affinities. What is most remarkable is that such nonequilibrium systems are functioning in regimes of rotations or oscillations, although there is no time-dependent external driving. The temporal periodicity is an intrinsic feature to the system. Both types of systems have important differences. The functionality of molecular motors finds its origin in their molecular architecture with proteins playing the roles of shaft and

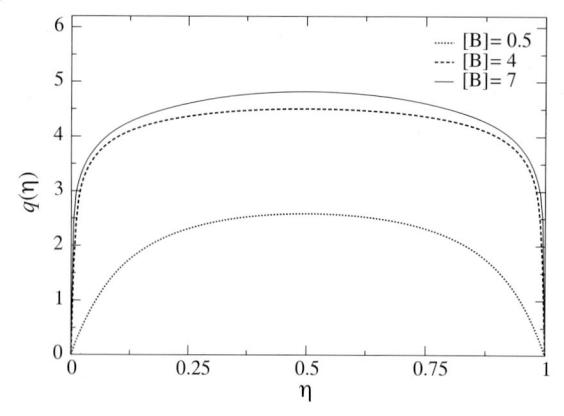

Figure 1.22 The generating function (1.146) numerically obtained for the Brusselator. The extensivity parameter takes the value $\Omega = 15$ while the control parameter takes the values $[B] = 0.5, 4, 7$. The reaction constants and the other concentrations are fixed at the values $k_{+1} = 0.5$, $k_{+2} = k_{+3} = 1$, $k_{-1} = k_{-2} = k_{-3} = 0.25$, and $[A] = [C] = 1$. For these parameter values, the equilibrium state is found at $[B]_{eq} = 0.0625$, the steady state is a stable node for $0 < [B] < 4.030\,24$ and a stable focus for $4.030\,24 < [B] < 6.366\,67$. The Hopf bifurcation happens at the critical concentration $[B]_{Hopf} = 6.366\,67$. Below this critical value, the steady state is an attractor. Above criticality for $6.366\,67 < [B]$, the attractor is the limit cycle of Figure 1.21. (Adapted from [122]).

barrel in the F_1 motor, for instance. On the other hand, chemical clocks are functioning by the time evolution of molecular populations. One must speculate as to which is most efficient in generating regular oscillations. The results above show that populations with several hundreds or thousands of molecules are required for chemical clocks to emerge from the molecular fluctuations [123]. These sizes are not significantly different from the numbers of atoms composing molecular motors, although the molecular architecture tends to confer to the latter well-defined shapes of their own [7–10].

1.5.4
Chemical Clocks Observed in Field Emission Microscopy

Nanometric chemical clocks have been experimentally observed thanks to field emission microscopy [124–129]. The principle of this microscopy is the magnification provided by an electric field extending from the nanometric tip of a metallic needle to a fluorescent screen [130].

Under a negative voltage, electrons are emitted by the needle and move along the lines of the electric field, arriving at the fluorescent screen at the points corresponding to the emission points at the surface of the needle tip. As a consequence, an image of the surface of the needle is projected on the screen with a magnification factor equal to the ratio of the curvature radii of the screen and the needle tip. This method is called field electron microscopy (FEM).

An alternative method is field ion microscopy (FIM). In this case, the needle is subjected to a positive voltage and the vacuum chamber is filled with a gas such as neon. When the neutral neon atoms collide with the surface of the needle, they are ionized and the resulting positive ion is projected to the screen along the line of electric field corresponding to the locus of ionization. The image seen on the fluorescent screen is the magnification of the surface of the needle. Since the needle is crystalline, its surface presents terraces and steps. The electric field is higher at the steps where the ionization rate is enhanced and, therefore, appear more clearly on the screen. Invented by Erwin Müller in the fifties, this was the first microscopy method that achieved atomic resolution [131].

In the nineties, chemical clocks were first observed in FEM [124, 125] and later in FIM with a higher (close to atomic) resolution [126, 127]. An example is provided by the reaction of catalytic water formation from hydrogen and oxygen on rhodium [128, 129]. The electric field at the tip of the rhodium needle is about 12 V/nm. The rhodium field emitter tip is exposed to a gaseous mixture of hydrogen and oxygen at fixed partial pressures. The radius of curvature of the tip is of the order of 10 nm. Since the reaction is concentrated at the tip because of the enhancement of the partial pressures by high electric fields, the field emitter tip constitutes a nanoreactor. Regular oscillations with a period of 30–40 seconds are observed around the partial pressures $P_{H_2} = 2 \times 10^{-3}$ Pa, $P_{O_2} = 2 \times 10^{-3}$ Pa, $P_{H_2O} = 0$, and temperature $T = 550$ K. These oscillations are self-sustained because the partial pressures of hydrogen, oxygen, and water are not in their chemical equilibrium ratios and the system is driven far from equilibrium.

The oscillations can be explained by the following reaction network [132]

adsorption-desorption of hydrogen:

$$H_2 \text{ (gas)} + 2\emptyset \text{(ad)} \underset{k_{dH}}{\overset{k_{aH}}{\rightleftharpoons}} 2H \text{ (ad)} \tag{1.158}$$

diffusion of hydrogen:

$$H \text{ (ad)} + \emptyset \text{(ad)} \overset{k_{diff}}{\rightleftharpoons} \emptyset \text{(ad)} + H \text{ (ad)} \tag{1.159}$$

adsorption-desorption of oxygen precursor:

$$O_2 \text{ (gas)} + \text{surface} \underset{\tilde{k}_d}{\overset{\tilde{k}_a}{\rightleftharpoons}} O_2 \text{ (pre)} + \text{surface} \tag{1.160}$$

dissociation of molecular oxygen and recombination:

$$O_2 \text{ (pre)} + 2\emptyset \text{(ad)} \underset{k_d}{\overset{k_a}{\rightleftharpoons}} 2O \text{ (ad)} \tag{1.161}$$

oxidation and reduction of rhodium:

$$O \text{ (ad)} + \emptyset \text{(sub)} \underset{k_{red}}{\overset{k_{ox}}{\rightleftharpoons}} \emptyset \text{(ad)} + O \text{ (sub)} \tag{1.162}$$

reaction of water formation:

$$2H \text{ (ad)} + O \text{ (ad)} \overset{k_r}{\rightarrow} 3\emptyset \text{(ad)} + H_2O \text{ (gas)} \tag{1.163}$$

Both hydrogen and oxygen diatomic molecules undergo a dissociative adsorption on the rhodium surface. The hydrogen atoms are highly mobile on rhodium. On the other hand, the oxygen atoms are strongly bounded to the surface and some of them move below the surface, forming a surface rhodium oxide layer with the stoichiometry of RhO_2. This surface oxide modifies the rate of adsorption of oxygen on the surface, which is the feedback mechanism at the origin of the oscillations. Water is formed from the combination of hydrogen and oxygen atoms and desorbs from the surface. Most of the water molecules leave the surface in a neutral form, but a fraction is ionized and contributes to the imaging of the surface by FIM.

Remarkable oscillatory patterns are observed in FIM [128, 129, 132]. These patterns have a length scale of tens of nanometers, which is much smaller than the typical length scales of ten to hundred micrometers for the patterns of reaction-diffusion processes in heterogeneous catalysis [133, 134]. The nanopatterns of the reactions observed in FIM can be explained in terms of the structural anisotropy of the crystalline tip, which results in different catalytic powers for the various exposed nanofacets [132]. Indeed, the activation energy $E_x^{\ddagger}$ and the prefactor k_x^0 of each rate coefficient depend on the crystalline orientation of the nanofacet where the reaction occurs. Each crystalline nanofacet is characterized by its Miller indices (h, k, l) or, equivalently, by the unit vector $\mathbf{n}$ perpendicular to the corresponding crystalline plane $\mathbf{n} = (h, k, l)/\sqrt{h^2 + k^2 + l^2}$ (see Figure 1.23). Moreover, the activation energy and the prefactor also depend on the magnitude F of the electric field normal to the metallic surface. If the tip is supposed to have the geometry of a paraboloid with a radius of curvature R at its apex, the electric field is known [130, 135] to vary

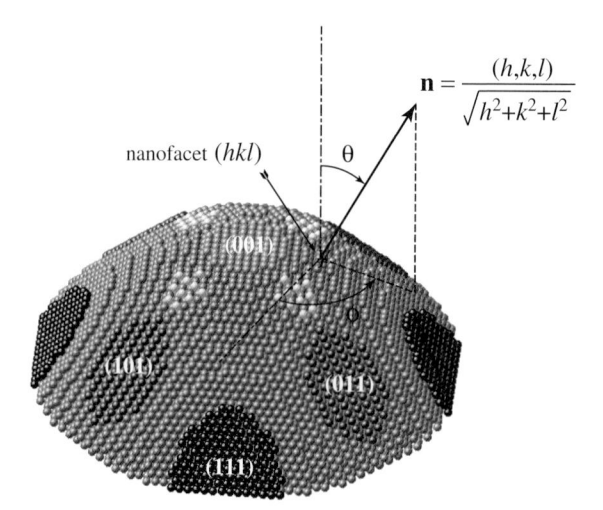

Figure 1.23 Ball model of the field emitter tip with the unit vector $\mathbf{n} = (\sin\theta\cos\phi, \sin\theta\sin\phi, \cos\theta)$ perpendicular to the nanofacet of Miller indices (h, k, l) of an underlying FCC crystal. All the balls inside a paraboloid are retained in this model of the field emitter. We notice that the mean electric field points in the same direction $\mathbf{F} = F\mathbf{n}$ because the electric field is always perpendicular to the surface of a conductor such as the field emitter tip. The (001) nanofacet is at the tip's apex. (Adapted from [132]).

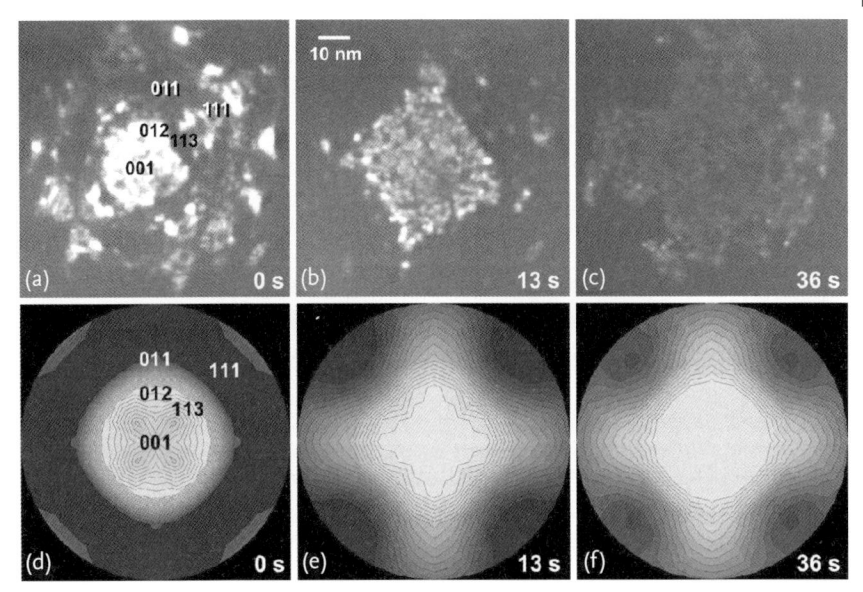

Figure 1.24 Series of FIM micrographs covering the complete oscillatory cycle as well as the corresponding time evolution of the subsurface oxygen distribution on a logarithmic scale as obtained within a kinetic model of the field emitter tip. Starting from a surface in the quasi-metallic state (a) and (d), an oxide layer invades the topmost plane and grows along the {011} facets, forming a nanometric cross-like structure (b) and (e). The oxide front finally spreads to the whole visible surface area (c) and (f). The temperature, electric field and partial pressures of oxygen in panels (a), (b) and (c) are $T = 550$ K, $F_0 = 12$ V/nm, $P_{O_2} = 2 \times 10^{-3}$ Pa, respectively. On the other hand, the hydrogen pressure in panels (c), (d) and (e) is $P_{H_2} = 2 \times 10^{-3}$ Pa in the FIM experiments and 4×10^{-3} Pa in the kinetic model (1.158)–(1.163). For the subsurface site occupation, the white areas indicate a high site occupation value while the dark areas indicate a low site occupation value. (Adapted from [132]).

as

$$F = \frac{F_0}{\sqrt{1 + (r/R)^2}}, \qquad (1.164)$$

where r is the radial distance with respect to the symmetry axis of the paraboloid and F_0 is the magnitude of the electric field at the apex of the tip. According to Arrhenius' law, the rate coefficient of each reaction can thus be written as

$$k_x = k_x^0(\boldsymbol{n}, F) \exp\left[-\frac{E_x^{\ddagger}(\boldsymbol{n}, F)}{k_B T}\right], \qquad (1.165)$$

giving the spatial dependence describing the anisotropy of the crystalline tip, which is necessary to explain the nanopatterns observed in the experiment (see Figure 1.24) [132].

In spite of the nanometric size of the chemical clocks, the oscillations are regular. The reason is that the system contains several thousands of adsorbed atoms, which is above the minimum number of a few hundred required for regular oscillations to emerge [123]. Consequently, chemical clocks can exist at the nanoscale.

1.5.5
Single-Copolymer Processes

The theorem 1.2 of nonequilibrium temporal ordering shows that dynamical order may appear in temporal sequences of events if the system is driven away from equilibrium. If the system had the ability to record the temporal sequence on a spatial support, the dynamical order would result into spatial order. The idea of coupling the dynamical order predicted by (1.51) with a spatial support of information has been developed in [32] to explain the possibility of information generation or information processing in nonequilibrium systems such as biosystems. Indeed, the theorem of nonequilibrium temporal ordering suggests that a nonequilibrium system can process information thanks to the directionality of its movements.

At the nanoscale, a natural spatial support of information is provided by random copolymers where information can be coded in the covalent bonds between the different monomers composing the copolymer chain. This is the idea of the aperiodic crystal that Erwin Schrödinger proposed in his well-known book, published in 1944, *What is Life?* [136]. As discovered in 1953 by Watson and Crick [137], the copolymer that codes genetic information is DNA. In DNA coding, a pair of nucleotides composed of about 64 atoms codes for two bits of information at the nanometer scale. DNA is but one among various types of copolymers in chemical and biological systems. Such copolymers are synthesized either with or without a template (see Figure 1.25). Styrene-butadiene is an example of a random copolymer grown without a template. Examples of copolymerizations with a template

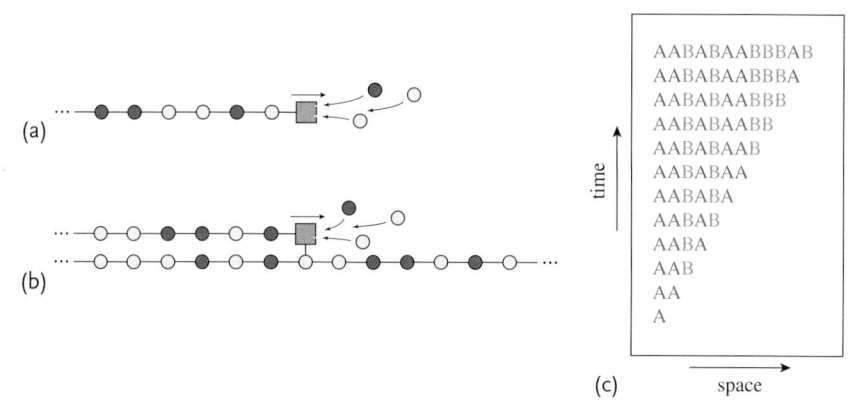

Figure 1.25 Schematic representations of (a) a copolymerization process without a template, (b) a copolymerization process with a template. The circles depict the monomers and the square predicts the catalyst of polymerization (adapted from [32]). (c) Schematic space-time plot of the growth process of a random copolymer composed of monomers A and B. The spatial sequence of monomers in the grown copolymer is directly determined by the temporal sequence of random attachments of A or B monomers at each time step.

are provided by the processes of DNA replication, DNA-mRNA transcription, and mRNA-protein translation [5].

Copolymerization necessarily proceeds away from equilibrium so that the growth of copolymers is controlled by the nonequilibrium conditions fixed, in particular by the chemical concentrations of the monomers in the solution surrounding the growing copolymer. Since copolymers are of nanometer dimensions, copolymerization processes are affected by the molecular fluctuations and should be described as stochastic processes. At equilibrium, the principle of detailed balancing prevents the ordering of temporal events and the possibility to generate or transmit information. Out of equilibrium, the ordering of temporal events becomes possible thanks to energy supply. Under this condition, the molecular motions acquire a directionality, allowing information generation or transmission (see Figure 1.25c).

1.5.5.1 Copolymerization without a Template

The stochastic growth of a single copolymer $\omega = m_1 m_2 m_3 \cdots m_l$ composed of monomers $m_i \in \{1, 2, \ldots, M\}$ can be described in terms of a master equation for the probability $P(\omega, t)$ in order to find the copolymer ω at time t [32, 33]

$$\frac{dP(\omega, t)}{dt} = \sum_{\omega'} \left[P(\omega', t) W(\omega'|\omega) - P(\omega, t) W(\omega|\omega') \right], \quad (1.166)$$

where $W(\omega|\omega')$ is the rate of the transition $\omega = m_1 m_2 m_3 \cdots m_l \rightarrow \omega' = m_1 m_2 m_3 \cdots m_{l'}$. During this transition, the length of the copolymer may change as $l \rightarrow l' = l \pm 1$ because of the attachment or detachment of a monomer. For many processes at fixed pressure and temperature T, the ratio of forward to backward transition rates can be expressed as [6]

$$\frac{W(\omega|\omega')}{W(\omega'|\omega)} = \exp \frac{G(\omega) - G(\omega')}{k_B T} \quad (1.167)$$

in terms of the free enthalpy $G(\omega)$ of a single copolymer chain ω surrounded by the solution. This Gibbs free energy is related to the enthalpy $H(\omega)$ and the entropy $S(\omega)$ of the copolymer ω in its environment at the temperature T by

$$G(\omega) = H(\omega) - T S(\omega). \quad (1.168)$$

Since the system is described by the statistical distribution $P(\omega, t)$ giving the probability in order to find the particular copolymer ω at time t, the overall entropy of the system is given by (1.21) and varies in time according to (1.22) because of the exchange of entropy (1.23) between the copolymer and its surrounding and the entropy production (1.24).

The growth may proceed in a regime described by the stationary statistical distribution $\mu_l(\omega)$ giving the composition of the copolymer chain ω, provided that its length is equal to l [138, 139]. This distribution is normalized as $\sum_\omega \mu_l(\omega) = 1$. In the regime of stationary growth, the probability distribution of the system becomes

$$P(\omega, t) = p(l, t) \mu_l(\omega), \quad (1.169)$$

where the time dependence is included in the statistical distribution $p(l, t)$ of the lengths l of the chains. The mean length of the chains is defined by $\langle l \rangle_t = \sum_l l \times p(l, t)$ and the mean growth velocity

$$v = \frac{d\langle l \rangle_t}{dt} \tag{1.170}$$

is supposed to be constant. Since the statistical composition of the copolymer is stationary, it is characterized by the mean entropy, enthalpy and free enthalpy per monomer defined as

$$s \equiv \lim_{l \to \infty} \frac{1}{l} \sum_\omega \mu_l(\omega) S(\omega) \tag{1.171}$$

$$h \equiv \lim_{l \to \infty} \frac{1}{l} \sum_\omega \mu_l(\omega) H(\omega) \tag{1.172}$$

$$g \equiv \lim_{l \to \infty} \frac{1}{l} \sum_\omega \mu_l(\omega) G(\omega) = h - Ts . \tag{1.173}$$

By substituting (1.169) in (1.21), the entropy of the system can be shown to have a dominant linear dependence on the mean chain length over long time intervals and its time derivative can be written as [32]

$$\frac{dS}{dt} = v\left[s + D(\text{polymer})\right] \tag{1.174}$$

in terms of the mean entropy per monomer s and the spatial disorder per monomer defined by the Shannon entropy per monomer [140–142]

$$D(\text{polymer}) = \lim_{l \to \infty} -\frac{1}{l} \sum_\omega \mu_l(\omega) \ln \mu_l(\omega) . \tag{1.175}$$

On the other hand, the entropy exchange (1.23) can be expressed in terms of the enthalpy per monomer as

$$\frac{d_e S}{dt} = v \frac{h}{T} \tag{1.176}$$

so that the thermodynamic entropy production is given by

$$\frac{d_i S}{dt} = v A \geq 0 \tag{1.177}$$

in terms of the affinity per monomer

$$A \equiv -\frac{g}{T} + D(\text{polymer}) = \varepsilon + D(\text{polymer}) , \tag{1.178}$$

where $\varepsilon = -g/T$ is the driving force of the copolymer growth [32]. This driving force is positive if the Gibbs free energy decreases as the copolymer grows, in which case the growth is driven by energetic effect. If the copolymer is random, its spatial disorder is positive so that the driving force can take negative values down to its equilibrium value $\varepsilon_{eq} = -D(\text{polymer})$ where the affinity (1.178) is vanishing. Consequently, a random copolymer can grow by entropic effects in an adverse free-energy landscape [32, 143].

1.5.5.2 Copolymerization with a Template

Similar considerations apply to the case of copolymerization processes taking place with a template given, for instance, by another copolymer [32]. The latter is characterized by the statistical distribution $\nu_l(\alpha)$ of the sequences α of length l, which is normalized as $\sum_\alpha \nu_l(\alpha) = 1$. The growing copolymer ω now acquires a composition which depends on the template α. In the stationary regime, the statistical distribution of the system can be written as

$$P(\omega, t) = p(l, t)\mu_l(\omega|\alpha), \tag{1.179}$$

where $\mu_l(\omega|\alpha)$ gives the conditional probability of the copy ω given the composition α of the template over the length l of the copy [32]. The joint probability to find the copy ω and the template α is defined as

$$\mu_l(\omega, \alpha) \equiv \nu_l(\alpha)\mu_l(\omega|\alpha), \tag{1.180}$$

and the probability of the copy ω for all the possible templates α is given by

$$\mu_l(\omega) \equiv \sum_\alpha \nu_l(\alpha)\mu_l(\omega|\alpha) = \sum_\alpha \mu_l(\omega, \alpha). \tag{1.181}$$

The Shannon conditional disorder of the copy grown on a given template is defined as [140–142]

$$D(\text{polymer}|\text{template}) = \lim_{l \to \infty} -\frac{1}{l} \sum_{\alpha, \omega} \nu_l(\alpha)\mu_l(\omega|\alpha) \ln \mu_l(\omega|\alpha), \tag{1.182}$$

while the Shannon disorder of all the possible copies is still defined by (1.175) with the probability distribution (1.181). The mutual information per monomer between the copy and the template is thus defined as [142]

$$I(\text{polymer}, \text{template}) \equiv D(\text{polymer}) - D(\text{polymer}|\text{template})$$
$$= \lim_{l \to \infty} \frac{1}{l} \sum_{\alpha, \omega} \mu_l(\omega, \alpha) \ln \frac{\mu_l(\omega, \alpha)}{\nu_l(\alpha)\mu_l(\omega)}. \tag{1.183}$$

The mutual information is always nonnegative and bounded as

$$0 \leq I(\text{polymer}, \text{template}) \leq \text{Min}\{D(\text{polymer}), D(\text{template})\}. \tag{1.184}$$

Following a similar reasoning, as in the case without a template, thermodynamic entropy production can be written as (1.177) with the affinity per monomer given by

$$A \equiv \varepsilon + D(\text{polymer}|\text{template}) = \varepsilon + D(\text{polymer}) - I(\text{polymer}, \text{template}) \tag{1.185}$$

with the driving force $\varepsilon = -g/T$ [32]. This fundamental result shows that positive mutual information becomes possible away from equilibrium where the thermodynamic entropy production and the affinity are positive: The larger the mutual information, the better the transmission of information between the template and the copy. This phenomenon can be illustrated for the case of DNA replication [32].

1.5.5.3 DNA Replication

In vivo, DNA replication is a nonequilibrium process which has a free energy cost of two ATP molecules for the attachment of one nucleotide [5]. DNA replication is performed by a whole machinery which separates the two strands of DNA and catalyzes the growth of two new strands by DNA polymerases. Moreover, an exonuclease performs proofreading for the correction of possible errors [144, 145].

The influence of the nonequilibrium constraints has been studied in the case of the DNA polymerase Pol γ, which replicates the human mitochondrial DNA [32]. The human mitochondrial DNA is 16.5 kb long and can be obtained from GenBank [146]. Forward kinetic constants k_{+mn} for the incorporation of both correct and incorrect nucleotides are available [147]. The reversed kinetic constants are taken as $k_{-mn} = k_{+mn}\, e^{-\varepsilon}$ in terms of the driving force ε, which is the control parameter of the nonequilibrium constraints. The thermodynamic equilibrium corresponds to the value $\varepsilon_{eq} = -\ln 4$. The replication process has been simulated by Gillespie's algorithm [57, 58].

Figure 1.26a depicts the mean velocity of replication in nucleotide per second as a function of the driving force. The velocity vanishes at equilibrium and increases towards a maximum value of about 34 nucleotides per second for a large and positive driving force. On the other hand, the percentage of errors in the repli-

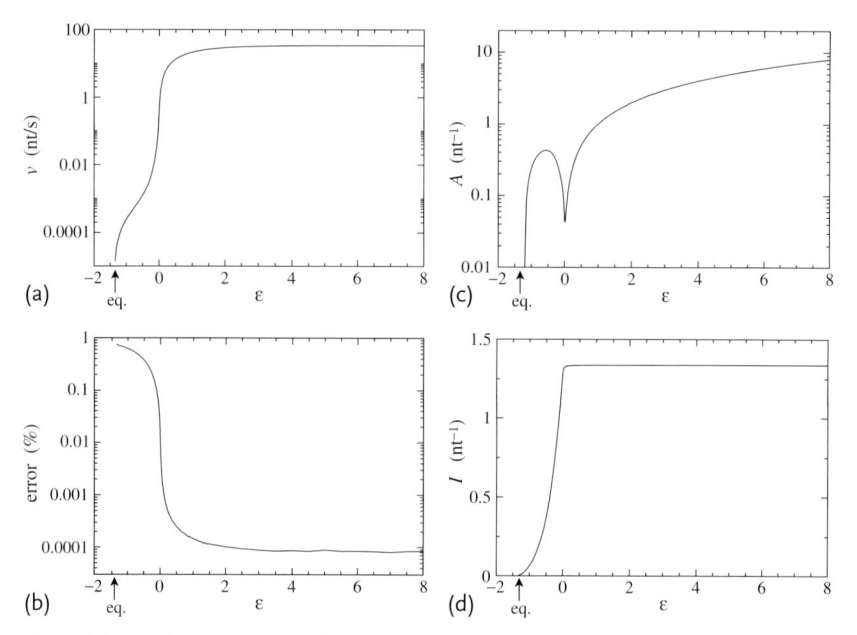

Figure 1.26 Replication process of human mitochondrial DNA by polymerase Pol γ: (a) Velocity versus the driving force ε. (b) Percentage of DNA replication errors versus the driving force ε. (c) Affinity per copied nucleotide versus the driving force ε. (d) Mutual information between the copied DNA strand and the original strand serving as template, versus the driving force ε. (Adapted from [32]).

cation process takes the large value of 75% at equilibrium and drops by several orders of magnitude away from equilibrium (see Figure 1.26b). The percentage of replication errors does not vanish far from equilibrium and constitutes a source of genetic mutations. Accordingly, the analysis shows that the thermal and molecular fluctuations cause replication errors and, thus, mutations.

Since the growth velocity is positive, the thermodynamic entropy production per copied nucleotide is given by the affinity (1.178) depicted in Figure 1.26c. The local minimum around $\varepsilon \simeq 0.015$ marks the transition between the regime driven by entropic effect and the one driven by energy effect. On the other hand, Figure 1.26d shows the mutual information per nucleotide (1.183) between the copy and the template. This mutual information vanishes at equilibrium and reaches a plateau at the maximum value $I_{\max} \simeq 1.337$ nats far from equilibrium. Therefore, the transmission of information between the template and the copy is not possible at equilibrium, but requires the process to be pushed far enough from equilibrium for replication to be accurate. The fidelity of replication is characterized by the percentage of errors or by the mutual information between the copy and the template [32].

If the copolymerization process was running too close to the thermodynamic equilibrium, the mutations would be too frequent to allow replication and self-reproduction. Therefore, the self-reproduction of biological systems is closely connected to their metabolism, that is, to their nonequilibrium nature. This connection finds its origin in the aforementioned phenomenon of nonequilibrium temporal ordering. It is remarkable that the ingredients of Darwinian evolution are so closely related to the basic physico-chemical laws of nonequilibrium nanosystems. The experimental study of copolymerization processes under tunable nonequilibrium conditions awaits the development of new single-molecule techniques such as nanopore sequencing [33, 148, 149].

1.6
Conclusions and Perspectives

Many nanosystems play an important role because they function out of equilibrium. The nonequilibrium constraints allow useful motions to be sustained in nanosystems as it is the case for molecular motors, electronic nanosystems, or catalytic nanodevices.

Although nanosystems are affected by thermal and molecular fluctuations, thermodynamic considerations continue to apply, thanks to recent advances in nonequilibrium statistical thermodynamics, which have led to the discovery of new fundamental relationships valid not only close to, but also far from equilibrium.

Nanosystems may be isolated or in contact with one or several reservoirs. Because of their intermediate size between the atoms and the macroscopic objects, their study requires the connection between different levels of description. Their microscopic dynamics are ruled by Newtonian or Hamiltonian equations for the motions of all the atoms. Often, a few degrees of freedom are relevant to the specific property of interest in a nanosystem. These few degrees of freedom are typically slower than the other ones, which results into a separation of time scales justifying a description in terms of stochastic processes, as explained in Section 1.2. Different stochastic processes can be envisaged depending on the level of coarse graining of the relevant quantities. These quantities may be the work performed on a nanosystem by some external force or the currents flowing across the nanosystem. At the nanoscale, these quantities are fluctuating in time so that their recording over some time interval generates random temporal sequences called paths or histories. Each path has a certain probability to occur in a long time series, which defines the probability distribution characterizing the stochastic process. If the stochastic process is stationary, the probability distribution is invariant under time evolution. This is the case at thermodynamic equilibrium where the microcanonical, canonical or grand-canonical probability distributions describe the statistical averages as well as the fluctuations of the relevant quantities. This concept of invariant probability distribution has the subtle feature of remaining a stationary solution of Liouville equation of time evolution while describing individual systems which are highly dynamical with incessant temporal fluctuations. The conceptual advantage of probability distribution introduces two levels of descriptions: (1) the single-system level which is dynamical and (2) the statistical-ensemble level in terms of a probability distribution which can remain stationary and thus invariant in time. Since the aforementioned equilibrium distributions are functions of the Hamiltonian, they are symmetric under time reversal if the Hamiltonian is.

Now, the concept of stationary probability distribution extends to nonequilibrium nanosystems in which heat or particle currents are flowing across the system between reservoirs at different temperatures or chemical potentials. Herein, the quantities of interest may fluctuate and be highly dynamical at the single-system level and, yet, be described by a stationary probability distribution for the possible random paths or histories followed by the system. The bonus provided by the probabilistic description is that nonequilibrium states such as chemical clocks, which are considered as being time-dependent at the macroscale, can nevertheless be described by a stationary probability distribution at the mesoscale in the presence of fluctuations. Indeed, the stationary probability distribution of the paths may lead to time correlation functions which present (damped) oscillations as illustrated in Figure 1.21. Thus, there is no incompatibility to describe a system with non-trivial time evolutions in terms of a stationary probability distribution.

Typically, the mean currents are non-vanishing in such nonequilibrium steady states as they flow from one reservoir to another in a well-defined direction. Although currents flowing in the opposite direction are possible, both nonequilibrium steady states are physically distinct. At the level of the stationary probability distribution, the paths in the direction of the mean currents are more probable than their time reversals. This fundamental remark shows that the stationary probability distributions of nonequilibrium steady states break the time-reversal symmetry. This symmetry breaking happens at the statistical level of description and, therefore, is perfectly compatible with microreversibility. The latter property only states

1.6 Conclusions and Perspectives

that if Newton's or Liouville's equations admit a solution, they also admit its time reversal as a solution. However, microreversibility does not mean that the solution and its time reversal should coincide. In the case where they are physically distinct, which most often occurs, the selection of one out of the pair is breaking the time-reversal symmetry. This phenomenon is well known in condensed-matter physics as spontaneous symmetry breaking. Nevertheless, this concept has not been considered until very recently for the time-reversal symmetry in nonequilibrium statistical mechanics [34–38].

Remarkably, the relationship (1.51) shows that the thermodynamic entropy production is an order parameter for the breaking of the time-reversal symmetry in nonequilibrium steady states. Indeed, (1.51) gives the entropy production as the difference between the temporal disorders of the time-reversed and forward paths. At equilibrium, the forward and reversed temporal disorders are equal because of the principle of detailed balancing. Out of equilibrium, the time reversals are less probable than the typical paths so that the time-reversed temporal disorder becomes larger than the forward one, which results into a positive thermodynamic entropy production. Accordingly, a directionality manifests itself away from equilibrium, which is expressed by the theorem 1.2 of nonequilibrium temporal ordering. Most remarkably, the breaking of time-reversal symmetry has been experimentally verified down to the nanoscale [30, 31]. In this way, the property of irreversibility that was previously envisaged for macrosystems containing about 10^{23} atoms is nowadays considered in small systems containing a few hundred or thousand atoms.

The time-reversal symmetry breaking of the stationary probability distribution concerns all of the large-deviation properties of nonequilibrium nanosystems, as reported in Section 1.2. Amazingly, these properties obey universal relationships which are the consequence of microreversibility. Such relationships have been discovered in different types of dynamical systems and stochastic processes, and are commonly called fluctuation theorems [11–27]. Recently, a fluctuation theorem has been proved for all the independent currents flowing across a nonequilibrium system thanks to graph theory [21–23]. This theory allows one to identify the thermodynamic forces, also called the affinities [69, 70], as well as the corresponding random currents by using the cycles of the graph associated with the stochastic process. To some extent, these cycles play a similar role as the periodic orbits in dynamical systems theory [38, 47]. Once the affinities are identified in a stochastic process, the symmetry (1.36) can be proved for the generating function of the statistical cumulants of the fluctuating currents, which is the content of the fluctuation theorem. This generating function provides us with the full counting statistics of the particles flowing across a nonequilibrium system such as an electronic, photonic, or chemical nanodevices [24–27]. Moreover, the symmetry of the fluctuation theorem for the currents allows us to deduce not only the Onsager–Casimir reciprocity relations for the linear response coefficients, but also the generalizations (1.44)–(1.46) of these relations to the nonlinear response coefficients [72]. These generalizations relate the nonlinear response coefficients to the statistical cumulants (1.33)–(1.35) characterizing the fluctuations. The fluctuation theorem for currents has also been proved for open quantum systems and apply to boson and fermion transport through mesoscopic junctions in electronic, photonic, or ultracold-atom devices [26, 27].

Additionally, fluctuation theorems have been obtained for Hamiltonian systems driven by time-dependent forces within Jarzynski's framework [11, 12]. The work performed on the system is a random variable similar to the number of particles exchanged between reservoirs and their fluctuations also obey a symmetry relation. In this way, the equilibrium free energy of conformation changes can be experimentally measured by folding and unfolding biomolecules [4]. These new fundamental results present promising perspectives in our understanding of nonequilibrium nanosystems, as revealed by the case studies presented in this review.

The mechanical nanosystems considered in Section 1.3 are Hamiltonian and isolated, and possibly driven by external time-dependent forces.

The double-walled carbon nanotubes (DWCNT) can slide relative to one another in a telescopic motion [83–86]. Systems containing about 1300 atoms can be studied by molecular dynamics simulations, showing that the energy of the sliding motion is dissipated among the vibrational degrees of freedom of each nanotube. This dissipation is caused by the friction coming from the van der Waals interaction between the nanotubes. The methods of Brownian motion theory extends from the micrometer down to the nanometer. Accordingly, the translational and rotational sliding motions are described by Langevin stochastic models with dynamic friction coefficients given by the Kirkwood formula of nonequilibrium statistical mechanics [60]. If the DWCNT system is isolated, it undergoes a relaxation towards a microcanonical equilibrium state with fluctuations in the sliding motions described by Langevin equations. However, the internal rotation between the two nanotubes can be driven by external forces as is the case in nanomechanical devices using DWCNTs for the shaft of rotary motors. In such devices, the energy continuously supplied by the external driving is dissipated by the property of rotational friction described in Section 1.3.

The other example presented in Section 1.3 is the heating of a microplasma by electromagnetic waves. This is a time-dependent Hamiltonian system to which Crooks fluctuation theorem applies for the work performed by the time-dependent electric force on the microplasma. This work represents the energy supplied to the system and is a random variable depending on the initial conditions. Heating corresponds to a positive value of the work and cooling to a negative value. As described by the fluctuation theorem, the work is statistically distributed around a positive most probable value, which corresponds to the heating of the system.

Section 1.4 presents the F_1-ATPase motor, which is an example of a nanosystem powered by a continuous supply of chemical energy [100, 101, 105, 106]. Both the shaft and the barrel of this nanomotor are composed of proteins. The barrel is a hexamer of proteins, three of which can bind adesonine triphosphate (ATP). ATP hydrolysis is the source of energy allowing the active rotation of the shaft in a mean unidirectional motion. The ATP molecules are coming from the aqueous solution surrounding the protein and they bind in the catalytic sites of the motor at random arrival times. These arrival times form a stationary random process for constant values of ATP concentration. In this regard, the nanomotor does not need an external cyclic driving, but has its own autonomous cycle as a car engine at constant gas supply. In the F_1 nanomotor, ATP hydrolysis is catalyzed by the conformational change of the protein, inducing the rotation of the shaft by a mechanism similar to a camshaft [102]. In this way, the chemical energy is transformed into mechanical motion at the expense of some dissipation which necessarily reduces the efficiency of energy transduction. This process can be investigated in detail thanks to continuous-state or discrete-state stochastic models [107, 108]. The regimes of tight or loose coupling between the chemistry and the mechanics of the motor can be identified. Although the chemical reaction and the mechanical rotation constitute a priori two independent dissipative processes leading to entropy production, they combine in the tight-coupling regime in such a way that a single independent dissipative process remains. In this tight-coupling regime, the mechanical efficiency can reach its maximum possible value (1.105). Remarkably, the rotation rate of the nanomotor has a highly nonlinear dependence on the thermodynamic force or affinity driving the system out of equilibrium, contrary to what is usually supposed. This nonlinear dependence allows the rotation to be much faster than would be the case if the nanomotor was functioning in the regime of linear response, whereupon a rotation rate of 130 rev/sec can be obtained under physiological conditions [100, 101]. Consequently, the rotation rate drops to extremely slow rates close to the equilibrium state and random backward rotations are very rare, although the fluctuation theorem remains valid [111]. As discussed at the end of Section 1.4, the directionality of the rotation is directly related to the fact that the motor is functioning out of equilibrium, which can be understood as the consequence of the theorem of nonequilibrium temporal ordering [29].

Further examples of nonequilibrium nanosystems evolving along an autonomous cycle are provided by the nanometric chemical clocks described in Section 1.5. In these systems, the time evolution concerns the populations composed of many identical molecules of small size, instead of the motion of mechanical pieces formed by large rigid molecules as carbon nanotubes or proteins. If the molecular architecture is instrumental to the mechanical rotation of molecular motors, it plays a secondary role in chemical clocks where the chemical concentrations of some intermediate species undergo oscillations. At the macroscale, these concentrations obey kinetic ordinary differential equations established by the laws of chemical kinetics [54, 70]. These equations are nonlinear if the reaction network contains autocatalytic steps. Far from equilibrium, their solutions may undergo bifurcations leading to bistability, limit cycles, or even chaotic attractors. At the nanoscale, the populations contain hundred or thousand molecules so that the reactive events induce random jumps in the concentrations, whereupon their time evolution is stochastic. Such stochastic processes can be driven out of equilibrium if the reactants and the products are supplied in proportions different from their chemical equilibrium values, in which case a source of chemical free energy maintains the matter fluxes from the reactants to the products. These fluxes are similar to the currents across an electronic circuit and obey a fluctuation theorem which is remarkably valid far from equilibrium, as shown in Section 1.5 for bistability in the

Schlögl trimolecular model [20] and for oscillations in the Brusselator model [122]. As illustrated with the chemical transistor, the fluctuation theorem allows us to recover the Onsager reciprocity relations for the linear response coefficients and to verify their generalizations to the nonlinear response coefficients [21, 72, 114]. The discovery of these new relations in the nonlinear response regime constitutes one of the most important advances in nonequilibrium statistical thermodynamics since Onsager's paper published in 1931 [71].

Field emission microscopy reveals the existence of nanometric chemical clocks [132]. These nonequilibrium nanosystems are the stage of catalytic reactions on metallic surface at the field emitter tip. By the localization of a high electric field, the tip constitutes a nanoreactor of a few dozen nanometers where patterns are observed in the coverage of the surface by adsorbed species. In some regimes, these nanopatterns may oscillate as it is the case in the reaction of water formation from hydrogen and oxygen on rhodium [132]. In spite of the nanometric size of the tip, several thousand atoms are adsorbed on the surface so that the system is large enough to sustain correlated oscillations and behave as a chemical clock [123]. The observed nanopatterns can be understood in terms of the spatial dependence of the reaction coefficients on the orientation of each nanofacet composing the tip with respect to the underlying metallic crystal [132].

Further nonequilibrium nanosystems where chemical reactions play a fundamental role are those composed of a single copolymer which is growing by the attachment of monomers coming from the surrounding solution [32, 33]. In this case, the process is stochastic since the single copolymer is of nanometric size and the reactive events occur at random with either attachment or detachment of monomers. These copolymerization processes can take place freely or with a template. The latter case is fundamental for biology since DNA replication, DNA-mRNA transcription, and mRNA-protein translation are examples of copolymerization processes. They are powered by a chemical energy supply and, therefore, proceed in nonequilibrium regimes. For instance, DNA replication requires two ATP for the attachment of each nucleotide [5]. In this regard, the self-reproduction closely depends on metabolism. This close connection can be further established by considering the nonequilibrium statistical thermodynamics of such copolymerization processes. In this way, the thermodynamic entropy of a single copolymer can be shown to depend on the Shannon disorder in the sequence of monomers composing the copolymer [32]. This spatial disorder contributes to the thermodynamic entropy production of copolymerization. Accordingly, the growth of the copolymer can be driven by the entropic effect of this spatial disorder besides the energetic effect due to the Gibbs free energy of monomer attachment. In the case of copolymerization with a template, the thermodynamic entropy production also depends on the mutual Shannon information between the template and the copy, which shows that nonequilibrium thermodynamics plays a fundamental role in the control of information processing at the molecular level [32, 33].

These new results pave the way for the statistical thermodynamics of nonequilibrium nanosystems. They introduce new perspectives in our understanding of the motions and processes that nanosystems can perform. The thermodynamic forces or affinities driving the fluxes and currents can be identified and related to thermodynamic quantities such as energy and entropy, allowing us to study energy dissipation in nanosystems and their efficiency. Moreover, a new light is shed on the possible bridges between biology and the physico-chemical laws. Indeed, biological systems present structures on all scales from the macroscale down to the nanoscale. The hierarchical organization of living systems often appears in contrast with textbook physico-chemical systems which are typically homogeneous such as gases, liquids, and other continuous media. Therefore, the investigation of nonequilibrium biological processes at the nanoscale is very new. In particular, the new advances provide conceptual methods to study the metabolism of biological systems at the molecular level and to shift from 3D to 4D molecular biology.

The new results also concern the dynamical aspects of information and establish the possibility of temporal ordering in nonequilibrium nanosystems. Indeed, recent results suggest that the dynamical order characteristic of biological systems can be understood on the basis of the second law of thermodynamics thanks to the appreciation of the importance of path probabilities and the breaking of time-reversal symmetry in the statistical description of nonequilibrium processes.

Acknowledgments

This research is financially supported by the Belgian Federal Government (IAP project "NOSY"), the "Communauté française de Belgique" (contract "Actions de Recherche Concertées" No. 04/09-312), and the F.R.S.-FNRS Belgium (contract F.R.F.C. No. 2.4577.04).

References

1 Einstein, A. (1956) *Investigations on the theory of the Brownian movement*, Dover, New York.
2 de Heer, W.A. (1993) *Rev. Mod. Phys.*, **65**, 611.
3 Schmidt, M., Kusche, R., Hippler, T., Donges, J., Kronmüller, W., von Issendorff, B., and Haberland, H. (2001) *Phys. Rev. Lett.*, **86**, 1191.
4 Collin, D., Ritort, F., Jarzynski, C., Smith, S.B., Tinoco, I. Jr., and Bustamante, C. (2005) *Nature*, **437**, 231.
5 Alberts, B., Bray, D., Johnson, A., Lewis, J., Raff, M., Roberts, K., and Walter, P. (1998) *Essential Cell Biology*, Garland Publishing, New York.
6 Hill, T.L. (2005) *Free Energy Transduction and Biochemical Cycle Kinetics*, Dover, New York.
7 Schienbein, M. and Gruler, H. (1997) *Phys. Rev. E*, **56**, 7116.
8 Lerch, H.-P., Rigler, R., and Mikhailov, A.S. (2005) *Proc. Natl. Acad. Sci. USA*, **102**, 10807.
9 Togashi, Y. and Mikhailov, A.S. (2007) *Proc. Natl. Acad. Sci. USA*, **104**, 8697.
10 Cressman, A., Togashi, Y., Mikhailov, A.S., and Kapral, R. (2008) *Phys. Rev. E*, **77**, 050901.
11 Jarzynski, C. (1997) *Phys. Rev. Lett.* **78**, 2690.
12 Crooks, G.E. (1999) *Phys. Rev. E*, **60**, 2721.
13 Evans, D.J., Cohen, E.G.D., and Morriss, G.P. (1993) *Phys. Rev. Lett.*, **71**, 2401.
14 Tasaki, S. and Gaspard, P. (1995) *J. Stat. Phys.*, **81**, 935.
15 Gallavotti, G. and Cohen, E.G.D. (1995) *Phys. Rev. Lett.*, **74**, 2694.

16 Gallavotti, G. (1996) *Phys. Rev. Lett.*, **77**, 4334.
17 Kurchan, J. (1998) *J. Phys. A: Math. Gen.*, **31**, 3719.
18 Lebowitz, J.L. and Spohn, H. (1999) *J. Stat. Phys.*, **95**, 333.
19 Maes, C. (1999) *J. Stat. Phys.*, **95**, 367.
20 Gaspard, P. (2004) *J. Chem. Phys.*, **120**, 8898.
21 Andrieux, D. and Gaspard, P. (2004) *J. Chem. Phys.*, **121**, 6167.
22 Andrieux, D. and Gaspard, P. (2007) *J. Stat. Phys.*, **127**, 107.
23 Andrieux, D. (2008) *Nonequilibrium Statistical Thermodynamics at the Nanoscale*, Ph.D. thesis, Université Libre de Bruxelles.
24 Andrieux, D. and Gaspard, P. (2006) *J. Stat. Mech.: Th. Exp.*, P01011.
25 Harbola, U., Esposito, M., and Mukamel, S. (2007) *Phys. Rev. B*, **76**, 085408.
26 Andrieux, D., Gaspard, P., Monnai, T., and Tasaki, S. (2009) *New J. Phys.*, **11**, 043014.
27 Esposito, M., Harbola, U., and Mukamel, S. (2009) *Nonequilibrium fluctuations, fluctuation theorems, and counting statistics in quantum systems*, arXiv:0811.3717, to appear in *Rev. Mod. Phys.*
28 Gaspard, P. (2004) *J. Stat. Phys.*, **117**, 599.
29 Gaspard, P. (2007) *C.R. Phys.*, **8**, 598.
30 Andrieux, D., Gaspard, P., Ciliberto, S., Garnier, N., Joubaud, S., and Petrosyan, A. (2007) *Phys. Rev. Lett.*, **98**, 150601.
31 Andrieux, D., Gaspard, P., Ciliberto, S., Garnier, N., Joubaud, S., and Petrosyan, A. (2008) *J. Stat. Mech.*, P01002.
32 Andrieux, D. and Gaspard, P. (2008) *Proc. Natl. Acad. Sci. USA*, **105**, 9516.
33 Andrieux, D. and Gaspard, P. (2009) *J. Chem. Phys.*, **130**, 014901.
34 Gaspard, P. (2005) *Chaotic Dynamics and Transport in Classical and Quantum Systems* (eds P. Collet et al.), Kluwer, Dordrecht, p. 107.
35 Gaspard, P. (2006) *Physica A*, **369**, 201.
36 Gaspard, P. (2007) *Adv. Chem. Phys.*, **135**, 83.
37 Gaspard, P. (2008) *Physics of Self-Organization Systems* (eds S. Ishiwata and Y. Matsunaga), World Scientific, New Jersey, p. 67.
38 Gaspard, P. (1998) *Chaos, Scattering and Statistical Mechanics*, Cambridge University Press, Cambridge UK.
39 Arnold, V.I. and Avez, A. (1968) *Ergodic Problems of Classical Mechanics*, W.A. Benjamin, New York.
40 Cornfeld, I.P., Fomin, S.V., and Sinai, Ya.G. (1982) *Ergodic Theory*, Springer, Berlin.
41 Eckmann, J.-P. and Ruelle, D. (1985) *Rev. Mod. Phys.*, **57**, 617.
42 Dellago, Ch., Posch, H.A., and Hoover, W.G. (1996) *Phys. Rev. E*, **53**, 1485.
43 van Beijeren, H., Dorfman, J.R., Posch, H.A., and Dellago, Ch. (1997) *Phys. Rev. E*, **56**, 5272.
44 van Zon, R., van Beijeren, H., and Dellago, Ch. (1998) *Phys. Rev. Lett.*, **80**, 2035.
45 Yang, H.-L. and Radons, G. (2005) *Phys. Rev. E*, **71**, 036211.
46 Gaspard, P. and Wang, X.-J. (1993) *Phys. Rep.*, **235**, 291.
47 Cvitanović, P. and Eckhardt, B. (1991) *J. Phys. A: Math. Gen.*, **24**, L237.
48 Boltzmann, L. (1964) *Lectures on Gas Theory*, Dover, New York.
49 Langevin, P. (1908) *C. r. Acad. Sci.*, **146**, 530.
50 Fokker, A.D. (1914) *Ann. Phys.*, **43**, 810.
51 Planck, M. (1917) *Sitzungsber. Preuss. Akad. Wissens.*, p. 324.
52 Pauli, W. (1928) *Festschrift zum 60. Geburtstage A. Sommerfelds*, Hirzel, Leipzig, p. 30.
53 McQuarrie, D.A. (1967) *Suppl. Rev. Ser. Appl. Prob.*, **8**, 1.
54 Nicolis, G. and Prigogine, I. (1977) *Self-Organization in Nonequilibrium Systems*, Wiley, New York.
55 van Kampen, N.G. (1981) *Stochastic Processes in Physics and Chemistry*, North-Holland, Amsterdam.
56 Gardiner, C.W. (2004) *Handbook of Stochastic Methods for Physics, Chemistry and the Natural Sciences*, 3rd ed., Springer, Berlin.
57 Gillespie, D.T. (1976) *J. Comput. Phys.*, **22**, 403.
58 Gillespie, D.T. (1977) *J. Phys. Chem.*, **81**, 2340.
59 Chandrasekhar, S. (1943) *Rev. Mod. Phys.*, **15**, 1.

60 Kirkwood, J.G. (1946) *J. Chem. Phys.*, **14**, 180.
61 Green, M.S. (1952) *J. Chem. Phys.*, **20**, 1281.
62 Green, M.S. (1954) *J. Chem. Phys.*, **22**, 398.
63 Kubo, R. (1957) *J. Phys. Soc. Japan*, **12**, 570.
64 Esposito, M. (2004) *Kinetic theory for quantum nanosystems*, Ph.D. thesis, Université Libre de Bruxelles.
65 Schnakenberg, J. (1976) *Rev. Mod. Phys.*, **48**, 571.
66 Luo, J.-L., van den Broeck, C., and Nicolis, G. (1984) *Z. Phys. B – Condens. Matter*, **56**, 165.
67 Lewis, G.N. (1925) *Proc. Natl. Acad. Sci. USA*, **11**, 179.
68 Bridgman, P.W. (1928) *Phys. Rev.*, **31**, 101.
69 De Donder, T. and Van Rysselberghe, P. (1936) *Affinity*, Stanford University Press, Menlo Park CA.
70 Prigogine, I. (1967) *Introduction to Thermodynamics of Irreversible Processes*, Wiley, New York.
71 Onsager, L. (1931) *Phys. Rev.*, **37**, 405.
72 Andrieux, D. and Gaspard, P. (2007) *J. Stat. Mech.: Th. Exp.*, P02006.
73 Andrieux, D. and Gaspard, P. (2008) *Europhys. Lett.*, **81**, 28004.
74 Hummer, G. and Szabo, A. (2005) *Acc. Chem. Res.*, **38**, 504.
75 Mukamel, S. (2003) *Phys. Rev. Lett.*, **90**, 170604.
76 Andrieux, D. and Gaspard, P. (2008) *Phys. Rev. Lett.*, **100**, 230404.
77 Kawai, R., Parrondo, J.M.R., and van den Broeck, C. (2007) *Phys. Rev. Lett.*, **98**, 080602.
78 Maes, C. and Netočný, K. (2003) *J. Stat. Phys.*, **110**, 269.
79 Iijima, S. (1991) *Nature*, **354**, 56.
80 Cumings, J. and Zettl, A. (2000) *Science*, **289**, 602.
81 Fennimore, A.M., Yuzvinsky, T.D., Han, W.-Q., Fuhrer, M.S., Cumings, J., and Zettl, A. (2003) *Nature*, **424**, 408.
82 Bourlon, B., Glattli, D.C., Miko, C., Forró, L., and Bachtold, A. (2004) *Nano Lett.*, **4**, 709.
83 Servantie, J. and Gaspard, P. (2003) *Phys. Rev. Lett.*, **91**, 185503.
84 Servantie, J. and Gaspard, P. (2006) *Phys. Rev. B*, **73**, 125428.
85 Servantie, J. and Gaspard, P. (2006) *Phys. Rev. Lett.*, **97**, 186106.
86 Servantie, J. (2006) *Dynamics and friction in double walled carbon nanotubes*, Ph.D. thesis, Université Libre de Bruxelles.
87 Saito, R., Dresselhaus, G., and Dresselhaus, M.S. (1998) *Physical Properties of Carbon Nanotubes*, Imperial College Press, London.
88 Brenner, D.W. (1990) *Phys. Rev. B*, **42**, 9458.
89 Lu, J.P., Li, X.P., and Martin, R.M. (1992) *Phys. Rev. Lett.*, **68**, 1551.
90 Buldum, A. and Lu, J.P. (1999) *Phys. Rev. Lett.*, **83**, 5050.
91 Zheng, Q. and Jiang, Q. (2002) *Phys. Rev. Lett.*, **88**, 045503.
92 Tangney, P., Cohen, M.L., and Louie, S.G. (2006) *Phys. Rev. Lett.*, **97**, 195901.
93 Zhang, X.H., Santoro, G.E., Tartaglino, U., and Tosatti, E. (2009) *Phys. Rev. Lett.*, **102**, 125502.
94 Bollinger, J.J. and Wineland, D.J. (1990) *Sci. Am.*, **262**, 114.
95 Walther, H. (1993) *Adv. At. Mol. Opt. Phys.*, **31**, 137.
96 Ghosh, P.K. (1995) *Ion Traps*, Clarendon Press, Oxford.
97 Gaspard, P. (2003) *Phys. Rev. E*, **68**, 056209.
98 Abrahams, J.P., Leslie, A.G.W., Lutter, R., and Walker, J.E. (1994) *Nature*, **370**, 621.
99 Ian Menz, R., Walker, J.E., and Leslie, A.G.W. (2001) *Cell*, **106**, 331.
100 Noji, H., Yasuda, R., Yoshida, M., and Kinosita, K. Jr. (1997) *Nature*, **386**, 299.
101 Yasuda, R., Noji, H., Yoshida, M., Kinosita, K. Jr., and Itoh, H. (2001) *Nature*, **410**, 898.
102 Kinosita, K. Jr., Adachi, K., and Itoh, H. (2004) *Annu. Rev. Biophys. Biomol. Struct.*, **33**, 245.
103 Jülicher, F., Ajdari, A., and Prost, J. (1997) *Rev. Mod. Phys.*, **69**, 1269.
104 Oosawa, F. and Hayashi, S. (1986) *Adv. Biophys.*, **22**, 151.
105 Elston, T., Wang, H., and Oster, G. (1998) *Nature*, **391**, 510.
106 Wang, H. and Oster, G. (1998) *Nature*, **396**, 279.

107 Gaspard, P. and Gerritsma, E. (2007) *J. Theor. Biol.*, **247**, 672.
108 Gerritsma, E. and Gaspard, P. *Discrete-versus continuous-state descriptions of the F_1-ATPase molecular motor*, arXiv:0904.4218.
109 Nicolis, G. (1972) *J. Stat. Phys.*, **6**, 195.
110 Malek-Mansour, M. and Nicolis, G. (1975) *J. Stat. Phys.*, **13**, 197.
111 Andrieux, D. and Gaspard, P. (2006) *Phys. Rev. E*, **74**, 011906.
112 McAdams, H.H. and Arkin, A. (1997) *Proc. Natl. Acad. Sci. USA*, **94**, 814.
113 Gonze, D., Halloy, J., and Gaspard, P. (2002) *J. Chem. Phys.*, **116**, 10997.
114 Andrieux, D. and Gaspard, P. (2008) *Phys. Rev. E*, **77**, 031137.
115 Schlögl, F. (1971) *Z. Phys.*, **248**, 446.
116 Schlögl, F. (1972) *Z. Phys.*, **253**, 147.
117 Belousov, B.B. (1958) *Sb. Ref. Radiats. Med.*, Medgiz, Moscow, p. 145.
118 Zhabotinskii, A.M. (1964) *Biophysika*, **9**, 306.
119 Goldbeter, A. (1996) *Biochemical Oscillations and Cellular Rhythms*, Cambridge University Press, Cambridge UK.
120 Lefever, R., Nicolis, G., and Borckmans, P. (1988) *J. Chem. Soc., Faraday Trans. 1*, **84**, 1013.
121 Baras, F., Pearson, J.E., and Malek Mansour, M. (1990) *J. Chem. Phys.*, **93**, 5747.
122 Andrieux, D. and Gaspard, P. (2008) *J. Chem. Phys.*, **128**, 154506.
123 Gaspard, P. (2002) *J. Chem. Phys.*, **117**, 8905.
124 van Tol, M.F.H., Gilbert, M., and Nieuwenhuys, B.E. (1992) *Catal. Lett.*, **16**, 297.
125 Gorodetskii, V., Block, J.H., and Ehsasi, M. (1992) *Appl. Surf. Sci.*, **67**, 198.
126 Gorodetskii, V., Drachsel, W., and Block, J.H. (1993) *Catal. Lett.*, **19**, 223.
127 Voss, C. and Kruse, N. (1996) *Appl. Surf. Sci.*, **94/95**, 186.
128 Visart de Bocarmé, T., Bär, T., and Kruse, N. (2001) *Ultramicroscopy*, **89**, 75.
129 Visart de Bocarmé, T., Beketov, G., and Kruse, N. (2004) *Surf. Interface Anal.*, **36**, 522.
130 Miller, M.K., Cerezo, A., Hetherington, M.G., and Smith, G.D.W. (1996) *Atom Probe Field Ion Microscopy*, Clarendon Press, Oxford.
131 Müller, E.W. (1951) *Z. Physik*, **131**, 136.
132 McEwen, J.-S., Gaspard, P., Visart de Bocarmé, T., and Kruse, N. (2009) *Proc. Natl. Acad. Sci. USA*, **106**, 3006.
133 Ertl, G., Norton, P., and Rustig, J. (1982) *Phys. Rev. Lett.*, **49**, 177.
134 Cox, M.P., Ertl, G., and Imbihl, R. (1985) *Phys. Rev. Lett.*, **54**, 1725.
135 Visart de Bocarmé, T., Kruse, N., Gaspard, P., and Kreuzer, H.J. (2006) *J. Chem. Phys.*, **125**, 054704.
136 Schrödinger, E. (1944) *What is Life?*, Cambridge University Press, Cambridge UK.
137 Watson, J.D. and Crick, F.H.C. (1953) *Nature*, **171**, 737.
138 Coleman, B.D. and Fox, T.G. (1963) *J. Polym. Sci. A*, **1**, 3183.
139 Coleman, B.D. and Fox, T.G. (1963) *J. Chem. Phys.*, **38**, 1065.
140 Shannon, C.E. (1948) *Bell System Tech. J.*, **27**, 379.
141 Shannon, C.E. (1948) *Bell System Tech. J.*, **27**, 623.
142 Cover, T.M. and Thomas, J.A. (2006) *Elements of Information Theory*, Wiley, Hoboken.
143 Jarzynski, C. (2008) *Proc. Natl. Acad. Sci. USA*, **105**, 9451.
144 Hopfield, J.J. (1974) *Proc. Natl. Acad. Sci. USA*, **71**, 4135.
145 Bennett, C.H. (1979) *Biosystems*, **11**, 85.
146 Homo sapiens DNA mitochondrion, 16569 bp, locus AC 000021, version GI:115315570, http://www.ncbi.nlm.nih.gov.
147 Lee, H. and Johnson, K. (2006) *J. Biol. Chem.*, **281**, 36236.
148 Astier, Y., Braha, O., and Bayley, H. (2006) *J. Am. Chem. Soc.*, **128**, 1705.
149 Zwolak, M. and Di Ventra, M. (2008) *Rev. Mod. Phys.*, **80**, 141.

2
Thermodynamics of Small Systems

Denis J. Evans, Stephen R. Williams, and Debra J. Searles

2.1
Introduction

Thermodynamics is the study of the flow and transformation of heat into work. Until recently, our understanding of thermodynamics was largely confined to equilibrium states. Linear irreversible thermodynamics is a simple extension of the nineteenth century concepts of equilibrium thermodynamics to systems that are sufficiently *close to* equilibrium, that intensive thermodynamic variables can be approximated by the same functions of *local* state variables as would be the case if the entire system was in complete thermodynamic equilibrium. Classical thermodynamics was limited in application to large systems where intensive thermodynamic functions do not change their values if the system size is increased. This is often referred to as the "thermodynamic limit".

In spite of these restrictions, thermodynamics is arguably the most widely applicable theory in physics. Its First and Second Laws are probably held with greater conviction that any other statements in physics.

In the last fifteen years, three new theorems have been proven that revolutionize our understanding of thermodynamics. Firstly, these new theorems remove the need to take the thermodynamic limit. This allows the application of thermodynamic concepts to finite, and even "nano" systems. Secondly, these new theorems can be applied to systems that are arbitrarily far from equilibrium. Thirdly, and for the first time, these theorems explain how *macroscopic* irreversibility appears naturally in systems that obey time reversible *microscopic* dynamics. Resolution of the Loschmidt (Irreversibility) Paradox had defied our best efforts for more than 100 years. These theorems remove the need for the Second *Law* of thermodynamics. That "law" now becomes a limiting (thermodynamic limit) consequence of the laws of mechanics and the Axiom of Causality: that an event A, can only influence event B, if A occurred prior to B.

Historically, the first of these theorems, the Fluctuation Theorem (FT), generalizes the Second Law of Thermodynamics so that it applies to small systems, including those that evolve far from equilibrium. It refers to a precisely defined mathe-

Nonlinear Dynamics of Nanosystems. Edited by Günter Radons, Benno Rumpf, and Heinz Georg Schuster
Copyright © 2010 WILEY-VCH Verlag GmbH & Co. KGaA, Weinheim
ISBN: 978-3-527-40791-0

matical function, namely, the dissipation function, and gives a precise statement of the probability ratio that time-averaged values of this function take on opposite values. In systems close to equilibrium, this dissipation function is what linear irreversible thermodynamics terms the rate of spontaneous entropy production.

Typically, the second theorem, the Nonequilibrium Work Relation (WR), provides a method of predicting *equilibrium* free energy differences from experimental information taken from *nonequilibrium* path integrals of "work" functions. These "work" functions turn out to be the sum of the equilibrium free energy difference and the time integrated dissipation function for the path.

The third of these theorems, the Dissipation Theorem, shows how the linear and the nonlinear nonequilibrium response of systems are related to temporal correlations in the dissipation function. The nonlinear response theory, known as the Transient Time Correlation Function formalism, is just a special case of the Dissipation Theorem, as is the more well known Green–Kubo linear response theory and the Fluctuation Dissipation Theorem. Each of these theorems refers to the dissipation function in some way. Thus, it is clear that the dissipation function is the central function in nonequilibrium statistical mechanics and thermodynamics.

Each of these theorems is at odds with a traditional understanding of nineteenth century thermodynamics. Indeed, conventional thermodynamics would have been better named thermostatics rather than thermodynamics. Furthermore, these theorems are essential for the application of thermodynamic concepts to nanotechnology systems which are currently of interest to biologists, physical scientists and engineers.

2.2
Thermostated Dynamical Systems

Consider a classical system of N interacting particles in a volume V. The microscopic state of the system is represented by a phase space vector of the coordinates and momenta of all the particles, in phase space – $(q_1, \ldots q_N, p_1, \ldots p_N) \equiv (q, p) \equiv \Gamma$ where q_i, p_i are the position and momentum of particle i, and the internal energy is given by $H_0(\Gamma) = \sum_{i=1}^{N} p_i^2/(2m) + \Phi(q)$ where $\Phi(q)$ is the interparticle potential energy. Initially (at $t = 0$), the microstates of the system are distributed according to a normalized probability distribution function $f(\Gamma, 0)$. While the results in this review are generally applicable, in order to demonstrate its application to realistic systems, we separate the N particle system into a system of interest and a wall region containing N_W particles. We assume that the wall region contains many more particles than the system of interest, $N_W \gg (N - N_W)$, and write the equations of motion for the composite N-particle system as,

$$\dot{q}_i = \frac{p_i}{m} + C_i(\Gamma) \cdot F_e$$

$$\dot{p}_i = F_i(q) + D_i(\Gamma) \cdot F_e - S_i \alpha(\Gamma) p_i , \qquad (2.1)$$

where $\boldsymbol{F}_e$ is the dissipative external field that couples to the system via the phase functions $C_i(\Gamma)$ and $D_i(\Gamma)$, $\boldsymbol{F}_i(\boldsymbol{q}) = -\partial\Phi(\boldsymbol{q})/\partial\boldsymbol{q}_i$ is the interatomic force on particle i, and the last term $-S_i\alpha(\Gamma)\boldsymbol{p}_i$ is a deterministic time reversible thermostat used to add or remove heat from the particles in a reservoir region [1–4]. The values of $C_i(\Gamma)$ and $D_i(\Gamma)$ can be set so that particles in the walls do not interact with the dissipative field. The thermostat employs a switch, S_i, which controls how many and which particles are thermostated, $S_i = 0$; $1 \le i \le (N - N_{\text{therm}})$, and $S_i = 1$; $(N - N_{\text{therm}} + 1) \le i \le N$ where $N_{\text{therm}} \le N_w$, and $N_{\text{therm}} = \sum_{i=1}^{N} S_i$.

The thermostat multiplier can be chosen in a number of ways, such as using Gauss' Principle of Least Constraint, to fix some thermodynamic constraint (e.g., temperature or energy), or using a Nosé–Hoover thermostat where an equation of motion is introduced for α [1–4]. Thus, although the equations of motion for the particles in the thermostating region are modified by the thermostating term, the equations of motion for the particles in the system of interest are quite natural. This construction has been applied in various studies (see, e.g., [5–8]). Of course, if $S_i = 1$ for all i, we obtain a homogeneously thermostated system that has been studied in detail [3]. We assume that in the absence of the thermostating terms, the adiabatic equations of motion preserve the phase space volume, $(\partial/\partial\Gamma)\cdot\dot{\Gamma}^{\text{ad}} = 0$. This is a condition known as the adiabatic incompressibility of phase space, or $AI\Gamma$ [3]. All Hamiltonian systems satisfy this condition. It is worth pointing out that for constant $\boldsymbol{F}_e$, and appropriate choices of $C_i(\Gamma)$ and $D_i(\Gamma)$, (2.1) is time reversible and heat can be either absorbed or given out by the thermostat.

One should not confuse a real thermostat composed of a very large (in principle, infinite) number of particles with the purely mathematical, albeit convenient, term α. In writing (2.1), it is assumed that the reservoir momenta $\boldsymbol{p}_i$ are peculiar (i.e., measured relative to the local streaming velocity of the fluid or wall). When a Gaussian thermostat is used, the thermostat multiplier is chosen to fix the peculiar kinetic energy of the thermostated wall particles

$$K_{\text{therm}} \equiv \sum_{i=1}^{N} S_i \frac{\boldsymbol{p}_i \cdot \boldsymbol{p}_i}{2m_i} = (d_C N_{\text{therm}} - 1) k_B T_{\text{therm}}/2 . \tag{2.2}$$

The quantity T_{therm} defined by this relation is called the kinetic temperature of the wall, k_B is Boltzmann's constant and d_C is the Cartesian dimension of the system. It is assumed that $N_W, N_{\text{therm}} \gg (N - N_{\text{therm}}) \ge (N - N_W)$. This means that the entire wall region can be assumed to be arbitrarily close to equilibrium at the thermodynamic temperature, therefore, $T_W = \partial E_W/\partial S_W$.

Under adiabatic conditions (i.e., in the absence of the thermostating term) we have assumed that the phase space is incompressible. Introduction of the thermostating results in phase space compression with a rate given by $\Lambda = (\partial/\partial\Gamma)\cdot\dot{\Gamma}$. This leads to a change in the energy due to the thermostating term. For dynamics described by (2.1) and thermostated by a Gaussian thermostat, $\dot{H}_0^{\text{therm}}(\Gamma) = \dot{Q}(\Gamma) = -2K_{\text{therm}}\alpha(\Gamma) = -(d_C N_{\text{therm}} - 1) k_B T_{\text{therm}}\alpha(\Gamma)$ and $\Lambda(\Gamma) = -d_C N_{\text{therm}}\alpha(\Gamma) + O(1)$. Therefore, the connection between the phase space compression factor and the heat exchange is given by

$$\Lambda(\Gamma) = \beta \dot{Q}(\Gamma) + O(1), \qquad (2.3)$$

where $\beta = 1/(k_B T_{therm})$. When a Nosé–Hoover thermostat is employed, a similar analysis yields $\Lambda(\Gamma) = \beta \dot{Q}(\Gamma)$ [3, 9].

One might object that our analysis is compromised by our use of artificial (time reversible) thermostats. However, the artificial thermostat region can be made arbitrarily remote from the system of interest by ensuring that the particles with $S_i = 1$ are far from the system of interest [5, 6, 10]. If this is the case, the system cannot 'know' the precise details of how heat was removed at such a remote distance. This means that the results obtained for the system using our simple mathematical thermostat must be the same as those we would infer for the same system surrounded (at a distance) by a real physical thermostat (e.g., with a huge heat capacity). We introduce the thermostat to simplify the bookkeeping of tracking changes to the phase space volume in open systems that exchange heat with their surroundings. In open Hamiltonian systems, phase space volumes are not preserved. This mathematical thermostat may be unnatural, however, in the final analysis it is a very convenient, but physically irrelevant device [6].

In [6], a mathematical proof is given showing that when the thermostating region has a significantly larger number of degrees of freedom than the unthermostated system of interest, the Fluctuation Theorem is independent of the mathematical details of how the thermostating is accomplished. The proof is for an infinite family of so-called μ-thermostats.

The exact equation of motion for the N-particle distribution function is the time-reversible Liouville equation [3]

$$\frac{\partial f(\Gamma, t)}{\partial t} = -\frac{\partial}{\partial \Gamma} \cdot [\dot{\Gamma} f(\Gamma, t)] \equiv -iL(\Gamma) f(\Gamma, t), \qquad (2.4)$$

where $iL(\Gamma)$ is the distribution function (or $f-$) Liouvillian and appears in the propagator for the phase space distribution function ($f(\Gamma, t) = \exp[-iL(\Gamma)t] f(\Gamma, 0)$). The Liouville equation can also be written in Lagrangian form [11],

$$\frac{df(\Gamma, t)}{dt} = -f(\Gamma, t) \frac{d}{d\Gamma} \cdot \dot{\Gamma} \equiv -\Lambda(\Gamma) f(\Gamma, t). \qquad (2.5)$$

The presence of the thermostat is reflected in the phase space expansion factor, $\Lambda(\Gamma) \equiv \partial/\partial \Gamma \cdot \dot{\Gamma}$, which is $\Lambda(\Gamma) = -d_C N_{therm} \alpha$ to first order in N_{therm}, assuming $\Lambda \textit{l} \Gamma$. The equation of motion for an arbitrary phase function $B(\Gamma)$, is [11],

$$\dot{B}(\Gamma) = \dot{\Gamma} \cdot \frac{dB}{d\Gamma} \equiv iL(\Gamma) B(\Gamma), \qquad (2.6)$$

where $iL(\Gamma)$ is the phase variable (or $p-$) Liouvillian and appears in the propagator for phase variables ($B(\Gamma(t)) = \exp[-iL(\Gamma)t] B(\Gamma(0))$). The difference between the f-Liouvillian and the p-Liouvillian is $iL(\Gamma) - iL(\Gamma) = \Lambda(\Gamma)$. The time-reversibility condition implies that there exists a time-reversal mapping, M^T such that $\Gamma = M^T \exp(iLt) M^T \exp(iLt) \Gamma$. Typically, $M^T(\mathbf{q}, \mathbf{p}) = (\mathbf{q}, -\mathbf{p})$.

The solution of (2.5) gives the phase space density at a phase point after evolution for a period t:

$$f(\Gamma(t), t) = \exp\left[-\int_0^t ds\, \Lambda(\Gamma(s))\right] f(\Gamma(0), 0). \tag{2.7}$$

If one considers an infinitesimal co-moving phase volume, $\delta\Gamma$, centered on Γ, for which the number of ensemble members is conserved, then (2.7) can be used to determine the time evolution of the phase volume:

$$\frac{\delta\Gamma(t)}{\delta\Gamma(0)} = \frac{f(\Gamma(0), 0)}{f(\Gamma(t), t)} = \exp\left[\int_0^t ds\, \Lambda(\Gamma(s))\right]. \tag{2.8}$$

2.3
The Transient Fluctuation Theorem

The first proof of any fluctuation theorem was for a special case of what is now known as the Evans–Searles Transient Fluctuation Theorem (ESFT). Here, we give a very general proof. Consider the response of a system initially in some known but arbitrary distribution,

$$f(\Gamma', 0) = \frac{\exp[-F(\Gamma')]}{\int d\Gamma' \exp[-F(\Gamma')]}, \tag{2.9}$$

where $F(\Gamma')$ is some arbitrary single-valued real function for which $f(\Gamma', 0) = f(M^T \Gamma', 0)$. Γ' is the extended phase space vector which includes the phase space vector Γ and may include additional dynamical variables such as the volume or those associated with the thermostat. In the following, we drop the prime in cases where the treatment is not altered and note where consideration of an extended phase space is important.

Consider any system whose dynamics is described by deterministic, time-reversible equations of motion. The equations of motion may have an applied dissipative field as in (1), or the field may be zero. If the field is zero, then in order to see anything interesting, the initial distribution should not be preserved by the equations of motion (if it is preserved, then the ESFT is completely trivial). On the other hand, if a dissipative field is applied, then it is often useful to consider the case where the initial distribution is the equilibrium distribution for the field free dynamics.

We assume the unthermostated equations of motion satisfy the $AI\Gamma$ condition. A thermostat may be added as in (2.1), but again, this is not absolutely essential. However, the equations of motion *must* be time-reversal symmetric.

2 Thermodynamics of Small Systems

The *dissipation function*, $\Omega(\Gamma)$, is defined as [12, 13]:

$$\int_0^t ds\, \Omega(\Gamma(s)) \equiv \ln \frac{p(\Gamma(0), \delta\Gamma)}{p(\Gamma^*(0), \delta\Gamma^*)}$$

$$= \ln\left(\frac{f(\Gamma(0), 0)}{f(\Gamma(t), 0)}\right) - \int_0^t \Lambda(\Gamma(s))\, ds$$

$$\equiv \bar{\Omega}_t t \tag{2.10}$$

where $\Gamma^*(0) \equiv M^T \Gamma(t)$ is the time-reversal mapped image of $\Gamma(t)$ (the endpoint of the trajectory starting at $\Gamma(0)$), as shown in Figure 1, and $p(\Gamma(0), \delta\Gamma) = f(\Gamma(0), 0)\delta\Gamma$ is the probability of observing ensemble members inside the infinitesimal phase volume $\delta\Gamma$, centered on the phase vector $\Gamma(0)$, according to the initial equilibrium distribution function, $f(\Gamma(0), 0)$[1].

In order for the dissipation function to be well-defined, for any $f(\Gamma(0), 0) \neq 0$, then $f(\Gamma(t), 0) \neq 0$, and vice versa. This is known as the *ergodic consistency* condition for the dissipation function [12]. There are systems that fail to satisfy this condition. For example, if we let the initial distribution be microcanonical and further assume that the dynamics do not preserve the energy (there may be no thermostat or the thermostat may fix the kinetic temperature or so), then the ergodic consistency obviously breaks down.

The ESFT [11, 12, 14] states that under the conditions given above, the dissipation function satisfies the following time-reversal symmetry:

$$\frac{p(\bar{\Omega}_t = A)}{p(\bar{\Omega}_t = -A)} = \exp[At], \tag{2.11}$$

where the notation $p(\bar{\Omega}_t = A)$ gives the probability that the time-averaged dissipation function takes on a value $A \pm \delta A$. Once the concepts, dynamics and definitions have been given, the proof of the ESFT is trivial. The probability, $p(\bar{\Omega}_t = A)$, is given by the integral of the phase space density over the set of all initial phase points that have a specified value (within the small tolerance) for the time-average of the dissipation function: $p(\bar{\Omega}_t = A, dA) = \int_{\bar{\Omega}_t(\Gamma)=A} d\Gamma\, f(\Gamma, 0)$. Next, we compute the probability ratio required in (2.11),

$$\frac{p(\bar{\Omega}_t = A)}{p(\bar{\Omega}_t = -A)} = \frac{\int_{\bar{\Omega}_t(\Gamma(0))=A} d\Gamma(0)\, f(\Gamma(0), 0)}{\int_{\bar{\Omega}_t(\Gamma^*(0))=-A} d\Gamma^*(0)\, f(\Gamma^*(0), 0)}$$

---

1) The time-integral of the dissipation function can therefore be considered to be the difference in surprise of observing trajectories and their time-reverse, according to the initial distribution function, and its ensemble of average is the relative entropy, or Kullback–Leibler divergence. This has recently been discussed for an adiabatic system [111].

$$= \frac{\int_{\bar{\Omega}_t(\Gamma(0))=A} d\Gamma(0)\, f(\Gamma(0),0)}{\int_{\bar{\Omega}_t(\Gamma(0))=A} d\Gamma(t)\, f(\Gamma(t),0)}$$

$$= \frac{\int_{\bar{\Omega}_t(\Gamma(0))=A} d\Gamma(0)\, f(\Gamma(0),0)}{\int_{\bar{\Omega}_t(\Gamma(0))=A} d\Gamma(0)\, \exp[\Lambda_t(\Gamma(0))]\, f(\Gamma(t),0)}$$

$$= \frac{\int_{\bar{\Omega}_t(\Gamma(0))=A} d\Gamma(0)\, f(\Gamma(0),0)}{\int_{\bar{\Omega}_t(\Gamma(0))=A} d\Gamma(0)\, \exp[-\Omega_t(\Gamma(0))]\, f(\Gamma(0),0)} = \exp[At] , \qquad (2.12)$$

where we apply the fact that $\Gamma^*(0) \equiv M^T \Gamma(t)$ is the time-reversal mapped image of $\Gamma(t)$. For time-reversible dynamics, if $\bar{\Omega}_t(\Gamma(0)) = A$, then $\bar{\Omega}_t(\Gamma^*(0)) = -A$ (see Figure 2.1). The third equality is obtained using (2.8), and the final equality using (2.10). We note that in order for this ratio to be well defined, $\int_{\bar{\Omega}_t(\Gamma^*(0))=-A} d\Gamma^*(0)\, f(\Gamma^*(0),0)$ must be nonzero. That is, the time-reversal mapping of the end-points of the trajectory that meet the condition $\bar{\Omega}_t(\Gamma^*(0)) = -A$, must be observable according to the equilibrium phase space density. This is the ergodic consistency condition for the fluctuation theorem, and is a weaker condition than the ergodic consistency condition mentioned above. An equivalent relation

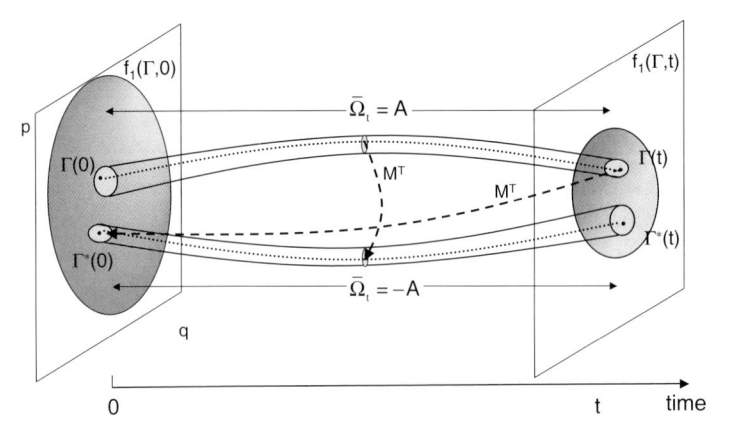

Figure 2.1 Schematic diagram showing the construction required to derive the transient ESFT. Two trajectories that are related by time-reversal mappings are shown as dotted lines, and the time evolution of a small phase volume centered on these trajectories is also shown. The dashed line indicates points related by time-reversal mappings (M^T).

to (2.12) is given for the time-integral of the dissipation function:

$$\frac{p(\Omega_t = A)}{p(\Omega_t = -A)} = \exp[A]. \tag{2.13}$$

The existence of the dissipation function (2.10) requires that the initial distribution is normalizable, that ergodic consistency holds and that the initial distribution is invariant under the time-reversal map. This assumption was used in going from line 1 to line 2 in (2.12). By definition, all equilibrium distributions satisfy this requirement.

The instantaneous dissipation function can be determined by the differentiation of (2.10) as

$$\begin{aligned}\Omega(\Gamma) &= -\frac{\partial}{\partial \Gamma} \cdot \dot{\Gamma}(\Gamma) - \frac{\dot{\Gamma}(\Gamma)}{f(\Gamma,0)} \cdot \frac{\partial}{\partial \Gamma} f(\Gamma,0) \\ &= -\frac{\partial}{\partial \Gamma} \cdot \dot{\Gamma}(\Gamma) - \dot{\Gamma}(\Gamma) \cdot \frac{\partial}{\partial \Gamma} \ln f(\Gamma,0)\end{aligned} \tag{2.14}$$

and, therefore, $\Omega(\Gamma) f(\Gamma,0) = -\frac{\partial}{\partial \Gamma} \cdot \left(\dot{\Gamma}(\Gamma) f(\Gamma,0) \right)$, the divergence of $\dot{\Gamma}(\Gamma) f(\Gamma,0)$ (i.e., the dissipation function weighted by the initial distribution is the weighted divergence of the phase space flow field).

The ESFT has generated much interest, as it shows how irreversibility emerges from the deterministic, reversible equations of motion[2], and is arbitrarily valid far from equilibrium. It provides a generalized form of the Second Law of Thermodynamics that can be applied to small systems observed for short periods of time. It also resolves the longstanding Loschmidt Paradox. The ESFT has been verified experimentally [15–22] (See Section 2.9).

The *form* of the above equation applies to any valid ensemble/dynamics combination, provided the distribution function is invariant with respect to time-reversal. However, the precise *expression* for $\bar{\Omega}_t$ given in (2.10) is dependent on both the initial distribution and the dynamics. This result is extremely general. The ESFT is so general because its proof requires so few assumptions: the ergodic consistency and time reversibility of the dynamics.

2.4
Thermodynamic Interpretation of the Dissipation Function

Although the definition of the dissipation function in (2.10) seems quite abstract, the dissipation function always takes on a physically significant form, which for systems close to equilibrium has average values which are equal to the spontaneous entropy production one meets in linear irreversible thermodynamics. Consider the

---

2) By time-reversible equations of motion, we mean that there exists a time reversal mapping M^T such that $\Gamma = M^T \exp(iLt) M^T \exp(iLt) \Gamma$.

special case where the kinetic energy $K_{therm}(\Gamma)$ of the thermostated particles is fixed and the initial distribution is isokinetic for the thermostating region, but canonical elsewhere. The distribution function is then

$$f(\Gamma,0) \equiv f_K(\Gamma,0)$$
$$= \frac{\delta(2K_{therm} - (d_C N_{therm} - 1)k_B T_{therm}) \exp[-\beta H_0(\Gamma)]}{\int d\Gamma \, \delta(2K_{therm} - (d_C N_{therm} - 1)k_B T_{therm}) \exp[-\beta H_0(\Gamma)]}, \quad (2.15)$$

where $H_0(\Gamma)$ is the internal energy of the entire system, and we recall $\beta = 1/(k_B T_{therm})$. In this case, it is evident [9, 12] that the dissipation function is related to the generalized entropy production $\Sigma(\Gamma)$,

$$\Omega(\Gamma) \equiv \Sigma(\Gamma) = -\beta J(\Gamma) V \cdot \mathbf{F}_e . \quad (2.16)$$

Here, V is the volume of the system of interest and $J(\Gamma)$ is the dissipative flux in the system of interest,

$$-J(\Gamma) V \cdot \mathbf{F}_e \equiv \sum_{i=1}^{N-N_W} \left[\frac{\mathbf{p}_i}{m} \cdot \mathbf{D}_i - \mathbf{F}_i \cdot \mathbf{C}_i \right] \cdot \mathbf{F}_e$$
$$= \left. \frac{dH_0}{dt} \right|^{tot} - \left. \frac{dH_0}{dt} \right|^{therm} = \left. \frac{dH_0}{dt} \right|^{ad}, \quad (2.17)$$

where we have assumed that the field only acts directly on the particles in the system of interest. The dissipative flux is thus the work performed on the system by the dissipative field. It is the "work" because it is the total change in the energy minus the change due to the thermostat.

Although we assumed a special dynamics where the kinetic energy of the thermostated particles is fixed, the form of (2.16) must be true for other "thermostated" dynamics (e.g., Nosé–Hoover or constant energy etc., see Appendix 1 of [9]). Furthermore, if the reservoir region does not directly interact with the field and N_{therm} is large, and much larger than the number of degrees of freedom in the system of interest, the form of (2.16) is generally true (e.g., for thermostats where higher order moments of the momenta are constrained, stochastic thermostats etc.) [6]. The dissipative flux, volume and field are properties of the system of interest and the only relevant property taken from the thermostated region is its temperature.

One might think that (2.16) is at odds with conventional linear irreversible thermodynamics in which it might be expected that the entropy production would be given by (2.16), except that the temperature appearing there would be the local thermodynamic equilibrium temperature rather than, as in (2.16), the equilibrium wall temperature. Close to equilibrium, the difference between these two temperatures is $O(\mathbf{F}_e^2)$. This further implies that the difference between the dissipation function and the entropy production conjectured for irreversible thermodynamics is in fact $O(\mathbf{F}_e^4)$. This goes beyond the domain of applicability of linear irreversible thermodynamics which is at most a $O(\mathbf{F}_e^2)$ theory. Thus, to the order that linear irreversible thermodynamics can be trusted (i.e., $O(\mathbf{F}_e^2)$) close to equilibrium, the dissipation function is equal to the spontaneous entropy production.

2.5
The Dissipation Theorem

We now derive the *Dissipation Theorem*, which shows that, as well as being the subject of the ESFT, the dissipation function is the central argument of both linear response theory (i.e., Green–Kubo theory) and nonlinear response theory. This theorem was first derived in 2008 [23, 24].

Taking the solution of the Lagrangian form of the Liouville equation (i.e., $f(\Gamma(t), t) = \exp\left[-\int_0^t ds\Lambda(\Gamma(s))\right] f(\Gamma(0), 0)$ as given in (2.7)), we can substitute for $f(\Gamma(0), 0)$ using the definition of the time-integrated dissipation function (2.10), thus obtaining

$$f(\Gamma(t), t) = \exp\left[-\int_0^t \Lambda(\Gamma(s))\right] f(\Gamma(t), 0)$$
$$\times \exp\left[\int_0^t ds\,\Omega(\Gamma(s)) + \int_0^t ds\,\Lambda(\Gamma(s))\right]$$
$$= f(\Gamma(t), 0) \exp\left[\int_0^t ds\,\Omega(\Gamma(s))\right]. \quad (2.18)$$

This is valid for any $\Gamma(t)$, therefore we select $\Gamma(t) = \Gamma^*$ and note that this implies $\Gamma(s) = \Gamma^*(s-t)$. Then,

$$f(\Gamma^*, t) = f(\Gamma^*, 0) \exp\left[\int_0^t ds\,\Omega(\Gamma^*(s-t))\right]$$
$$= f(\Gamma^*, 0) \exp\left[-\int_0^{-t} ds'\,\Omega(\Gamma^*(s'))\right], \quad (2.19)$$

where the second equality is obtained by introducing $s' = s - t$. Replacing the dummy variables gives

$$f(\Gamma, t) = f(\Gamma, 0) \exp\left[-\int_0^{-t} ds\,\Omega(\Gamma(s))\right]. \quad (2.20)$$

This result shows that the propagator for the N-particle distribution function, $\exp[-iL(\Gamma)t]$, has a very simple relation to exponential time integrals of the dissipation function. As shown below, in the case of isokinetic nonequilibrium dynamics, this equation reduces to (7.2.17) of [23][3]. In the case of adiabatic (i.e., unthermostated) dynamics for an ensemble that is initially a canonical ensemble, the result is equivalent to (7.2.8) of [23], which is the distribution function derived by Yamada and Kawasaki in 1967 [25]. However, (2.20) is much more general and, like the ESFT, can be applied to any initial ensemble and any time-reversible, and possibly thermostated dynamics that satisfies $AI\Gamma$.

[3] Note that an alternative derivation of (20), that more closely resembles the approach used in [3], is given in [23, 24].

From (2.20), we can calculate nonequilibrium ensemble averages in the Schrödinger representation

$$\langle B(t) \rangle_{F_e, f(\Gamma,0)} = \left\langle B(0) \exp\left[-\int_0^{-t} ds\, \Omega\left(\Gamma(s)\right)\right] \right\rangle_{F_e, f(\Gamma,0)} \quad (2.21)$$

and by differentiating and integrating (2.21) with respect to time, we can write the averages in the Heisenberg representation as

$$\langle B(t) \rangle_{F_e, f(\Gamma,0)} = \langle B(0) \rangle_{f(\Gamma,0)} + \int_0^t ds\, \langle \Omega(0) B(s) \rangle_{F_e, f(\Gamma,0)} . \quad (2.22)$$

On both sides of (2.20)–(2.22), the time evolution is governed by the field-dependent thermostated equations of motion (2.1). The derivation of (2.21) and (2.22) from the definition of the dissipation function (2.10), is called the *Dissipation Theorem*. This Theorem is extremely general and allows the determination of the ensemble average of an arbitrary phase variable under very general conditions. Like the ESFT, it is valid arbitrarily far from equilibrium. Equations (2.20) and (2.21) can be obtained for time-dependent fields by including the explicit time-dependence of Ω, but (2.22) cannot [26]. As in the derivation of the ESFT, the only unphysical terms in the derivation are the thermostating terms within the wall region. However, because these thermostating particles can be moved arbitrarily far from the system of interest, the precise mathematical details of the thermostat are unimportant. Since the number of degrees of freedom in the reservoir is assumed to be much larger than that of the system of interest, the reservoir can always be assumed to be in thermodynamic equilibrium. Therefore, there is no difficulty in defining the thermodynamic temperature of the walls. This is in marked contrast with the system of interest, which may be very far from equilibrium where the thermodynamic temperature cannot be defined.

For the special case of isokinetic dynamics where the kinetic energy $K_{therm}(\Gamma)$ of the thermostated particles is fixed and if the initial distribution is isokinetic (2.15), (2.22) can be written as the Transient Time Correlation function expression [23] for the thermostated nonlinear response of the phase variable B to the dissipative field F_e:

$$\langle B(t) \rangle_{F_e, f_K(\Gamma,0)} = \langle B(0) \rangle_{f_K(\Gamma,0)} - \beta V \int_0^t ds\, \langle J(0) B(s) \rangle_{F_e, f_K(\Gamma,0)} \cdot F_e . \quad (2.23)$$

In the weak field limit, this reduces to the well known Green–Kubo expression [23] for the linear response

$$\lim_{F_e \to 0} \langle B(t) \rangle_{F_e, f_K(\Gamma,0)} = \langle B(0) \rangle_{f_K(\Gamma,0)} - \beta V \int_0^t ds\, \langle J(0) B(s) \rangle_{F_e=0, f_K(\Gamma,0)} \cdot F_e , \quad (2.24)$$

where the right-hand side is given by the integral of an *equilibrium* (i.e., $F_e = 0$) time correlation function. The Transient Time Correlation Function (TTCF) in

(2.23) has been used frequently to compute the nonlinear transport behavior of systems over extremely wide values of the applied field [27–33]. For small fields, the values of field-dependent properties of the system are often swamped by noise from naturally occurring fluctuations making direct calculation of the left-hand side of (2.23) or (2.24) problematic. This is particularly relevant regarding the calculation of the transport coefficient which can be obtained from the ratio of the flux to the field. The TTCF can be applied at any field strength, even zero, where it reduces to the Green–Kubo expression for the linear response. Equation 2.24 has the form of a susceptibility equation and, because the correlation function is the equilibrium one, this equation is valid for time dependent fields.

It is interesting to compare a number of different relationships between the distribution function, the dissipation function and the phase space expansion factor. The first such relations are (2.7) and (2.18) above. We note that although the time argument in (2.20) is negative, the dynamics must still be governed by the field dependent, thermostated equations of motion (2.1). By rewriting (2.10), we have

$$f(\Gamma(t),0) = \exp\left[-\int_0^t ds \Omega\left(\Gamma(s)\right) + \Lambda(\Gamma(s))\right] f(\Gamma(0),0) . \qquad (2.25)$$

In a nonequilibrium steady state (SS), $\langle \Omega(t) \rangle_{ss} = -\langle \Lambda(t) \rangle_{ss}$. We also note that if the initial ensemble is microcanonical (has a uniform density) and the dynamics are such that the total energy (system of interest *plus* walls and thermostat) is constant, then $\Omega(t) = -\Lambda(t), \forall t$.

Rather obviously, the results of the Dissipation Theorem (2.22) can also be used to obtain a Fluctuation Dissipation Theorem as described in [34] by considering the case where the phase function $B(\Gamma) = J(\Gamma)$. Furthermore, following [34], we find that when the equilibrium dissipative flux autocorrelation function is δ-correlated, $\langle J(t_1)J(t_2)\rangle_{F_e=0} = \langle J(t_1)J(t_2)\delta(t_2-t_1)\rangle_{F_e=0}$, and we obtain the fluctuation dissipation relation, and $\lim_{F_e \to 0} \langle J(t)\rangle_{F_e} = -\frac{1}{3}\beta V \langle J(0)\cdot J(0)]\rangle_{F_e=0} \cdot F_e$.

2.6
Nonequilibrium Work Relations

Traditionally, free energy differences between two equilibrium states have been determined by methods based on measuring the work performed along a quasistatic (equilibrium, reversible) pathway between the two states, or by considering the ratio of the partition functions. The Jarzynski Equality (JE) [35, 36] and the Crooks Fluctuation Theorem (CFT) [37, 38] provide alternative approaches whereby the work is measured along an ensemble of nonequilibrium pathways.

The JE and CFT were originally developed for determining the difference in free energy of canonical equilibrium states at the same temperature; however more recently, they have been extended to other systems [10, 39–46]. Here, we present a very general formalism [40] for obtaining the nonequilibrium free energy theorems that can be applied in all these cases. We then show how it leads the usual canonical JE and CFT under appropriate conditions (see (2.34) and (2.35) below). In Section 2.7, the general approach will be employed in order to consider the case of the physical problem of a varying temperature.

We consider two closed N-particle systems: 1 and 2. These systems may have the same or different Hamiltonians, temperatures or volumes; it does not matter. They may have the same or different temperatures or volumes which are again, irrelevant. In addition, the type of ensemble does not matter, whether microcanonical, canonical, isothermal, or isobaric and so on. A protocol and the corresponding time-dependent dynamics are then defined that will transform system 1 to system 2. The systems are distinguished by introducing a parameter, λ, which takes on a value λ_1 in system 1, and λ_2 in system 2. The transformation is also parameterized through $\lambda(t)$ with $\lambda(0) = \lambda_1$ and $\lambda(\tau) = \lambda_2$. We define a generalized dimensionless "work", $\Delta X_\tau(\Gamma)$, for a trajectory of duration τ, under these dynamics as in [40],

$$\exp[\Delta X_\tau(\Gamma)] = \frac{p_1(\Gamma(0), \delta\Gamma(0))Z(\lambda_1)}{p_2(\Gamma(\tau), \delta\Gamma(\tau))Z(\lambda_2)}$$

$$= \frac{f_1(\Gamma(0), 0)\delta\Gamma(0)Z(\lambda_1)}{f_2(\Gamma(\tau), 0)\delta\Gamma(\tau)Z(\lambda_2)} \quad (2.26)$$

where $Z(\lambda_i)$ is the partition function for the system with system i. If the system is canonically distributed, then $Z(\lambda_i)$ is related to the Helmholtz free energy, $A(\lambda_i) = -k_B T \ln(Z(\lambda_i))$. For other ensembles, the partition functions are well known. $p_i(\Gamma, \delta\Gamma) = f_i(\Gamma)\delta\Gamma$ is the probability of observing the infinitesimal phase volume $\delta\Gamma$, centered on the phase vector Γ, according to the initial equilibrium distribution function, f_i. In order for ΔX_τ to be well defined, for any $f_1(\Gamma(0), 0) \neq 0$, then $f_2(\Gamma(\tau), 0) \neq 0$, and vice versa. This is known as the ergodic consistency for generalized work [47].

Although the physical significance of the variable X might seem obscure at this point, we will show that for particular choices of dynamics and ensemble, it is related to important physical properties. We identify $\partial\Gamma(\tau)/\partial\Gamma(0)$ as the Jacobian matrix and note that

$$\left\|\frac{\partial\Gamma(\tau)}{\partial\Gamma(0)}\right\| = \frac{\delta\Gamma(\tau)}{\delta\Gamma(0)}. \quad (2.27)$$

Since the distribution function is normalized and by means of (2.27), it is obvious that

$$\langle\exp[-\Delta X_\tau(\Gamma)]\rangle_1 = \int d\Gamma(0) f(\Gamma(0), 0) \frac{f_2(\Gamma(\tau), 0)\delta\Gamma(\tau)Z(\lambda_2)}{f_1(\Gamma(0), 0)\delta\Gamma(0)Z(\lambda_1)}$$

$$= \frac{Z(\lambda_2)}{Z(\lambda_1)} \quad (2.28)$$

where the brackets $\langle\ldots\rangle_1$ denote an equilibrium ensemble average over the initial distribution.

This relationship, called a generalized Jarzynski Equality, is widely applicable [40, 47]. It relates the ensemble average of the exponential of a nonequilibrium path integral to equilibrium thermodynamic free energy differences. The validity of (2.28) requires that there is an integrable region in the phase space of the final equilibrium distribution for which $f_2(\Gamma(\tau), 0) \neq 0$, that is, $\int d\Gamma(\tau) f_2(\Gamma(\tau), 0) \neq 0$. We call this the ergodic consistency condition for the generalized JE. The relationship is very general and even applies to stochastic dynamics (see [40]). The paths do not need to be quasistatic paths as in traditional thermodynamics. Additionally, other nonequilibrium (even dissipative) processes can be carried out during the period $0 < t < \tau$, as is the case with the CFT and JE [48, 49].

The CFT considers the probability, $p_f(\Delta X_\tau = B)$, of observing values of $\Delta X_\tau = B \pm dB$ for forward trajectories starting from the initial equilibrium distribution $f_1(\Gamma, 0)$, and the probability $p_r(\Delta X_\tau = -B)$ of observing $\Delta X_\tau = -B \pm dB$ for reverse trajectories, though starting from the equilibrium given by $f_2(\Gamma, 0)$, as in Figure 2.2. Proof of the generalized CFT closely resembles the proof of the ESFT (2.12):

$$\frac{p_f(\Delta X_\tau = B)}{p_r(\Delta X_\tau = -B)} = \frac{\int_{\Delta X_\tau(\Gamma(0))=B} d\Gamma(0) f_1(\Gamma(0), 0)}{\int_{\Delta X_\tau(\Gamma^*(0))=-B} d\Gamma^*(0) f_2(\Gamma^*(0), 0)}$$

$$= \frac{\int_{\Delta X_\tau(\Gamma(0))=B} d\Gamma(0) f_1(\Gamma(0), 0)}{\int_{\Delta X_\tau(\Gamma(0))=B} d\Gamma(\tau) f_2(\Gamma(\tau), 0)}$$

$$= \frac{\int_{\Delta X_\tau(\Gamma(0))=B} d\Gamma(0) f(\Gamma(0), 0)}{\int_{\Delta X_\tau(\Gamma(0))=B} d\Gamma(0) \exp[-\Delta X_\tau(\Gamma)] f_1(\Gamma(0), 0) Z(\lambda_1)/Z(\lambda_2)}$$

$$= \exp[B] \frac{Z(\lambda_2)}{Z(\lambda_1)}. \tag{2.29}$$

Here, we use the fact that $\Gamma^*(0) \equiv M^T \Gamma(\tau)$, and for time-reversible dynamics, if $\Delta X_\tau(\Gamma(0)) = B$, then $\Delta X_\tau(\Gamma^*(0)) = -B$, as shown in Figure 2.2.

The derivation of the JE, as shown in (2.28), is trivial once the definitions have been made. However, a more instructive approach is to obtain it by integration of the CFT, (2.29):

$$\langle \exp[-\Delta X_\tau] \rangle = \int_{-\infty}^{\infty} dB\, p_f(\Delta X_\tau = B) \exp(-B)$$

$$= \int_{-\infty}^{\infty} dB\, p_r(\Delta X_\tau = -B) \frac{Z(\lambda_2)}{Z(\lambda_1)}$$

$$= \frac{Z(\lambda_2)}{Z(\lambda_1)}. \tag{2.30}$$

From the first line of (2.30), it is clear that trajectories for which the value of ΔX_τ is negative have a contribution to the ensemble average that is exponentially en-

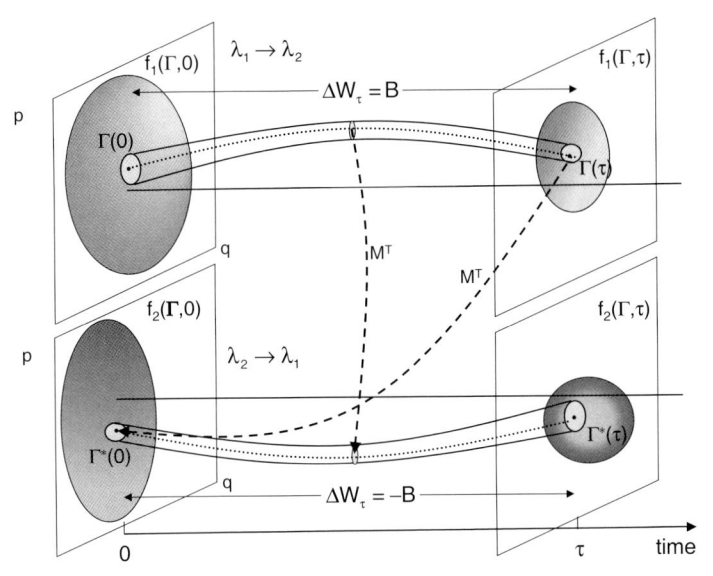

Figure 2.2 Schematic diagram showing the construction required to derive the Crooks fluctuation relation. Two trajectories related by time-reversal mappings are shown as dotted lines, and sampled from different initial distributions. The time evolution of a small phase volume centered on these trajectories is also shown. The dashed line indicates points related by time-reversal mappings (M^T). It is assumed that the time-evolution of λ for the trajectories sampled from f_1 is $\lambda_1 \to \lambda_2$, and it is reversed for the trajectories that start in f_2 (i.e., $\lambda_2 \to \lambda_1$).

hanced. Therefore, in order to obtain numerical convergence of the ensemble average, it is important that these trajectories are well sampled. Many recent studies have addressed this issue and have developed algorithms in order to improve convergence [50–62]. If the averaging process is not sufficiently exhaustive for these possibly extremely rare events to be observed, (2.29) and (2.30) will yield incorrect results. This observation has an immediate impact on the calculation of free energy differences in the thermodynamic limit. This difference must be calculated in finite systems for a series of system sizes and then extrapolation must be employed in order to obtain the thermodynamic limit. If you apply the CFT or JE to extremely large systems, one will never observe the required fluctuations and incorrect estimates will be inferred.

We now show that these general results lead to the usual canonical forms of the JE and CFT. The relevant distribution function is the canonical distribution function,

$$f(\Gamma, 0) = \frac{\exp\left[-\beta H_0(\Gamma)\right]}{Z}. \tag{2.31}$$

The thermodynamic potential is the Helmholtz free energy, A, which is related to the phase space integral of the negative exponential of the Hamiltonian $H_0(\Gamma)$ of the system[4],

$$A(\lambda) = -k_B T \ln Z(\lambda)$$
$$= -k_B T \ln \left(\int d\Gamma \exp(-\beta H_0(\Gamma, \lambda)) \right) . \quad (2.32)$$

In order to transform from the initial equilibrium state, with $\lambda = \lambda_1 = \lambda(0)$, to the final equilibrium state with $\lambda = \lambda_2 = \lambda(\tau)$, the functional form of the system's Hamiltonian may vary parametrically over the period $0 < t < \tau$. For example, $H_0(\Gamma, \lambda(t)) = \sum_{i=1}^{N} p_i^2/(2m) + \Phi(q, \lambda(t))$ where $\Phi(q, \lambda(t))$ is the interparticle potential. For $t > \tau$, the Hamiltonian's parametric dependence is fixed at $H_0(\Gamma, \lambda(\tau))$. Over the times $0 < t < \tau$, the ensemble is driven away from equilibrium, and if the transformation is halted at $t = \tau$, the system will eventually relax to a new equilibrium state.

Using (2.26), the generalized "work" becomes:

$$\Delta X_\tau = \beta \left(H_0(\Gamma(\tau), \lambda(\tau)) - H_0(\Gamma(0), \lambda(0)) + \ln \left[\frac{\delta \Gamma(0)}{\delta \Gamma(\tau)} \right] \right)$$
$$= \beta (H_0(\Gamma(\tau), \lambda(\tau)) - H_0(\Gamma(0), \lambda(0)) - \int_0^\tau ds \Lambda(\Gamma(s)))$$
$$= \beta (H_0(\Gamma(\tau), \lambda(\tau))) - H_0(\Gamma(0), \lambda(0) - \Delta Q_\tau)$$
$$= \beta \Delta W_\tau . \quad (2.33)$$

The final equality is obtained from the First Law of Thermodynamics, and the equations of motion must satisfy $\Lambda I \Gamma$. We note that if $t = \tau$ at the end of the protocol, then the system is not in equilibrium and it does not matter. Any subsequent relaxation processes will have no effect on ΔW. Furthermore, at the end of the protocol, the system cannot "know" how long the final relaxation process will take [47]. Analogous statements apply for ΔX in general, and stem from the fact that ΔX is defined in terms of the ratio of the partition functions of the two equilibrium states regardless of the relaxation that takes place after the protocol has ceased ($t > \tau$).

For a system where the phase space is extended due to the introduction of additional dynamical variables such as the volume or those associated with the thermostat (such as in the case of Nosé–Hoover dynamics [63], as detailed below), the work becomes $\Delta W_\tau = H_E(\Gamma'(\tau), \lambda(\tau)) - H_E(\Gamma'(0), \lambda(0)) - \Delta Q_\tau$, where H_E is the Hamiltonian of the extended system [46].

Using (2.29) and (2.33), the CFT is given as

$$\frac{p_F(\Delta W_\tau = B)}{p_R(\Delta W_\tau = -B)} = \exp[\beta B] \frac{Z_2}{Z_1} = \exp[-\beta(\Delta A - B)] \quad (2.34)$$

---

[4] We assume that the system's center of mass motion is zero.

where $\Delta A = A(\lambda_\tau) - A(\lambda_0)$, and using (2.28), the JE is

$$\langle \exp(-\beta \Delta W_\tau)\rangle = \frac{Z_2}{Z_1} = \exp(-\beta \Delta A) \,. \tag{2.35}$$

The same results are obtained for the canonical distribution when the dynamics are thermostated by a Gaussian thermostat [64], a Nosé–Hoover thermostat [46], or the dynamics are adiabatic (i.e., unthermostated). For other ensembles and transformations, (2.28) does not necessarily refer to a work (e.g., see [39, 40, 43]).

2.7
Nonequilibrium Work Relations for Thermal Processes

To obtain experimentally applicable forms of these theorems which are valid arbitrarily far from equilibrium, it is necessary to introduce a thermal reservoir that is large and remote enough from the system of interest to effectively remain in equilibrium. As in Section 2.2, we surround the system of interest with a large synthetic thermostating region. We wish to consider a realistic model of a system that is driven away from equilibrium by a reservoir whose temperature is changing (e.g., see [43]). For this case, the simple parametric change in Hamiltonian or external field, usually employed in the derivation of the JE or the CFT, is not applicable and care is needed in developing the physical assumptions, as in [43, 65].

Here, we address the issue by considering a system of interest containing some very slowly relaxing constituents, such as soft matter or pitch [66], in contact with a rapidly relaxing reservoir. The reservoir may be formed from a copper block or another highly thermally conductive material. Changing the temperature of the reservoir (e.g., with a thermostatically controlled heat exchanger) then drives the system of interest out of equilibrium. The change in temperature is slow enough that the reservoir may be treated to high accuracy, as in undergoing a quasistatic temperature change. The slowly relaxing system of interest is far from equilibrium. We have developed generalized versions of the CFT and the JE applied to this system. Importantly, the quantities that appear in the theory are physically measurable variables.

Since we choose the thermostat to be large and remote, details of how it is implemented will not affect the way the system behaves. For convenience, from a theoretical perspective, we choose the Nosé–Hoover thermostating mechanism and the equations (based on (2.1)), including that of the thermostat multiplier, are thus:

$$\dot{q}_i = \frac{p_i}{m}$$
$$\dot{p}_i = F_i(q) - S_i \alpha(\Gamma) p_i$$
$$\dot{\alpha} = \left(\frac{\sum_{i=1}^N S_i p_i \cdot p_i / m}{d\, N_{\text{therm}} k_B T(t)} - 1\right) \frac{1}{\tau_\alpha^2}, \tag{2.36}$$

where τ_α is the Nosé–Hoover time constant. The value of $T(t)$ is the target temperature of the thermostat. The extended, time-dependent internal energy is $H_E(\Gamma, \alpha, t) = H_0(\Gamma) + \frac{d}{2} N_{\text{therm}} k_B T(t) \alpha^2 \tau_\alpha^2$ and the extended phase space of the system is $\Gamma' = (\Gamma, \alpha)$. The Liouville equation states: $df/dt = -\Lambda f$, [3] and using (2.36) it is easy to show that $k_B T \Lambda = k_B T \left(\frac{\partial}{\partial \Gamma} \cdot \dot{\Gamma} + \frac{\partial}{\partial \alpha} \dot{\alpha} \right) = -d N_{\text{therm}} k_B T \alpha$ $\equiv \dot{Q}$, where $\dot{Q}$ is the rate of increase in H_E at constant T due to the thermostating alone. The equilibrium distribution function for this system is then easily shown to be

$$f_{\text{eq}}(\Gamma, T, \alpha) = \frac{\tau_\alpha \sqrt{d N_{\text{therm}}/(2\pi)}}{Z(T)} \exp(-\beta H_E(\Gamma, T, \alpha)), \tag{2.37}$$

where $Z(T)$ is given in (2.32). In this case, the parameter that is varied in time is the temperature of the wall, $\lambda(t) \equiv T(t)$.

We now consider applying the generalized JE, (2.28) when a thermal rather than a mechanical process occurs. Consider a thermostated system of N particles whose target temperature is changed from T_1 to T_2 over a period $0 < t < \tau$. We do not change the Hamiltonian during this process. For simplicity, we consider a canonical ensemble for the two equilibrium states (2.37), and use the equations of motion (2.36). The temperature dependence of the reservoir is achieved by making the Nosé–Hoover target temperature $T(t)$ in (2.36) a time dependent variable.

The change in temperature is slow enough that the reservoir may be treated as changing quasi-statically at the target temperature $T(t)$, while the slowly relaxing system of interest is driven out of equilibrium: that is, it changes irreversibly. However, if one is only interested in the synthetic dynamics, this restriction may be lifted and the temperature can be changed at an arbitrary rate. Either way, the system of interest will approach the temperature T_2 in the limit $t/\tau \to \infty$. We use (2.26) with f_1 and f_2 given by (2.37) at the two different temperatures to obtain

$$\Delta X_\tau(\Gamma'; 0, \tau) = \beta_2 H_E(\Gamma'(\tau)) - \beta_1 H_E(\Gamma'(0)) - \int_0^\tau dt \, \dot{\beta}(t) \dot{Q}(\Gamma'(t)), \tag{2.38}$$

where $\beta(t) = 1/(k_B T(t))$ is the inverse, time-dependent target temperature. Now, if we take the derivative of the extended Hamiltonian while the temperature is changing, but with no other external agent acting on the system, by using (2.36) we obtain

$$\frac{d}{dt} H_E(\Gamma'(t)) = \dot{Q}(\Gamma'(t)) + \frac{d}{2} N_{\text{therm}} k_B \dot{T}(t) \alpha^2(t) \tau_\alpha^2. \tag{2.39}$$

We then obtain

$$\frac{d}{dt}[\beta(t) H_E(\Gamma'(t))] = -\beta(t) \left[H_0(\Gamma(t)) \frac{\dot{T}(t)}{T(t)} - \dot{Q}(\Gamma'(t)) \right], \tag{2.40}$$

and combining these, the generalized "power" for a change in the target temperature with time is

$$\dot{X}(\Gamma(t)) = \dot{\beta}(t) H_0(\Gamma(t)). \tag{2.41}$$

2.7 Nonequilibrium Work Relations for Thermal Processes

Note that (2.41) only depends upon Γ and not the thermostat multiplier α. Equation 2.28 then becomes

$$\left\langle \exp\left(-\int_0^\tau dt\, \dot{\beta}(t) H_0(\Gamma(t))\right)\right\rangle_1 = \frac{Z^{(2)}}{Z^{(1)}} = \exp[-\beta_2 A_2 + \beta_1 A_1]. \quad (2.42)$$

One can see that this equation is consistent with thermodynamics because in the quasistatic limit, equilibrium thermodynamics yields the relation

$$\dot{\beta}(t)\langle H_0\rangle_{eq} = \dot{T}\frac{d}{dT}[\beta(t) A(t)]. \quad (2.43)$$

The Hamiltonian of the total system may be split in parts representing the system of interest, H_{si}, the reservoir H_r and the interaction between the reservoir particles and the system of interest particles H_{sir}, yielding, in rather obvious notation, $H_0 = H_{si}(\Gamma_{si}) + H_r(\Gamma_r) + H_{sir}(\Gamma)$. Now, by construction, we have set up our system such that the changes to $\langle H_r \rangle$ and $\langle H_{sir} \rangle$ are quasistatic. This allows us to take the contributions of these parts of the Hamiltonian through the average appearing in (2.42),

$$\left\langle \exp\left(-\int_0^\tau dt\, \dot{\beta}(t) H_{si}(\Gamma_{si}(t))\right)\right\rangle_1 \times \exp\left(-\int_0^\tau dt\, \dot{\beta}(t) \langle H_r + H_{sir}\rangle_{T(t),eq}\right)$$
$$= \exp[-\beta_2 A_2 + \beta_1 A_1] \quad (2.44)$$

and obtain,

$$\left\langle \exp\left(-\int_0^\tau ds\, \dot{\beta}(t) H_{si}(\Gamma_{si}(t))\right)\right\rangle_1 = \exp[-\beta_2 A_{si,2} + \beta_1 A_{si,1}], \quad (2.45)$$

where β and T are given by the temperature of the reservoir, and

$$\int_0^\tau ds\, \dot{\beta}(t) \langle H_{si}(\Gamma_{si}(t))\rangle_{eq} = \beta_2 A_{si,2} - \beta_1 A_{si,1}. \quad (2.46)$$

For temperature changes at finite rates, the thermodynamic temperature of the system of interest can not be defined and the kinetic temperature of the system of interest may not be equal to the temperature of the thermal reservoir. Nonetheless, (2.45) can still be used to compute changes in the free energy of the system of interest, as specified by (2.46), because the reservoir is being changed approximately quasistatically.

From the above, one observes that the function appearing in the quasistatic thermodynamic path integral (2.46) is the same as that which appears in the nonequilibrium free energy relation. One could conjecture that any correct microscopic expression for the thermodynamic path integral derived using classical statistical thermodynamics would yield a correct Nonequilibrium Free Energy Relation for some protocol. All that is required is sufficient ingenuity to design a protocol consistent with the microscopic expression for the generalized work. To be absolutely sure that your microscopic expression and protocol are consistent, one should simply check that when substituted into (2.28), that the protocol generates the required generalized "work". However, if the Nonequilibrium Free Energy Relation is to be used beyond the synthetically thermostated dynamics, care is required. It must be ensured that the system is controlled by a thermal reservoir which remains in equilibrium.

If one constructs an algorithm (2.36) in order to accomplish some thermal transformation $(N_1, V_1, T_1) \rightarrow (N_1, V_1, T_2)$, then (2.28) gives a precise microscopic form for the generalized "work" appearing in the classical thermodynamic path integral for the free energy change. Although the quasistatic path integral expression is unique, the nonequilibrium expression is certainly not. This is because there are infinitely many protocols that accomplish the required change. Nonetheless, each of these expressions gives identical values for the free energy difference.

2.8
Corollaries of the Fluctuation Theorem and Nonequilibrium Work Relations

In this section, we describe some of the results that can be derived using the fluctuation theorem and nonequilibrium work relations, and provide references for further details.

2.8.1
Generalized Fluctuation Theorem

For an arbitrary phase function $\phi(\Gamma)$, one can derive an equality for the conjugate probability ratio [12, 67]:

$$\frac{p[\bar{\phi}_t = B]}{p[\bar{\phi}_t = -B]} = \langle \exp[-\Omega_t] \rangle^{-1}_{\bar{\phi}_t = B} . \tag{2.47}$$

In this equation, $\langle \ldots \rangle_{\bar{\phi}_t = B}$ denotes an average over those ensemble members for which $\bar{\phi}_t = B$. A derivation can be found in references [12, 67].

2.8.2
Integrated Fluctuation Theorem

In experimental tests of the Fluctuation Theorem, it is almost always easier to test the Integrated Fluctuation Theorem (IFT), rather than the distinct Fluctuation Theorem [5, 15]. The IFT simply asks what the ratio between positive and negative time-averaged values of the dissipation function is. The IFT states that

$$\frac{p[\Omega_t > 0]}{p[\Omega_t < 0]} = \langle \exp[-\Omega_t] \rangle_{\Omega_t < 0} = \langle \exp[-\Omega_t] \rangle^{-1}_{\Omega_t > 0} . \tag{2.48}$$

This can be obtained by integration of (2.13),

$$\frac{p(\Omega_t > 0)}{p(\Omega_t < 0)} = \frac{\int_0^{+\infty} dB\, p(\Omega_t = B)}{\int_{-\infty}^0 dB\, p(\Omega_t = B)}$$

$$= \frac{\int_0^{+\infty} dB\, p(\Omega_t = B)}{\int_0^{+\infty} dB\, p(\Omega_t = -B)}$$

$$= \frac{\int_0^{+\infty} dB\, p(\Omega_t = B)}{\int_0^{+\infty} dB\, p(\Omega_t = B)\exp[-B]}$$

$$= \langle \exp[-\Omega_t] \rangle_{\Omega_t > 0}^{-1}. \tag{2.49}$$

2.8.3
Second Law Inequality

Linear irreversible thermodynamics asserts that the instantaneous local spontaneous entropy production must always be nonnegative. However, for a viscoelastic fluid, this is not always the case. Given the fundamental status of the Second Law, this presents a problem. The derivation of the Second Law Inequality (SLI) from the FT provides new insight into this problem. The SLI shows that *time averages* (rather than instantaneous values) of the entropy production are nonnegative. This Second Law Inequality is valid for the appropriately time-averaged entropy production, though the instantaneous entropy production may be negative for various ranges of times.

The Second Law Inequality states that [68]

$$\langle \Omega_t \rangle \geq 1, \tag{2.50}$$

and is obtained by integration of (2.13)

$$\langle \Omega_t \rangle = \int_{-\infty}^{+\infty} dB\, p(\Omega_t = B)\, B$$

$$= \int_0^{+\infty} dB\, p(\Omega_t = B)\, B + \int_{-\infty}^0 dB\, p(\Omega_t = B)\, B$$

$$= \int_0^{+\infty} dB\, p(\Omega_t = B)\, B - \int_0^{+\infty} dB\, p(\Omega_t = -B)\, B$$

$$= \int_0^{+\infty} dB\, p(\Omega_t = B)\, B(1 - \exp[-B]) \geq 0. \tag{2.51}$$

We note that the Second Law Inequality is a macroscopic consequence of the Fluctuation Theorem. All previously derived consequences of ESFT and JE were microscopic in nature. This finding should have important consequences in widely varied applications such as atmospheric physics and aerodynamics.

If one applies the Second Law Inequality to the case of periodic time dependent fields, then ergodic consistency limits its application to a discrete set of times that are multiples of the period or in some cases, the half period. However, one can also derive the Second Law Equality from the Crooks Fluctuation theorem for the case where there is no change in the free energy $\Delta A = 0$ (e.g., straining a liquid) and consequently, there is no distinction between forward and reverse processes. Because the ergodic consistency condition is much weaker for CFT, this more general form of the Second Law Inequality applies for all integration times and for arbitrary waveforms (i.e., periodic waveforms are not required).

The Second Law Inequality derived from CFT resolves a long-standing paradox in linear irreversible thermodynamics [69]. In this theory, it is frequently stated that the entropy production rate is nonnegative. This statement is manifestly false, as any casual analysis of electric circuits in which time dependent voltages cause currents to flow in circuits with a complex impedance shows. Similar systems include viscoelastic fluids. In these systems, there is a phase lag between the applied field and the induced flux which guarantees that for short intervals of time, the product of the force and the flux will be negative. The ensemble-averaged entropy production *can* be negative. However, the Second Law Inequality must always be satisfied.

2.8.4
Nonequilibrium Partition Identity

This identity (also referred to as the Kawasaki identity, Kawasaki normalisation factor, Kawasaki function, or integral fluctuation theorem) was first derived for Hamiltonian systems by Yamada and Kawasaki in 1967, and for thermostated dynamical systems by Morriss and Evans in 1984 [11, 70, 71]. The Nonequilibrium Partition Identity (NPI) is stated as:

$$\langle \exp[-\Omega_t] \rangle = 1 . \qquad (2.52)$$

A very simple proof can be obtained using the ESFT given in (2.13):

$$\langle \exp[-\Omega_t] \rangle = \int_{-\infty}^{+\infty} dB\, p(\Omega_t = B) \exp[-B]$$

$$= \int_{-\infty}^{+\infty} dB\, p(\Omega_t = -B) = 1 . \qquad (2.53)$$

It is quite extraordinary that although the Second Law Inequality says the exponent of the NPI is negative on average, the rare instances when the dissipation function has a negative time average occur with such frequency that their exponentially enhanced effect insures the average of the exponential is always unity.

If one applies the Jarzynski Equality to a situation where the free energy difference of the two states is zero, one derives (2.53). Although the ESFT, JE and CFT each imply the NPI, the converse is not true [71].

2.8.5
The Steady State Fluctuation Theorem

In many nonequilibrium systems, if the system starts at equilibrium but is driven away from equilibrium by a dissipative field and is thermostated in some way, then after the relaxation of initial transients, the system relaxes to a *nonequilibrium steady state*. Nonequilibrium steady states are curious states. All macroscopic thermophysical properties are time independent. There is a balance between the work done on the system by the dissipative field and the energy dissipated to the thermostat. Not all thermostated nonequilibrium systems relax to steady states. Some evolve into periodic or quasi-periodic nonequilibrium systems, others suffer from thermal run-away (explosions) because the thermostated process cannot dissipate sufficient heat, while other systems become turbulent. Other systems evolve into multiple steady states with different macroscopic properties depending on the initial conditions. In what follows, we assume we are dealing with systems that evolve into a unique steady state. The necessary and sufficient conditions for a thermostated nonequilibrium system to evolve into a unique nonequilibrium steady state are unknown.

Deterministic nonequilibrium steady states display many fascinating properties. The steady state distribution function collapses onto a strange attractor of lower dimension than the phase space in which it is embedded. Indeed, the dimensional reduction can be used to compute the entropy production and the associated transport coefficients [72]. The fine-grained Gibbs entropy of deterministic steady state systems diverges towards negative infinity. An extensive literature is devoted to this subject [4, 73].

The transient fluctuation theorem (2.12) applies to systems where the trajectories are sampled from a known initial ($t = 0$) distribution function. It is exact, applies at all times, and averages of the dissipation function are taken from the start of each trajectory ($t = 0$) for a period of time, t. The same is true for the generalized ESFT form (2.47).

However, these transient fluctuation theorems can be extended so that the time averaging is carried out for a duration, t, but starting at a time t_0 (see Figure 2.3). As t_0 increases, the statistics of these delayed averages approach those of a nonequilibrium steady state. For a given t, t_0 the (exact) transient fluctuation theorem gives the following result [9, 13]:

$$\frac{p[\bar{\Omega}_{t_0,t_0+t} = B]}{p[\bar{\Omega}_{t_0,t_0+t} = -B]} = \exp[tB] + \langle \exp[-\Omega_{0,t_0}\Omega_{,t_0,2t_0+t}]\rangle^{-1}_{\bar{\Omega}_{t_0,t_0+t}=B} \quad (2.54)$$

where we use the notation $\bar{\Omega}_{t_1,t_2} = \frac{1}{t_2-t_1}\int_{t_1}^{t_2} \Omega(s)\,ds = \frac{1}{t_2-t_1}\Omega_{t_1,t_2}$. If fluctuations in the dissipation function had no serial correlations, the second term on the right hand side of (2.54) would be unity and, hence, would become insignificant in the limit $t \to \infty$.

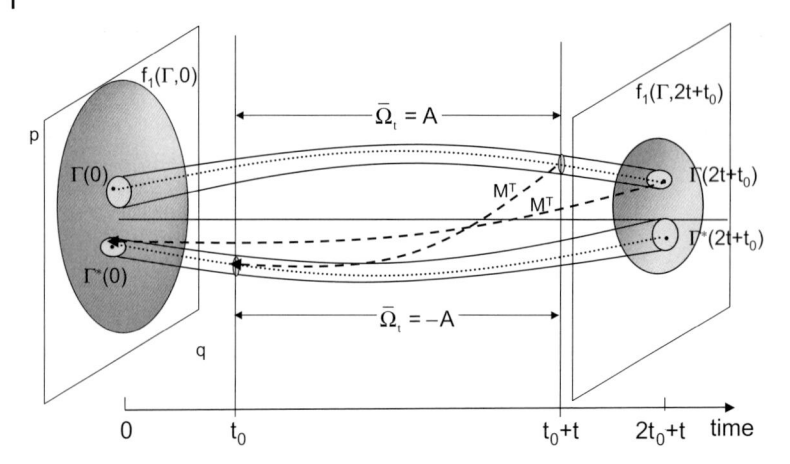

Figure 2.3 Schematic diagram showing the construction required to derive the steady state ESFT. Two trajectories that are related by time-reversal mappings are shown as dotted lines, and the time evolution of a small phase volume centered on these trajectories is also shown. The trajectory segments over which the dissipation function is averaged are considered, and are considered in the steady state fluctuation relation start at t_0 and end at $t_0 + t$. The dashed line indicates points related by time-reversal mappings (M^T).

For an arbitrary phase function ϕ we have:

$$\frac{p[\bar{\phi}_{t_0,t_0+t} = B]}{p[\bar{\phi}_{t_0,t_0+t} = -B]} = \langle \exp[-\Omega_{0,2t_0+t}]\rangle^{-1}_{\bar{\phi}_{t_0,t_0+t}=B} . \tag{2.55}$$

If a steady state exists, and we choose t_0 to be much larger than the Maxwell time so that the system has reached a steady state, these expressions will apply to the statistics of steady state trajectories sampled from an initial distribution. In the case of large t_0, the second term on the right hand side of (2.54) might be expected to be bounded with respect to t. In that case, we can write [9]

$$\lim_{t \to \infty} \frac{1}{t} \ln\left[\frac{p[\bar{\Omega}_{t_0,t_0+t} = B]}{p[\bar{\Omega}_{t_0,t_0+t} = -B]}\right] = B \tag{2.56}$$

and

$$\lim_{t \to \infty} \frac{1}{t} \ln\left[\frac{p[\bar{\phi}_{t_0,t_0+t} = B]}{p[\bar{\phi}_{t_0,t_0+t} = -B]}\right] = \frac{1}{t} \ln \langle \exp[-\Omega_{t_0,t_0+t}]\rangle^{-1}_{\bar{\phi}_{t_0,t_0+t}=B} . \tag{2.57}$$

We note that these equations do not explicitly refer to the behavior of properties of the system before t_0, and therefore only refer to the steady state portion of the dynamics. They will be independent of t_0 if it is much larger than the Maxwell time. Furthermore, the statistics of the phase variables Ω and ϕ will be the same as that of trajectories selected from points along a single steady state trajectory, provided there is only one steady state. Therefore, under these conditions, we can drop the reference to t_0, and write the steady state relations that will apply to both

2.8 Corollaries of the Fluctuation Theorem and Nonequilibrium Work Relations

a set of trajectory segments generated from an initial ensemble or from a single steady state trajectory

$$\lim_{t \to \infty} \frac{1}{t} \ln \left[\frac{p[\bar{\Omega}_t = B]}{p[\bar{\Omega}_t = -B]} \right] = B \tag{2.58}$$

and

$$\lim_{t \to \infty} \frac{1}{t} \ln \left[\frac{p[\bar{\phi}_t = B]}{p[\bar{\phi}_t = -B]} \right] = \frac{1}{t} \ln \langle \exp[-\Omega_t] \rangle^{-1}_{\bar{\phi}_t = B} . \tag{2.59}$$

We call (2.58) the steady state ESFT and (2.59) the generalized ESFT.

We emphasize that there are a number of conditions required to obtain (2.56)–(2.59) that are not required for the transient relations or (2.54), (2.55) (for a full description see [9]). A steady state must exist which implies that the correlation time is finite and the system must be chaotic; and if the phase function is singular, then it must be integrable. For the relation to apply to samples from a single trajectory, only one steady state must exist. Furthermore, they will not apply at all times, only asymptotically in the limit of long t. However, they do apply at arbitrary field strengths [74].

An alternative way of obtaining (2.56) and (2.57) is to assume that the average from time 0 to $t_0 + t$ will approach that of t_0 to $t_0 + t$ when t_0 is fixed and t is made large [12]. Therefore, the statistics of these sets will be indistinguishable. Then, from (2.12) we can directly obtain (2.58) in this limit. The assumptions stated above are still required.

Using the steady state fluctuation relations, the Einstein and Green–Kubo relations can be derived for systems close to equilibrium (see Section 2.5).

We note here that another type of Steady State Fluctuation Relation has been developed by Gallavotti, Cohen and co-workers [75–77]. It has been proven for Anosov systems [78], but is anticipated to apply to a wider range of systems and can be written as

$$\lim_{t \to \infty} \frac{1}{t} \ln \left[\frac{p[\bar{\Lambda}_t = B]}{p[\bar{\Lambda}_t = -B]} \right] = B \quad \text{for} \quad |B| \leq B^* , \tag{2.60}$$

where Λ is the phase space expansion rate, and B^* is some constant[5]. For isoenergetic systems $\Lambda \equiv \Omega$, and therefore, the relations (2.60) and (2.58) become identical, implying for this circumstance that $B^* = \infty$.

Application of the Gallavotti–Cohen Fluctuation Theorem (GCFT) to systems that are not isoenergetic has recently been discussed [77, 79], and it has been found that there are serious limitations to its utility. For instance, for many common systems the value of $B^* = O(F_e^2) \to 0$ as $F_e \to 0$, and it must be modified if the phase

---

5) The existence of the bound to the range of fluctuations in phase space compression [78] was not mentioned in the original Gallavotti–Cohen papers [75, 76]. This has caused confusion in the literature.

space contraction is not bounded. Perhaps even more difficult is the fact that for these systems the time required for convergence of the GCFT diverges as $O(F_e^{-2})$. Since much of the interest in fluctuation theorems arises from the fact that they are exact arbitrarily far from equilibrium, the bound on the range of fluctuations means that the GCFT is of limited use in large deviation theory.

2.8.6
Minimum Average Work Principle

From the Jarzynski Equality, it is easy to compute a bound on the work for a thermodynamic process [36]:

$$\begin{aligned}
\exp[-\beta \Delta A] &= \langle \exp[-\beta \Delta W] \rangle \\
&= \exp[-\beta \langle \Delta W \rangle] \langle \exp[-\beta \Delta W + \beta \langle \Delta W \rangle] \rangle \\
&\geq \exp[-\beta \langle \Delta W \rangle] \langle 1 - \beta \Delta W + \beta \langle \Delta W \rangle \rangle \\
&\geq \exp[-\beta \langle \Delta W \rangle]
\end{aligned} \qquad (2.61)$$

In deriving this results, we have used the fact that $e^x \geq 1 + x$, $\forall x$. The above equation implies that the ensemble average thermodynamic work is never less than the free energy difference:

$$\langle \Delta W \rangle \geq \Delta A. \qquad (2.62)$$

This is called the Minimum Work Principle (MWP). Naturally, if one considers the purely dissipative work, namely, $\Delta W - \Delta A$, the Second Law Identity (2.50) for this quantity is consistent with MWP (2.62).

If one studies a cyclic process, the total change in free energy around the cycle is zero. Application of the MWP implies that the ensemble-averaged work for the cyclic process is nonnegative.

2.9
Experiments

Until quite recently, these theorems were explored theoretically and numerically using computer simulation. Numerical results can provide insight into practical issues associated with application of the theorems such as the degree of sampling required. In cases where fluctuation theorems have been obtained for a coarse-grained or stochastic models of a system [22, 80], or there is uncertainty in the initial distribution [47], numerical calculation provides information on the validity of the models. It is only in the last few years that the practicality of these theorems, applied to the small dynamical systems, has been experimentally explored. It is important to note that experimentation is in no way a "proof" of the theorems. Instead, experiments verify that the conditions (e.g., ergodic consistency, time reversibility, synthetic thermostats at a distance) required for satisfaction of the theorem are actually present under experimental conditions. If this is so, we can then conclude that the mathematical theorem is actually *relevant* to natural systems. Experiments also give us information that may be presently theoretically unknown. An example of this is the information provided by experiments (both laboratory and computer) regarding B^* in the GCFT. At the time of the experiments, almost nothing was known about B^*. The scaling properties for B^* were only discovered theoretically after the experiments had indicated the information.

Here, we briefly mention some of the experiments that have been carried out and refer the reader to recent reviews including discussions on experiments [49, 81–84], and a literature review by Searles and Evans on recent experiments [85].

The first conclusive experimental tests of the fluctuation relations were on a colloidal particle in water held by an optical trap [15, 16]. The trap was either translated by movement of the optical trap relative to the system, or used to 'capture' the particle by increasing the strength of the optical trap [16]. Carberry *et al.* [71] later experimentally demonstrated the NPI using this approach, and Wang *et al.* [86, 87] verified the *steady state* version of the ESFT. In early studies, the particle was in a viscous fluid and therefore the equations of motion of the particle were well approximated by a white noise stochastic Langevin equation. In 2007, a capture experiment was carried out in a viscoelastic solvent where this approximation no longer applies [22]. It was shown that despite this, the experiments validated the ESFT, and therefore could not be considered to be just a special property of Brownian dynamics. Blickle *et al.* [88] verified the fluctuation relation for the work (or dissipation function) for a system where the trap potential was not harmonic.

Narayan and Dhar [89] demonstrated the importance of choosing the correct expression for the entropy production (see discussion in Section 2.5) by demonstrating that an FT for heat (corresponding to the GCFT, (2.60)) is not obeyed in their experimental studies, whereas the FT for work (corresponding to the ESFT, (2.11)) is.

Garnier and Ciliberto [90] have studied fluctuations in the power injected to an electrical dipole that is subject to a current and verified the steady state FT (2.58) and a heat fluctuation relation as predicted by van Zon *et al.* [91–93]. This group has more recently studied stochastic nonequilibrium steady states, verifying FTs that are valid at all times [94].

Douarche *et al.* [95] verified the transient ESFT and steady state ESFT for a harmonic oscillator (a brass pendulum in a water-glycerol solution that is driven out of equilibrium by an applied torque). They also developed a steady state relation for a system with a sinusoidal forcing and showed that the convergence time for the steady state relation was considerably longer in this case.

Other systems used for verification of the ESFT have included a diamond with a single defect periodically excited by a laser [21, 96], an electric circuit [97] and particles undergoing diffusion [98].

The first tests of the JE and CFR were by Liphardt *et al.* [19], who used optical tweezers to extend a DNA-RNA hybrid chain, measuring the work required as the extension proceeded. As well as demonstrating the ability of observing fluctuations that would allow the JE and CFR to be applied, it led to the use of the JE as an experimental tool for studying protein folding and for generating free energy landscapes.

More recently, Collin *et al.* [18] carried out an experiment using the CFR to determine the difference in free energies of an RNA molecule and a mutant that differs by one base pair. The CFR was shown to be useful far from equilibrium where insufficient sampling hampers convergence of the JE.

Hummer and Szabo [99] demonstrated that in single molecule stretching experiments, the JE provides an expression for the work at different times, whereas from an experimental point of view, it is of more interest to know the free energy difference between states at different extensions of the molecule. They show how this can be obtained and apply it in experiments.

Douarche *et al.* [100] have verified the CFR and JE for fluctuations in the work of a mechanical oscillator that is in contact with a reservoir and driven by a large external field.

In the future, it will be interesting to see how the relationships can be beneficially used in experimental studies or interpretation of experimental results. In this vein, Noy has used [101] the JE to benefit in interpretation of experimental results of chemical force microscopy where the probes of atomic force microscopy are functionalized.

2.10
Conclusion

At first sight, the definitions of the dissipation function and of the generalized work function may seem a little obscure. However, we can give a more physical explanation of why they take the form they do. If you look at the dissipation function defined in (2.10), you can see that on the second line of (2.10) we have two terms. Consider the first term. That term will be zero for microcanonical ensembles and it will be a difference in energies or enthalpies for canonical or isothermal isobaric ensembles.

Now, look at the second term. The integral of the phase space compression factor is just related to the energy lost to the thermostat (which would be zero for unthermostated dynamics). Thus, the sum of the two terms on the right hand side of (2.10) reduces to either the heat loss in the microcanonical case or a generalized dissipative work (energy or enthalpy) for the other ensembles.

For the microcanonical ensemble and constant energy dynamics, (the dynamics required to satisfy the ergodic consistency condition), the heat loss and the dissipated work are exactly equal, so the dissipation function can be clearly interpreted as the dissipated work[6].

The same form of analysis can be carried out on the generalized work defined in (2.26), however in this case, the probabilities are determined with respect to different ensembles. Because the distribution functions, free energies and associated partition functions may be different for the two states, the only way we can ob-

---

6) If the initial distribution is microcanonical, adiabatic dynamics cannot be used because this would violate ergodic consistency.

tain a generalized work is to multiply the distribution functions by their respective partition functions, precisely as in (2.26). In the case of canonical dynamics, this then leads to an expression based on the difference in the internal energies and the energy lost to the thermostat. There is a second difference between generalized work and the dissipation function. Because the two equilibrium states may have different free energies, the resulting generalized work is *not* purely dissipative. This observation leads to a rather simple derivation of the generalized Jarzynski Equality (2.28). We decompose the generalized work into its purely dissipative component and its reversible component: $\Delta X_\tau \equiv \Delta X_{\text{rev},\tau} + \Delta X_{\text{diss},\tau}$ where $\exp(-\Delta X_{\text{rev},\tau}) = Z(\lambda_2)/Z(\lambda_1)$, we immediately see from the NPI that

$$\langle \exp(-\Delta X_\tau) \rangle = \frac{Z(\lambda_2)}{Z(\lambda_1)} \langle \exp(-\Delta X_{\text{diss},\tau}) \rangle = \frac{Z(\lambda_2)}{Z(\lambda_1)}. \qquad (2.63)$$

This most simple and general derivation of the generalized JE, (also given by (2.28)) shows what is required for the Equality to work in practice. One needs to see the anti-trajectories that correspond to the most probable trajectories in order for the generalized JE to yield reliable averages in practice. These trajectories are required for the Nonequilibrium Partition Identity to be unity. The derivation also points out the intimate relationship between the Nonequilibrium Work Relations and the Fluctuation Theorem.

Using the approach applied in Section 2.5, Williams and Evans have rederived an exact expression for the time dependent nonlinear response [26]:

$$f(\Gamma, t) = f(\Gamma, 0) \exp\left[-\beta \int_0^t ds\, J(\Gamma(-s)) V F_e(t-s)\right]. \qquad (2.64)$$

For the case of time independent fields, this equation is consistent with the dissipation Theorem (2.20). This equation was previously derived [102] by assuming that the nonequilibrium fine-grained Gibbs entropy, $S(t) = -k_B \int d\Gamma\, f(\Gamma, t) \ln[f(\Gamma, t)]$, is a minimum, subject to the constraints that the distribution is normalized, that the average energy is fixed and that the average dissipative flux is fixed at any time and also is a continuous function of time. This result is now confirmed without assumptions regarding the entropy or the free energy.

Recently, there has been some interest in what has become known as the Maximum Entropy Production (MEP) approach [103, 104] to nonlinear dynamical systems. This approach asserts that nonequilibrium systems arrange themselves in a way that maximizes the rate at which entropy is produced, subject to a set of constraints. As Hoover pointed out in 1986 [105], the problem with these theories is that there is no objective way to comprehensively list the set of such constraints. Combining (2.64) with the work reported in [102] shows that the nonequilibrium distribution function for a dissipative system does indeed minimize the fine-grained entropy of the nonequilibrium system. Subject to the constraints placed on the system (the distribution is normalized, the initial average energy is fixed, and the average dissipative flux is constrained at all times and is a continuous function of time) the exact distribution function (2.64) minimizes the Gibbs entropy for all times, including the transient behavior of the system. This is in spite of the fact that for a thermostated deterministic steady state, the entropy diverges towards negative infinity as time increases [73]. This is a consequence of the fact that the steady state distribution (which is only approached and never actually reached) is a strange attractor of lower dimension than the ostensible dimension of phase space [4, 72]. It can be seen in [102] that at any finite time, assuming that the entropy to be a minimum (subject to the correct constraints of course), results in the same expression, (2.64), which is rigorously derived in [26]. In the steady state, the average rate of change of the entropy is constant. Thus, the rate of decrease of the Gibbs entropy is maximal. However, the approach in [102] is far more complex than the MEP approximations. The MEP only employs sets of Lagrange multipliers for the corresponding sets of constraints. In order to achieve temporal continuity in the average dissipative flux, [102] employs a Lagrange multiplier functional (i.e., the set of constraints becomes infinite). This functional becomes a memory function that determines what happens in the future by what has happened in the past, ensuring that the dissipative flux is a continuous function of time.

If we take the logarithm of (2.64) and assume the initial ensemble is canonical, we see that

$$\ln[f(\Gamma, t)] = \beta A(t = 0) - \beta H_0(\Gamma) - \beta \int_0^t ds\, J(\Gamma(-s))\, V F_e(t - s)]. \quad (2.65)$$

Since the logarithm is a monotonic increasing function of its argument, we see that the probability of observing a phase Γ at time t is increased if the dissipation integrated along the phase space trajectory that terminates at Γ at time t, is large and positive. What is important here is that the probability is influenced by the path integral and not just the instantaneous value. As we have seen in time dependent systems (e.g., viscoelastic systems), the instantaneous entropy production is not always positive. However, the Second Law Inequality [69] guarantees that the average time integral is positive and, in the long time limit, subject to the constraints, is actually maximal.

We also see that the probability is increased if the value of the Hamiltonian at the current phase point, namely, $H_0(\Gamma)$, is also low. In the long time limit, we expect that the integrated dissipation will dominate over the Hamiltonian term. This is probably why MEP provides a reasonable approximation at long times in some circumstances [103]. We note, as always, that the reciprocal temperature β is not directly related to the temperature of the system of interest (this temperature may not be well defined far from equilibrium), but rather to that of the large, effectively equilibrium, heat bath to which the heat eventually dissipates.

The FT is a rigorous analytical result from which the SLI and then the Second Law of Thermodynamics can be directly proven, despite the fact that the underlying equations of motion are microscopically reversible. It therefore explains the Loschmidt Paradox, which remained a paradox until the FT was obtained in the early 1990s (i.e., for about 100 years).

On page 33 of the Landau and Lifshitz's textbook [106], they state, "The question of the physical foundations of the law of monotonic increase of entropy thus remains open: it may be of cosmological origin and related to the general problem of initial conditions in cosmology ... the violation of symmetry under time reversal in some weak interactions between elementary particles may play some part ... in all closed systems which occur in Nature, the entropy never decreases..." (for a discussion on the cosmological approach see Chapter 7 of [107]). It is interesting that in 1980, Landau and Lifshitz still talked of the *monotonic increase of* entropy as though the Second Law was absolute rather than statistical. This same mistake was made by Einstein in 1903 when he believed he had given a proof of the Second Law. His proof does lead to a monotonic increase in entropy, but the proof rested on an erroneous assumption, namely, that more probable distributions *always* follow less probable ones!

Another quite frequent "explanation" of the origin of the Second Law is that the Second Law is satisfied in our region of the Universe because we as complex biological organisms are here to observe this part of the Universe. This is the anthropomorphic theory of the Second Law.

These views are each quite different from those of J. C. Maxwell, who wrote in 1878 ([108], p. 280), "Hence, the Second Law of thermodynamics is continually being violated, and that to a considerable extent, in any sufficiently small group of molecules belonging to a real body. As *the number of molecules in the group is increased*, the deviations from the mean of the whole become smaller and less frequent; and when the number is increased till the group includes a sensible portion of the body, the probability of a measurable variation from the mean occurring *in a finite number of years* becomes so small that it may be regarded as practically an impossibility". Thus, it is quite clear that Maxwell would be completely unsurprised by the Fluctuation Theorem. He understood that the Second Law is not absolute and that it is "continually being violated" in small systems for short times.

If all the laws of mechanics and quantum mechanics are time-reversal symmetric, then clearly, you cannot prove an asymmetric result like the Fluctuation Theorem. In the first proof given by Evans and Searles in 1994 [14], this time symmetry was indeed broken, though it was broken in such a natural way that many people who have analyzed these proofs fail to see where the time-reversal symmetry is broken. The assumption made was that processes are *causal* [109].

Once again, Landau and Lifshitz are quoted [106] (p. 32), "In quantum mechanics ... The fundamental equation is itself symmetrical under time reversal ... However, despite this symmetry, quantum mechanics does involve an important non-equivalence of the two directions of time ... the probability of any particular result of process B is determined by the result of process A, can only be valid if process A occurred earlier than process B."

This is the Axiom of *Causality*. It is used frequently in quantum mechanics and (unrecognized by Landau and Lifshitz), it is also required in classical mechanics. In the proof of the ESFT and the CFT, the probabilities of observing particular values of time integrals of the dissipation function or of the generalized work are computed from the probabilities of observing the initial states from which those sets of trajectories began, $f(\Gamma, 0)d\Gamma$. We never used the probabilities of the endpoints; indeed, had we done so we would have proved the anti-Fluctuation Theorem and an anti-Second Law [109]. The Axiom of Causality is so natural that people fail to observe that they have indeed made this assumption. Landau and Lifshitz failed to notice that it is constantly used in classical mechanics. This is evidenced by the simple fact that Laplace transforms are only defined by $(0, \infty)$ time integrals rather than $(-\infty, \infty)$ time integrals as for spatial Fourier transforms. Consequently, this leads to *memory* functions rather than anti-memory functions. For an extensive discussion of causality and thermodynamics, see [109].

The Mori–Zwanzig theory of thermal transport processes involves an exact manipulation of propagators and leads to exact expressions for linear transport coefficients at zero wave-vector. It is not completely clear what the sign of these expressions is (there is no Second Law Inequality), but it is clear that time-reversal symmetry has been broken because these expressions are odd under time-reversal symmetry. Time reversal symmetry was broken within the Mori–Zwanzig formalism by invoking the Axiom of Causality and employing memory functions rather than anti-memory functions. Another instance of the application of the Axiom of Causality is provided by the classical theory of electromagnetism where one finds both "advanced" and "retarded" vector potentials as solutions of Maxwell's time symmetric field equations [110]. Ignoring the "advanced" potentials is just another example of the application of the Axiom of Causality. This Axiom is so natural that most physicists don't even recognize instances of its application.

If one is prepared to accept the Axiom of Causality without proof, then the Fluctuation Theorem and the Second Law are inescapable. In fact, the Second Law ceases to be a "Law" and becomes, instead, the limiting case ($N \to \infty$) of a very generally applicable theorem.

Acknowledgements

We would like to thank the Australian Research Council for their support. DJE acknowledges many useful discussions with Chris Jarzynski at the Isaac Newton Institute for Mathematical Sciences in Cambridge, UK, during 2006. DJE also acknowledges the very generous support from the Volkswagen Foundation. DJE, DJS and SRW acknowledge support from the Australian Research Council. DJS thanks Suresh Bhatia at the University of Queensland for hosting her sabbatical leave while this manuscript was written.

References

1 Hoover, W.G., Ladd, A.J.C., Moran, B. (1982) *Phys. Rev. Lett.*, **48**, 1818.
2 Evans, D.J. (1983) *J. Chem. Phys.*, **78**, 3297.
3 Evans, D.J., Morriss, G.P. (2008) *Statistical Mechanics of Nonequilibrium Liquids*, 2nd edn, Cambridge Univ. Press, London.
4 Klages, R. (2007) *Microscopic Chaos, Fractals and Transport in Nonequilibrium Statistical Mechanics*, World Scientific, Singapore.
5 Ayton, G., Evans, D.J., Searles, D.J. (2001) *J. Chem. Phys.*, **115**, 2033.
6 Williams, S.R., Searles, D.J., Evans, D.J. (2004) *Phys. Rev. E*, **70**, 066113.
7 Jarzynski, C. (2000) *J. Stat. Phys.*, **98**, 77.
8 Gallavotti, G. (2006) *Chaos*, **16**, 043114.
9 Searles, D.J., Rondoni, L., Evans, D.J. (2007) *J. Stat. Phys.*, **128**, 1337.
10 Williams, S.R., Searles, D.J., Evans, D.J. (2007) *Mol. Phys.*, **105**, 1059.
11 Evans, D.J., Searles, D.J. (1995) *Phys. Rev. E*, **52**, 5839.
12 Evans, D.J., Searles, D.J. (2002) *Adv. Phys.*, **51**, 1529.
13 Searles, D.J., Evans, D.J. (2000) *J. Chem. Phys.*, **113**, 3503.
14 Evans, D.J., Searles, D.J. (1994) *Phys. Rev. E*, **50**, 1645.
15 Wang, G.M., Sevick, E.M., Mittag, E., Searles, D.J., Evans, D.J. (2002) *Phys. Rev. Lett.*, **89**, 050601.
16 Carberry, D.M., Reid, J.C., Wang, G.M., Sevick, E.M., Searles, D.J., Evans, D.J. (2004) *Phys. Rev. Lett.*, **92**, 140601.
17 Reid, J.C., Carberry, D.M., Wang, G.M., Sevick, E.M., Evans, D.J., Searles, D.J. (2004) *Phys. Rev. E*, **70**, 016111.
18 Collin, D., Ritort, F., Jarzynski, C., Smith, S.B., Tinoco, I., Bustamante, C. (2005) *Nature*, **437**, 231.
19 Liphardt, J., Dumont, S., Smith, S.B., Tinoco, I., Bustamante, C. (2002) *Science*, **296**, 1832.
20 Trepagnier, E.H., Jarzynski, C., Ritort, F., Crooks, G.E., Bustamante, C.J., Liphardt, J. (2004) *Proc. Natl. Acad. Sci. U.S.A.*, **101**, 15038.
21 Schuler, S., Speck, T., Tietz, C., Wrachtrup, J., Seifert, U. (2005) *Phys. Rev. Lett.*, **94**, 180602.
22 Carberry, D.M., Baker, M.A.B., Wang, G.M., Sevick, E.M., Evans, D.J. (2007) *J. Opt. A*, **9**, S204.
23 Evans, D.J., Searles, D.J., Williams, S.R. (2008) *J. Chem. Phys.*, **128**, 014504.
24 Evans, D.J., Searles, D.J., Williams, S.R. (2008) *J. Chem. Phys.*, **128**, 249901.
25 Yamada, T., Kawasaki, K. (1967) *Progr. Theor. Phys.*, **38**, 1031.
26 Williams, S.R., Evans, D.J. (2008) *Phys. Rev. E*, **78**, 021119.
27 Morriss, G.P., Evans, D.J. (1987) *Phys. Rev. A*, **35**, 792.
28 Evans, D.J., Morriss, G.P. (1988) *Phys. Rev. A*, **38**, 4142.
29 Borzsak, I., Cummings, P.T., Evans, D.J. (2002) *Mol. Phys.*, **100**, 2735.
30 Todd, B.D., Daivis, P.J. (1999) *Comput. Phys. Commun.*, **117**, 191.
31 Delhommelle, J., Cummings, P.T., Petravic, J. (2005) *J. Chem. Phys.*, **123**, 114505.
32 Delhommelle, J., Cummings, P.T. (2005) *Phys. Rev. B*, **72**, 172201.
33 Pan, G., McCabe, C. (2006) *J. Chem. Phys.*, **125**, 194527.
34 Kubo, R. (1966) *Rep. Prog. Phys.*, **29**, 255.
35 Jarzynski, C. (1997) *Phys. Rev. Lett.*, **78**, 2690.
36 Jarzynski, C. (1997) *Phys. Rev. E*, **56**, 5018.
37 Crooks, G.E. (1998) *J. Stat. Phys.*, **90**, 1481.
38 Crooks, G.E. (1999) *Phys. Rev. E*, **60**, 2721.
39 Adib, A.B. (2005) *Phys. Rev. E*, **71**, 056128.
40 Reid, J.C., Sevick, E.M., Evans, D.J. (2005) *Europhys. Lett.*, **72**, 726.
41 Chatelain, C. (2007) *J. Stat. Mech.-Theory Exp.*, P04011.
42 Chelli, R., Marsili, S., Barducci, A., Procacci, P. (2007) *J. Chem. Phys.*, **126**, 044502.
43 Chelli, R., Marsili, S., Barducci, A., Procacci, P. (2007) *Phys. Rev. E*, **75**, 050101.
44 Cuendet, M.A. (2006) *J. Chem. Phys.*, **125**, 144109.
45 Cleuren, B., Van den Broeck, C., Kawai, R. (2006) *Phys. Rev. Lett.*, **96**, 050601.
46 Williams, S.R., Searles, D.J., Evans, D.J. (2008) *Phys. Rev. Lett.*, **100**, 250601.

47 Williams, S.R., Searles, D.J., Evans, D.J. (2008) *J. Chem. Phys.*, **129**, 134504; also available at arXiv:0806.2918v1.

48 Horowitz, J., Jarzynski, C. (2007) *J. Stat. Mech.-Theory Exp.*, P11002.

49 Sevick, E.M., Prabhakar, R., Williams, S.R., Searles, D.J. (2008) *Ann. Rev. Phys. Chem.*, **59**, 603.

50 Adjanor, G., Athènes, M., Calvo, F. (2006) *Eur. Phys. J. B*, **53**, 47.

51 Kofke, D.A. (2006) *Mol. Phys.*, **104**, 3701.

52 Wu, D., Kofke, D.A. (2005) *J. Chem. Phys.*, **123**, 054103.

53 Wu, D., Kofke, D.A. (2005) *J. Chem. Phys.*, **123**, 084109.

54 Wu, D., Kofke, D.A. (2005) *J. Chem. Phys.*, **122**, 204104.

55 MacFadyen, J., Andricioaei, I. (2005) *J. Chem. Phys.*, **123**, 074107.

56 Lechner, W., Dellago, C. (2007) *J. Stat. Mech.-Theory Exp.*, P04001.

57 Lua, R.C., Grosberg, A.Y. (2005) *J. Phys. Chem. B*, **109**, 6805.

58 Shirts, M.R., Pande, V.S. (2005) *J. Chem. Phys.*, **122**, 144107.

59 Ytreberg, F.M., Swendsen, R.H., Zuckerman, D.M. (2006) *J. Chem. Phys.*, **125**, 184114.

60 Lechner, W., Oberhofer, H., Dellago, C., Geissler, P.L. (2006) *J. Chem. Phys.*, **124**, 044113.

61 Schmiedl, T., Seifert, U. (2007) *Phys. Rev. Lett.*, **98**, 108301.

62 Vaikuntanathan, S., Jarzynski, C. (2008) *Phys. Rev. Lett.*, **100**, 190601.

63 Hoover, W.G. (1985) *Phys. Rev. A*, **31**, 1695.

64 Evans, D.J. (2003) *Mol. Phys.*, **101**, 1551.

65 Jarzynski, C., Wójcik, D.K. (2004) *Phys. Rev. Lett.*, **92**, 230602.

66 Edgeworth, R., Dalton, B.J., Parnell, T. (1984) *Eur. J. Phys.*, **5**, 198.

67 Searles, D.J., Ayton, G., Evans, D.J. (2000) *AIP Conf. Ser.*, **519**, 271.

68 Searles, D.J., Evans, D.J. (2004) *Aust. J. Chem.*, **57**, 1119.

69 Williams, S.R., Evans, D.J., Mittag, E. (2007) *Comptes Rendus Phys.*, **8**, 620.

70 Morriss, G.P., Evans, D.J. (1985) *Mol. Phys.*, **54**, 629.

71 Carberry, D.M., Williams, S.R., Wang, G.M., Sevick, E.M., Evans, D.J. (2004) *J. Chem. Phys.*, **121**, 8179.

72 Evans, D.J., Cohen, E.G.D., Searles, D.J., Bonetto, F. (2000) *J. Stat. Phys.*, **101**, 17.

73 Evans, D.J., Rondoni, L. (2002) *J. Stat. Phys.*, **109**, 895.

74 Williams, S.R., Searles, D.J., Evans, D.J. (2006) *J. Chem. Phys.*, **124**, 194102.

75 Gallavotti, G., Cohen, E.G.D. (1995) *Phys. Rev. Lett.*, **74**, 2694.

76 Gallavotti, G., Cohen, E.G.D. (1995) *J. Stat. Phys.*, **80**, 931.

77 Bonetto, F., Gallavotti, G., Giuliani, A., Zamponi, F. (2006) *J. Stat. Phys.*, **123**, 39.

78 Gallavotti, G. (1995) *Math. Phys. Electron. J.*, **1**, 12 p.

79 Evans, D.J., Searles, D.J., Rondoni, L. (2005) *Phys. Rev. E*, **71**, 056120.

80 Rahav, S., Jarzynski, C. (2007) *J. Stat. Mech.-Theory Exp.*, P09012.

81 Ritort, F. (2006) *J. Phys.-Condens. Matter*, **18**, R531.

82 Bustamante, C., Liphardt, J., Ritort, F. (2005) *Phys. Today*, **58**, 43.

83 Bustamante, C. (2005) *Q. Rev. Biophys.*, **38**, 291.

84 Tinoco, I., Li, P.T.X., Bustamante, C. (2006) *Q. Rev. Biophys.*, **39**, 325.

85 Searles, D.J., Evans, D.J. (2008) Fluctuation Relations, Free Energy Calculations and Irreversibility, in *Specialist Periodic Reports of the Rsc "Chemical Modelling: Applications and Theory"* (ed. A. Hinchliffe) Vol. 5; pp. 182–207.

86 Wang, G.M., Reid, J.C., Carberry, D.M., Williams, D.R.M., Sevick, E.M., Evans, D.J. (2005) *Phys. Rev. E*, **71**, 046142.

87 Wang, G.M., Carberry, D.M., Reid, J.C., Sevick, E.M., Evans, D.J. (2005) *J. Phys.-Condens. Matter*, **17**, S3239.

88 Blickle, V., Speck, T., Helden, L., Seifert, U., Bechinger, C. (2006) *Phys. Rev. Lett.*, **96**, 070603.

89 Narayan, O., Dhar, A. (2004) *J. Phys. A-Math. Gen.*, **37**, 63.

90 Garnier, N., Ciliberto, S. (2005) *Phys. Rev. E*, **71**, 060101.

91 van Zon, R., Ciliberto, S., Cohen, E.G.D. (2004) *Phys. Rev. Lett.*, **92**, 130601.

92 van Zon, R., Cohen, E.G.D. (2004) *Phys. Rev. E*, **69**, 056121.

93 van Zon, R., Cohen, E.G.D. (2004) *Physica A*, **340**, 66.

94 Joubaud, S., Garnier, N.B., Ciliberto, S. (2008) *Europhys. Lett.*, **82**, 30007.
95 Douarche, F., Joubaud, S., Garnier, N.B., Petrosyan, A., Ciliberto, S. (2006) *Phys. Rev. Lett.*, **97**, 140603.
96 Tietz, C., Schuler, S., Speck, T., Seifert, U., Wrachtrup, J. (2006) *Phys. Rev. Lett.*, **97**, 050602.
97 Andrieux, D., Gaspard, P., Ciliberto, S., Garnier, N., Joubaud, S., Petrosyan, A. (2007) *Phys. Rev. Lett.*, **98**, 150601.
98 Seitaridou, E., Inamdar, M.M., Phillips, R., Ghosh, K., Dill, K. (2007) *J. Phys. Chem. B*, **111**, 2288.
99 Hummer, G., Szabo, A. (2005) *Accounts Chem. Res.*, **38**, 504.
100 Douarche, F., Ciliberto, S., Petrosyan, A. (2005) *J. Stat. Mech.-Theory Exp.*, P09011.
101 Noy, A. (2006) *Surf. Interface Anal.*, **38**, 1429.
102 Evans, D.J. (1985) *Phys. Rev. A*, **32**, 2923.
103 Lorenz, R. (2003) *Science*, **299**, 837.
104 Kleidon, A., Lorenz, R.D. (eds) (1986) *Non-Equilibrium Thermodynamics and the Production of Entropy, Life, Earth and Beyond*, Springer, Berlin/Heidelberg, 2005.
105 Hoover, W.G. (1986) *J. Stat. Phys.*, **42**, 587.
106 Landau, L.D., Lifshitz, E.M. (1980) *Statistical Physics, Part 1*, Butterworth-Heinemann, Oxford.
107 Penrose, R. (1989) *The Emperor's New Mind*, Oxford University Press, Oxford.
108 Maxwell, J.C. (1878) *Nature*, **17**, 257.
109 Evans, D.J., Searles, D.J. (1996) *Phys. Rev. E*, **53**, 5808.
110 Panofsky, W.K.H., Phillips, M. (1962) *Classical Electricity and Magnetism*, 2nd edn, Addison & Wesley, USA.
111 Kawai, R., Parrondo, J.M.R., Van den Broeck, C. (2007) *Phys. Rev. Lett.*, **98**, 080602.

3
Quantum Dissipative Ratchets
Milena Grifoni

3.1
Introduction to Microscopic Ratchets

This contribution is dedicated to quantum dissipative ratchets, which are a special class of nano- or mesoscopic engines that employ asymmetric devices (ratchets) to direct particle motion in one specified direction (for reviews see [1–4]). After a brief introduction about the state of the art in the fields of theoretical and experimental investigation of quantum ratchet systems, I will discuss some concrete examples, placing more emphasis on the underlying physical principles and observable phenomena than on the mathematical treatment. The reader interested in the mathematical details can consult the scientific articles cited in this manuscript.

In macroscopic objects, an example of the ratchet principle is demonstrated in windmills. The issue of whether random microscopic fluctuations, such as those due to thermal motion, can act as a random energy source that can cause particles to flow in a single direction, however, is much subtler. In particular, Richard Feynman showed that no work can be extracted from a microscopic ratchet acting at thermal equilibrium, in agreement with the second law of thermodynamics (for a detailed discussion see [2, 5]). The second law of thermodynamics places clear constraints upon the attainable efficiencies of heat engines, devices which use energy in the form of heat to do work. Typically, the engine extracts energy in the form of heat from a hot reservoir and uses part of it to do work, with the remaining energy given to a colder reservoir. In this context, the second law says that one cannot create a heat engine that extracts heat and converts it all to useful work. In particular, the maximal or Carnot efficiency is determined by the temperature difference between the two reservoirs. If the reservoirs are at the same temperature, no work can be extracted. In other words, a global *out of thermal equilibrium situation* has to be maintained in order to extract work.

Following Feynman's ideas several quantum ratchet devices, that is, periodic structures with broken spatial symmetry, have been investigated over the last several years. As a minimal model, one considers a quantum particle moving in a ratchet potential that simultaneously interacts with one or more equilibrium reser-

Nonlinear Dynamics of Nanosystems. Edited by Günter Radons, Benno Rumpf, and Heinz Georg Schuster
Copyright © 2010 WILEY-VCH Verlag GmbH & Co. KGaA, Weinheim
ISBN: 978-3-527-40791-0

voirs and is subject to unbiased forces. The latter insure that the whole system is driven out of equilibrium. From the theoretical point of view one distinguishes between dissipative ratchets [6–14] where the tunneling particle continuously interacts with the bath as it moves along the periodic structure, and coherent ratchets [15–17] in which the dissipation originates from the coupling between a ballistic device and fermionic reservoirs. In dissipative ratchets, the interplay of dissipative tunneling [18] with unbiased driving enriches the quantum ratchet effect with features absent in its classical counterpart, such as temperature-dependent current reversals [6].

After the pioneering semi-classical work [6], further progress towards a full quantum description was made in [9], in which the role of the band structure in ratchet potentials sustaining only a few bands below the barrier was investigated. Such a situation does not have a classical counterpart because of the different conditions for the classical case. In the latter, many bands contribute to transport in such a way that the band structure of the asymmetric periodic potential should be of no importance. The classical situation is achieved if the temperature is large enough that particles can be easily excited across the barrier separating two nearby energy wells.

In [9] it was established that a ratchet state of particle transport can only be achieved when at least the two lowest Bloch bands contribute to transport. On the other hand, at least two harmonics of the potential should enter the dynamical equations in order to obtain the ratchet effect in systems with weak periodic potentials [11, 12]. Interestingly, in [11, 12] an expression for the ratchet current valid for weak dissipation was found upon generalization of a duality relation put forward in [19] for a cosine potential. The relation was generalized to an arbitrary ratchet potential and a time-dependent driving.

The growing interest in spintronics, the field which exploits the spin degrees of freedom for new applications in magneto-electronic devices, has stimulated the first proposal for the realization of a dissipative spin ratchet [13, 14]. Such a device exploits the spin-orbit interaction present in semiconductor heterostructures, including dissipation and spatial asymmetry, in order to create a device where a net spin current is produced while no net charge current occurs.

As mentioned above, the ratchet effect has also been discussed in quantum transport across ballistic quantum wires. Here rectification is a result of the dissipative coupling of the wire to fermionic baths. Coherent charge ratchets based on molecular wires with an asymmetric level structure of the orbital energies were proposed in [15]. The spin ratchet effect in coherent quantum wires with Rashba spin-orbit interaction was first investigated in [16]. The Zeeman ratchet effect which occurs in the presence of a non-uniform static magnetic field was studied in [17] for coherent quantum wires formed in a two-dimensional electron gas (2DEG).

Despite the increasing number of theoretical works on the subject, quantum ratchet systems have been realized in only a few experiments [20, 21]. In [20] the experimental realization of a coherent quantum ratchet has been demonstrated for electrons moving in nanopatterned asymmetric semiconductor heterostructures. In the experiment an alternating bias voltage is applied. A net current is generated despite the fact that the time-averaged electric field is zero. A current reversal was detected by sweeping the temperature while keeping all the other parameters fixed, including potential shape and bias voltage. This was interpreted as a crossover from the classical to the quantum tunneling regime in accordance with the prediction in [6]. In [21] the ratchet effect in the deep quantum realm has been reported in which only a few energy bands below the potential barrier contribute to transport. In the experiment of Majer *et al.*, the tunneling particles are topological excitations present in superconducting Josephson junction arrays and are termed vortices.

3.2
The Feynman Ratchet

The Feynman ratchet is illustrated in Figure 3.1 and consists of an axle with a sawtoothed wheel or ratchet at one end and a paddle at the other.

The whole device is surrounded by a gas of molecules in thermal equilibrium at some temperature T. The gas of molecules hitting the paddle cause it to turn, but the question of which direction arises. In a ratchet-pawl system, motion in one direction is allowed by the ratchet and motion in the opposite direction is prevented by a pawl such that a small load can be lifted. This, however, would violate the second law of thermodynamics. Feynman demonstrated that as the pawl will be of a size comparable to the paddle, it will also undergo similar thermal fluctuations. In particular, thermal noise causes the pawl to release so that the ratchet can move "backwards". As shown by Feynman, forward and backward motion compensate on average and no net work is produced. No contradiction with the second principle occurs if the ratchet-pawl system is operated out of equilibrium which, for example, would be the case if the ratchet and paddle would be in contact with thermal reservoirs at different temperatures.

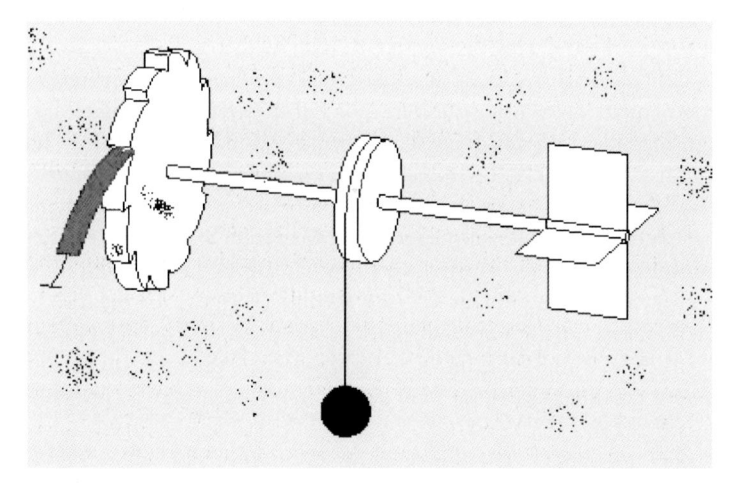

Figure 3.1 Feynman ratchet and pawl system. In thermal equilibrium no load can be lifted.

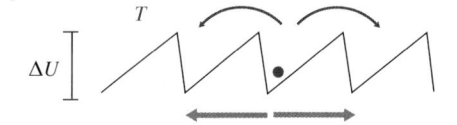

Figure 3.2 One-dimensional schematization of a Feynman ratchet. A particle (dot) which sits in one of the minima of the periodic ratchet potential can with equal probability be thermally activated over the left or right barrier (upper arrow processes). Hence, no preferred direction of motion exists (as indicated by the equal length of the lower arrows).

Feynman's analysis is schematized in Figure 3.2 which shows a particle (spot) in a one dimensional asymmetric potential in the presence of equilibrium noise at a temperature T. According to the laws of classical statistical mechanics, an energy barrier can be overcome only if a particle acquires enough energy, for example by thermal activation (upper arrows). For large enough barriers, the probabilities per unit time $\Gamma_{L/R}$ that the particle has to overcome the barrier to the left or to the right, respectively, are equal. The probabilities depend exponentially on the height ΔU of the barrier to be overcome and on the magnitude of the thermal fluctuations ($\Gamma_{L/R} \approx \exp(-\Delta U/k_B T)$). This implies that the larger the temperature/barrier ratio, the larger the escape probability. Hence, no preferred direction of motion exists and the ratchet current, being the difference between the probabilities to overcome the barrier to the left and to the right (indicated by the lower arrows pointing towards the left and right, respectively), is zero. This brings us to the general statement that in order for a Brownian motor to produce useful work, the system has to be driven permanently out of equilibrium.

3.3
Tunneling Ratchets: Temperature Driven Current Reversal

In the remainder of this article we focus on rocked ratchets, where the temperature and the asymmetric potential stay constant in time but the system is subjected to an external time dependent driving force that disrupts thermal equilibrium. The external driving force is chosen to be unbiased, for example with zero time average for any of its odd moments such that it does not induce any additional asymmetry. Moreover, we shall consider the regime of low enough temperatures such that quantum tunneling provides an alternative mechanism to thermal activation to overcome energy barriers and obtain directed motion. In Figure 3.3 the case is shown in which the force can assume two opposite values. As we shall discuss below, a classical particle will preferably move towards the right rather than to the left due to the different potential slopes (a). The situation, however, may be reversed for a quantum particle (b).

In a seminal work, Reimann et al. [6] predicted a current inversion with decreasing temperatures in rocked quantum ratchets due to a transition to a tunneling

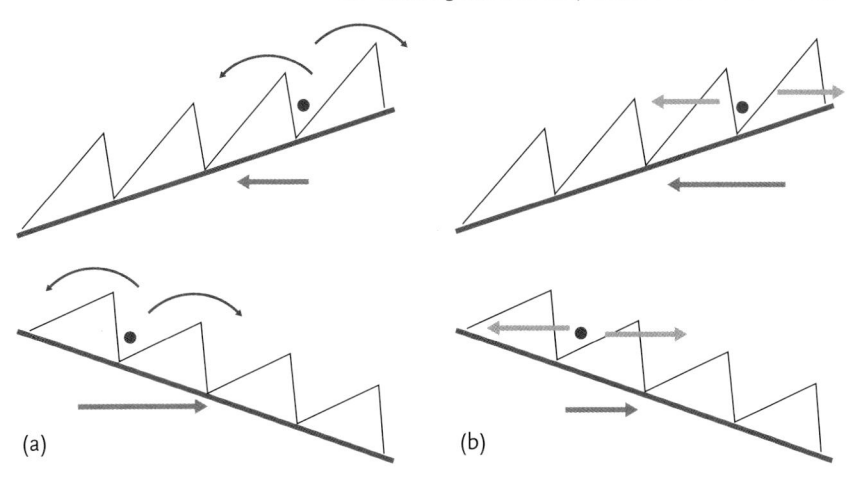

Figure 3.3 Current reversal in a rocked ratchet. Depending on the temperature, thermal activation over the barrier (upper arrows) or tunneling through the barrier (gray arrows) are the dominant escape mechanisms. For classical particles the net motion, as determined by the lower arrows, is towards the right (a), while for quantum particles is towards the left (b).

dominated regime. As explained in Figure 3.3, the occurrence of a current reversal as the temperature is decreased can be understood as follows. Consider first a classical particle in a periodic potential (a). Classically, a potential barrier can be overcome only if a particle acquires enough energy to jump over the barrier, for example by thermal fluctuations. If $\Delta U_{L/R}$ is the energy barrier seen by the particle on its left/right side and T is the temperature, the probability per unit time $\Gamma_{L/R}$ to overcome the barrier to the left/right is given by the Arrhenius rate $\Gamma_{L/R} \approx \exp(-\Delta U_{L/R}/k_B T)$ for large enough barriers, where k_B is the Boltzmann constant. In the tilted potential shown in the top (bottom) panel of Figure 3.3a, the potential barrier to the right (left) is less than that to the left (right). By summing the contributions from the two tilting situations (lower arrows), a net ratchet current to the right is expected.

In the quantum case, however, a finite probability occurs to go from one well to the other via quantum tunneling, even if the energy of the particle is much less than the potential height. This phenomenon can also occur at zero temperature as it relies only on the wave nature of quantum particles. For a quantum particle moving against a potential hill, the wave function describing the particle can extend to the other part of the hill. In particular, it turns out that the tunneling probability exponentially depends not only on the barrier height, but also on the distance to be traveled through the potential barrier. For the potential shown in Figure 3.3b, tunneling in the tilted potential favors a net motion towards the left. Due to the fact that the Arrhenius rate becomes exponentially smaller as the temperature is lowered, a transition temperature is expected below which tunneling through the barrier dominates over thermal activation above the barrier, and hence a current reversal is expected. This prediction was verified few years later in the first experimental demonstration by Linke et al. [20] of the quantum ratchet effect in rocked asymmetric semiconductor heterostructures.

3.4
Rocked Ratchets in the Deep Quantum Regime

One of the fundamental predictions of quantum mechanics for particles moving in a periodic potential is that only discrete sets of energies forming "energy bands" are allowed.

Significantly, in the temperature and driving regime in which the dynamics is effectively restricted to the lowest band of the periodic potential, no current rectification occurs as shown in [9]. In fact, a reduction to the lowest band of the ratchet potential retains only information about the periodicity of the original Hamiltonian, but not about its reflection properties. At least two bands should contribute in order to take into account the vibrational motion within the well and thus the asymmetry of the ratchet potential leading to the ratchet effect. Quantitative calculations of ratchet currents in the limit in which only the three lowest bands contribute to the dynamics have been performed in [9]. In this limit, a tight-binding model with tight binding parameters related to the intraband and interband energies was derived and solved in the limit of moderate to strong dissipation.

To date only one experimental realization of the ratchet effect in the deep quantum realm has been reported, with only a few energy bands below the potential barrier contributing to transport [21]. In the experiment of Majer et al., the tunneling particles are topological excitations present in superconducting Josephson junction arrays named vortices (Figure 3.4).

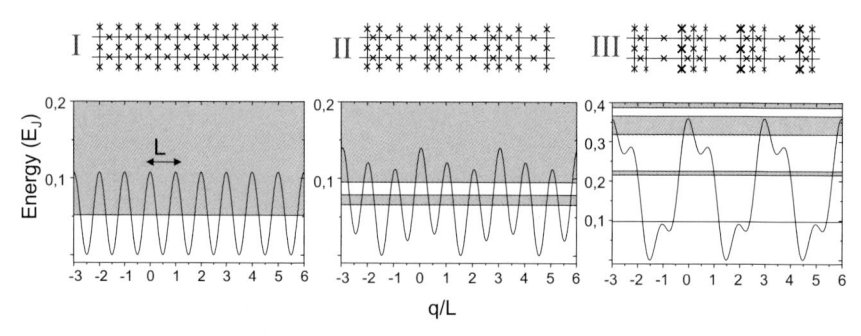

Figure 3.4 Ratchet potentials for vortices. Top: Schematic layout of the devices, with one cross denoting a Josephson junction. The device I serves as a reference and yields a symmetric periodic potential for the vortices. Bottom: Calculation of the potential seen by the vortices for the regular sample I and the samples II, III [21].

The upper part of Figure 3.4 shows a schematic layout where the Josephson junctions are represented by crosses. Cells are areas enclosed by four junctions.

The peculiarity of the device is that junctions of three different sizes, large, medium, and small, are periodically repeated along the length of the quasi-one-dimensional array. Applying a magnetic field perpendicular to the array induces vortices in the system. Vortices have lower energy in cells with larger area and smaller junctions. Hence, by properly choosing junction sizes and cell areas, different energy potentials felt by the vortices can be designed. In the lower part of Figure 3.4 three different quasi-one-dimensional Josephson junction arrays and the corresponding potentials felt by a single vortex are shown in units of the Josephson energy E_J with $E_J/k_B = 5$ K and at temperatures of $T = 12$ mK. In particular, the device denoted sample III yields the most asymmetric ratchet potential. Sample I yields a regular periodic potential, and serves to check that accidental asymmetries in the measured voltage current characteristics are not present. Finally, sample II also gives rise to a ratchet profile, but with only one band below the barrier.

If a current is applied vertically and homogeneously along the length of the array, the vortices start to move. Such motion can be detected as a voltage drop across the array as shown in the bottom panels of Figure 3.4 and in Figure 3.5. As expected, samples I and II do not exhibit ratchet behavior, while sample III shows clear rectification. The power law dependence $V \propto I^\delta$ with $\delta > 1$ of the voltage on the applied current is noteworthy. This is a quantum effect, as classical dynamics and a zero temperature Ohm's law ($\delta = 1$) is expected above the critical current. Indeed, such a power law behavior characterizes the incoherent dynamics at low temperatures and large biases of tunneling particles in Ohmic environments [18]. However, a quantitative explanation of such behavior was not possible in [21], where it is suggested that a realistic description of the experiment might require the aban-

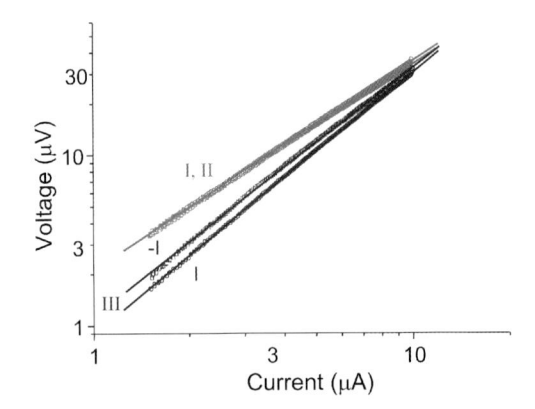

Figure 3.5 Experimental demonstration of the quantum ratchet effect in superconducting Josephson junctions arrays. The ratchet effect is observed in sample III where $V(I) \neq -V(-I)$. Astonishingly, sample I and II exhibit identical behavior and no rectification [21].

3.5
Rocked Shallow Ratchets

A striking feature of the experiment shown in Figure 3.5 is that despite the fact that samples I and II have seemingly different shapes, the resulting $V - I$ curves are the same. The qualitative understanding of this feature lies in the theoretical investigations carried out in [11, 12]. A first observation is that a ratchet potential can, in general, always be expanded in a Fourier series because it is periodic in space, for example $V(q) = V_0 + \sum_n V_n \cos(2nq/L - n)$ for the one dimensional case, where L is the periodicity length. For the two ratchets reported in [21], only the first three harmonics were relevant. Moreover, the second harmonic was essentially absent in the less asymmetric sample. In the theory developed in [11, 12], it is shown that for the case in which an expansion of the tunneling current in the amplitudes of the potential's harmonics is rapidly converging, the terms that dominate the ratchet current are those linear in the second harmonic. While these contributions are present for the potential characterizing sample III, they vanish for the potential characterizing sample II. Thus, sample II behaves in a similar fashion as the symmetric sample I at low temperatures. Indeed, a deeper look at the potentials in Figure 3.4 shows that while the potential in sample III possesses narrow bands below the barrier, this is not the case for the shallow potentials in samples I and II, which support no bands and one band below the barrier, respectively.

In [11, 12] a duality transformation was used to map quantum Brownian motion in a tight-binding description, with Ohmic damping characterized by a viscosity coefficient η. In this description weak dissipation in the original model corresponds to strong dissipation in the tight-binding model and vice versa. Moreover, the periodicity of the tight-binding model depends on the viscosity coefficient η and the tight-binding matrix elements are proportional to the Fourier coefficients V_n.

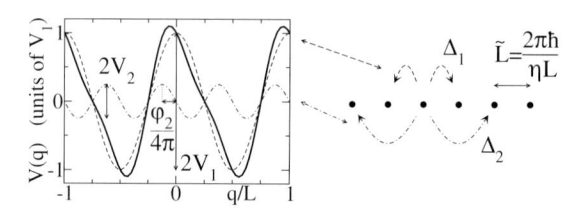

Figure 3.6 Dual relation between a dissipative ratchet system and a tight-binding (TB) model sketched for a two-harmonics ratchet potential (thick curve). Each harmonic (thin curves) generates couplings to different neighbors in the TB system. The periodicity $\tilde{L}$ of the TB model is determined by the viscosity η in the original model [11, 12].

By means of the duality relation, various quantities such as the mobility or the ratchet current can be calculated as a correction to the dynamics in the absence of the periodic potential. Due to the Eherenfest theorem, in the absence of the potential the classical and quantum expectation values are the same, such that in the limit of vanishing potential the duality relation yields the correct classical limit. The quantum corrections due to the presence of the ratchet potential can be calculated, for example, as a series expansion in the Fourier amplitudes V_n. For shallow potentials, that is, small amplitudes, a truncation of the series yields a good approximation to the ratchet current.

3.6
Spin Ratchets

During the last decade the new research field of spintronics has emerged, in which one makes use of the spin degree of freedom of a particle for transport and storage of information. One essential difference between spin and charge is that a particle can have more than one spin state, while it has only one charge state. In the context of transport, it is significant that the spin state of a particle can depend on the transport conditions, as it happens, for example, in systems with spin-orbit interaction. This fact has stimulated research on spintronics devices made up of nonmagnetic materials but still exploiting the spin-orbit interaction. Among the various different spin-orbit mechanisms, the Rashba spin-orbit interaction (RSOI) plays a distinguished role because the spin-orbit coupling strength can be controlled by an external electric field.

The possibility to transfer the spin separately from charge plays an important role. This can be implemented by so-called pure spin currents that are not accompanied by charge currents. In a recent seminal work, Smirnov *et al.* [13] have addressed the challenging question of how to make use of the Rashba spin-orbit interaction and ratchet geometry to generate pure spin currents in dissipative quasi-one-dimensional systems. In periodic quasi-one-dimensional systems with a periodic potential along the longitudinal direction, the RSOI removes the spin degeneracy of the one dimensional Bloch bands and couples different transverse branches. When only a few M transverse modes are relevant, each Bloch band splits into $2M$ sub-bands which may carry different spin [22]. Following [9], the authors restricted their attention to the lowest Bloch band in order to block the generation of charge currents and investigate the possibility of pure spin current generation. The Rashba Hamiltonian is not invariant under reflection of a transport direction. Thus the Rashba Hamiltonian itself already has a built in spatial asymmetry. In coherent quantum wires this asymmetry turns out to be sufficient to generate pure spin currents [16]. As shown in [13], however, this is no longer the case for dissipative wires. Despite the intrinsic Rashba asymmetry, the system must additionally lack the spatial inversion symmetry and its orbital degrees of freedom must be coupled in order to generate pure spin currents.

Acknowledgments

I thank all of the students and colleagues with whom I have worked and discussed the theme of quantum ratchets along the years. This work was supported by the SFB 689.

References

1 Linke, H. (Ed.) (2002) *Appl. Phys. A*, **75**, 167; special issue on *Ratchets and Brownian motors*.
2 Reimann, P. (2002) *Phys. Rep.* **361**, 57.
3 Astumian, R.D. and Hänggi, P. (2002) *Phys. Today*, **55** (11), 33.
4 Hänggi, P. and Marchesoni, F. (2009) *Rev. Mod. Phys.*, **81**, 387.
5 Feynman, R.P., Leighton, R.B., and Sands, M. (1963) *The Feynman Lectures on Physics*, Addison-Wesley, Reading, MA.
6 Reimann, P., Grifoni, M., and Hänggi, P. (1997) *Phys. Rev. Lett.*, **79**, 10.
7 Roncaglia, R. and Tsironis, G.P. (1998) *Phys. Rev. Lett.*, **81**, 10.
8 Scheidl, S. and Vinokur, V.M. (2002) *Phys. Rev. B*, **65**, 195305.
9 Grifoni, M. et al. (2002) *Phys. Rev. Lett.*, **89**, 146801.
10 Machura, L., Kostur, M., Hänggi, P., Talkner, P., Luczka, J. (2004) *Phys. Rev. E*, **70**, 031107.
11 Peguiron, J. and Grifoni, M. (2005) *Phys. Rev. E*, **71**, 010101(R).
12 Peguiron, J. and Grifoni, M. (2006) *Chem. Phys.*, **322**, 169.
13 Smirnov, S., Bercioux, D., Grifoni, M., and Richter, K. (2008) *Phys. Rev. Lett.*, **100**, 230601.
14 Flatté, M.E. (2008) *Nat. Phys.*, **4**, 587.
15 Lehmann, J., Kohler, S., Hanggi, P., and Nitzan, A. (2002) *Phys. Rev. Lett.*, **88**, 228305.
16 Scheid, M., Pfund, A., Bercioux, D., and Richter, K. (2007) *Phys. Rev. B*, **76**, 195303.
17 Scheid, M., Bercioux, D., and Richter, K. (2007) *New J. Phys.*, **9**, 401.
18 Weiss, U. (2008) *Quantum Dissipative Systems*, 3rd ed., World Scientific, Singapore.
19 Fisher, M.P.A. and Zwerger, W. (1985) *Phys. Rev. B*, **32**, 6190.
20 Linke, H., Humphrey, T.E., Lofgren, A., Suhskov, A.O. Newbury, R., Taylor, R.P., and Omling, P. (1999) *Science*, **286**, 2314.
21 Majer, J.B., Peguiron, J., Grifoni, M., Tusveld, M., and Mooij, J.E. (2003) *Phys. Rev. Lett.*, **90**, 056802.
22 Smirnov, S., Bercioux, D., and Grifoni, M. (2007) *Europhys. Lett.*, **80**, 27003.

Part II Surface Effects

4
Dynamics of Nanoscopic Capillary Waves

Klaus Mecke, Kerstin Falk, and Markus Rauscher

With the advent of nanofluidics in the last several years it has become evident that thermal noise may play an important role in all hydrodynamic processes occurring at free interfaces on small scales. Although in bulk fluids hydrodynamic Navier–Stokes equations are proven to be valid down to the nanometer scale, in free interface flow stochastic forces induced by molecular motion can significantly alter the behavior even on a micrometer scale. Moseler and Landman, for instance, found that the deterministic lubrication approximation for axial-symmetric free boundary flow is not applicable for the description of nanoscopic cylindrical jets [1]. The lack of thermally triggered fluctuations in the classical hydrodynamic continuum modeling was identified as the most likely source for deviations of Navier–Stokes solutions from molecular dynamics simulations. They derived a stochastic differential equation that includes thermal noise, whose influence on the dynamics increases as the radius of the nanojet becomes smaller, leading finally to the emergence of symmetric double cone neck shapes during the breakup, instead of a long thread solution as expected in the absence of noise. In [2], path integral methods were applied to confirm that thermal noise indeed induces qualitative changes in the breakup of a liquid nanometer jet. Thermal fluctuations speed up the dynamics and make surface tension an irrelevant force for the breakup. Very recently, the importance of thermal noise for drop formation was observed in a colloidal dispersion with an ultra-low surface tension [3].

The hydrodynamics of liquid interfaces are particularly poorly understood on microscopic length scales where the standard capillary wave theory [4, 5] is not applicable. In contrast to solid surfaces, relevant experimental information is absent on the nanometer scale even for the simplest liquid–vapor interfaces. Recent developments in grazing incidence X-ray scattering experiments has removed this uncomfortable situation for the equilibrium structure, but has yet to do so for the dynamics. The theoretical results and X-ray experiments reported in the last several years in [6–8] give the first complete determination of the structure and the equilibrium fluctuations of a liquid–vapor interface, and represent a significant improvement in the understanding of fluid interfaces on a molecular level. In particular, it has been demonstrated that the dominant effect below a few nanometers

Nonlinear Dynamics of Nanosystems. Edited by Günter Radons, Benno Rumpf, and Heinz Georg Schuster
Copyright © 2010 WILEY-VCH Verlag GmbH & Co. KGaA, Weinheim
ISBN: 978-3-527-40791-0

is a large decrease of the surface energy due to dispersion forces. This calls for a reexamination of all small scale interfacial processes. For such a reexamination, the hydrodynamics of capillary waves have to be studied on length scales comparable to the range of molecular interactions. If the wavelengths of undulations are below 50 nm, one expects new phenomena due to direct non-local interactions of molecules which cannot be described by local hydrodynamics. The final theoretical goal would be a derivation of the time dependence of capillary waves on nanometer length scales and, in particular, the dependence of dispersion relations and damping factors on molecular interactions. Experimentally, the dynamics of capillary waves may be measured on liquid–vapor interfaces by X-ray photon correlation spectroscopy and scattering techniques.

4.1
Stochastic Hydrodynamics

The effect of thermal noise has already been introduced phenomenologically into hydrodynamics by Landau and Lifšic [9] and further discussed by Fox and Uhlenbeck [10, 11]. A microscopic justification for the noisy hydrodynamical equations has been provided by showing that the form proposed can be derived from the deterministic Boltzmann equation by a long wave approximation [12]. The noisy hydrodynamical equations have been discussed, for example, in the context of turbulence in randomly stirred fluids [13, 14] as well as for the onset of instabilities in Rayleigh–Bénard convection [15] and Taylor–Couette flow [16]. Introductions can be found in [9, 17, 18].

4.1.1
Stochastic Interfaces

We consider an incompressible Newtonian liquid with a free fluid boundary as sketched in Figure 4.1. We assume that the liquid–vapor interface can be parameterized by a single-valued function $h(\vec{R}, t)$ of the lateral coordinates $\vec{R} = (x, y)$, which defines the plane of the averaged position $\langle h \rangle = 0$ of the interface. For later use we introduce the surface normal vector

$$\vec{n} = \frac{1}{\sqrt{1 + (\nabla h)^2}} \begin{pmatrix} -\nabla h \\ 1 \end{pmatrix}$$

and also the tangent vectors $\vec{t} = (1, \nabla h)/\sqrt{1 + (\nabla h)^2}$ and $\vec{t}' = \vec{n} \times \vec{t}$ to the surface. To the lowest order of the interface position, the mean curvature reads

$$H(\vec{R}, t) = \frac{1}{2}(\kappa_1 + \kappa_2) = \frac{1}{2}\Delta h(\vec{R}, t) .$$

4.1 Stochastic Hydrodynamics

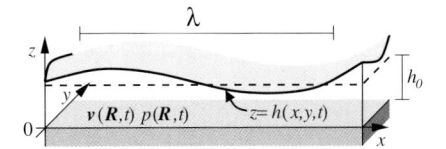

Figure 4.1 A fluid surface is parameterized by the single-valued function $h(\vec{R}, t)$ of the lateral coordinates $\vec{R}$. The flow is characterized by the flow velocity $\vec{v} = (v_x, v_y, v_z)$ and the pressure p. Capillary waves of wave vector $\vec{q} = \frac{2\pi}{\lambda}$ are thermally excited which leads to moving boundary for the fluid.

The incompressibility condition and momentum conservation within the fluid are given by the stochastic Navier–Stokes equations [9]

$$\nabla \cdot \vec{v} = 0 \tag{4.1}$$

$$\rho \left(\frac{\partial \vec{v}}{\partial t} + (\vec{v} \cdot \nabla)\vec{v} \right) = \eta \vec{\nabla}^2 \vec{v} - \vec{\nabla}(p + V_{\text{ext}}) + \vec{\nabla} \cdot \mathcal{S}. \tag{4.2}$$

By $\vec{v}$ and p, we denote velocity and pressure fields, respectively. An external potential $V_{\text{ext}}(\vec{r})$ may be given by gravity, $V_{\text{ext}} = \rho g z$, or a substrate potential $V_{\text{ext}}(z) \sim -A/z^6$ due to dispersion forces. The mass density ρ is constant within the fluid and η is the shear viscosity (kinematic viscosity $\nu = \eta/\rho$). The random stress fluctuations $\mathcal{S}$ represent the effect of molecular motion. $\mathcal{S}$ is symmetric, has zero mean

$$\langle \mathcal{S} \rangle = 0$$

and the correlator is given as

$$\langle \mathcal{S}_{ij}(\vec{r}, t) \mathcal{S}_{lm}(\vec{r}', t') \rangle = 2\eta k_B T \delta(\vec{r} - \vec{r}') \delta(t - t') (\delta_{il} \delta_{jm} + \delta_{im} \delta_{jl}) \tag{4.3}$$

with the thermal energy $k_B T$. $\mathcal{S}$ is spatially uncorrelated, and therefore the divergence of $\mathcal{S}$ in (4.2) poses mathematical questions we do not want to enter into at this point. From a physical point of view, hydrodynamical equations are only valid at a scale large compared to the molecular scale. Therefore, $\delta(\vec{r} - \vec{r}')$ in (4.3) might be replaced by a correlation function of small but finite width. In order to show that equilibria are characterized by Gaussian velocity distributions as required by thermodynamics, we need spatially uncorrelated noise. For this reason, we keep the notation commonly used in physical literature.

We assume that at the free surface $z = h(x, t)$ the normal and tangential stresses are balanced, so that the boundary condition reads

$$(\sigma - \sigma' + \mathcal{S}) \cdot \vec{n} = 2\gamma H \vec{n} \tag{4.4}$$

with the surface tension coefficient γ and the stress tensor for an incompressible fluid

$$\sigma_{ij} = \eta(\partial_i v_j + \partial_j v_i) - p \delta_{ij}.$$

In the vapor phase one finds $\sigma' = -p'\delta_{ij}$. Finally, assuming that the fluid is non-volatile, the component of the flow velocity normal to the surface is identical to the surface normal velocity and we get the kinematic condition

$$\vec{n} \cdot \vec{e}_z \frac{\partial h}{\partial t} = \vec{n} \cdot \vec{v} , \qquad (4.5)$$

that is,

$$\frac{\partial h}{\partial t} = v_z - v_x \frac{\partial h}{\partial x} - v_y \frac{\partial h}{\partial y} \quad \text{at} \quad z = h$$

$$= -\nabla \cdot \vec{j} , \qquad (4.6)$$

with the total flow current in the film at position $\vec{R} = (x, y)$

$$\vec{j}(\vec{R}, t) = \int_{-L}^{h(\vec{R},t)} v_{x,y}(\vec{R}, z, t) \, dz , \qquad (4.7)$$

where $-L$ denotes the bottom of our container. It is often convenient to use the Fourier transformation of a quantity $a(\vec{r}, z, t)$ parallel to the interface, for instance

$$\tilde{a}(\vec{q}, z, \omega) := \int_{\mathbb{R}^2} d^2 R \int_{\mathbb{R}} dt \, a(\vec{R}, z, t) \, e^{-i(\vec{q} \cdot \vec{R} + \omega t)} , \qquad (4.8)$$

to define the surface modes

$$\tilde{h}(\vec{q}, \omega) := \int_{\mathbb{R}^2} d^2 R \int_{\mathbb{R}} dt \, h(\vec{R}, t) \, e^{-i(\vec{q} \cdot \vec{R} + \omega t)} .$$

Before we study capillary waves on a liquid interface, it is instructive to derive the dispersion relation and damping of acoustic waves in the bulk of a compressible liquid.

4.1.2
Acoustic Waves

Assuming that the liquid in equilibrium is at rest with $\vec{v}_0 = 0$, $\varrho_0 = \text{const.}$, and $p_0 = \text{const.}$, the thermal noise causes a velocity profile $\delta \vec{v}(\vec{R}, t) \equiv \vec{v}$ and fluctuations in density $\delta \varrho(\vec{R}, t) \equiv \delta \varrho$ and pressure $\delta p(\vec{R}, t) \equiv \delta p$. On may use the thermodynamic relation

$$\delta \varrho = \frac{1}{c^2} \delta p$$

with the adiabatic sound velocity c, so that the linearized compressible Navier–Stokes equation and continuity equation,

$$\varrho_0 \partial_t \vec{v} = -\nabla \delta p + \eta \Delta \vec{v} + \left(\zeta + \frac{\eta}{3} \right) \nabla (\nabla \cdot \vec{v}) + \nabla \cdot \mathcal{S}$$

$$\partial_t \delta \varrho = -\varrho_0 \nabla \cdot \vec{v} \qquad (4.9)$$

can be written as

$$\left(\partial_t^2 - \frac{4}{3}\nu\partial_t\Delta - c^2\Delta\right)\delta p = -c^2\nabla\cdot(\nabla\cdot\mathcal{S}),\tag{4.10}$$

with the second viscosity ζ is set to zero. After Fourier transformation this is written as

$$\left(\omega^2 - \frac{4}{3}\nu i\omega k^2 - c^2 k^2\right)\delta\tilde{p}(\vec{k},\omega) = -c^2\vec{k}\cdot\left(\vec{k}\cdot\tilde{\mathcal{S}}(\vec{k},\omega)\right).\tag{4.11}$$

Non-trivial solutions for the averaged pressure $\langle\delta\tilde{p}\rangle \neq 0$ are only possible if

$$0 = \omega^2 - i\omega\frac{4}{3}\nu k^2 - c^2 k^2 \tag{4.12}$$

because $\langle S_{ij}\rangle = 0$. From this we can determine the dispersion relation $\omega = ck$ for acoustic waves in an ideal liquid as well as the damping coefficient $\frac{2}{3}\nu k^2$. In the following we assume an incompressible fluid for convenience.

4.1.3
Capillary Waves

For an ideal liquid with vanishing viscosity η in a gravitational field, one may assume a constant density ϱ_0 and introduce a potential φ so that the velocity is given by the gradient $\vec{v} = \nabla\varphi$. Incompressibility of the fluid leads to the equation $\Delta\varphi = 0$. Ignoring the stochastic stress tensor and linearizing Euler's equation, that is, the inviscid Navier–Stokes equation, one finds for the pressure

$$p = -\varrho_0\partial_t\varphi - \varrho_0 gz \tag{4.13}$$

with the boundary condition

$$p|_{z=0} = -\gamma\Delta_{xy}h. \tag{4.14}$$

Using the pressure $-\varrho_0\partial_t\varphi - \varrho_0 gh$ at the free surface h and the kinematic boundary condition $\partial_t h = \partial_z\varphi$ one obtains

$$\partial_t^2\varphi + \left(g - \frac{\gamma}{\varrho_0}\Delta_{xy}\right)\partial_z\varphi\bigg|_{z=0} = 0 \tag{4.15}$$

and after Fourier transformation

$$\varphi(\vec{R},t) = \varphi_0\mathrm{Re}\left\{e^{i(\vec{k}\cdot\vec{R}+\omega t)}\right\} \tag{4.16}$$

one obtains the dispersion relation

$$k_x^2 + k_y^2 + k_z^2 = 0$$
$$\omega^2 - \left(g + \frac{\gamma}{\varrho_0}\left(k_x^2 + k_y^2\right)\right)ik_z = 0 \tag{4.17}$$

Table 4.1 Density ϱ, surface tension γ, adiabatic sound velocity c, kinematic viscosity ν, capillary length l_c (4.19) and λ_0 for water, ethanol, glycerol and mercury at 20°C [19].

	$\varrho\left[\frac{g}{cm^3}\right]$	$\gamma\left[10^{-3}\frac{N}{m}\right]$	$c\left[10^3\frac{m}{s}\right]$	$\nu\left[\frac{cm^2}{s}\right]$	l_c [cm]	λ_0 [nm]
Water	1.00	72.8	1.48	0.01	0.27	0.21
Ethanol	0.79	22.6	1.16	0.015	0.17	0.14
Glycerol	1.26	63.4	1.90	11.8	0.23	0.09
Mercury	13.5	476	1.45	0.0012	0.19	0.10

with the parallel wave vector $\vec{q} := (k_x, k_y)$ and the perpendicular decay length $k_z = \pm iq$. Thus, one finds waves with wavelengths $\lambda = \frac{2\pi}{q}$ and frequency

$$\omega_0^2(q) = gq + \frac{\gamma q^3}{\varrho_0}. \tag{4.18}$$

The waves within the liquid are exponentially damped at a distance z from the free surface. The capillary length

$$l_c := \sqrt{\frac{\gamma}{g\varrho_0}} \tag{4.19}$$

separates gravitational waves with $\lambda > l_c$ from capillary waves with wavelengths $\lambda < l_c$. In the following, we are interested in nanoscopic wavelengths for which the gravitational term can be neglected. Comparing the dispersion relation for capillary waves with acoustic modes $\omega = cq$, one finds a crossover at a characteristic frequency and wave vector

$$q_0 = \frac{\rho c^2}{\gamma}$$
$$\omega_0 = q_0 c = \frac{\rho c^3}{\gamma}. \tag{4.20}$$

Resulting values are in the range $\lambda_0 = \frac{2\pi}{q_0} \approx 0.2$ nm given the typical fluid parameters in Table 4.1. At these small length scales one cannot neglect the damping due to viscous forces. Without derivation we mention that for wavelength

$$q > q_{max} = 1.69 \frac{\gamma \rho}{\eta^2}$$

capillary waves are overdamped [19].

4.1.4
Linearized Stochastic Hydrodynamics

We now want to study an alternative approach to capillary waves which is based on solving the stochastic Navier–Stokes equation (4.2). It is convenient to introduce the perpendicular v_z and parallel velocities $\vec{v}_\| := (v_x, v_y)$. Decomposing the tensor $\mathcal{S}$ accordingly yields

$$\mathcal{S} = \begin{pmatrix} \mathbb{S}_\| & \vec{S}_{\|z} \\ \vec{S}_{\|z}^T & S_{zz} \end{pmatrix} \quad \text{with} \quad \mathbb{S}_\| := \begin{pmatrix} S_{xx} & S_{xy} \\ S_{yx} & S_{yy} \end{pmatrix}$$

$$\text{and} \quad \vec{S}_{\|z} := \begin{pmatrix} S_{xz} \\ S_{yz} \end{pmatrix} = \begin{pmatrix} S_{zx} \\ S_{zy} \end{pmatrix}. \tag{4.21}$$

Linearizing leads to

$$\left(\partial_z^2 - q_{p\perp}^2\right)\delta\tilde{p}(z) = \frac{\partial_z^2 \tilde{S}_{zz}(z) + 2i\partial_z\vec{q}\cdot\tilde{\vec{S}}_{\|z}(z) - \vec{q}\cdot(\vec{q}\cdot\tilde{\mathbb{S}}_\|(z))}{1 + i\omega\frac{4\nu}{3c^2}}$$

$$\eta\left(\partial_z^2 - q_{v\perp}^2\right)\tilde{v}_z(z) = \partial_z\tilde{p}(z) - \partial_z\tilde{S}_{zz}(z) - i\vec{q}\cdot\tilde{\vec{S}}_{\|z}(z)$$

$$\eta\left(\partial_z^2 - q_{v\perp}^2\right)i\vec{q}\cdot\tilde{\vec{v}}_\|(z) = -q^2\tilde{p}(z) - i\partial_z\vec{q}\cdot\tilde{\vec{S}}_{\|z}(z) + \vec{q}\cdot(\vec{q}\cdot\tilde{\mathbb{S}}_\|(z))$$

$$\tag{4.22}$$

with

$$q_{v\perp}^2(q,\omega) = q^2 + \frac{i\omega}{\nu}$$

$$q_{p\perp}^2(q,\omega) = q^2 - \frac{\omega^2}{c^2 + i\omega\frac{4}{3}\nu}. \tag{4.23}$$

The solution has to fulfill the continuity equation

$$\frac{i\omega}{c^2\rho_0}\delta\tilde{p}(z) + i\vec{q}\cdot\tilde{\vec{v}}_\|(z) + \partial_z\tilde{v}_z(z) = 0 \tag{4.24}$$

and the Fourier transformed linearized boundary conditions

$$0 = \eta(\partial_z\tilde{\vec{v}}_\|(z) + i\vec{q}\tilde{v}_z) + \tilde{\vec{S}}_{\|z}\Big|_{z=0}$$

$$-\gamma\vec{q}^2\tilde{h} = -\tilde{p} + 2\eta\partial_z\tilde{v}_z + \tilde{S}_{zz}\Big|_{z=0}$$

$$i\omega\tilde{h} = \tilde{v}_z|_{z=0} \tag{4.25}$$

which determines the velocities $\tilde{v}_z(z)$ and $\tilde{\vec{v}}_\|(z)$, the pressure $\tilde{p}(z)$, and the interface position $\tilde{h}$. For an ideal liquid with $\eta = 0$ one finds

$$i\omega\rho\tilde{v}_z(z) + \partial_z\tilde{p}(z) = \partial_z\tilde{S}_{zz}(z) + i\vec{q}\cdot\tilde{\vec{S}}_{\|z}(z)$$

$$\omega\rho\vec{q}\cdot\tilde{\vec{v}}_\|(z) + q^2\tilde{p}(z) = -i\partial_z\vec{q}\cdot\tilde{\vec{S}}_{\|z}(z) + \vec{q}\cdot(\vec{q}\cdot\tilde{\mathbb{S}}_\|(z))$$

$$0 = i\vec{q}\cdot\tilde{\vec{v}}_\|(z) + \partial_z\tilde{v}_z(z) \tag{4.26}$$

with the boundary conditions

$$0 = \tilde{\vec{S}}_{\|z}\Big|_{z=0}$$

$$-\gamma\vec{q}^2\tilde{h} = -\tilde{p} + \tilde{S}_{zz}\Big|_{z=0}$$

$$i\omega\tilde{h} = \tilde{v}_z|_{z=0}. \tag{4.27}$$

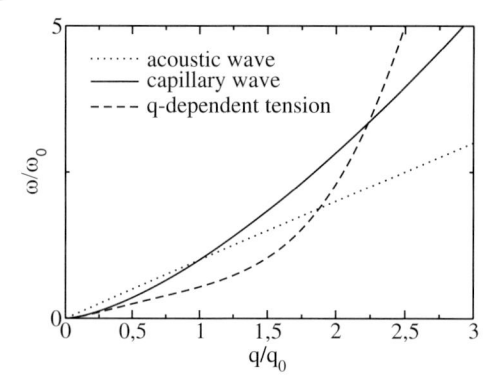

Figure 4.2 Nanostructure and dynamics of capillary waves: the sketch of dispersion relations demonstrates the possible transition between acoustic modes and capillary waves. Due intermolecular interaction potential $V(r)$ the surface tension $\gamma(q)$ depends on the wave vector leading to a dispersion relation $\omega^2 \sim q^3 \gamma(q)$; assuming that an extension from hydrodynamics to smaller time and length scales is provided by $\gamma(q)$. Does one find a similar signature of the microscopic forces $-V'(r)$ on dynamical properties of capillary waves, for instance, on damping factors or viscosities?

Finally, one obtains the interface dynamics

$$\frac{\rho_0}{q^2}\left[\omega^2 - \sqrt{q^2 - \omega^2/c^2}\,\frac{\gamma}{\rho_0} q^2\right] \tilde{h}(\vec{q},\omega)$$

$$= \int_{-\infty}^{0} e^{q_{p\perp} z} \left(\frac{q_{p\perp}^2}{q^2} \tilde{S}_{zz}(z) - 2\mathrm{i}\frac{q_{p\perp}}{q}\vec{e}_q \cdot \tilde{\vec{S}}_{\|z}(z) - \vec{e}_q \cdot (\vec{e}_q \cdot \tilde{S}_{\|}(z))\right) dz$$

$$+ \frac{q_{p\perp}}{q}\int_{-\infty}^{0} e^{q_{v\perp} z} 2\mathrm{i}\vec{e}_q \cdot \tilde{\vec{S}}_{\|z}(z)\, dz\;, \tag{4.28}$$

which reduces to $D(q,\omega)\langle \tilde{h}(\vec{q},\omega)\rangle = 0$ for the average position with the dispersion term

$$D(q,\omega) := \omega^2 - \sqrt{q^2 - \omega^2/c^2}\,\frac{\gamma}{\rho_0} q^2\;. \tag{4.29}$$

For $q \to 0$ one recovers the dispersion relation of (4.18) for an ideal incompressible fluid. In contrast to the previous results, the coupling of acoustic and capillary waves avoids a crossover of the dispersion relations but does lead to a crossover of capillary waves to acoustic waves for $q \to \infty$. For a more detailed analysis one has to take into account the viscosity of a compressible fluid which is done in [19].

Next we want to focus on another important problem in the dynamics of nanoscopic capillary waves: on these small length scales one can no longer neglect the molecular interaction potentials. Recent X-ray scattering measurements of the static structure function showed significant deviations of the surface tension from its macroscopic value due to molecular interactions [7, 8]. From the stochastic hydrodynamics one obtains the dynamic structure function

$$\tilde{S}(\vec{q},\omega) := \int_{\mathbb{R}^2} d\vec{q}' \int_{\mathbb{R}} d\omega' \left\langle \tilde{h}(\vec{q},\omega)\tilde{h}^*(\vec{q}',\omega') \right\rangle$$

$$= \frac{2k_BT}{\omega\varrho_0} \mathcal{I}m \left\{ \frac{q_{p\perp}}{D(q,\omega)} \right\} \tag{4.30}$$

and the static structure function by

$$\tilde{C}(\vec{q}) = \int_{\mathbb{R}^2} \frac{d\vec{q}'}{(2\pi)^2} \left\langle \tilde{h}(\vec{q},t)\tilde{h}^*(\vec{q}',t) \right\rangle$$

$$= \int_{-\infty}^{\infty} \frac{d\omega}{2\pi} \tilde{S}(\vec{q},\omega) , \tag{4.31}$$

which is proportional to the cross section $\frac{d\sigma}{d\Omega}$ of scattered X-rays with the difference $\vec{q}$ of in- and outcoming wavefronts. In the following section this relation is used to determine the surface energy γ experimentally.

4.2
Surface Tension at Nanometer Length Scales:
Effect of Long Range Forces and Bending Energies

Liquid interfaces are of fundamental importance in many areas of science and technology and have been the subject of continuous attention since van der Waals [20]. It is only in recent years, however, that a continuous effort in theory [6], experimental methods [7], and numerical simulations [21, 22] has given us a more complete picture of their microscopic structures.

In the approach initiated by van der Waals [20, 23], the liquid–vapor interface was described as a region of smooth transition (intrinsic profile) from the density of the liquid to that of the gas over approximately the bulk correlation length ξ. Conversely, the 1965 capillary-wave model of Buff, Lovett, and Stillinger [24] describes a wandering, step-like interface whose structure is determined by the height correlation spectrum (see (4.31))

$$\tilde{C}(\vec{q}) = \langle h(\vec{q})h(-\vec{q}) \rangle \propto \frac{k_BT}{\gamma \vec{q}^2} ,$$

where $2\pi/q$ is the wavelength of the capillary excitation, in good agreement with experiments [25, 26]. This description is necessarily expected to fail at small length scales, at least for wavelengths $\lambda = 2\pi/q \approx \xi$ [27]. Since the interfacial structure is determined by the surface energy associated with the deformation modes, the problem of the small scale structure can be addressed by considering corrections to the surface energy through an effective Hamiltonian or wave vector-dependent surface energy $\gamma(q)$. Following Helfrich [28], the surface free energy can be expanded in powers of the mean curvature H and of the Gaussian curvature. Fourier transforming and applying the theorem of equipartition of energy, one obtains

$$\gamma(q) = \gamma + \kappa q^2, \tag{4.32}$$

where $\gamma \equiv \gamma(q=0)$ is the macroscopic surface tension and κ is the bending rigidity constant. A large reduction in the surface energy was predicted with increasing wavevectors in a density functional theory [6], which was in strong contradiction to an expected positive bending rigidity. The results could, however, be successfully measured for water and other liquids using X-ray scattering [7, 8]. Thus, the analytic expansion in (4.32) cannot be performed in the presence of long range forces, but the surface tension should reach its macroscopic value from below as $\sim q^2 \ln q$ in the limit $q \to 0$.

The density functional theory is constructed by describing the liquid using the Carnahan–Starling equation of state and long range interactions with a potential

$$V(r) = -\frac{w_0 r_0^6}{(r^2 + r_0^2)^3}.$$

An effective interfacial Hamiltonian is constructed as the difference between the grand potential minimized with the constraint of a given density on a given deformed surface, and that for the flat interface. This leads to a momentum dependent surface tension:

$$\gamma(q) = 4\frac{\tilde{h}(0) - \tilde{h}(q)}{q^2} + 2\left[\tilde{\kappa}^H(q) - \tilde{\kappa}^H(0)\right]$$
$$+ \kappa q^2 - \tilde{\kappa}^{HH}(q)q^2 + \mathcal{O}(q^4). \tag{4.33}$$

The density distribution at the fluctuating interface is different from the flat intrinsic density profile because there is a displacement of the average interface position due to capillary waves, and also because curvature induces density changes in the intrinsic profile. The first term in (4.33) gives the contribution of long range forces due to interface displacement, neglecting the distortion in the intrinsic profile, and the other terms are bending terms, either local (κq^2 as in (4.32)), or nonlocal ($\tilde{\kappa}^{HH}(q)q^2$ and $\tilde{\kappa}^H$ due to long range interactions). For convenience one may use a product approximation [6], which is valid for very short ranged intermolecular potentials with $r_0 \ll \xi$, but remains accurate to approximately 10% even for $\xi \approx r_0$ [6]. Within this approximation,

$$\tilde{h}(q) = \frac{\gamma}{2r_0^2}(1 + qr_0)e^{-qr_0},$$

$$\tilde{\kappa}^H(q) = \frac{\gamma}{2} C_H \frac{\xi^2}{r_0^2}(1 + qr_0)e^{-qr_0},$$

$$\tilde{\kappa}^{HH}(q) = 0.74\gamma C_H^2 \frac{\xi^4}{r_0^2}(1 + qr_0)e^{-qr_0},$$

$$\kappa = 0.74\gamma C_H^2 \xi^2 \left(\frac{1}{2} + \frac{\xi^2}{r_0^2}\right). \tag{4.34}$$

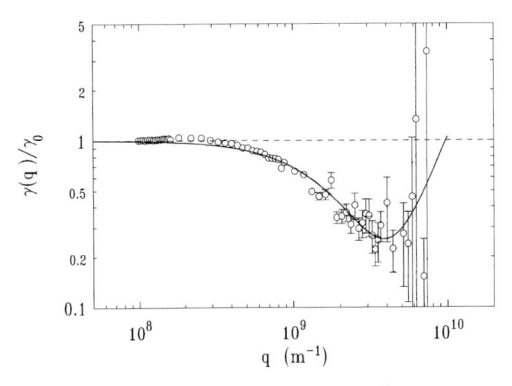

Figure 4.3 The wavevector dependent surface energy $\gamma(q)$ of water normalized to the macroscopic surface tension γ_0 [7, 8]. The line is the analytical result of the density functional theory described in [6] and given by (4.33) and (4.34).

C_H is the "susceptibility" of the density profile to curvature (see [6]), with the curvature corrections to the profile being proportional to C_H. Landau theory gives $C_H = 0.25$, which yields a good description of the experimental data (see Figure 4.3) when used in (4.33).

Although the equilibrium properties of capillary waves seem to be well understood, there has been little work done on their dynamical properties at the nanometer scale. What is the molecular nature of the flow field close to the fluctuating interface? Do capillary waves in the liquid–vapor interface play a significant role for the fluid flow in open nanochannels? As far as we know, these questions are as yet unanswered and unexplored. The Navier–Stokes equations correctly describe the dynamics of Newtonian liquids in all cases where the time and spatial scales on which the dynamics are investigated are well separated from the scales of the microscopic dynamics of the molecules. In such cases the conservation laws for mass, energy, and translational momentum, together with appropriate boundary conditions, completely determine the dynamics of the fluids [9]. This is not the case for capillary waves with wavelengths below 50 nm, where features of the molecular dynamics become relevant for the spatiotemporal description of the fluid dynamics.

Extensive measurements of dynamic properties of interfaces such as the spectrum of surface waves have been undertaken with dynamic light scattering [25, 29]. Lateral length scales below 100 µm are unattainable, however, because an unambiguous distinction between surface scattering and scattering from the bulk liquid underneath becomes impossible. The use of X-ray photon correlation spectroscopy [30] allows improvement of the lateral resolution by two decades, but the surface physics at wavelengths of about 1 µm is nevertheless still dominated by classical capillary waves. Consequently, the spectrum of the surface ripples can be described by a linear response analysis of the equations of hydrodynamics (see previous section and [31]). Thus, the theoretical analysis is unambiguous for these wavelengths that are much larger than intermolecular spacing.

Further progress in experimental techniques, especially with high brilliance sources for synchrotron radiation, opens up the possibility to study the dynamics of surfaces at wavelengths close to intermolecular distances [32–37]. For these length scales the hydrodynamic limit, which is always the long wavelength limit of the dynamics of a statistical system, will certainly break down, and another description is called for. The main theoretical challenge here lies not so much in exhaustive simulations of the dynamical behavior of interfaces, but rather in identifying the relevant dynamical variables which describe the fluctuations of the surface. Nevertheless, it is indispensable to judge the viability of a theoretical model by careful comparison to molecular dynamics data and, hopefully in a few years, to new experiments.

In the following section a first example is presented in which thermal fluctuations and capillary waves play an important role for the nonlinear dynamics of thin liquid films.

4.3
Thermal Noise Influences Fluid Flow in Nanoscopic Films

Thin liquid films are ubiquitous in nature and the understanding of their dynamical behavior is important for many technological applications. The fabrication of electronic chips now requires thicknesses of insulating layers or photoresists on the order of a few nanometers. Reliable predictions of the dynamics of ultra-thin films play an important role in guaranteeing stability during production and use of such devices. In bulk fluids, hydrodynamic Navier–Stokes equations are proven to be valid down to the nanometer scale. Until recently thin film flow has been studied solely by deterministic equations [38], although thermal noise plays an increasingly important role the smaller the system size becomes, and may play an important role in thin liquid films with thicknesses of a few nanometers. For instance, in [39, 40] a stochastic version of the thin film equation was derived based on the lubrication approximation for stochastic hydrodynamic equations. To a linear approximation, this treatment predicted that the spectrum of capillary waves changes from an exponential decay to a power law for large wave vectors due to thermal fluctuations. Consequently, the time evolution of the film thickness $h(\vec{r}, t)$, that is of the film roughness $\sigma^2(t) = \langle h^2 \rangle$, and also of the typical wavelengths of the maximum of the power spectrum are found to change qualitatively. Whereas the deterministic equation predicts a constant wavelength in the linear regime, the stochastically evolving structures coarsen in time and $\sigma^2(t)$ is expected to increase much faster due to thermal noise. These consequences of the stochastic nature of the thin film dynamics are robust. The failure of the deterministic hydrodynamic description due to thermal fluctuations is already expected for small noise amplitudes in thin liquid films and for a large class of substrate interactions. Recent numerical studies of thin film evolution also indicate that thermal noise might influence characteristic time scales of the dewetting process [40].

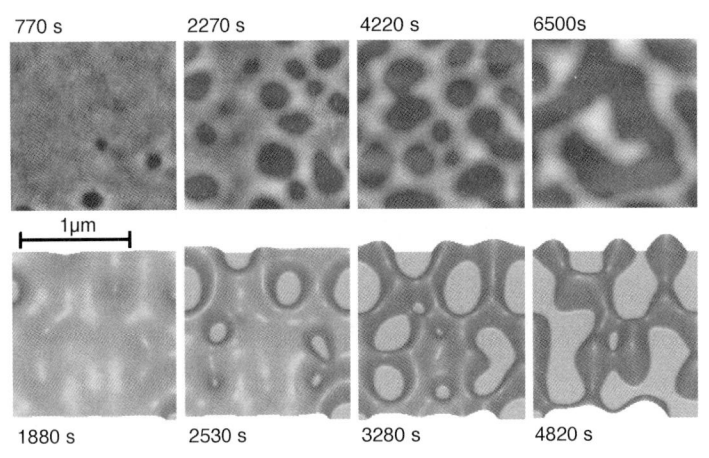

Figure 4.4 A 3.9 nm polystyrene film beads off an oxidized Si wafer [41]: temporal series of experimental scanning force microscopy of the dewetting process (top) can be simulated by the Navier–Stokes equation in lubrication approximation, (4.37), with identical system parameters. The temporal evolution of the dewetting morphology can be modeled quantitatively by stochastic Navier–Stokes equation (4.35).

In [42] these predictions of noisy hydrodynamics have been confirmed experimentally by AFM measurements of the dewetting of thin polymer films [41]. The experimental system consists of a polystyrene film with a molecular weight of 2 kg/mol (PS(2k)) prepared on a silicon substrate with a 191 nm thick amorphous oxide layer on top. The film thickness is chosen to be 3 to 4 nm so that the system is unstable in the spinodal regime. Heating the sample above T_g leads to capillary waves at the PS surface. The amplitudes of the waves increase exponentially with time, finally reaching the order of the film thickness and causing holes that grow in the further stages of dewetting. The whole process of spinodal dewetting is scanned in situ by SPM (Figure 4.4). This process involves the emergence and amplification of capillary waves, the appearance, growth, and coalescence of holes, and finally the formation of droplets.

4.3.1
Dynamics of the Film Thickness

The film can be described as an incompressible Newtonian liquid with a constant mass density ρ on an infinite flat solid substrate, that is by the Navier–Stokes equation (4.2)

$$\rho\left(\frac{\partial \vec{v}}{\partial t} + (\vec{v} \cdot \nabla)\vec{v}\right) = \eta \vec{\nabla}^2 \vec{v} - \vec{\nabla}(p - \Pi) + \vec{\nabla} \cdot \mathcal{S}, \tag{4.35}$$

but with an additional disjoining pressure

$$\Pi(h) = -\frac{\partial \Phi(h)}{\partial h} \qquad (4.36)$$

which is given by the effective interface potential (see (4.2))

$$\Phi(h) = -\frac{A}{12\pi h^2}$$

and determines the dewetting properties of a substrate. Thus, the Hamaker constant A determines the disjoining pressure in (4.35) if we neglect the short ranged part of the potential. We also assume that there is no slip between the fluid and the substrate and that at the free surface $z = h(\vec{r}, t)$ the normal and tangential stresses are balanced in the polymer film.

For a smooth thin film, where the ratio of the characteristic film height h_0 is much smaller than the length scale over which the film thickness varies laterally, one can find an approximate solution for the free boundary flow. This approach is well described in [38–40], so that we give here only the resulting nonlinear Langevin equation

$$\eta \frac{\partial h}{\partial t} = \vec{\nabla} \cdot \left\{ \frac{h^3}{3} \vec{\nabla} \left[\Phi'(h) - \gamma \vec{\nabla}^2 h \right] + \sqrt{\frac{2 k_B T \eta}{3} h^3} \vec{\mathcal{N}}(t) \right\} \qquad (4.37)$$

for the film thickness $h(\vec{r}, t)$, with a single multiplicative conserved noise vector $\vec{\mathcal{N}}(\vec{r}, t)$ obeying $\langle \vec{\mathcal{N}}(\vec{r}, t) \rangle = 0$ and the correlator

$$\langle \mathcal{N}_i(\vec{r}, t) \mathcal{N}_j(\vec{r}', t') \rangle = \delta_{ij} \delta(\vec{r} - \vec{r}') \delta(t - t') \,.$$

The polystyrene (PS) film of thickness $h_0 \approx 4$ nm on silicon dioxide is linearly unstable and the characteristic lateral length scale is given by the dispersive capillary length

$$\frac{2\pi}{q_0} = \sqrt{\frac{-32\pi^2 \gamma}{\Pi'(h_0)}} = 4 h_0^2 \sqrt{\pi^3 \gamma / A} \,.$$

With the Hamaker constant $A \approx 2 \cdot 10^{-20}$ N m and the surface tension coefficient $\gamma \approx 3 \cdot 10^{-2}$ N/m, we see that $\frac{2\pi}{q_0} \approx 400$ nm. The viscosity is $\eta \approx 1200$ N s/m². In the deterministic part of (4.37) there are two terms which can drive the flow, the disjoining pressure and the surface tension. The flow associated with each part is of the order of $h_0 U$ with two characteristic velocities, namely

$$U_\Pi = \frac{A}{6\pi h_0 \eta} \frac{q_0}{2\pi} \approx 0.6 \text{ nm/s}$$

$$U_\gamma = \frac{h_0^3 \gamma}{3\eta} \left(\frac{q_0}{2\pi}\right)^3 \approx 8 \cdot 10^{-3} \text{ nm/s} \,. \qquad (4.38)$$

Taking the larger of the two velocities, the dimensionless noise amplitude is

$$\tilde{T} = \frac{k_B T h_0}{\eta \, U_\Pi} \left(\frac{q_0}{2\pi}\right)^3 = \frac{3 k_B T}{8\pi^2 h_0^2 \gamma} \, .$$

This result is, in fact, independent of the form of the disjoining pressure and also holds if the short ranged part is included. The experiments were performed at 53 °C which leads to the dimensionless amplitude $\tilde{T} \approx 4 \cdot 10^{-4}$ of the noise. The noise induced current is therefore about two orders of magnitude smaller than the current induced by the disjoining pressure.

4.3.2
Comparison with Experiments

Quantitative information about the dewetting process can be obtained by measuring the variance $\sigma^2(t) = \langle \delta h(\vec{r}, t) \, \delta h(\vec{r}, t) \rangle$ of the film height $h(\vec{r}, t)$, and the variance $k^2(t) = \langle (\vec{\nabla} h(\vec{r}, t))^2 \rangle / (2\pi \sigma^2(t))$ of the local slope of the film height, which gives information about the preferred wave vector within the surface. Here, the brackets represent an integration over all positions $\vec{r}$ of the image. We analyze only the early stage of spinodal dewetting where capillary waves are amplified until the first holes appear (up to about 1000 s). During the intermediate stage when holes emerge and grow, typically between 1000 s and 4000 s, as well as during the late stage of dewetting after 4000 s when the coalescence of holes and forming of droplets take place, we analyze the prepared film only around the initial height h_0 and ignore the emerging holes (details of the technique are given in [43]). Thus, we restrict the integral $\langle \cdot \rangle$ on regions without holes. The results for σ^2 and k^2 are shown in Figure 4.5.

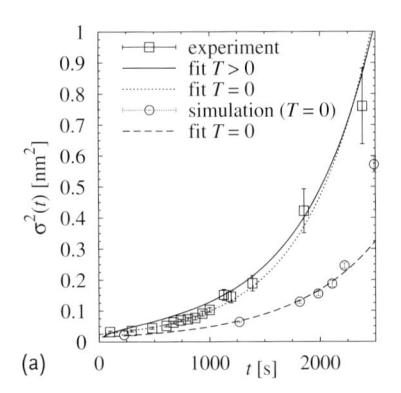

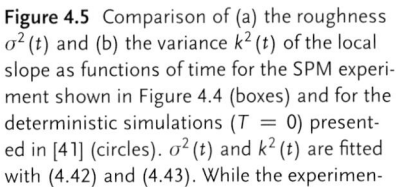

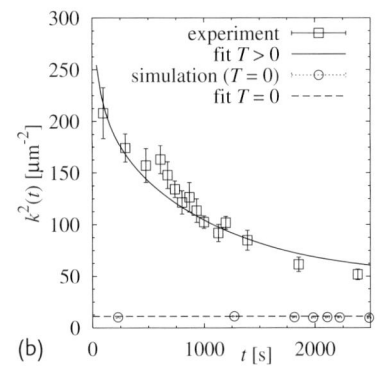

Figure 4.5 Comparison of (a) the roughness $\sigma^2(t)$ and (b) the variance $k^2(t)$ of the local slope as functions of time for the SPM experiment shown in Figure 4.4 (boxes) and for the deterministic simulations ($T = 0$) presented in [41] (circles). $\sigma^2(t)$ and $k^2(t)$ are fitted with (4.42) and (4.43). While the experimental $\sigma^2(t)$ can be fitted with the deterministic theory by adjusting the initial roughness σ_0^2 and the characteristic time scale t_0 ($T = 0$, dashed line), this is not possible for the experimental $k^2(t)$. The deterministic $k^2(t)$ is always constant in time during the linear regime – independent of σ_0^2 and t_0.

Since the noise induced current is smaller than the current induced by the disjoining pressure, let us first neglect the thermal noise and solve (4.37) numerically with $T = 0$ and random initial conditions. This was done previously in [41] and shows excellent agreement in the spatial structure. However, the time scales and the time dependence of $\sigma^2(t)$ and $k^2(t)$ do not match at all, as can be seen in Figure 4.5. The discrepancy is assumed to be caused by thermal noise in the experimental system, which was thus far neglected in the simulations. However, the preliminary numerical results for (4.37) at finite temperature $T \neq 0$ presented in [40] indicates that the influence of thermal fluctuations on the dewetting dynamics may fix the discrepancy in the early stages of dewetting, where the effect of noise is largest. In this case noise can accelerate the initial dynamics of thin polymer films by at least a factor five, if realistic values are chosen for surface tension, substrate potential and viscosity [40].

4.3.3
Linearized Stochastic Thin Film Equation

We can study these early stages of dewetting further in a linear approximation of (4.37). In the beginning of the dewetting process, the deviations $\delta h(\vec{r}, t) = h(\vec{r}, t) - h_0$ from the initial film height h_0 are small. By expanding (4.37) to the first order of δh and $\vec{\mathcal{N}}$, assuming that the noise amplitude is small as well, and applying a Fourier transformation $\delta h(\vec{r}, t) = \int \frac{d^2 q}{(2\pi)^2} \tilde{\delta h}(\vec{q}, t) e^{i\vec{q}\cdot\vec{r}}$, we obtain the linear stochastic equation

$$\frac{\partial \tilde{\delta h}(\vec{q}, t)}{\partial t} = \omega(q) \tilde{\delta h}(\vec{q}, t) + i\sqrt{\frac{2k_B T h_0^3}{3\eta}} \vec{q} \cdot \tilde{\vec{\mathcal{N}}}(\vec{q}, t) . \qquad (4.39)$$

The dispersion relation

$$\omega(q) = \left[1 - \left(q^2/q_0^2 - 1\right)^2\right]/t_0 \qquad (4.40)$$

has a maximum at the wave vector

$$q_0^2 = -\frac{\Phi''(h_0)}{2\gamma}$$

with $\Phi''(h_0) < 0$ and a characteristic time

$$t_0 = \frac{3\eta}{\gamma h_0^3 q_0^4} .$$

Note that the multiplicative noise in (4.37) becomes additive in the linear approximation in (4.39). Then the spectrum of the height reads

$$\left\langle \tilde{\delta h}(\vec{q}, t) \tilde{\delta h}(\vec{q}\,', t) \right\rangle = (2\pi)^2 \delta(\vec{q} + \vec{q}\,') \tilde{C}(|\vec{q}|; t)$$

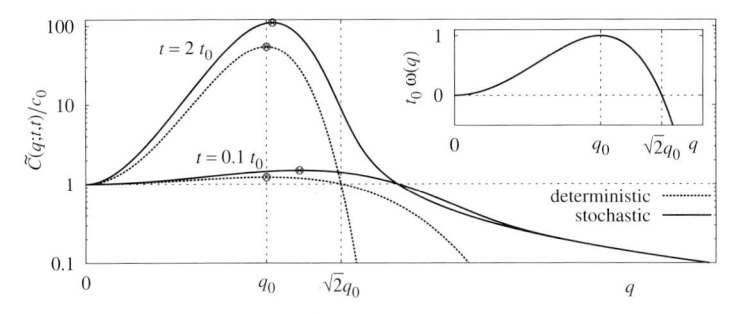

Figure 4.6 Power spectrum of a spinodally dewetting film with (solid line) and without noise (dashed line, for $T = \gamma q_0^4 c_0$) for two different times $t = 0.1 t_0$ and $t = 2.0 t_0$, cf. (4.41). The initial spectrum at $t = 0$ is $\tilde{C}(q; 0, 0) = c_0$. The inset shows the dispersion relation $\omega(q)$.

with the static structure function (see (4.31))

$$\tilde{C}(q; t) = \tilde{C}_0(q) e^{2\omega(q)t} + \frac{k_B T h_0^3}{3\eta} \frac{q^2}{\omega(q)} \left[e^{2\omega(q)t} - 1 \right] \qquad (4.41)$$

and the initial power spectrum

$$\tilde{C}_0(q) = \left\langle \left| \delta h(\vec{q}, 0) \right|^2 \right\rangle \quad \text{at} \quad t = 0.$$

The time evolution of the power spectrum with and without noise for a white initial spectrum $\tilde{C}(q; 0, 0) = c_0$ is shown in Figure 4.6. Note that in the case of a spinodally unstable film ($\Phi''(h_0) < 0$) the dispersion relation $\omega(q)$ is negative for $q > \sqrt{2} q_0$ (see inset in Figure 4.6). Thus, for $t \to \infty$ one finds exponentially decaying height–height correlations

$$\tilde{C}(q; t) \to \tilde{C}_0(q) e^{-2|\omega(q)|t}$$

for the deterministic dynamics ($T = 0$) but we recover the the algebraic capillary wave spectrum

$$\tilde{C}(q; t) \to \frac{k_B T h_0^3}{3\eta} \frac{q^2}{|\omega(q)|} \to \frac{k_B T}{\gamma q^2}$$

for any finite temperature T. Note that the maximum of the deterministic spectrum stays at q_0 for all times, but the maximum of the stochastic spectrum approaches q_0 from above as $t \to \infty$. This noise generated coarsening process can last until nonlinearities become important, effectively masking the typical feature of the linear deterministic regime, namely that the maximum of the power spectrum stays at a fixed wave number.

At this point we note that the spectrum necessarily has a microscopic cutoff $q_{max} = 2\pi/r_0 \gg q_0$ at the scale r_0 of the fluid particles. For simplicity we assume (4.41) holds up to this point and $\tilde{C}(q; t) = 0$ for $q > q_{max}$. In order to illustrate further the spatial features of the dynamics, we calculate the roughness of the film

$$\sigma^2(t) = \int \frac{d^2 q}{(2\pi)^2} \tilde{C}(q; t)$$

$$= \sigma^2_{T=0}(t) + \frac{k_B T}{4\pi \gamma} \int_{-1}^{\frac{q^2_{max}}{q_0^2}-1} d\theta \frac{1 - e^{-2\frac{t}{t_0}(\theta^2-1)}}{\theta - 1} \quad (4.42)$$

with the deterministic evolution

$$\sigma^2_{T=0}(t) = \sigma_0^2 \sqrt{\frac{\pi t_0}{8t}} \, e^{2\frac{t}{t_0}} \, \mathrm{erf}\left(\sqrt{\frac{2t}{t_0}}\right)$$

of the roughness, the initial roughness $\sigma^2(0) = \sigma_0^2$, and the initial spectrum $\tilde{C}_0(q) = \frac{2\pi}{q_0^2} \sigma_0^2$ for $q < \sqrt{2}\, q_0$ and $\tilde{C}_0(q) = 0$ for $q > \sqrt{2}\, q_0$. Due to a rapid increase of thermal fluctuations on an atomistic time scale t_m, the initial spectrum is irrelevant for the film evolution on the characteristic time scale t_0. For large t we can calculate the ratio $\sigma^2(t)/\sigma^2_{T=0}(t) \to 1 + \Xi + \mathcal{O}(t^{-1})$ with $\Xi = \frac{k_B T}{2\pi \gamma \sigma_0^2} > 0$. Note that Ξ is given by the ratio of the thermal (capillary) roughness T/γ to the initial roughness σ_0^2. It is this ratio which determines the importance of thermal fluctuations for the dynamics of the film. One may argue that the initial roughness is due to thermally equilibrated capillary waves before the dewetting process starts, so that one may expect Ξ to be of order unity. A numerical integration of (4.42) shown in Figure 4.7a illustrates that thermal noise is most important in the beginning of the process. One finds a fast linear increase $\sigma^2(t)/\sigma^2_{T=0}(t) = 1 + \frac{\Xi}{2} t/t_m + \mathcal{O}(t^2)$ of the thermal roughness with the characteristic (microscopic) time $t_m = \frac{q_0^4}{q_{max}^4} t_0$ due to a rapid build up of a thermal spectrum for $q > \sqrt{2} q_0$, followed by a slower increase for $t_m < t$ and up to t_0 due to the linear dewetting process. However, for times $t \gg t_0$ thermal fluctuations become less important compared to the exponential increase of the unstable mode q_0 and one reaches a 'quasi'-deterministic behavior $\sigma^2_{T=0}(t)$, but with a renormalized initial roughness $\sigma_0^2 + \frac{k_B T}{2\pi \gamma}$.

For the variance of the local slope

$$2\pi \sigma^2(t) k^2(t) = \left\langle \left[\vec{\nabla} \delta h(\vec{r}, t)\right]^2 \right\rangle = \int \frac{d^2 q\, q^2}{(2\pi)^2} \tilde{C}(q; t)$$

which is a measure for the characteristic wavelength of fluctuations, we find

$$\frac{k^2(t)}{k_0^2} = 1 + \frac{k_B T}{4\pi \gamma \sigma^2(t)} \int_{-1}^{\frac{q^2_{max}}{q_0^2}-1} d\theta \, \theta \, \frac{1 - e^{-\frac{2t}{t_0}(\theta^2-1)}}{\theta - 1}, \quad (4.43)$$

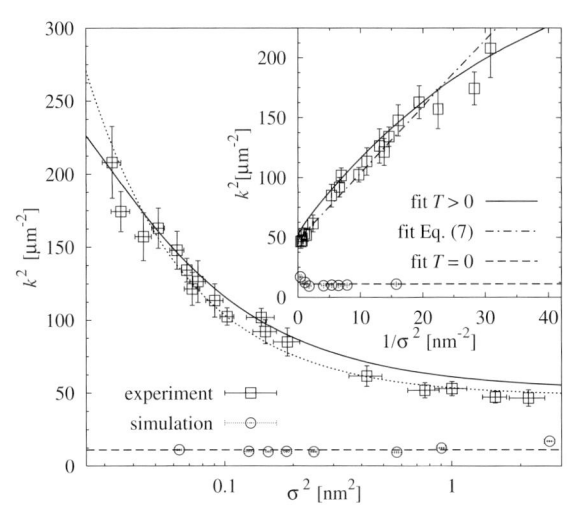

Figure 4.7 Variance k^2 of the local slope in the experiment (boxes) and the deterministic simulation (circles) plotted versus the roughness σ^2. Due to thermal noise one finds $k^2(t)/k^2(\infty) - 1 \sim 1/\sigma^2(t)$ for any noise amplitude T as long as the characteristic wavelength $2\pi/q_0$ is much larger than the molecular cutoff $2\pi/q_{max}$. In contrast, simulations for $T = 0$ are consistent with a constant k^2.

with the initial and final value $k^2(0) = k^2(\infty) = \frac{q_0^2}{2\pi}$. Note that for the deterministic dynamics $k^2_{T=0}(t) = \frac{q_0^2}{2\pi}$ is constant in time for the chosen initial spectrum, and that the position of the maximum in the structure function $\tilde{C}(q, t)$ does not change during the dewetting process. In contrast, in the stochastic dynamics thermal noise induces a time dependence of $k^2(t)$ that starts at the deterministic value at $t = 0$ and increases linearly in time until it reaches a maximum at $t \approx t_m$, before approaching the deterministic value $k^2(t)/k^2(\infty) \to 1 + \frac{\Xi}{1+\Xi}\frac{t_0}{4t} + \mathcal{O}(t^{-2})$ from above for $t \gg t_0$. If the microscopic cutoff q_{max} is much larger than q_0, one obtains an intermediate time regime $t_m < t$ up to $t \approx t_0$, where the integral in (4.43) is approximately constant. One then gets

$$k^2(t) \approx k^2(\infty) + \frac{k_B T}{2\gamma r_0^2}\frac{1}{\sigma^2(t)} \tag{4.44}$$

for intermediate times up to $t \approx t_0$, independent of the initial conditions. Thus, thermal noise generates coarsening even in the linear regime for which the deterministic linear dynamics predicts a fixed characteristic wave vector $k^2(t) = k^2(\infty) = \frac{q_0^2}{2\pi}$. The second term in (4.44) can easily be derived by calculating the variance of the local slope

$$\left\langle \left[\vec{\nabla}\delta h(\vec{r}, t)\right]^2 \right\rangle_{eq} = \int \frac{d^2 q\, q^2}{(2\pi)^2} \tilde{C}_{eq}(q) = \frac{k_B T}{4\pi\gamma}q_{max}^2,$$

with the equilibrium capillary wave spectrum $\tilde{C}_{eq}(q) = \frac{k_B T}{\gamma q^2}$. Thus, the variance

$$2\pi\sigma^2(t)k^2(t) = \left\langle\left[\vec{\nabla}\delta h(\vec{r},t)\right]^2\right\rangle = q_0^2\sigma^2(t) + \left\langle\left[\vec{\nabla}\delta h(\vec{r},t)\right]^2\right\rangle_{eq}$$

consists of two terms: the equilibrium contribution due to thermal noise and the growing peak at q_0 due to the dewetting process $\int d^2q\,\vec{q}\vec{q}^2\,\tilde{C}_0(q)\,e^{2\omega(q)t} \to q_0^2\int d^2q\,\tilde{C}_0(q)\,e^{2\omega(q)t}$, which is independent of the initial spectrum and proportional to $\sigma^2(t)$ in the limit $t \gg t_0$.

For a glass-forming liquid such as polystyrene, one may argue that the equilibrium variance $\left\langle\left[\vec{\nabla}\delta h(\vec{r},t)\right]^2\right\rangle_{eq} = 2\pi\alpha(T - T_g)$ is proportional to the temperature difference $T - T_g$ in the vicinity of the glass transition at T_g, so that (4.44) actually reads $k^2(t) \approx k^2(\infty) + \alpha(T - T_g)/\sigma^2(t)$.

A separation of length scales $q_{max} \gg q_0$ also leads to a separation of time scales $t_m = \frac{q_0^4}{q_{max}^4}\,t_0 \ll t_0$, so that the algebraic decrease of $k^2(t)$ with $\sigma^2(t)$ is visible before the exponentially growing peak in the structure function causes a crossover to an algebraic behavior in time $k^2(t)/k_0^2 - 1 \sim 1/t$ for $t > t_0$. We expect the linear approximation in (4.39) to hold at least for the fast initial ($t < t_m$) formation of the thermal spectrum for $q > q_0$, as well as for the noise dominated spinodal dewetting process up to $t \approx t_0$.

Finally, one can conclude that the noise term in the structure function $\tilde{C}(q;t)$ is relevant for any value of the noise amplitude T, as long as the dispersion relation $\omega(q)$ becomes negative for large wave vectors $q_{max} > q > \sqrt{2}q_0$. For realistic values of surface tension and substrate potentials one finds $2\pi/q_0 \approx 0.1\ldots 1$ µm, which is much larger than the size of molecules and provides an upper cutoff q_{max} for allowed wave vectors. Thus, the time evolution of $\sigma^2(t)$ and $k^2(t)$ given by (4.42) and (4.43), respectively, are always dominated by the thermal noise term for times $t < t_0$ up to the characteristic time t_0 of the fastest growing mode q_0.

During the dewetting process of liquid films of nanometer thickness, the interplay of substrate potentials and thermal noise may result in qualitatively different lateral behavior on scales up to microns. In particular, for the further development of efficient tools to be used in the design of microfluidic devices or electronic components whose function relies on thin film properties, it is essential to gain a quantitative understanding of thermal fluctuations in thin film flow and its interplay with molecular interactions. In the course of miniaturization of microfluidic devices, a fully quantitative description of Newtonian liquids at surfaces is essential and requires quantitative stochastic modeling of ultrathin film dynamics as well as mathematically well controlled numerical schemes.

References

1 Moseler, M. and Landman, U. (2000) *Science*, **289**, 1165.
2 Eggers, J. (2002) *Phys. Rev. Lett.*, **89**, 084502.
3 Hennequin, Y. et al. (2006) *Phys. Rev. Lett.*, **97**, 244502.
4 Rowlinson, J. and Widom, B. (1982) *Molecular theory of capillarity*, Clarendon, Oxford.
5 Buff, F.P., Lovett, R.A., and Stillinger, F.H. Jr. (1965) Interfacial density profile for fluids in the critical region. *Phys. Rev. Lett.*, **15**, 621.
6 Mecke, K. and Dietrich, S. (1999) Effective Hamiltonian for liquid–vapor interfaces. *Phys. Rev. E*, **59**, 6766.
7 Fradin, C., Braslau, A., Luzet, D., Smilgies, D., Alba, M., Boudet, N., Mecke, K., and Daillant, J. (2000) Liquid interfaces beyond capillarity. *Nature*, **403**, 871.
8 Mora, S., Mecke, K., Daillant, J., Luzet, D., and Struth, B. (2003) X-ray synchrotron study of liquid–vapor interfaces at shor length scales: effect of long-range forces and bending energies. *Phys. Rev. Lett.*, **90**, 216101.
9 Landau, L.D. and Lifschitz, J.M. (1991) Hydrodynamik, Vol. VI, *Lehrbuch der Theoretischen Physik*, Akademie Verlag, 5th edn.
10 Fox, R.F. and Uhlenbeck, G.E. (1970) Contributions to Non-equilibrium Thermodynamics. I. Theory of Hydrodynamical Fluctuations. *Phys. Fluids*, **13**, 1893.
11 Fox, R.F. and Uhlenbeck, G.E. (1970) Contributions to Nonequilibrium Thermodynamics. II. Fluctuation Theory for the Boltzmann Equation. *Phys. Fluids*, **13**, 2881.
12 Mashiyama, K.T. and Mori, H. (1978) Origin of the Landau–Lifshitz hydrodynamic fluctuations in nonequilibrium systems and a new method for reducing the Boltzmann equation. *J. Stat. Phys.*, **18**, 385.
13 Forster, D., Nelson, D.R., and Stephen, M.J. (1976) Long-time tails and the large-eddy behavior of a randomly stirred fluid. *Phys. Rev. Lett.*, **36**, 867.
14 Forster, D., Nelson, D.R., and Stephen, M.J. (1977) Large-distance and long-time properties of a randomly stirred fluid. *Phys. Rev. A*, **16**, 732.
15 Hohenberg, P.C. and Swift, J.B. (1992) Effects of additive noise at the onset of Rayleigh–Bénard convection. *Phys. Rev. A*, **46**, 4773.
16 Swift, J.B., Babcock, K.L., and Hohenberg, P.C. (1994) Effects of thermal noise in Taylor–Couette flow with corotation and axial through flow. *Physica A*, **204**, 625.
17 Hansen, J.-P. and McDonald, I (1986) *Molecular Hydrodynamics*, Academic.
18 Boon, J.P. (1990) *Theory of Simple Liquids*, Academic.
19 Falk, K. (2008) *Stochastische Hydrodynamik von Oberflächen kompressibler Flüssigkeiten*, Diplomarbeit, Universität Erlangen-Nürnberg.
20 van der Waals, J.D. (1893) *Verhandel. Konink. Akad. Weten. Amsterdam*, **1** (8).
21 Stecki, J. (1998) *J. Chem. Phys.*, **109**, 5002–5007.
22 van Giessen, A.E. and Blokhuis, E.M. (2002) *J. Chem. Phys.*, **116**, 302–310.
23 Cahn, J.W. and Hilliard, J.E. (1958) *J. Chem. Phys.*, **28**, 258–267.
24 Buff, F.P., Lovett, R.A., and Stillinger, R.H. (1965) *Phys. Rev. Lett.*, **15**, 621–623.
25 Loudon, R. (1984) *Modern problems in condensed matter sciences* (eds V.M. Agranovich and R. Loudon), Vol. 9, pp. 589–638.
26 Sanyal, M.K. et al. (1991) *Phys. Rev. Lett.*, **66**, 628–631.
27 Sengers, J.V. and van Leeuwen, J.M.J. (1989) *Phys. Rev. A*, **39**, 6346–6355.
28 Helfrich, W. (1973) *Z. Naturforsch.*, **28c**, 693–703.
29 Langevin, D. (ed.) (1992) *Light Scattering by Liquid Surfaces and Complementary Techniques*, Marcel Dekker, New York.
30 Seydel, T. et al. (2001) Capillary waves in slow motion. *Phys. Rev. B*, **63**, 073409.
31 Jäckle, J. (1998) The spectrum of surface waves on viscoelastic liquids of arbitrary depth. *J. Phys.: Condens. Matter*, **10**, 7121.
32 Gutt, C., Tolan, M. et al. (2003) *Phys. Rev. Lett.*, **91**, 076104.
33 Kim, H., Rühm, A. et al. (2003) *Phys. Rev. Lett.*, **90**, 068302.

34 Weber, R., Tolan, M. et al. (2001) *Phys. Rev. E*, **64**, 061508.

35 Tolan, M., Seeck, O.H. et al. (1998) *Phys. Rev. Lett.*, **81**, 2731.

36 Tolan, M. (1999) X-Ray Scattering from Soft-Matter Thin Films – Materials Science and Basic Research. *Springer Tracts in Modern Physics*, Vol. 148, Springer, Berlin, Heidelberg.

37 Wang, J., Tolan, M. et al. (1999) *Phys. Rev. Lett.*, **83**, 564.

38 Oron, A., Davis, S.H., and Bankoff, S.G. (1997) Long-scale evolution of thin liquid films. *Rev. Mod. Phys.*, **69**, 931.

39 Mecke, K. and Rauscher, M. (2005) On thermal noise in thin film flow. *J. Phys.: Condens. Matter*, **17**, S3515.

40 Grün, G., Mecke, K., and Rauscher, M. (2006) Thin-film flow influenced by thermal noise. *J. Stat. Phys.*, **122**, 1261.

41 Becker, J., Grün, G., Seemann, R., Mantz, H. Jacobs, K., Mecke, K., and Blossey, R. (2003) Complex dewetting scenarios captured by thin film models. *Nat. Mater.*, **2**, 59.

42 Fetzer, R., Rauscher, M., Seemann, R., Jacobs, K., and Mecke, K. (2007) Thermal noise influences fluid flow in thin films during spinodal dewetting. *Phys. Rev. Lett.*, **99**, 114503.

43 Mantz, H., Jacobs, K., and Mecke, K. (2008) Utilising Minkowski Functionals for Image Analysis: a marching square algorithm. *J. Stat. Mech.: Theory Exp. (JSTAT)*, P12015.

5
Nonlinear Dynamics of Surface Steps

Joachim Krug

5.1
Introduction

Surface steps are key elements in the dynamics of a crystal surface below its thermodynamic roughening transition because they constitute long-lived structural defects that are nevertheless highly mobile and prone to strong fluctuations [1]. The description of surface morphology evolution in terms of the thermodynamics and kinetics of steps goes back at least half a century [2]. During the past few decades, the subject has experienced a significant revival due to the availability of imaging methods such as scanning tunneling microscopy that allow for a direct visualization of step conformation and step motion on the nanoscale (see [3–7] for recent reviews). In this chapter I will focus specifically on cases where steps have been found to display complex *dynamic* behavior, such as oscillatory shape evolution under constant driving.

The examples to be discussed below can be naturally organized according to the underlying topology of the step configurations: I first consider driven single layer islands (closed step loops), and then vicinal surfaces (arrays of parallel steps). A certain familiarity with the basic thermodynamics and kinetics of crystal surfaces is assumed. For an elementary introduction the reader may consult [8].

5.2
Electromigration-Driven Islands and Voids

Electromigration is the directed transport of matter in a current carrying material, which is primarily caused by the scattering of conduction electrons off defects such as interstitials or atoms adsorbed on the surface. The latter are henceforth referred to as *adatoms* (Figure 5.1). Much of the work on electromigration has been motivated by its importance as a damage mechanism limiting the lifetime of integrated circuits [9]. Because electromigration forces are small compared to the typical energy barriers involved in the thermal diffusion of atoms, the direct observation of

---

Nonlinear Dynamics of Nanosystems. Edited by Günter Radons, Benno Rumpf, and Heinz Georg Schuster
Copyright © 2010 WILEY-VCH Verlag GmbH & Co. KGaA, Weinheim
ISBN: 978-3-527-40791-0

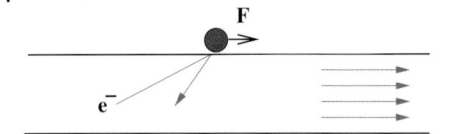

Figure 5.1 Schematic of the microscopic origin of the electromigration force: conduction electrons scattering off an adatom give rise to a transfer of momentum in the direction of the current flow.

electromigration effects in real time on atomistic length scales is difficult (see [10] for recent progress in this direction). In this chapter the electromigration force will be used as a conceptually simple way of driving a system of surface steps out of equilibrium, giving rise to surprisingly complex dynamical behavior.

5.2.1
Electromigration of Single Layer Islands

Two-dimensional single layer islands are the simplest nanoscale structures that appear on a surface during the early stages of thin film growth, when the amount of deposited material is a small fraction of a monolayer [5]. Because of their small size, such islands already display considerable shape fluctuations in thermal equilibrium that may cause diffusive motion of the island as a whole [4]. The electromigration-induced drift of single layer islands on the Si(111) surface was observed experimentally by Métois and collaborators in 1999 [11]. In the following, I summarize recent theoretical work on this problem that is based on a continuum formulation due to Pierre-Louis and Einstein [12].

I focus here on the simplest case in which the motion of atoms is restricted to the boundary of the islands, such that the island area is conserved.[7] The local normal velocity v_n of the island boundary then satisfies a continuity equation,

$$v_n = -\frac{\partial}{\partial s} j = \frac{\partial}{\partial s} \sigma \left[\frac{\partial}{\partial s} (\tilde{\gamma} \kappa) - F_t \right], \tag{5.1}$$

where s denotes the arc length measured along the island contour. The mass current j along the island boundary is proportional to the step edge mobility σ, and it is driven by capillary forces and the component F_t of the electromigration force tangential to the boundary. The capillary force, in turn, is given by the tangential gradient of the edge chemical potential, which is the product of the edge stiffness $\tilde{\gamma}$ and the edge curvature κ. The stiffness $\tilde{\gamma}$ is derived from the edge free energy per unit length γ according to $\tilde{\gamma} = \gamma + \gamma''$, where primes denote derivatives with respect to the orientation angle of the edge. In the absence of external forces ($F_t = 0$), (5.1) guarantees the relaxation of the island to its equilibrium shape characterized by $\tilde{\gamma}\kappa = $ const. [5]. Throughout this section the electromigration force is assumed

---

7) A nonconserved situation where the step exchanges atoms with the terrace is treated below in Section 5.2.4.

to be constant in magnitude and direction. This implies that

$$F_t = F_0 \cos \theta , \tag{5.2}$$

where θ denotes the angle between the boundary and the direction of the force.

In the absence of crystalline anisotropy the material parameters σ and $\tilde{\gamma}$ in (5.1) are constants and it is straightforward to check that (5.1), (5.2) are solved by a circle of arbitrary radius R moving at constant speed $V = \sigma F_0/R$ [13]. Linear stability analysis of the circular solution shows that it becomes unstable at a critical radius [14]

$$R_c \approx 3.26 l_E , \tag{5.3}$$

where the characteristic length scale obtained by nondimensionalizing (5.1) reads

$$l_E = \sqrt{\tilde{\gamma}/F_0} . \tag{5.4}$$

Beyond the linear instability of the circular solution one finds a family of stationary shapes that are elongated in the direction of the force and become increasingly sensitive to breakup with increasing size [15, 16].

The effect of crystalline anisotropy in the mobility σ was explored, mostly numerically, in [15, 17]. Using the expression [18]

$$\sigma(\theta) = \sigma_0[1 + S \cos^2(n\theta)] , \tag{5.5}$$

where $2n$ denotes the number of symmetry axes, a surprisingly rich phase diagram of migration modes was obtained in the plane spanned by the anisotropy strength S and the dimensionless island radius $R_0 = R/l_E$ for the case of sixfold anisotropy ($n = 3$) (Figure 5.2). In these calculations the force was oriented along a direction of maximal mobility.

For small R_0 the dynamics is dominated by capillarity and the island shape is close to the equilibrium shape. The island moves at constant speed in the direction of the applied force (**ss** = straight stationary motion). With increasing size a bifurcation to a regime of oblique stationary (**os**) motion occurs, in which the symmetry with respect to the force direction is spontaneously broken. A suitable order parameter for this bifurcation is the angle between the direction of force and the direction of motion (Figure 5.3). Increasing the radius further, another bifurcation occurs to a phase in which the obliquely moving island displays periodic shape oscillations (the **oo** phase). At smaller values of S the island performs an oscillatory zig-zag motion that is, on average, directed along the applied force (Figure 5.4).

A clear signature of the transition from stationary oblique to oscillatory behavior shows up in the angle of island migration (Figure 5.3). In addition, it is observed that the period τ of the shape oscillation diverges as the critical radius R_0^{oo} of the transition is approached from above (Figure 5.5). Although the data show some dependence on the number of discretization points, a power law fit indicates that the period diverges as

$$\tau \sim (R_0 - R_0^{oo})^{-2.5} . \tag{5.6}$$

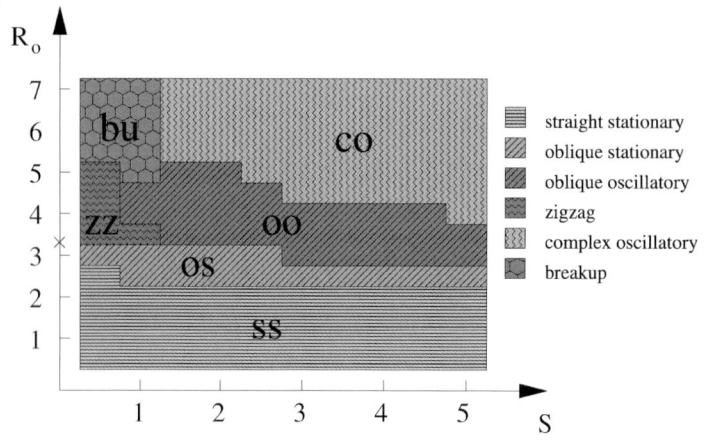

Figure 5.2 Numerically generated phase diagram of island migration modes as a function of the anisotropy strength S, as defined in (5.5), and the dimensionless island radius $R_\sigma = R/l_E$. In the regions denoted by **zz** and **oo** the island shape oscillates periodically, while in the **co** region the behavior is irregular, possibly chaotic. The cross on the R_0-axis indicates the bifurcation from circular to elongated shapes in the isotropic case at the critical radius (5.3). The phase diagram is based on a grid of resolution 0.5×0.5 in the S–R_0-plane.

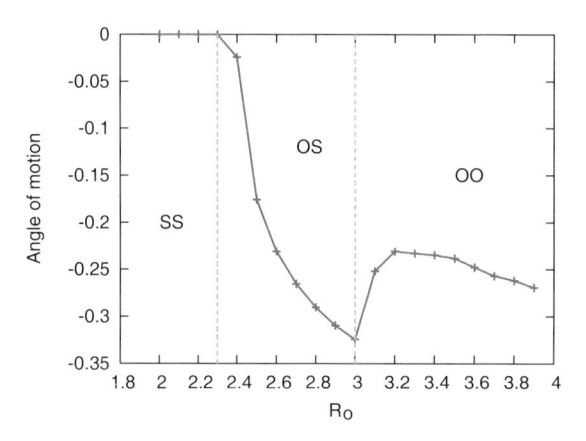

Figure 5.3 Angle enclosed by the direction of island motion and the direction of the applied force as a function of the scaled island radius for $S = 2$. The transitions between the different phases in Figure 5.2 are manifest as slope discontinuities in this graph.

Increasing the island size further, the oscillations become increasingly irregular. This is illustrated in Figure 5.6 by the time series of the island perimeter. The uppermost curve in the figure displays large scale fluctuations that can be traced back to reversals of the direction of island motion that occur at irregular intervals [17]. The Fourier spectrum of such a time series is broad and shows clear signatures of period doubling (Figure 5.7).

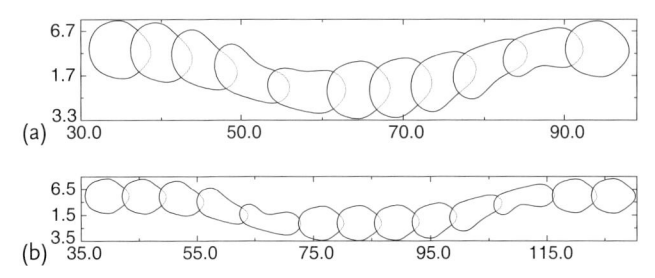

Figure 5.4 Oscillatory island motion in the zig-zag phase of the phase diagram. Parameters are $R_0 = 3.5$, $S = 0.5$ for (a) and $R_0 = 3.5$, $S = 1$ for (b). All lengths are measured in units of l_E.

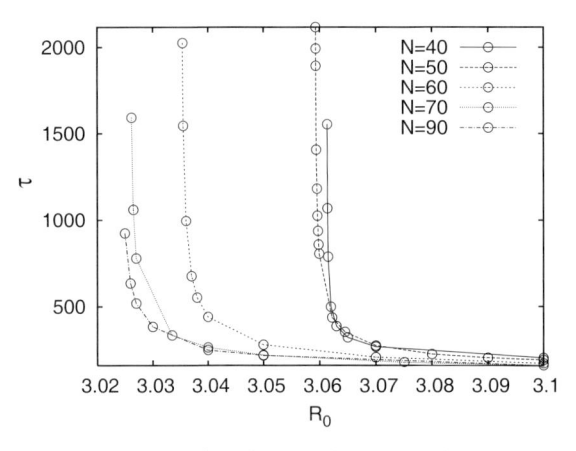

Figure 5.5 Period τ of the shape oscillation near the transition from the **oo** to the **os** phase. Different curves show results obtained for different numbers N of discretization points in the numerical solution, with N increasing from right to left (from [16]).

5.2.2
Continuum vs. Discrete Modeling

In the preceding section it was seen that electromigration-driven islands display a number of features that are consistent with the behavior of a low-dimensional, nonlinear dynamical system. This is remarkable since physically such an island consists of a large number of atoms that move stochastically under the influence of thermal fluctuations and a very small systematic force.

In order to determine whether the phenomena predicted on the basis of the deterministic continuum model given in (5.1) also persist under experimentally realistic conditions, extensive kinetic Monte Carlo (KMC) simulations were carried out using a lattice model that has been shown to provide an accurate representation

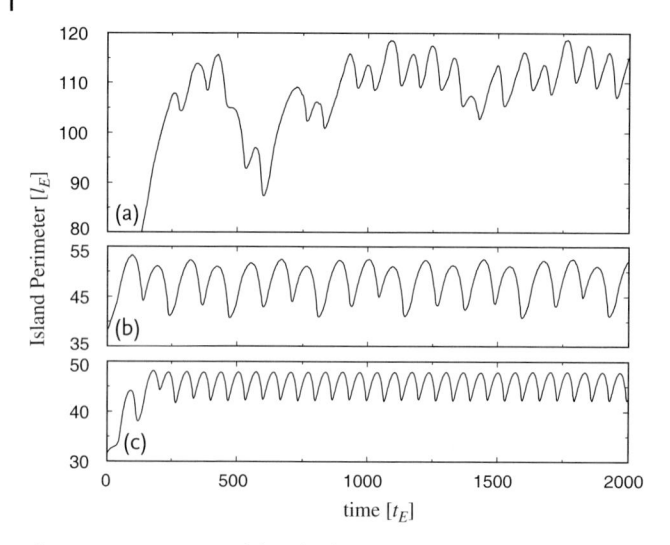

Figure 5.6 Time series of the island perimeter, measured in units of l_E. From (c) to (a), parameters are $S = 2$, $R_0 = 5$; $S = 5$, $R_0 = 5$; and $S = 5$, $R_0 = 6.5$. Time is measured in units of $t_E = l_E^4/(\sigma_0 \tilde{\gamma})$.

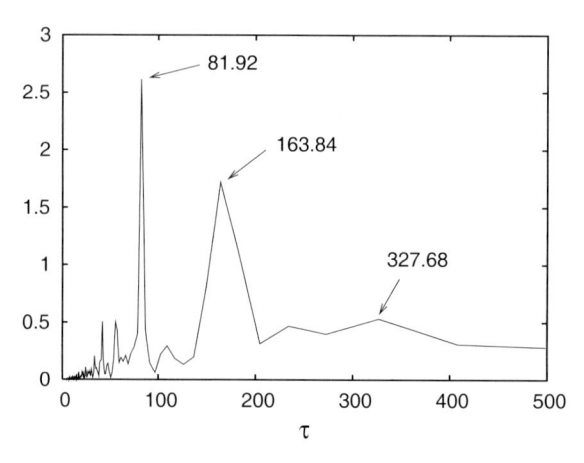

Figure 5.7 Fourier spectrum of the island perimeter time series for $S = 3$ and $R_0 = 6$, plotted as a function of the period $\tau = 2\pi/\omega$ (from [16]).

of metal surfaces[8] such as Cu(100) [19]. In a suitably chosen range of parameters, a regime of oscillatory motion could be identified which shows dynamic behavior in good, essentially quantitative agreement with the continuum model (Figure 5.8).

8) See [5, 7] for an overview of similar models, and [12, 20] for earlier KMC simulations of island electromigration.

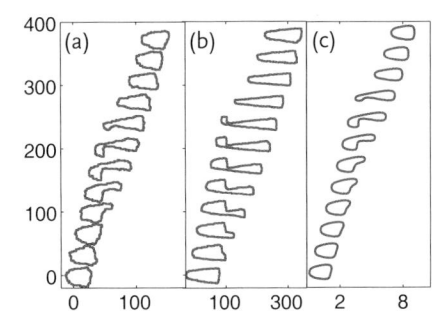

Figure 5.8 Comparison of island shape evolution obtained from KMC simulations (a and b) and numerical solution of the continuum model (c). The simulated islands consist of 1000 atoms in (a) and 4000 atoms in (b). (a) and (c) correspond to a temperature of $T = 700$ K, while in (b) $T = 500$ K. In (a) and (b) lengths are measured in units of the lattice constant, in (c) in units of l_E.

For the comparison to KMC simulations, realistic expressions for the step edge mobility σ and the stiffness $\tilde{\gamma}$ in (5.1) were derived and implemented. Both of these quantities display a fourfold anisotropy on the fcc(100) surface. A rough exploration of the full phase diagram conducted within the continuum model is depicted in Figure 5.9. Since the physical parameter controlling the anisotropy is the temperature T, with lower temperatures corresponding to more pronounced anisotropy, the temperature axis in Figure 5.9 replaces the anisotropy axis in Figure 5.2. The regions displaying oscillatory behavior without leading to island breakup are much

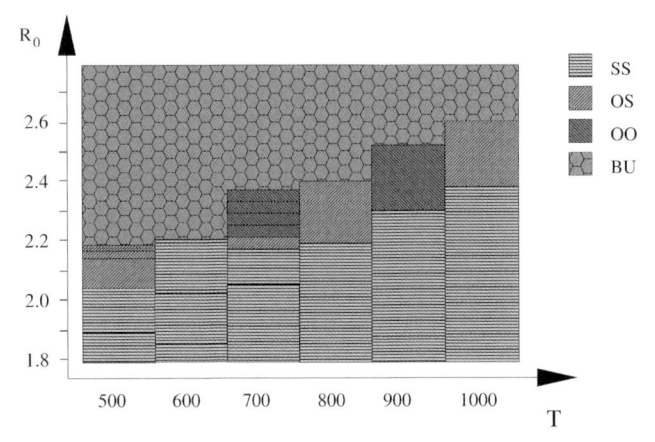

Figure 5.9 Phase diagram of island migration modes obtained by numerical solution of the continuum equations for a mobility and stiffness of fourfold crystalline anisotropy. The temperature is measured in Kelvin and the anisotropy increases with decreasing temperature. Temperature was varied in steps of 100 K. Each rectangle corresponds to a single value of R_0 and T, which is located in the center of the rectangle. The cases $T = 500$ K and $T = 700$ K, which correspond to the KMC simulations, were explored with higher resolution. The abbreviations used for the different phases are explained in Figure 5.2 (from [16]).

more limited than in the case of sixfold anisotropy. In particular, at $T = 500$ K no oscillatory regime was found in the continuum model, despite the fact that oscillations are seen in the KMC simulations at this temperature (Figure 5.8b). This is one of the indications of a breakdown of the continuum description at low temperatures which were reported in [19].

5.2.3
Nonlocal Shape Evolution: Two-Dimensional Voids

Formally, the island electromigration problem described in the preceding sections is largely equivalent to the problem of electromigration of cylindrical voids in a thin metallic film. The formation, migration, and shape evolution of such voids plays an important part in the failure of metallic interconnects in integrated electronic circuits [9]. In this context, the size scale of interest is usually in the range of micrometers rather than nanometers, but on the level of the continuum description on which (5.1) is based this difference is immaterial.

A more relevant distinction is illustrated in Figure 5.10. In the case of an island on top of a thick metallic substrate, the disturbance of the electric current distribution in the bulk due to the presence of the island can be neglected, and correspondingly the force F_t in (5.1) can be approximated by the simple constant expression in (5.2). On the other hand, in the presence of an insulating void in a current-carrying film, the current is obviously forced to flow around the void. As a consequence the current distribution and, hence, the distribution of electromigration forces is strongly dependent on the void shape itself, and the shape evolution becomes a non-local moving boundary value problem for the electric potential [18]. It is possible to interpolate between the two cases depicted in Figure 5.10 by considering a conducting void and varying the conductivity ratio between the interior and the exterior regions [14].

Oscillatory shape evolution of two-dimensional voids was first observed numerically by Gungor and Maroudas [21]. They considered edge voids located at the boundary of a two-dimensional conducting strip. In the presence of crystalline

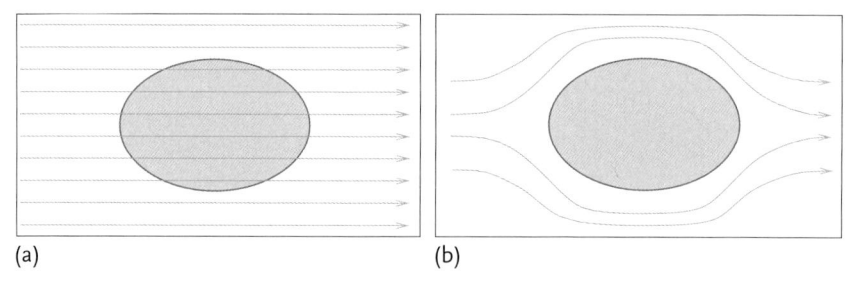

Figure 5.10 Comparison between the electromigration problem for islands (a) and voids (b). Arrows indicate the flow of the electric current. The shape evolution problem in (a) is *local*, whereas in (b) has to solve a *nonlocal* moving boundary value problem.

anisotropy in the mobility of adatoms along the inner void surface, a transition from stationary to oscillatory behavior occurs with increasing electromigration force or void area. Subsequent detailed analysis has shown that this transition has the character of a Hopf bifurcation [22]. The experimental signature of oscillatory void evolution is rapid oscillations in the resistance of the conductor, which have indeed been reported in the literature [23].

5.2.4
Nonlocal Shape Evolution: Vacancy Islands with Terrace Diffusion

The exchange of atoms between the step and the surrounding terraces is another source of nonlocality in the motion of the steps, as it necessitates the solution of a moving boundary value problem for the concentration of adatoms on the terraces [2, 8]. A particular case in this class of problems is the *interior model* for the electromigration of vacancy islands introduced in [12], and studied in detail in [16, 24].

As illustrated in Figure 5.11, one considers a vacancy island (i.e., a surface region which is one atomic height lower than the surrounding terrace) bounded by an ascending step. Atoms can detach from the step and diffuse across the island, but an energy barrier prevents atoms from entering the island from the exterior terrace. This leads to a moving boundary value problem in the bounded interior domain where the adatom concentration $\rho(\mathbf{r}, t)$ satisfies the drift-diffusion equation

$$\frac{\partial \rho}{\partial t} = D\nabla^2 \rho - \frac{D}{k_B T} \mathbf{F} \cdot \nabla \rho \tag{5.7}$$

with appropriate boundary conditions at the step edge (see [8] for a general discussion). If the exchange of atoms with the step edge is rapid, such that thermal equilibrium is maintained at the boundary at all times, a circular stationary solution drifting at constant speed against the force direction can be found [12].

From the perspective of nonlinear dynamics, an intriguing feature of this problem is that the circular solution is linearly stable, although numerical simulation of the fully nonlinear evolution shows that the circle develops an instability under finite perturbations that eventually leads to the pinching off of a small island [24].

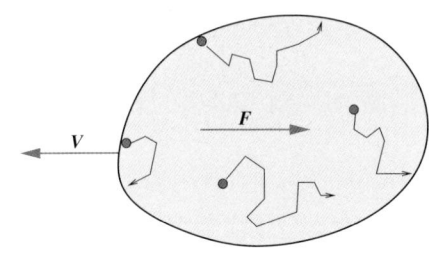

Figure 5.11 Sketch of a vacancy island migrating by internal terrace diffusion. The drift force leads to a net transport of material from the left to the right, which implies island migration in the opposite direction.

The critical perturbation strength needed to trigger the instability decreases as the dimensionless island size, defined in this case by

$$R_0 = \frac{R}{\xi}, \quad \xi = \frac{k_B T}{|F|} \tag{5.8}$$

is increased by increasing either the force or the island size. A similar scenario combining linear stability with nonlinear instability was previously found in the problem of two-dimensional void migration [18, 25] as well as in the dynamics of ionization fronts [26, 27].

The effects of crystalline anisotropy in this problem have not been explored thus far. However, in view of the results described in the preceding subsections, it seems likely that oscillatory and other modes of complex shape evolution may arise in this case as well.

5.3
Step Bunching on Vicinal Surfaces

A vicinal surface is obtained by cutting a crystal at a small angle relative to a high symmetry orientation, such that a staircase of well separated atomic height steps forms. When such an array of steps is set into motion by growing or sublimating the crystal or by applying an electromigration force on the adatoms, a variety of patterns emerges.

Quite generally, the pattern formation process can be understood as a competition between the destabilizing effects of the external forces and thermodynamic forces arising from the step free energy and repulsive step–step interactions, which act to restore the equilibrium state of straight, equidistant steps. The resulting instability scenarios have been studied extensively on the level of linear stability analysis, (e.g. [28]). The two basic modes of instability are illustrated in Figure 5.12. In *step bunching* the individual steps remain straight but the initially homogeneous step train breaks up into regions of high step density (bunches) separated by wide terraces. By contrast, in *step meandering* the individual steps become wavy; often the repulsive interactions between the steps then force the different steps to meander in phase, such that an overall periodic surface corrugation perpendicular to the

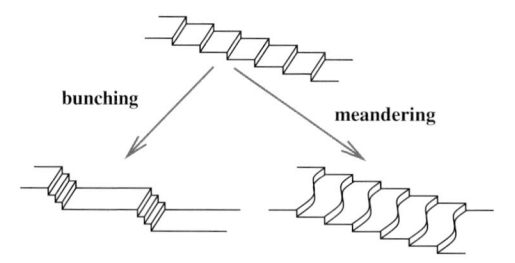

Figure 5.12 Schematic of the two main morphological instabilities of a vicinal surface.

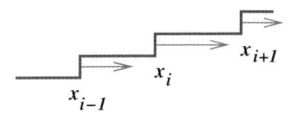

Figure 5.13 Sketch of a one-dimensional step train. Under sublimation, ascending steps move to the right.

direction of vicinality results. In some cases step bunching and step meandering have been observed to coexist [29, 30].

In the following some recent results on the *nonlinear* evolution of step bunches will be summarized, focusing again on instances of complex *temporal* behavior of the step configurations. For a discussion of the nonlinear dynamics of meandering steps we refer to [31].

When step bunching is the dominant instability, the steps, to a first approximation, can be assumed to be straight, and the problem reduces to the one-dimensional motion and interaction of point-like steps. Figure 5.13 illustrates the situation for the case of *sublimation*, where ascending steps move (on average) to the right. The equations of motion for the steps can be obtained from the solution of a one-dimensional moving boundary value problem for the adatom concentration on the terraces. This procedure has been reviewed in detail elsewhere [8]. Here we start the discussion directly from the nonlinear equations of motion, regarded as a physically motivated many-dimensional dynamical system.

5.3.1
Stability of Step Trains

As a first orientation, suppose the velocity $\dot{x}_i$ of the ith step is the sum of contributions f_+ and f_-, which are functions of the length of the leading terrace (in front of the step) and the trailing terrace (behind the step), respectively, such that

$$\frac{dx_i}{dt} = f_+(x_{i+1} - x_i) + f_-(x_i - x_{i-1}) \tag{5.9}$$

for the N steps $i = 1, \ldots, N$, and periodic boundary conditions are employed. Then a uniform step train of equally spaced steps

$$x_i^{(0)} = il + vt \tag{5.10}$$

is always a solution, with l denoting the step spacing and $v = f_+(l) + f_-(l)$ the step speed. A straightforward linear stability analysis of (5.9) reveals that the solution given in (5.10) is stable if

$$\frac{d}{dx}[f_+(x) - f_-(x)]|_{x=l} > 0, \tag{5.11}$$

and step bunching occurs when this condition is violated.

There are obviously different ways in which such an instability can be realized. One possibility is that both contributions on the right hand side of (5.9) are increasing functions of the terrace size, but the contribution from the trailing terrace is larger, that is the step motion is primarily driven from behind. This is the scenario first described by Schwoebel and Shipsey [32, 33], who pointed out that the preferential attachment/detachment of adatoms from/to the lower terrace bordering a step leads to step bunching during sublimation. The mechanism for electromigration-induced step bunching first described by Stoyanov [34] is of a similar nature. We will return to this case in the following sections.

A different scenario was investigated by Kandel and Weeks [35, 36], who considered a class of one-sided models with $f_- \equiv 0$ and a *nonmonotonic* function f_+ of the form

$$f_+(x) = cx(x_0 - x). \tag{5.12}$$

This work was motivated by the physics of impurity-induced step bunching during growth, where steps are slowed down by impurities that accumulate on the terraces [37, 38]. Larger terraces have been exposed to the impurity flux for longer times, which leads to a decrease of the step speed and ultimately to its vanishing when $x = x_0$. The equidistant step train is stable for $l < x_0/2$ and unstable for $l > x_0/2$. Perturbing a single step in an unstable equidistant step train leads to a disturbance wave which travels backwards because of the one-sided nature of the dynamics, leaving behind a frozen configuration of step bunches separated by terraces of size x_0. Varying the initial step spacing, one finds a sequence of spatial bunching patterns which can be periodic, intermittent, or chaotic[9].

5.3.2
Strongly and Weakly Conserved Step Dynamics

An important global characteristic of the step dynamics is the overall sublimation or growth rate of the crystal, which is given by

$$\mathcal{R} = \frac{1}{N} \sum_i \frac{dx_i}{dt}. \tag{5.13}$$

We distinguish between *strongly conserved* step dynamics in which $\mathcal{R} = 0$, and *weakly conserved* dynamics where $\mathcal{R}$ is nonzero but independent of the step configuration[10]. The latter case is realized during growth at relatively low temperature, where desorption of adatoms can be neglected and therefore the growth rate is completely determined by the external deposition flux [41].

A generic model that incorporates the strongly and weakly conserved situation is given by

$$\frac{dx_i}{dt} = \gamma_+ \cdot (x_{i+1} - x_i) + \gamma_- \cdot (x_i - x_{i-1}) + U \cdot (2f_i - f_{i+1} - f_{i-1}) \tag{5.14}$$

with

$$f_i = \frac{l^3}{(x_i - x_{i-1})^3} - \frac{l^3}{(x_{i+1} - x_i)^3}. \tag{5.15}$$

[9] A similar scenario has been found in a model for sand ripple formation in an oscillatory flow [39].

[10] In [40], only the *strongly* conserved case is referred to as "conserved". The reason for our choice of nomenclature will become clear below in Section 5.3.6.

These equations were first written down by Liu and Weeks [42] as a model for electromigration-induced step bunching in the presence of sublimation[11]. In contrast to (5.9), here $\gamma_\pm$ and U are constant coefficients multiplying the terms in parenthesis. Comparison with (5.9) shows that $f_\pm$ are linear functions with slopes $\gamma_\pm$, such that the stability condition reads $\gamma_+ > \gamma_-$. In addition to the linear terms depending on the nearest neighbor step positions, (5.14) contains nonlinear next-nearest neighbor contributions arising from repulsive thermodynamic step–step interactions of entropic and elastic origin [3, 8] that drive the relaxation of the step train to its equidistant or equilibrium shape.

The sublimation rate for the model given in (5.14) is $\mathcal{R} = (\gamma_+ + \gamma_-)l$, hence for strongly conserved dynamics one has to set $\gamma_+ = -\gamma_-$. This case is realized in electromigration-induced step bunching without growth or sublimation [43]. In the following we will focus on the weakly conserved case where $\mathcal{R} > 0$. It is then convenient to normalize the time scale such that $\gamma_+ + \gamma_- = 1$, and to introduce the asymmetry parameter b through [44]

$$\gamma_+ = \frac{1-b}{2}, \quad \gamma_- = \frac{1+b}{2}, \tag{5.16}$$

such that step bunching occurs for $b > 0$. Together (5.14), (5.15), (5.16) define a two parameter family of nonlinear many body problems which have been investigated in detail in [44–46]. In the following two sections some pertinent results of this study will be summarized.

5.3.3
Continuum Limit, Traveling Waves and Scaling Laws

The analysis of the nonlinear dynamics of step bunches is greatly simplified if it is possible to perform a continuum limit of the problem, thus passing from the discrete dynamical system of (5.14) to a partial differential equation [8, 47]. Coarse graining the discrete equations of motion given in (5.14), one arrives first at a "Lagrangian" continuum description for the step positions x_i or the terrace sizes $l_i = x_{i+1} - x_i$. This is accomplished by converting the layer index i into a continuous surface height $h = i h_0$, where h_0 denotes the height of an elementary step [43, 48]. In a second step this is transformed into an "Eulerian" evolution equation for the surface height profile $h(x, t)$ or, equivalently, the step density $m = \partial h/\partial x$, which reads, for the model of (5.14), [44, 45]

$$\frac{\partial h}{\partial t} + \frac{\partial}{\partial x}\left[-\frac{b}{2m} - \frac{1}{6m^3}\frac{\partial m}{\partial x} + \frac{3U}{2m}\frac{\partial^2(m^2)}{\partial x^2}\right] + 1 = 0. \tag{5.17}$$

To unburden the notation, we have normalized vertical and horizontal lengths by setting $h_0 = l = 1$. In the weakly conserved case the evolution law has the form

---

11) We will see below in Section 5.3.6 that the weakly conserved form of (5.14) is, in fact, not really appropriate in the presence of sublimation.

5 Nonlinear Dynamics of Surface Steps

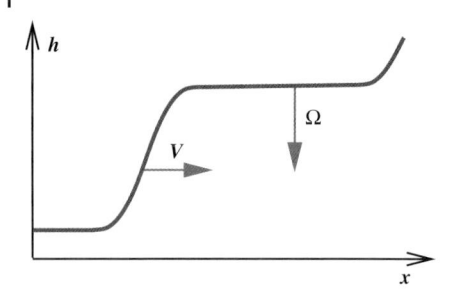

Figure 5.14 Sketch of a moving step bunch.

of a continuity equation, with the corresponding current given by the terms inside the square brackets.

The solution $h(x, t) = x/l - t$ of (5.17) is linearly unstable for $b > 0$. The physically relevant nonlinear solutions take the form of a generalized traveling wave

$$h(x, t) = f(x - Vt) - \Omega t, \tag{5.18}$$

as illustrated in Figure 5.14. The conserved nature of (5.17) implies the sum rule

$$\Omega + V = 1, \tag{5.19}$$

but the individual values of the vertical and horizontal speed are not fixed by the ansatz[12]. An analysis of periodic solutions of the discrete equations of motion shows that, under rather general conditions,

$$V \sim 1/N. \tag{5.20}$$

Since the mean velocity of a single step is unity in the present units, this implies that bunches move more slowly than steps. Similar to cars in a traffic jam, steps join the bunch from behind, move slowly through the bunch, and accelerate into the *outflow region* which separates one bunch from the next[13].

Inserting (5.18) into (5.17) one arrives at a third order nonlinear ODE, which can, to a large extent, be handled analytically [44]. A key result are scaling laws [50] for the shape of stationary bunches. As illustrated in Figure 5.15, the shape can be characterized by the bunch width W and the bunch spacing L, both of which are functions of the number N of steps in the bunch. The global constraint on the average slope of the surface implies that $L \sim N$, but the bunch width typically scales with a sublinear power of N, which implies that bunches become steeper as more steps are added. Related quantities of interest are the minimal terrace size l_{min} in the bunch and the size l_1 of the first terrace in the bunch. On the basis of

---

12) For the relation of this problem to the standard velocity selection problem for traveling waves moving into unstable states, see [49].

13) Note, however, that traffic jams generally move in the direction *opposite* to the traffic flow [52, 53].

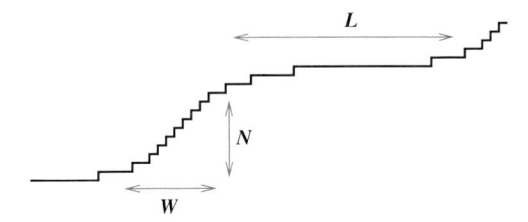

Figure 5.15 Quantities characterizing the shape of a step bunch.

the continuum model of (5.17) one finds that, asymptotically for large N [44],

$$W \approx 4.1(UN/b)^{1/3}, \quad l_{\min} \approx 2.4(U/bN^2)^{1/3}, \quad l_1 \approx (2U/bN)^{1/3}, \quad (5.21)$$

in good agreement with numerical simulations of the discrete model [45]. Note that $W \sim Nl_{\min}$, as one would expect, but $l_1 \gg l_{\min}$. An experimental study of the shapes of electromigration-induced step bunches on Si(111) is consistent with $l_{\min} \sim N^{-2/3}$ [51].

5.3.4
A Dynamic Phase Transition

As with any hydrodynamic description, the validity of the continuum limit passing from (5.14) to (5.17) is restricted to step configurations in which the step density is slowly varying on the scale of the mean step spacing. To check the consistency of this assumption, we consider the outflow region of the bunch, where the spacing between steps leaving the bunch becomes large and hence the nonlinear interaction terms on the right hand side of (5.14) can be neglected. We are thus left with the linear system

$$\frac{dx_i}{dt} = \frac{1-b}{2}(x_{i+1} - x_i) + \frac{1+b}{2}(x_i - x_{i-1}), \quad (5.22)$$

which can be solved by the exponential traveling wave ansatz

$$l_i \equiv x_{i+1} - x_i = A e^{Q(i + \Omega t)}. \quad (5.23)$$

Inserting (5.23) into (5.22) yields the relation

$$b = \frac{\sinh Q - \Omega Q}{\cosh Q - 1} \approx \frac{\sinh Q - Q}{\cosh Q - 1}, \quad (5.24)$$

where we have used the fact that $\Omega \to 1$ for large bunches according to (5.19) and (5.20).

The step spacing is slowly varying when $Q \ll 1$, which according to (5.24) requires $b \ll 1$. More strikingly, (5.24) has no solution when $b > 1$. At $b = 1$ the bunch undergoes a *dynamic phase transition* which is reflected, among other things, in the number of "crossing" steps between bunches. For $b < 1$ this number grows with N as $\ln N$, whereas for $b > 1$ at most a single step can reside between two bunches at one time [46].

The physical origin of this change of behavior can be traced back to the evolution equations (5.22). For a step about to leave the bunch, the leading terrace is much larger than the trailing terrace and $x_{i+1} - x_i \gg x_i - x_{i-1}$, such that the right hand side of (5.22) is dominated by the first term, which is *negative* for $b > 1$. The linear term thus pushes the step back into the bunch, and it can escape only thanks to the repulsive, nonlinear step–step interaction. Since the bunches become steeper with increasing size, the ability of a bunch to eject crossing steps also depends on the number of steps N that it contains.

The result of this interplay between linear and nonlinear effects is the phase diagram in the U–N-plane depicted in Figure 5.16. At moderate values of U it predicts a qualitative change in the behavior of bunches with increasing N. For small bunches the emission of steps ceases completely, such that all steps constituting the bunch move at the speed of the whole bunch and $V = 1$ in our units. Larger bunches emit one step at a time. Figure 5.17 shows the transition between the two regimes in a time-dependent situation. The initial condition consists of 4 small bunches of 16 steps each. These bunches initially merge in a hierarchical fashion without exchanging steps. This behavior is characteristic of *strongly* conserved step dynamics [40, 43], which in our units corresponds to $b \to \infty$. After the last merger, the bunch enters the region in the phase diagram of Figure 5.16 where step emission is possible, and, correspondingly, the overall bunch motion slows down. It can also be seen that the emission of steps is accompanied by a periodic "breathing" of the entire bunch [46].

A rough estimate of experimental parameters indicates that both regimes $b < 1$ and $b > 1$ can be accessed in experiments on electromigration-induced step bunching of the Si(111) surface by varying the temperature [44]. The identification of the

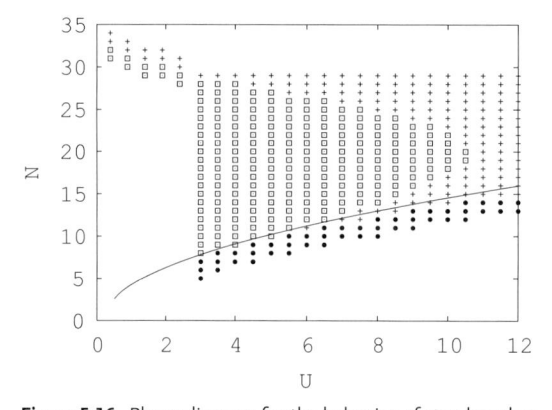

Figure 5.16 Phase diagram for the behavior of step bunches at $b = 11$. The line is the linear stability limit, below which the equidistant step train is stable (full circles). In the linearly unstable regime above this line, bunches either eject no steps (open squares) or they eject one step at a time (crosses).

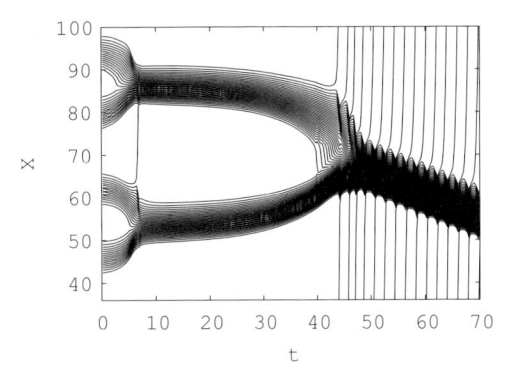

Figure 5.17 Trajectories of 64 step evolving under the weakly conserved dynamics (5.14) with $b = 20$ and $U = 12$. Step positions are shown in a frame moving with the mean step velocity. Initially, the trajectories are horizontal because the entire bunch moves at the mean step speed.

predicted phase transition is, however, not straightforward because real steps can bend [54], thus invalidating the one-dimensional approximation used throughout this section.

5.3.5
Coarsening

The time evolution depicted in Figure 5.17 is an example of *coarsening*, a term that is generally used to describe the unlimited increase of bunch size with time. In many cases coarsening proceeds according to a power law,

$$L \sim N \sim t^n, \tag{5.25}$$

defining the coarsening exponent n. Despite recent progress in the theory of coarsening dynamics for one-dimensional fronts [55], a quantitative analysis of coarsening dynamics based on nonlinear continuum equations such as (5.17) seems still out of reach. Nevertheless, heuristic arguments (to be explained below) in combination with numerical [40, 42] and experimental [56] evidence indicate that the coarsening exponent is

$$n = \frac{1}{2} \tag{5.26}$$

under a wide range of conditions, as far as the weakly conserved system of (5.14) is concerned, including its strongly conserved limit. In particular, the value of n does not seem to be affected by the phase transition at $b = 1$ [57].

The first heuristic argument goes back to Chernov [58], and it is based on the relation given in (5.20) for the bunch velocity. The key assumption is that V is the only velocity scale in the problem, such that the velocity *difference* between two bunches of similar size $\sim N$ is also of order $\Delta V \sim 1/N$. The time required for two bunches to merge is then of order $L/\Delta V \sim N^2$, and (5.26) follows. A weakness of this argument is that it assumes coarsening to proceed by the merging of bunches, which does not need to be true when bunches can exchange steps.

The second argument, due to Liu and Weeks [42], is based on the generally conserved form of the continuum equation for the height profile $h(x, t)$, which reads

$$\frac{\partial h}{\partial t} + \frac{\partial j}{\partial x} = 0 \tag{5.27}$$

in a frame where the constant rate of sublimation has been subtracted. Without further specifying the current j, Liu and Weeks assume the existence of a single lateral length scale $\sim t^n$, such that both the height profile and the current take on scaling forms

$$h(x, t) = t^n H(x/t^n), \quad j(x, t) = J(x/t^n). \tag{5.28}$$

Inserting (5.28) into (5.27) enforces the scaling in (5.26). Similar scaling arguments have been advanced by Pimpinelli and coworkers [50].

Like the argument of Chernov, the ansatz of (5.28) is problematic because the bunch spacing is *not* the only length scale in the system [31, 45]. For example, the bunch width W defines a second time-dependent scale which cannot obviously be ignored. An explicit counterexample where the existence of an additional length scale leads to coarsening exponents that differ from (5.26) was presented in [49].

5.3.6
Nonconserved Dynamics

In the presence of sublimation the rate of volume change given in (5.13) couples to the step configuration, and therefore the weakly conserved form of the discrete (5.14) and continuous (5.17) evolution equations is no longer appropriate [28]. The minimal modification of (5.14) which takes account of this fact reads [59]

$$\frac{dx_i}{dt} = (1 + g f_i) \left[\frac{1+b}{2}(x_i - x_{i-1}) + \frac{1-b}{2}(x_{i+1} - x_i) \right] \\ + U(2 f_i - f_{i+1} - f_{i-1}), \tag{5.29}$$

where the new dimensionless parameter g is proportional to the strength of the repulsive step–step interactions. On the linearized level the introduction of the new term shifts the instability condition, which now reads [48, 59]

$$b > 6g. \tag{5.30}$$

The nonlinear consequences of the new term are quite dramatic. Numerical simulations of (5.29) [60] as well as of a more complicated nonconserved model [40] show that the coarsening of step bunches is *arrested* when the bunches have reached a certain size. Correspondingly, a large initial step bunch evolving under the dynamics of (5.29) breaks up into smaller bunches, as illustrated in Figure 5.18.

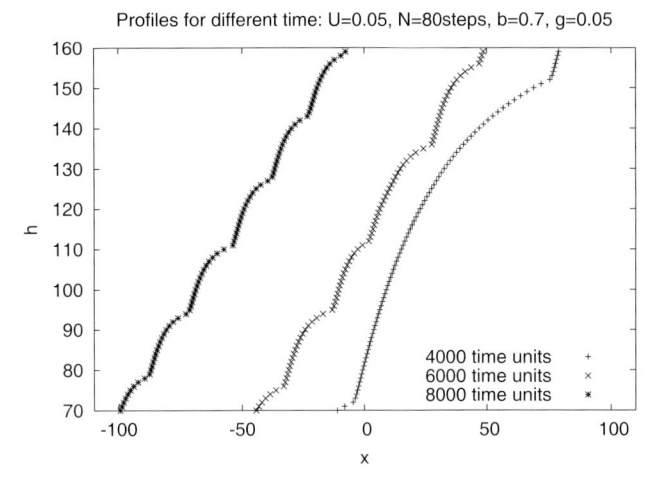

Figure 5.18 Surface profiles generated with the nonconserved discrete model of (5.29) with parameters $b = 0.7$ and $U = g = 0.05$. The initial condition is a single large bunch, which first relaxes into a quasi-stationary configuration and then breaks up into smaller bunches after 4000 time steps. Height profiles at different times have been shifted in the horizontal direction.

The absence of asymptotic coarsening in the nonconserved case is consistent with analyses in which *weakly* nonlinear continuum equations, in the sense of [61], are derived from the discrete step dynamics close to the instability threshold, that is for $1 - 6g/b \ll 1$ [62, 63]. These equations typically display spatiotemporal chaos or structure formation at a fixed length scale, but no coarsening [31]. However, for strongly nonlinear continuum equations similar to (5.17) that are expected to apply when $b \gg g$, such results are so far not available.

5.3.7
Beyond the Quasistatic Approximation

With few exceptions [64–66], most theoretical studies of step dynamics work in the *quasistatic* approximation, which implies that the dynamics of the diffusing adatoms on the terraces separating the steps are assumed to be much faster than the step motion. As a consequence, a step reacts instantaneously to the motion of its neighbors, which mathematically leads to coupled first order equations for the step positions such as those of (5.14).

A simple and conceptually appealing way of explicitly including the time scale of adatom dynamics was recently proposed by Ranguelov and Stoyanov, who derived and studied a coupled system of two sets of evolution equations, one for the terrace widths $l_i = x_{i+1} - x_i$ and one for the suitably parametrized adatom concentration profile on the terraces. Remarkably, in this setting the equidistant step train may undergo an instability into a new dynamic phase characterized by step compression waves [67], even if it would be completely stable in the quasistatic limit. The instability is caused solely by the time delay that is introduced into the interaction between steps by the finite time scale of the adatom dynamics, similar to the instabilities induced in follow the leader models of highway traffic by the finite reaction time of drivers [52, 53]. In the presence of electromigration and sublimation, the non-quasistatic model reproduces the main features of the phase transition described above in Section 5.3.4 [68].

5.4
Conclusions

The fact that the evolution of nanostructures is intrinsically noisy is by now widely appreciated [1]. In contrast, the role of deterministic nonlinear dynamics, in the sense of dynamical systems theory, as a source of complex behavior is largely unexplored in this context. Here I have presented the results of two case studies in which concepts from nonlinear dynamics appear naturally in the analysis of the evolution of surface nanostructures. In both cases surface steps constitute the relevant degrees of freedom which, despite satisfying simple equations of motion, can display a wide range of dynamic phenomena. Many other systems not discussed here fit into the same framework. An example of current interest is the thermal decay of nanoscale mounds, either through the periodic collapse of the top island [69] or through the jerky rotation of a spiral step emanating from a screw dislocation [70]. Hopefully it has become clear that much, perhaps most, of the work in this field remains to be done.

Acknowledgments

This chapter is based on joint work with Frank Hausser, Marian Ivanov, Philipp Kuhn, Vladislav Popkov, Marko Rusanen, and Axel Voigt. I am grateful to Dionisios Margetis, Olivier Pierre-Louis, Alberto Pimpinelli, Paolo Politi, Bogdan Ranguelov, Stoyan Stoyanov, Vesselin Tonchev and John D. Weeks for useful interactions, and to DFG for support within SFB 616 *Energy dissipation at surfaces* and project KR 1123/1-2.

References

1 Williams, E.D. (2004) *MRS Bull.*, p. 621.
2 Burton, W.K., Cabrera, N., and Frank, F.C. (1951) *Phil. Trans. Roy. Soc. A*, **243**, 299.
3 Jeong, H.-C. and Williams, E.D. (1999) *Surf. Sci. Rep.*, **34**, 171.
4 Giesen, M. (2001) *Prog. Surf. Sci.*, **68**, 1.
5 Michely, T. and Krug, J. (2004) *Islands, Mounds and Atoms. Patterns and Processes in Crystal Growth Far from Equilibrium*, Springer, Berlin.
6 Pierre-Louis, O. (2005) *C.R. Phys.*, **6**, 11
7 Evans, J.W., Thiel, P.A., and Bartelt, M.C. (2006) *Surf. Sci. Rep.*, **61**, 1.
8 Krug, J. (2005) *Multiscale Modeling in Epitaxial Growth*, (ed. A. Voigt), Birkhäuser, Basel, p. 69.

9. Tu, K.N. (2003) *J. Appl. Phys.*, **94**, 5451.
10. Bondarchuk, O., Cullen, W.G., Degawa, M., Williams, E.D., Bole, T., and Rous, P.J. (2007) *Phys. Rev. Lett.*, **99**, 206801.
11. Métois, J.-J., Heyraud, J.-C., and Pimpinelli, A. (1999) *Surf. Sci.*, **420**, 250.
12. Pierre-Louis, O. and Einstein, T.L. (2000) *Phys. Rev. B*, **62**, 13697.
13. Ho, P.S. (1970) *J. Appl. Phys.*, **41**, 64.
14. Wang, W., Suo, Z., and Hao, T.-H. (1996) *J. Appl. Phys.*, **79**, 2394.
15. Kuhn, P. and Krug, J. (2005) *Multiscale Modeling in Epitaxial Growth*, (ed. A. Voigt), Birkhäuser, Basel, p. 159.
16. Kuhn, P. (2007) *Nichtlineare Dynamik von Oberflächeninseln unter Elektromigrationseinfluss*, PhD Dissertation, Universität zu Köln.
17. Kuhn, P., Krug, J., Hausser, F., and Voigt, A. (2005) *Phys. Rev. Lett.*, **94**, 166105.
18. Schimschak, M. and Krug, J. (2000) *J. Appl. Phys.*, **87**, 695.
19. Rusanen, M., Kuhn, P., and Krug, J. (2006) *Phys. Rev. B*, **74**, 245423.
20. Mehl, H., Biham, O., Millo, O., and Karimi, M. (2000) *Phys. Rev. B*, **61**, 4975.
21. Gungor, M.R. and Maroudas, D. (2000) *Surf. Sci.*, **461**, L550.
22. Cho, J., Gungor, M.R., and Maroudas, D. (2008) *Surf. Sci.*, **602**, 1227.
23. Cho, J., Gungor, M.R., and Maroudas, D. (2005) *Appl. Phys. Lett.*, **86**, 241905.
24. Hausser, F., Kuhn, P., Krug, J., and Voigt, A. (2007) *Phys. Rev. E*, **75**, 046210.
25. Schimschak, M. and Krug, J. (1998) *Phys. Rev. Lett.*, **80**, 1674.
26. Meulenbroek, B., Ebert, U., and Schäfer, L. (2005) *Phys. Rev. Lett.*, **95**, 195004.
27. Ebert, U., Meulenbroek, B. and Schäfer, L. (2007) *SIAM J. Appl. Math.*, **68**, 292.
28. Pierre-Louis, O. (2003) *Surf. Sci.*, **529**, 114.
29. Néel, N., Maroutian, T., Douillard, L., and Ernst, H.-J. (2003) *Phys. Rev. Lett.*, **91**, 226103.
30. Yu, Y.-M. and Liu, B.-G. (2006) *Phys. Rev. B*, **73**, 035416.
31. Krug, J. (2005) *Collective Dynamics of Nonlinear and Disordered Systems*, (eds G. Radons, W. Just, P. Häussler), Springer, Berlin, p. 5.
32. Schwoebel, R.L. and Shipsey, E.J. (1966) *J. Appl. Phys.*, **37**, 3682.
33. Schwoebel, R.L. (1969) *J. Appl. Phys.*, **40**, 614.
34. Stoyanov, S. (1991) *Jpn. J. Appl. Phys.*, **30**, 1.
35. Kandel, D. and Weeks, J.D. (1992) *Phys. Rev. Lett.*, **69**, 3758.
36. Kandel, D. and Weeks, J.D. (1993) *Physica D*, **66**, 78.
37. van der Eerden, J.P. and Müller-Krumbhaar, H. (1989) *Phys. Scripta*, **40**, 337.
38. Vollmer, J., Hegedüs, J., Grosse, F., and Krug, J. (2008) *New J. Phys.*, **10**, 053017.
39. Krug, J. (2001) *Adv. Compl. Syst.*, **4**, 353.
40. Sato, M. and Uwaha, M. (1999) *Surf. Sci.*, **442**, 318.
41. Krug, J. (1997) *Adv. Phys.*, **46**, 139.
42. Liu, D.-J. and Weeks, J.D. (1998) *Phys. Rev. B*, **57**, 14891.
43. Chang, J., Pierre-Louis, O., and Misbah, C. (2006) *Phys. Rev. Lett.*, **96**, 195901.
44. Popkov, V. and Krug, J. (2005) *Europhys. Lett.*, **72**, 1025.
45. Krug, J., Tonchev, V., Stoyanov, S., and Pimpinelli, A. (2005) *Phys. Rev. B*, **71**, 045412.
46. Popkov, V. and Krug, J. (2006) *Phys. Rev. B*, **73**, 235430.
47. Krug, J. (1997) *Dynamics of Fluctuating Interfaces and Related Phenomena*, (eds Kim, D., Park, H., Kahng, B.), World Scientific, Singapore, p. 95.
48. Fok, P.-W., Rosales, R.R., and Margetis, D. (2007) *Phys. Rev. B*, **76**, 033408.
49. Slanina, F., Krug, J., and Kotrla, M. (2005) *Phys. Rev. E*, **71**, 041605.
50. Pimpinelli, A., Tonchev, V., Videcoq, A., and Vladimirova, M. (2002) *Phys. Rev. Lett.*, **88**, 206103.
51. Fujita, K., Ichikawa, M., and Stoyanov, S. (1999) *Phys. Rev. B*, **60**, 16006.
52. Chowdhury, D., Santen, L., and Schadschneider, A. (2000) *Phys. Rep.*, **329**, 199.
53. Helbing, D. (2001) *Rev. Mod. Phys.*, **73**, 1067.
54. Thürmer, K., Liu, D.-J., Williams, E.D., and Weeks, J.D. (1999) *Phys. Rev. Lett.*, **83**, 5531.
55. Politi, P. and Misbah, C. (2006) *Phys. Rev. E*, **73**, 036133.
56. Yang, Y.-N., Fu, E.S., and Williams, E.D. (1996) *Surf. Sci.*, **356**, 101.

57 Ranguelov, B. and Tonchev, V. (unpublished).
58 Chernov, A. (1961) *Sov. Phys. Usp.*, **4**, 116.
59 Ivanov, M. (2007) *Dynamik von Stufen auf vizinalen Kristalloberflächen*. Diploma thesis, Universität zu Köln.
60 Ivanov, M., Popkov, V., and Krug, J. (unpublished).
61 Pierre-Louis, O. (2005) *Europhys. Lett.*, **72**, 894.
62 Sato, M. and Uwaha, M. (1995) *Europhys. Lett.*, **32**, 639.
63 Misbah, C. and Pierre-Louis, O. (1996) *Phys. Rev. E*, **53**, R4318.
64 Ghez, R. and Iyer, S.S. (1988) *IBM J. Res. Dev.*, **32**, 804.
65 Ghez, R., Cohen, H.G., and Keller, J.B. (1993) *J. Appl. Phys.*, **73**, 3685.
66 Keller, J.B., Cohen, H.G., and Merchant, G.J. (1993) *J. Appl. Phys.*, **73**, 3694.
67 Ranguelov, B. and Stoyanov, S. (2007) *Phys. Rev. B*, **76**, 035443.
68 Ranguelov, B. and Stoyanov, S. (2008) *Phys. Rev. B*, **77**, 205406.
69 Margetis, D., Fok, P.-W., Aziz, M.J., and Stone, H.A. (2006) *Phys. Rev. Lett.*, **97**, 096102.
70 Ranganathan, M., Dougherty, D.B., Cullen, W.G., Zhao, T., Weeks, J.D., and Williams, E.D. (2005) *Phys. Rev. Lett.*, **95**, 225505.

6
Casimir Forces and Geometry in Nanosystems
Thorsten Emig

Casimir interactions, predicted in 1948 [14, 15] between atoms and macroscopic surfaces, and probed in a series of high precision experiments over the past decade [8, 46, 50], are particularly important at micrometer to nanometer length scales due to their strong power-law increase at short separations between particles. Therefore, in constructing and operating devices at these length scales, it is important to have an accurate understanding of the material, shape and geometry dependence of these forces. In particular, the observation of Casimir forces in devices on submicron scales has currently generated a great deal of interest regarding the exploration of the role of these forces for the development and optimization of micro- and nanoelectromechanical systems [9, 16, 17]. These systems can serve as on-chip fully integrated sensors and actuators with a growing number of applications. It was pointed out that Casimir forces can make an important contribution to the principal cause of malfunctions of these devices in form of stiction that results in permanent adhesion of nearby surface elements [10]. This initiated interest in repulsive Casimir forces by modifying material properties as well as the geometry of the interacting components [11, 43, 52].

The study of fluctuation induced forces has a long history. When these forces result from fluctuations of charges and currents inside particles or macroscopic objects, they are usually summarized under the general term, van der Waals forces [55]. This interaction appears at the atomic scale in the guise of Keesom, Debye, London, and Casimir–Polder forces. An important property of all these interactions is their non-additivity. The total interaction of macroscopic objects is generally not given by the sum of the interactions between all pairs of particles forming the objects. This inherent many-body character of the force leads to interesting and often unexpected behaviors, but makes studying these forces a difficult problem. Commonly used approximations as pairwise additivity assumptions become unreliable for systems of condensed atoms. The collective interaction of condensed macroscopic systems is better formulated in terms of their dielectric properties. Such a formulation was established by Lifshitz for two parallel and planar, infinitely extended dielectric surfaces [47], extending Casimir's original work for perfect metals. In practice, one encounters objects of finite size with curved surfaces and/or

Nonlinear Dynamics of Nanosystems. Edited by Günter Radons, Benno Rumpf, and Heinz Georg Schuster
Copyright © 2010 WILEY-VCH Verlag GmbH & Co. KGaA, Weinheim
ISBN: 978-3-527-40791-0

edges, like structured surfaces, spheres or cylinders. Also, in small-scale devices, often more than two objects are at close separation and one would like to know the collective effects resulting from non-additivity. In this chapter, we shall encounter a selection of examples for the interesting behavior of fluctuation forces that result from shape and material properties that have been obtained from a recently developed method that makes it possible to compute van der Waals–Casimir interactions for arbitrary compact objects based on their scattering properties for electromagnetic waves [30, 32, 33, 42].

6.1
Casimir Effect

The Casimir effect is the attraction between two uncharged, parallel and perfectly conducting plates [14]. For this simple geometry, the interaction can be obtained directly from the plate induced *change* of the energies of the quantum mechanical harmonics oscillators associated with the normal modes of the electromagnetic field. The derivation given here closely follows the one originally presented by Casimir. Consider two parallel and planar surfaces of size $L \times L$ and separation d. We assume that the system is at zero temperature so that the interaction is given by the ground state energies of harmonic oscillators. When we are interested in the pressure (force per plate area L^2) between large plates with $L \gg d$, we can ignore edge effects $\sim L$ and allow for a continuum of wave vectors parallel to the plates. For a perfect conductor, the tangential electric field has to vanish at the surface and the normal modes correspond to the allowed wave vectors $\mathbf{k} = (\mathbf{k}_\|, \pi n/d)$ where $\mathbf{k}_\|$ is the two-dimensional wave vector parallel to the plates. The linear dispersion of photons yields the eigenfrequencies $\omega_{n\mathbf{k}_\|} = c\sqrt{k_\|^2 + (\pi n/d)^2}$ so that the ground state energy becomes

$$E = \frac{\hbar}{2} \sum_{n=0}^{\infty}{'} \left(\frac{L}{2\pi}\right)^2 \int d^2 k_\| 2\omega_{n\mathbf{k}_\|} \, , \tag{6.1}$$

where we have included a factor of 2 since for each mode with $n \neq 0$, two polarizations exist. The primed summation assigns a weight of $1/2$ to the term for $n = 0$. Obviously, the expression of (6.1) is divergent. This is a consequence of the assumption that the surfaces behave as a perfect conductor for arbitrarily high frequencies. In practice, as pointed out by Casimir, for very high frequencies (X-rays, e.g.) the plates are hardly an obstacle for electromagnetic waves and therefore the ground state energy of these modes will not be changed by the presence of the plates. We implement this observation by introducing a cut-off function $\chi(z)$ that is regular at $z = 0$ with $\chi(0) = 1$ and vanishes, along with all its derivatives, for $z \to \infty$ sufficiently fast. After a change of variables, $\omega = c\sqrt{k_\|^2 + (\pi n/d)^2}$

6.1 Casimir Effect

with $c^2 k_\| dk_\| = \omega d\omega$, we obtain the finite expression for the energy

$$E = \frac{\hbar L^2}{2\pi c^2} \sum_{n=0}^{\infty}{}' f(n) \quad \text{with} \quad f(n) = \int_{\pi nc/d}^{\infty} \omega^2 \chi(\omega/\omega_c) \, d\omega, \tag{6.2}$$

where ω_c is a cut-off frequency. As mentioned before, we are interested in the change of the energy due to the presence of the plates. Let us imagine that we increase the separation d between the plates to infinity, thus creating empty space. When we subtract the energy of the latter configuration from the total energy of (6.2), we obtain the change in energy that is the relevant interaction potential between the plates. When the separation d tends to infinity, the sum in (6.2) can be replaced by an integral, yielding after the substitution $\Omega = \pi nc/d$ the energy

$$E_\infty = \frac{\hbar}{2\pi^2 c^3} L^2 d \int_0^{\infty} d\Omega \, \tilde{f}(\Omega) \quad \text{with} \quad \tilde{f}(\Omega) = \int_{\Omega}^{\infty} \omega^2 \chi(\omega/\omega_c) \, d\omega. \tag{6.3}$$

As expected, the energy E_∞ is proportional to the volume $L^2 d$ of empty space and to a cut-off dependent factor that is given by the integrals of (6.3). This factor describes the self-energy of the bounding surfaces. It is infinite for perfect conductors which correspond to $\omega_c \to \infty$. For a non-ideal conductor or any other material, this factor is finite but depends on material properties like the plasma wavelength for a metal. Now, we compute the change in energy when the plates are moved in from infinity,

$$\Delta E = E - E_\infty = \frac{\hbar L^2}{2\pi c^2} \left[\sum_{n=0}^{\infty}{}' f(n) - \int_0^{\infty} dn \, f(n) \right]. \tag{6.4}$$

The difference between the sum and the integral is given by the Euler–Maclaurin formula $\sum_{n=0}^{\infty}{}' f(n) - \int_0^{\infty} dn \, f(n) = -\frac{1}{12} f'(0) + \frac{1}{6!} f'''(0) + \mathcal{O}(f^v(0))$. This series of derivatives of odd order can be truncated in the limit of perfect conductors $\omega_c \to \infty$ since $f'(0) = 0$, $f'''(0) = -2(\pi c/d)^3$ and $f^{(v)}(0) \sim (c/d)^3 (c/d\omega_c)^{v-3}$. The Casimir potential hence becomes

$$\Delta E = -\frac{\pi^2}{720} \frac{\hbar c}{d^3} L^2 + \mathcal{O}(\omega_c^{-2}), \tag{6.5}$$

and the pressure for perfect metal plates is

$$\frac{F}{L^2} = -\frac{\pi^2}{240} \frac{\hbar c}{d^4}. \tag{6.6}$$

The interesting fact is that the amplitude of the interaction is *universal*, that is, independent of the cut-off that can be viewed as a simplified description of a real metal. This implies that for any pair of surfaces with metallic response in the limit of small frequencies $\omega \to 0$, the interaction at asymptotically large separations is described by the potential of (6.5). At a separation of $d = 100$ nm, (6.6) yields a pressure of $1.28 \cdot 10^{-4}$ atm or 13.00 Pa.

6.2
Dependence on Shape and Geometry

Casimir interactions result from a modification of the fluctuation spectrum of the electromagnetic field due to boundaries or coupling to matter. This suggest that these interactions strongly depend on the shape of the interacting objects and geometry, that is, relative position and orientation. The most commonly encountered geometry is a sphere-plate setup that was used in the first high-precision tests of the Casimir effect [46, 50]. Since then, this geometry has been successfully used in most of the experimental studies of Casimir forces between metallic surfaces [17, 19–21, 25, 27, 51, 60]. In order to keep the deviations from two parallel plates sufficiently small, a sphere with a radius much larger than the surface distance has been used. The effect of curvature has been accounted for by the "proximity force approximation" (PFA) [55]. This scheme is assumed to describe the interaction for sufficiently small ratios of radius of curvature to distance. However, this an uncontrolled assumption since PFA becomes exact only for infinitesimal separations, and corrections to PFA are generally unknown.

At the other extreme, the interaction between a planar surface and an object that is either very small or at an asymptotically large distance is governed by the Casimir–Polder potential that was derived for the case of an atom and a perfectly conducting plane [15]. There have been attempts to go beyond the two extreme limits of asymptotically large and small separations by measuring the Casimir force between a sphere and a plane over a larger range of ratios of sphere radius to distance [44].

Until very recently, no practical tools were available to compute the electromagnetic Casimir interaction between objects of arbitrary shape at all distances, including the important sphere-plate geometry. Progress in understanding the geometry dependence of fluctuation forces was hampered by the lack of methods that are applicable over a wide range of separations. Unlike the case of parallel plates, the eigenvalues of the Helmholtz equation in more complicated geometries are generally unknown, and a summation over normal modes, as in the original Casimir calculation of Section 6.1, is not practical. Conceptually, the effects of geometry and shape are difficult to study due to the non-additivity of fluctuation forces.

For decades, there has been considerable interest in the theory of Casimir forces between objects with curved surfaces. Two types of approaches have been pursued. Attempts to compute the force explicitly in particular geometries and efforts to develop a general framework which yields the interaction in terms of characteristics of the objects, such as polarizability or curvature. Within the second type of approach, Balian and Duplantier studied the electromagnetic Casimir interaction between compact and perfect metallic shapes in terms of a multiple reflection expansion and also derived explicit results to leading order at asymptotically large separations [4, 5]. For parallel and partially transmitting plates, a connection to scattering theory has been established which yields the Casimir interaction of the plates as a determinant of a diagonal matrix of reflection amplitudes [40]. For nonplanar, deformed plates, a general representation of the Casimir energy as a functional determinant of a matrix that describes reflections at the surfaces and free propagation between them has been developed in [31]. Later, an equivalent representation has been applied to perturbative computations in the case of rough and corrugated plates with finite conductivity [45, 53, 54].

Functional determinant formulas have been used also for open geometries that do not fall into the class of parallel plates with deformations. For the Casimir interaction between planar plates and cylinders, a partial wave expansion of the functional determinant has been employed [6, 36]. More recently, a new method based on a multipole expansion of fluctuating currents inside the objects has been developed [30, 32, 33]. This method allows for accurate and efficient calculations of Casimir forces and torques between compact objects of arbitrary shape and material composition in terms of the scattering matrices of the *individual* objects. A similar scattering approach has been developed in [42].

In this section, three examples for the strong geometry dependence of Casimir forces will be made explicit. First, an overview on forces between *deformed and structured surfaces* will be given. The interactions are obtained from both a perturbative and numerical evaluation of a functional determinant representation of Casimir interactions between ideal metal surfaces. As a second example, we describe the effects that occur in the interaction of *one-dimensional structures as cylinders and wires* and related non-additivity phenomena for more than two objects. Finally, as an example for the interaction between compact objects, the Casimir force between *metallic spheres* is presented for the full range of separations, covering the crossover from the asymptotic Casimir–Polder law to proximity approximations. The analysis of the last two examples is based on a scattering approach.

6.2.1
Deformed Surfaces

The dependence of the Casimir force on shape and material properties offers the opportunity to manipulate this interaction in a controlled way, for example, by imprinting patterns on the interacting surfaces. It has been shown that a promising route to this end is via modifications of the parallel plate geometry [34, 35, 38]. The corrections due to deformations, such as sinusoidal corrugations, of the metal plates can be significant. In searching for non-trivial shape dependences, Roy and Mohideen [61] measured the force between a sphere with large radius and a sinusoidally corrugated plate with amplitude $a \approx 60$ nm and wavelength $\lambda \approx 1.1$ µm. Over the range of separations $H \approx 0.1$–0.9 µm, the observed force showed clear deviations from the dependence expected on the basis of decomposing the Casimir force to a sum of pairwise contributions (in effect, an average over the variations in separations). Motivated by this experiment, the effect of corrugations on the Casimir force between surfaces has been studied without using pairwise additivity approximations. The analysis is based on a path integral quantization of the fluctuating field with appropriate boundary conditions which leads to a functional determinant representation of the Casimir energy [38] which can be evaluated perturbatively for a small deformation amplitude [34, 35]

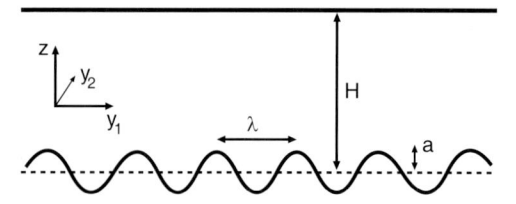

Figure 6.1 Configuration of a flat and a corrugated plate at mean separation H.

or numerically for general amplitudes [12, 13, 28]. In recent experiments [18] the Casimir force between a gold sphere and a silicon surface with an array of nanoscale, rectangular corrugations has been measured and the results were found to be consistent with the theory based on a numerical evaluation of functional determinants for ideal metals (see below). Qualitative agreement can be expected only when material properties are taken into account in addition to shape.

While we will be interested in the interaction between flat and corrugated surfaces as depicted in Figures 6.1 and 6.4, we must first consider the two surfaces with arbitrary *uniaxial* deformations without overhangs so that their profiles can be described by height functions $h_\alpha(y_1)$ ($\alpha = 1, 2$ for the two surfaces), with $\int dy_1 h_\alpha(y_1) = 0$. It is further assumed that the surfaces are perfectly conducting and infinitely extended along the plane spanned by $\mathbf{y}_\parallel = (y_1, y_2)$. As explained before, for two planar plates, the Casimir energy at zero temperature corresponds to the difference of the ground state energies of the quantized electromagnetic field for plates at distance H and at $H \to \infty$, respectively. To obtain this energy, we employ a path integral quantization method. For general, non-uniaxial deformations or objects of more general shape, it is necessary to consider the action for the electromagnetic field since the two polarizations (TM for transverse magnetic waves and TE for transverse electric waves) are coupled. However, for the uniaxial deformations under consideration here, we can develop a simpler quantization scheme, by a similar reasoning also used in the context of waveguides with constant cross-sectional shape [34]. In this case, the two polarizations are *independent* modes which do not couple under scattering between the surfaces. For TM waves, all field components are then fully specified by a scalar field corresponding to the electric field along the invariant direction,

$$\Phi_{\text{TM}}(t, y_1, y_2, z) = E_2(t, y_1, y_2, z), \tag{6.7}$$

with the Dirichlet boundary condition $\Phi_{\text{TM}}|_{S_\alpha} = 0$ on each surface S_α. The TE waves are analogously described by the scalar field

$$\Phi_{\text{TE}}(t, y_1, y_2, z) = B_2(t, y_1, y_2, z), \tag{6.8}$$

with the Neumann boundary condition $\partial_n \Phi_{\text{TE}}|_{S_\alpha} = 0$, where ∂_n is the normal derivative of the surface S_α pointing into the space between the two plates. After a Wick rotation to the imaginary time variable $X^0 = ict$, both fields Φ_{TM} and Φ_{TE}

can be quantized using the Euclidean action

$$S\{\Phi\} = \frac{1}{2} \int d^4 X (\nabla \Phi)^2 . \tag{6.9}$$

In order to obtain the change in the ground state energy that is associated with the presence of the plates, we now consider the partition functions $\mathcal{Z}_D$ and $\mathcal{Z}_N$ for the scalar field Euclidean action both with Dirichlet (D) and Neumann (N) boundary conditions at the surfaces. We implement the boundary conditions on the surfaces S_α using delta functions, which leads to the partition functions

$$\mathcal{Z}_D = \frac{1}{\mathcal{Z}_0} \int \mathcal{D}\Phi \prod_{\alpha=1}^{2} \prod_{X_\alpha} \delta[\Phi(X_\alpha)] \exp(-S\{\Phi\}/\hbar) , \tag{6.10}$$

$$\mathcal{Z}_N = \frac{1}{\mathcal{Z}_0} \int \mathcal{D}\Phi \prod_{\alpha=1}^{2} \prod_{X_\alpha} \delta[\partial_n \Phi(X_\alpha)] \exp(-S\{\Phi\}/\hbar) , \tag{6.11}$$

where $\mathcal{Z}_0$ is the partition function of the space without plates. Here, $X_1(y) = [y, h_1(y_1)]$ and $X_2(y) = [y, H + h_2(y_1)]$, where $y = (y_0, y_1, y_2) = (y_0, y_\parallel)$, and $y_0 = ict$, is a parametrization of the plates in 4-D Euclidean space. The Casimir energy $\mathcal{E}$ per unit area (at zero temperature) that results from moving the plates in from infinity is obtained from the partition function as

$$\mathcal{E}(H) = E(H) - \lim_{H \to \infty} E(H) , \tag{6.12}$$

with

$$E(H) = -\frac{\hbar c}{AL} [\ln \mathcal{Z}_D + \ln \mathcal{Z}_N] , \tag{6.13}$$

where A is the surface area of the plates and the limit where the overall Euclidean length in time direction, L, tends to infinity is implicitly assumed. The partition functions can be expressed as functional determinants, using auxiliary fields (for details see [35]),

$$\ln \mathcal{Z}_D = -\frac{1}{2} \ln \det \mathbb{M}_D , \quad \ln \mathcal{Z}_N = -\frac{1}{2} \ln \det \mathbb{M}_N . \tag{6.14}$$

The kernels $\mathbb{M}_D$ and $\mathbb{M}_N$ are given by

$$[\mathbb{M}_D]_{\alpha\beta}(y, y') = [g_\alpha(y_1)]^{1/4} G[X_\alpha(y) - X_\beta(y')][g_\beta(y_1')]^{1/4} , \tag{6.15}$$

$$[\mathbb{M}_N]_{\alpha\beta}(y, y') = [g_\alpha(y_1)]^{1/4} \partial_{n_\alpha(y_1)} \partial_{n_\beta(y_1')} G[X_\alpha(y) - X_\beta(y')]$$
$$\times [g_\beta(y_1')]^{1/4} , \tag{6.16}$$

where $g_\alpha(y_1) = 1 + [h'_\alpha(y_1)]^2$ is the determinant of the induced metric, and $n_\alpha(y_1) = (-1)^\alpha g_\alpha^{-1/2}(y_1)[h'_\alpha(y_1), 0, -1]$ is the normal vector to the surface S_α, while

$$G(x) = \frac{1}{4\pi^2} \frac{1}{x^2} \tag{6.17}$$

is the *free* Euclidean space Green's function with $\mathbf{x} = (\mathbf{y}, z)$. Equations 6.12 to (6.17) constitute the functional determinant representation of the Casimir interaction. This representation is exact. To proceed, the determinant has to be evaluated either by perturbation theory in the deformation amplitude or numerically for a specific shape of the surface profiles. First, we present the perturbative approach.

For both boundary conditions (X = D, N), we divide by the partition function $\mathcal{Z}_{X,\infty}$ for $H \to \infty$ and expand $\ln(\mathcal{Z}_X/\mathcal{Z}_{X,\infty})$ in a series $\ln(\mathcal{Z}_X/\mathcal{Z}_{X,\infty})|_0 + \ln(\mathcal{Z}_X/\mathcal{Z}_{X,\infty})|_1 + \ln(\mathcal{Z}_X/\mathcal{Z}_{X,\infty})|_2 + \cdots$, where the subscript indicates the corresponding order in h_α. The lowest order result is

$$\ln(\mathcal{Z}_X/\mathcal{Z}_{X,\infty})|_0 = \frac{AL}{H^3}\frac{\pi^2}{1440} \tag{6.18}$$

for both types of modes, corresponding to two flat plates, as in (6.5). The first order result $\ln(\mathcal{Z}_X/\mathcal{Z}_{X,\infty})|_1$ vanishes since we assume, without loss of generality, that the mean deformations are zero, $\int dy_1 h_\alpha(y_1) = 0$. The second order contribution is given by

$$\ln(\mathcal{Z}_X/\mathcal{Z}_{X,\infty})|_2 = \frac{\pi^2}{240}\frac{1}{H^5}\int d^3y \left\{[h_1(y_1)]^2 + [h_2(y_1)]^2\right\}$$
$$- \frac{1}{2}\int d^3y \int d^3y' K_X(\mathbf{y}-\mathbf{y}')\left\{\frac{1}{2}[h_1(y_1) - h_1(y_1')]^2 + \frac{1}{2}[h_2(y_1) - h_2(y_1')]^2\right\}$$
$$- \frac{1}{2}\int d^3y \int d^3y' Q_X(\mathbf{y}-\mathbf{y}')[h_1(y_1)h_2(y_1') + h_2(y_1)h_1(y_1')] \,. \tag{6.19}$$

The terms in the first row are local contributions which are identical for TM and TE modes. They also follow from a pairwise summation approximation (PWS) that sums a "renormalized" Casimir–Polder potential over the volumes of the interacting bodies [35]. The remaining terms are nonlocal and cannot be obtained in approximative schemes. For Dirichlet boundary conditions, the kernels depend only on $|\mathbf{y}-\mathbf{y}'|$ and are given by

$$K_D(y) = -\frac{1}{2\pi^4 y^8} + \frac{\pi^2}{128}\frac{1}{H^6 y^2}\frac{\cosh^2(s)}{\sinh^6(s)}, \tag{6.20}$$

$$Q_D(y) = \frac{\pi^2}{128}\frac{1}{H^6 y^2}\frac{\sinh^2(s)}{\cosh^6(s)}, \tag{6.21}$$

where $s = \pi y/(2H)$. The kernels for Neumann boundary conditions assume a more complicated form since the normal derivative breaks the equivalence of space and time directions. Hence, they depend separately on $|y_0 - y_0'|$ and $|\mathbf{y}_\| - \mathbf{y}_\|'|$. Their explicit form can be found in [35]. The results obtained thus far apply to general uniaxial deformations of both surfaces.

Now, we apply these results to the important case of corrugated plates. We begin with the geometry depicted in Figure 6.1 which is parametrized by

$$h_1(y_1) = a\cos(2\pi y_1/\lambda), \quad \text{and} \quad h_2(y_1) = 0\,. \tag{6.22}$$

For this profile, the computation of the partition function to second order in a reduces to the Fourier transforming of the kernels with respect to y_1. The corresponding expression for $\mathcal{E}$ in (6.12) can be written as

$$\mathcal{E} = \mathcal{E}_0 + \mathcal{E}_{cf}, \tag{6.23}$$

where $\mathcal{E}_0$ is the energy per unit area of two flat plates [see (6.5)] and

$$\mathcal{E}_{cf} = -\frac{\hbar c a^2}{H^5}\left[G_{TM}\left(\frac{H}{\lambda}\right) + G_{TE}\left(\frac{H}{\lambda}\right)\right] + \mathcal{O}(a^3), \tag{6.24}$$

where the index cf of $\mathcal{E}_{cf}$ stands for corrugated-flat geometry. The functions that describe the λ-dependence in this expression can be computed exactly [35]. They can be expressed in terms of the polylogarithm function $\mathrm{Li}_n(z) = \sum_{\nu=1}^\infty z^\nu/\nu^n$, leading to

$$G_{TM}(x) = \frac{\pi^3 x}{480} - \frac{\pi^2 x^4}{30}\ln(1-u) + \frac{\pi}{1920x}\mathrm{Li}_2(1-u) + \frac{\pi x^3}{24}\mathrm{Li}_2(u)$$
$$+ \frac{x^2}{24}\mathrm{Li}_3(u) + \frac{x}{32\pi}\mathrm{Li}_4(u) + \frac{1}{64\pi^2}\mathrm{Li}_5(u)$$
$$+ \frac{1}{256\pi^3 x}\left(\mathrm{Li}_6(u) - \frac{\pi^6}{945}\right) \tag{6.25}$$

$$G_{TE}(x) = \frac{\pi^3 x}{1440} - \frac{\pi^2 x^4}{30}\ln(1-u) + \frac{\pi}{1920x}\mathrm{Li}_2(1-u)$$
$$- \frac{\pi x}{48}(1+2x^2)\mathrm{Li}_2(u) + \left(\frac{x^2}{48} - \frac{1}{64}\right)\mathrm{Li}_3(u) + + \frac{5x}{64\pi}\mathrm{Li}_4(u)$$
$$+ \frac{7}{128\pi^2}\mathrm{Li}_5(u) + \frac{1}{256\pi^3 x}\left(\frac{7}{2}\mathrm{Li}_6(u) - \pi^2\mathrm{Li}_4(u) + \frac{\pi^6}{135}\right) \tag{6.26}$$

with $u \equiv \exp(-4\pi x)$. Figure 6.2 separately displays the contributions from G_{TM} and G_{TE} to the corrugation induced correction $\mathcal{E}_{cf}$ to the Casimir energy. While $G_{TM}(H/\lambda)$ is a monotonically increasing function of H/λ, $G_{TE}(H/\lambda)$ displays a minimum for $H/\lambda \approx 0.3$.

Examining the limiting behaviors of (6.24) is instructive. In the limit $\lambda \gg H$, the functions G_{TM} and G_{TE} approach constant values, and the total Casimir energy takes the λ-independent form

$$\mathcal{E} = -\frac{\hbar c}{H^3}\frac{\pi^2}{720}\left(1 + 3\frac{a^2}{H^2}\right) + \mathcal{O}(a^3). \tag{6.27}$$

Note that *only* in this case, both wave types provide the same contribution to the total energy and the result agrees with the pairwise summation approximation (see Figure 6.2). In the opposite limit of $\lambda \ll H$, both G_{TM} and G_{TE} grow linearly in H/λ. Therefore, in this limit the correction to the Casimir energy decays according to a *slower* power law in H, as

$$\mathcal{E} = -\frac{\hbar c}{H^3}\frac{\pi^2}{720}\left(1 + 2\pi\frac{a^2}{\lambda H}\right) + \mathcal{O}(a^3), \tag{6.28}$$

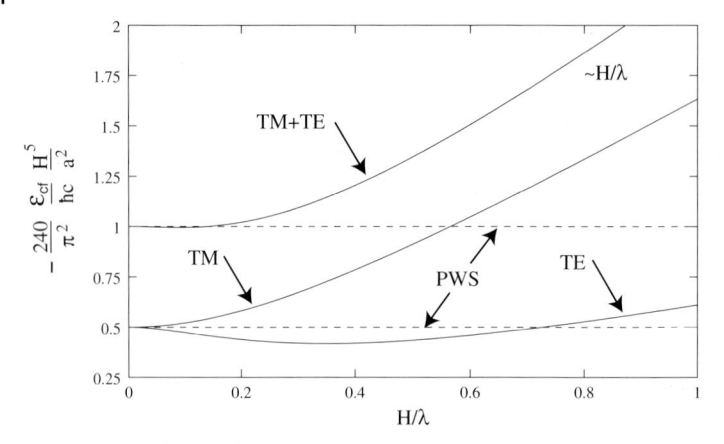

Figure 6.2 Rescaled correction $\mathcal{E}_{cf}$ to the Casimir energy due to the corrugation as given by (6.24) (upper curve). The lower curves show the separate contributions from TM and TE modes. The rescaling of $\mathcal{E}_{cf}$ is chosen such that the corresponding prediction of the pairwise summation (PWS) approximation [corresponding to the local terms of (6.19)] is a constant (dashed lines).

with an amplitude proportional to $1/\lambda$. Note that this behavior is completely missed by the pairwise summation approach which always yields a λ independent Casimir energy in the presence of modulations on one plate [35]. As we will discuss below, in the context of the numerical approach, the apparent divergence for $\lambda \to 0$ in (6.28) is an artifact of the perturbative expansion which assumes that the amplitude a is the smallest length scale.

Next we turn to a numerical approach for computing the functional determinants of (6.14). Such an approach has been developed for periodic surface profiles in [12, 28]. In this approach, it is convenient to directly compute the Casimir force $F = -\partial_H \mathcal{E}$ per unit area which is the sum of TM and TE contributions, $F = F_{TM} + F_{TE}$ that according to (6.14), are given by (for X = D, N)

$$F_X = -\frac{\hbar c}{2AL} \text{Tr}\left(\mathbb{M}_X^{-1} \partial_H \mathbb{M}_X\right). \tag{6.29}$$

The right-hand side of this expression is always finite, and no divergences due to self-energies have to be subtracted. The trace in (6.29) can be efficiently computed by Fourier transforming $\mathbb{M}$ with respect to $\mathbf{y}$, $\mathbf{y}'$. The transformed operator can then be transformed to block-diagonal form by making use of the periodicity of the surface profile along the y_1 direction. In this representation, the blocks can be numbered by the wave vector $q_1 \in [0, 2\pi/\lambda)$ along the y_1 direction. A block matrix with label q_1 couples only waves whose momenta differ from the Bloch momentum q_1 by integer multiples of $2\pi/\lambda$. The integers multiplying $2\pi/\lambda$ number the matrix elements within a block matrix. Hence, the problem of computing the total trace has been simplified to the computation of the trace of each block matrix with label q_1. Finally, integration over q_1 from 0 to $2\pi/\lambda$ and over the unrestricted momenta q_0, q_2 (along the time direction and invariant spatial direction of the surfaces, respectively) yields the force of (6.29). For the particular choice of a *rectangular* corrugation (see Figure 6.5a), analytic expressions for all matrix elements of $\mathbb{M}$ can be obtained. For details of the implementation of the numerical approach and expressions for the matrix elements see [12].

In comparison to the profile of Figure 6.1, we consider the corresponding situation of a flat plate and a plate with a rectangular corrugation profile parametrized by

$$h_1(y_1) = \begin{cases} +a & \text{for} \quad |y_1| < \lambda/4 \\ -a & \text{for} \quad \lambda/4 < |y_1| < \lambda/2 \end{cases}, \quad (6.30)$$

and continuation by periodicity $h_1(y_1) = h_1(y_1 + n\lambda)$ for any integer n. The numerical results for the total Casimir force between the two plates is shown in Figure 6.3 for different corrugation wavelengths λ. For all λ, the forces at a fixed separation H are bounded between a minimal force F_∞ and a maximal force F_0. For small λ/a, the upper bound F_0 is approached, whereas for asymptotically large λ/a, the force converges towards the lower bound F_∞. Analytic expressions can be derived for these bounds. For large λ, the corrugated surface is composed of large flat segments with a low density of edges. At sufficiently small surface sepa-

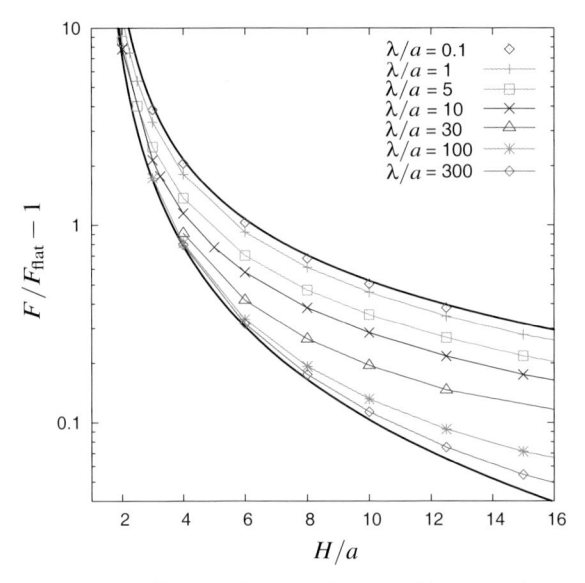

Figure 6.3 Total Casimir force as a function of the mean plate separation H. The relative change of the force compared to the total Casimir force F_{flat} between two flat plates is shown. The two bold curves enclosing the numerical data are the analytical results F_0 for $\lambda \to 0$ (upper curve) and F_∞ for $\lambda \to \infty$ (lower curve), see text.

rations $H \ll \lambda$, the main contribution to the force comes from wavelengths which are much smaller than the scale λ of the surface structure. Thus, in the dominant range of modes, diffraction can be neglected and a simple proximity force approximation [55] should be applicable. Such an approximation assumes that the total force can be calculated as the sum of local forces between opposite *flat* and *parallel* small surface elements at their local distance $H - h_1(y_1)$. No distinction is made between TM and TE modes. This procedure is rather simple for the rectangular corrugation considered here since the surface has no curvature (except for edges). There are only two different distances $H + a$, $H - a$ which each contribute one-half across the entire surface area, leading for $\lambda \to \infty$ to the proximity approximation for the force,

$$F_\infty/A = -\frac{\pi^2 \hbar c}{240} \frac{1}{2} \left[\frac{1}{(H-a)^4} + \frac{1}{(H+a)^4} \right]. \tag{6.31}$$

In the limit $\lambda \to 0$, the important fluctuations should not get into the narrow valleys of the corrugated plate. Even for small but finite λ, this picture should be a good, though approximate, description since it still effects the wavelengths of order H which give the main contribution to the force. Thus, one can expect that the plates feel a force which is equal to the force between two *flat* plates at the *reduced* distance $H - a$. Fortunately, this expectation can be checked by an explicit calculation since the leading part of determinant of $\mathbb{M}_X$ in the limit $\lambda \to 0$ can be computed. Indeed, this computation confirms the expectation, leading to the Casimir force per surface area [12]

$$F_0/A = -\frac{\pi^2}{240} \frac{1}{(H-|a|)^4} \tag{6.32}$$

with equal contributions from TM and TE modes. Notice that this result is not analytic in a/H and is *exact* in the limit $\lambda \to 0$. As we have seen before, perturbation theory for smoothly deformed surfaces always yields corrections to the interaction of order a^2. However, for small a/H, the result of (6.32) has the expansion

$$F_0/A = -\frac{\pi^2}{240} \frac{1}{H^4} \left[1 + 4\frac{|a|}{H} + \mathcal{O}\left(\left(\frac{a}{H}\right)^2\right) \right] \tag{6.33}$$

which indicates that perturbation theory is not applicable if $\lambda \ll a$. This implies that the apparent divergent behavior for $\lambda \to 0$ in (6.28) actually disappears for $\lambda \simeq a$.

6.2.2
Lateral Forces

As a natural generalization of the geometry of the previous section, we study the Casimir interaction between two sinusoidally corrugated plates. For direct correspondence to experiments for this type of configuration [22], we consider the specific profiles

$$h_1(y_1) = a\cos(2\pi y_1/\lambda), \quad \text{and} \quad h_2(y_1) = a\cos(2\pi(y_1+b)/\lambda), \qquad (6.34)$$

which are shifted relative to each other by the length b (see Figure 6.4). When these profiles are substituted into the general expression for the second order term of the partition function of (6.19), one finds for the Casimir energy

$$\mathcal{E} = \mathcal{E}_0 + 2\mathcal{E}_{cf} + \mathcal{E}_{cc}, \qquad (6.35)$$

with $\mathcal{E}_{cf}$ given in (6.24), and where the corrugation–corrugation interaction energy $\mathcal{E}_{cc}$ can be calculated in terms of the kernels $Q_X(y)$ in (6.19). Besides oscillating contributions to the normal Casimir force from $\mathcal{E}_{cc}(b)$, a *lateral* force

$$F_{\text{lat}} = -\frac{\partial \mathcal{E}_{cc}}{\partial b} \qquad (6.36)$$

is induced by the corrugation–corrugation interaction. This lateral force is better suited for experimental tests of the influence of deformations since there is no need for subtracting a larger baseline force (the contribution of flat plates) as in the case of the normal force. The lateral force can be also employed as a actuation mechanism in mechanical oscillators as we will see in Section 6.4. In analogy to the previous section, the corrugation–corrugation interaction can be expressed as

$$\mathcal{E}_{cc} = \frac{\hbar c a^2}{H^5} \cos\left(\frac{2\pi b}{\lambda}\right) \left[J_{\text{TM}}\left(\frac{H}{\lambda}\right) + J_{\text{TE}}\left(\frac{H}{\lambda}\right) \right] + \mathcal{O}(a^3) \qquad (6.37)$$

with

$$\begin{aligned}
J_{\text{TM}}(x) =\ & \frac{\pi^2}{120}(16x^4 - 1)\operatorname{arctanh}(\sqrt{u}) \\
& + \sqrt{u}\left[\frac{\pi}{12}\left(x^3 - \frac{1}{80x}\right) \Phi(u, 2, \tfrac{1}{2}) + \frac{x^2}{12}\Phi(u, 3, \tfrac{1}{2}) \right. \\
& \left. + \frac{x}{16\pi}\Phi(u, 4, \tfrac{1}{2}) + \frac{1}{32\pi^2}\Phi(u, 5, \tfrac{1}{2}) + \frac{1}{128\pi^3 x}\Phi(u, 6, \tfrac{1}{2}) \right],
\end{aligned} \qquad (6.38)$$

$$\begin{aligned}
J_{\text{TE}}(x) =\ & \frac{\pi^2}{120}(16x^4 - 1)\operatorname{arctanh}(\sqrt{u}) + \sqrt{u}\left[-\frac{\pi}{12}\left(x^3 + \frac{x}{2} + \frac{1}{80x}\right) \right. \\
& \times \Phi(u, 2, \tfrac{1}{2}) + \frac{1}{24}\left(x^2 - \frac{3}{4}\right)\Phi(u, 3, \tfrac{1}{2}) + \frac{5}{32\pi}\left(x - \frac{1}{20x}\right) \\
& \left. \times \Phi(u, 4, \tfrac{1}{2}) + \frac{7}{64\pi^2}\Phi(u, 5, \tfrac{1}{2}) + \frac{7}{256\pi^3 x}\Phi(u, 6, \tfrac{1}{2}) \right],
\end{aligned} \qquad (6.39)$$

where $u \equiv \exp(-4\pi x)$ and $\Phi(z, s, a) = \sum_{k=0}^{\infty} z^k/(a+k)^s$ is the Lerch transcendent. In the limit of large corrugation length, $H/\lambda \to 0$, this result agrees to lowest order with a pairwise summation approximation where $J_{\text{TM}}(0) + J_{\text{TE}}(0) = \pi^2/120$.

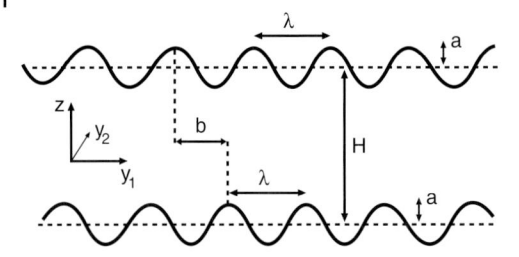

Figure 6.4 Geometry used for calculating the lateral Casimir force between two corrugated plates with lateral shift b. The equilibrium position is at $b = \lambda/2$.

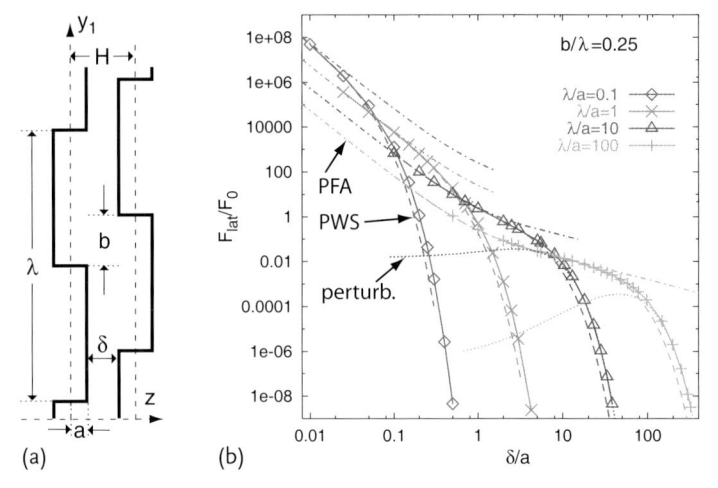

Figure 6.5 (a) Geometry consisting of two parallel plates with laterally shifted uniaxial rectangular corrugations. (b) Lateral force F_{lat} (in units of normal force F_0 between flat surfaces) at $b = \lambda/4$ for the geometry shown in (a) as a function of the gap δ (solid curves). The proximity force (PFA, dash-dotted curves) and pairwise summation (PWS, dashed curves) approximations, and the perturbative result F_{pt} that follows from a calculation for sinusoidal profiles (dotted curves) are also plotted.

At the other extreme of $\lambda \ll H$, $J_{\text{TM}}(x) + J_{\text{TE}}(x)$ decays *exponentially* fast. This decay distinguishes the lateral force from the normal force. In particular, for large $x = H/\lambda$, we arrive at the leading order

$$J_{\text{TM}}(x) + J_{\text{TE}}(x) = \frac{4\pi^2}{15} \left(x^4 + \mathcal{O}(x^2) \right) e^{-2\pi x}. \tag{6.40}$$

Since $J_{\text{TM}}(x) + J_{\text{TE}}(x)$ is positive for all values of x, the equilibrium position of two modulated surfaces is predicted at $b = \lambda/2$. This corresponds to aligning the maxima and minima of the two corrugations (see Figure 6.4).

The numerical approach for computing the functional determinant in the case of periodic surfaces can be also applied to the lateral force [13]. Once again, we consider a rectangular corrugation, though now, on both surfaces with a lateral shift

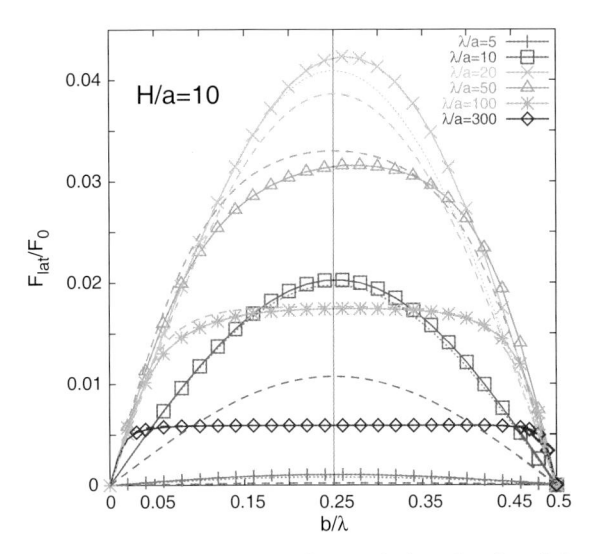

Figure 6.6 Shape dependence of F_{lat} on the lateral surface shift b at fixed distance $H = 10a$ for different corrugation lengths. The dashed and the dotted curves represent the PWS and the full perturbative result for sinusoidal profiles with arbitrary H/λ, respectively.

of b, as in Figure 6.5. The numerical results for the lateral force in this geometry are summarized in Figures 6.5 and 6.6. Figure 6.5 shows the numerical result for the lateral force for a shift $b = \lambda/4$ and different values of λ/a over more than four orders of magnitude for the gap $\delta = H - 2a$, together with two approximate results (PFA and PWS) and the perturbative result for sinusoidal profiles for $\lambda \ll H$. An exponential decay of the force as predicted by perturbation theory can be clearly observed.

The PFA yields a lateral force per unit area $F_{\text{lat,PFA}} = [2\mathcal{E}_0(H) - \mathcal{E}_0(H - 2a) - \mathcal{E}_0(H + 2a)]/\lambda$ for $0 < b < \lambda/2$ where $\mathcal{E}_0$ has the same meaning as before. $F_{\text{lat,PFA}}$ changes sign at $b = \lambda/2$ discontinuously which is an artifact of this approximation. The pairwise summation (PWS) of Casimir–Polder potentials is strictly justified for rarefied media only but it is often also applied to metals, using the two-body potential $U(r) = -(\pi/24)\hbar c/r^7$ with the amplitude chosen such as to reproduce the correct result for flat ideal metal plates [7]. It yields a lateral force $F_{\text{lat,PWS}} = -\frac{\partial}{\partial b} \int_{V_l} d^3\mathbf{x} \int_{V_r} d^3\mathbf{x}' \, U(|\mathbf{x} - \mathbf{x}'|)$ with V_l and V_r denoting the semi-infinite regions to the left and right of the two surfaces in Figure 6.5a, respectively. $F_{\text{lat,PWS}}$ can be obtained by numerical integration. For small gaps δ, both approximations agree and match the exact numerical results. Beyond $\delta \gtrsim \lambda/20$ the PFA starts to fail since it does not capture the exponential decay of F_{lat} for increasing δ. The PWS approach has a slightly larger validity range and reproduces the exponential decay. However it deviates by at least *one order of magnitude* from F_{lat} for $\delta \gtrsim 2.5\lambda$.

Although the perturbative result of (6.37) applies to sinusoidal surfaces, it is instructive to compare it to the numerical results for the rectangular profiles. Since the lateral force decays exponentially, $F_{\text{lat}} \sim e^{-2\pi H/\lambda}$, with the characteristic scale set by the modulation wavelength of the profile, the force at large H should be determined by the lowest harmonic of the periodic surface profile. This implies a *universal* lateral force for large $H \gg \lambda$ that is independent of the precise form of the surface corrugation. This universal force is the force between two sinusoidal surfaces where the amplitude follows from the projection of an arbitrary periodic profile of wavelength λ onto a sinusoidal profile with the same wavelength. The latter force follows from (6.37) and in the limit $\lambda \ll H$ is given by

$$F_{\text{pt}} = \frac{8\pi^3 \hbar c}{15} \frac{a_0^2 A}{\lambda^5 H} \sin\left(\frac{2\pi}{\lambda} b\right) e^{-2\pi H/\lambda}, \qquad (6.41)$$

where we assumed an amplitude a_0 for the sinusoidal profiles. For the rectangular corrugation of Figure 6.5, the lowest harmonic has the amplitude $a_0 = 4a/\pi$. When we compare F_{pt} and the numerical results of Figure 6.5b, we find excellent agreement for distances $\delta \gtrsim \lambda$.

The universal behavior of the lateral force is also clearly demonstrated by the dependence of the lateral force on the surface shift b. Corresponding numerical results together with PWS approximations and the force that follows from the *full* perturbative result of (6.37) for sinusoidal surfaces with arbitrary H/λ are shown in Figure 6.6 for fixed $H = 10a$ and varying λ/a. With decreasing λ, three regimes can be identified. For $\lambda \gg H$, the force profile nearly resembles the rectangular shape of the surfaces, and the PWS approximation yields consistent results. For smaller λ, yet larger than H, the force profile becomes asymmetric with respect to $b = \lambda/4$ and more peaked, signaling the crossover to the universal regime for $\lambda \lesssim H$ where the force profile becomes sinusoidal. In the latter case, for slightly small $\lambda/a \approx 10$, the numerical results for F_{lat} agree with the perturbative result for sinusoidal surfaces with arbitrary H/λ. We note that the PWS approach fails to predict the asymmetry of the force profile, and the PFA even predicts no variation with b for $0 < b < \lambda/2$. To observe this universal behavior of the lateral force experimentally, one should consider surfaces with very small corrugation wavelengths in the range of nanometers so that the exponential decay does not diminish the force for $H \gg \lambda$ too strongly.

6.2.3
Cylinders

In this section, we give examples for two central aspects of fluctuation forces: Effects resulting from the nonadditivity and the particular properties of systems with a codimension of two, which plays a special role as we will see below. These problems are considered in the context of interactions between cylinders and sidewalls. It has been demonstrated that Casimir forces in these geometries have only a weak logarithmic dependence on the cylinder radius [36] and can be nonmonotonic [56, 57, 59], consequences of codimension and nonadditivity. These forces between quasi-one-dimensional structures could be probed in mechanical oscillators that are composed of nanowires or carbon nanotubes. Exact results for the interaction can be obtained by employing a recently developed scattering approach for Casimir forces [30, 32]. This approach is based on the concept that electromagnetic Casimir interactions result from fluctuating currents inside the bodies. It is possible to formulate an effective action for the multipole moments $\mathbf{Q}_{a,X}$ of the currents inside the bodies where a labels the bodies and X is a multi-index that numbers polarizations (electric and magnetic multipoles) and the elements of the basis for the multipole expansion, for example, cylindrical waves. The effective action can then be written as the quadratic form

$$S = \sum_{a,a'} \sum_{X,X'} \mathbf{Q}^*_{a,X} \mathbb{M}_{aa',XX'} \mathbf{Q}_{a',X'} , \qquad (6.42)$$

with the matrix kernel

$$\mathbb{M}_{aa',XX'} = \kappa \left\{ \left[(\mathbb{T}_a)^{-1} \right]_{XX'} \delta_{aa'} - \mathbb{U}_{aa',XX'}(1 - \delta_{aa'}) \right\} , \qquad (6.43)$$

where κ is the Wick-rotated frequency, $\omega = ic\kappa$, the matrix $\mathbb{T}_a$ is the so-called T-matrix of object a that relates incoming and scattered waves and $\mathbb{U}_{aa'}$ is a "translation" matrix that relates the incoming wave at object a to the outgoing wave at object a'. The T-matrix is related to the scattering matrix of the object, $\mathbb{S}_a$, by the relation $\mathbb{T}_a = (\mathbb{S}_a - 1)/2$. Analytic results for all elements of the scattering matrix are available for symmetric shapes such as cylinders and spheres. The $\mathbb{S}_a$ matrix contains all information about shape and material composition of the object that is relevant to the Casimir interaction. The translation matrices $\mathbb{U}_{aa'}$ are independent of the properties of the interacting bodies and depend only on the relative position (separation vector) of the objects a and a', and the properties of the fluctuating field. For the electromagnetic field, the translation matrices are known in many bases, for example, for cylindrical and spherical waves [30]. To obtain the Casimir energy, the multipole fluctuations are integrated out, leading to the determinant of the infinite dimensional matrix $\mathbb{M}$. Integration over all frequencies κ yields the interaction energy

$$\mathcal{E} = \frac{\hbar c}{2\pi} \int_0^\infty d\kappa \ln \frac{\det \mathbb{M}}{\det \mathbb{M}_\infty} , \qquad (6.44)$$

where the division by the determinant of the matrix $\mathbb{M}_\infty$ accounts for the subtraction of the residual energy of the configuration where the separations between all objects tend to infinity. Since the translation matrices decay to zero with increasing separation, the matrix $\mathbb{M}_\infty$ is given by (6.43) with the $\mathbb{U}_{aa'}$ set to zero. In the special case of two objects, the energy can be simplified to [32]

$$\mathcal{E}_2 = \frac{\hbar c}{2\pi} \int_0^\infty d\kappa \ln \det(1 - \mathbb{N}) , \qquad (6.45)$$

where $\mathbb{N} = \mathbb{T}_1 \mathbb{U}_{12} \mathbb{T}_2 \mathbb{U}_{21}$.

First, the scattering approach is applied to two parallel, infinitely long, perfectly conducting cylinders of equal radius R and center-to-center separation d, see Figure 6.7. For this geometry, it is most convenient to use cylindrical vector waves

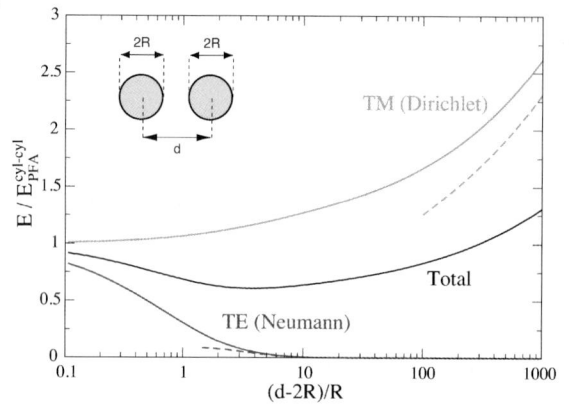

Figure 6.7 Casimir energy for two cylinders of equal radius R as a function of surface-to-surface distance $d - 2R$ (normalized by the radius). The energy is divided by the PFA estimate $E_{\text{PFA}}^{\text{cyl-cyl}} = -\frac{\pi^3}{1920}\hbar c L \sqrt{R/(d-2R)^5}$ for the energy which is applicable in the limit $d \to 2R$ only. The solid curves show numerical results; the dashed lines represent the asymptotic results of (6.53). The inverse logarithmic correction to the leading order result for TM modes cause very slow convergence. For the parameter range shown here, it was sufficient to consider $m = 40$ partial waves in order to obtain convergence.

for the multipole expansion. This basis consists of the vector fields $\mathbf{M}_{k_z m}^{i(o)}(\mathbf{x}) = \frac{1}{q}\nabla \times \mathbf{V}^{i(o)}(\mathbf{x})$ for magnetic (M) multipoles and $\mathbf{N}_{k_z m}^{i(o)}(\mathbf{x}) = \frac{c}{q\omega}\nabla \times \nabla \times \mathbf{V}^{i(o)}(\mathbf{x})$ for electric (E) multipoles where $q = \sqrt{(\omega/c)^2 - k_z^2}$ and incoming (i) and outgoing (o) waves differ in the definition of the vector fields $\mathbf{V}^i(\mathbf{x}) = \hat{\mathbf{z}}J_m(qr)e^{im\phi}e^{ik_z z}$, $\mathbf{V}^o(\mathbf{x}) = \hat{\mathbf{z}}H_m^{(1)}(qr)e^{im\phi}e^{ik_z z}$. Here, (r, ϕ, z) denote cylindrical coordinates and J_m, $H_m^{(1)}$ are Bessel and Hankel functions of the first kind. In this basis, the matrices of (6.43) assume a simple form where the multi-index X now represent the polarization (M or E), the wave vector k_z along the cylinder axis and the partial wave index m. The T-matrix is diagonal in polarizations k_z and m, with diagonal elements

$$T_{Mk_z m} = (-1)^m \frac{\pi}{2i} \frac{I'_m(qR)}{K'_m(qR)} \tag{6.46}$$

$$T_{Ek_z m} = (-1)^m \frac{\pi}{2i} \frac{I_m(qR)}{K_m(qR)}, \tag{6.47}$$

where we have applied a Wick rotation $\omega = ic\kappa$ which leads to modified Bessel functions of the first (I_n) and second (K_n) kind. The translation matrices are diagonal in polarization and k_z, and the elements are identical for both polarizations,

$$U_{12,Mk_z nm} = U_{12,Ek_z nm} = \frac{2}{i\pi}(-i)^{m-n} K_{m-n}(pd) \tag{6.48}$$

$$U_{21,Mk_z nm} = U_{21,Ek_z nm} = \frac{2}{i\pi} i^{m-n} K_{m-n}(pd) \tag{6.49}$$

with $p = \sqrt{\kappa^2 + k_z^2}$. Due to the decoupling of electric and magnetic multipoles, corresponding to transverse magnetic (TM) and transverse electric (TE) field modes, respectively, the Casimir energy of (6.45) has two independent contributions, $\mathcal{E}_2 = \mathcal{E}_{TM} + \mathcal{E}_{TE}$, with

$$\mathcal{E}_{TM(TE)} = \frac{\hbar c L}{4\pi} \int_0^\infty p \, dp \, \ln \det(1 - \mathbb{N}_{TM(TE)}) \,, \tag{6.50}$$

where $L \, (\to \infty)$ is the cylinder length and the determinant only runs over the partial wave indices m, m' of the matrix elements

$$\mathbb{N}_{TE,mm'} = i^{m'-m} \sum_n \frac{I'_m(pR)}{K'_m(pR)} K_{n+m}(pd) \frac{I'_n(pR)}{K'_n(pR)} K_{m'+n}(pd) \tag{6.51}$$

$$\mathbb{N}_{TM,mm'} = i^{m'-m} \sum_n \frac{I_m(pR)}{K_m(pR)} K_{n+m}(pd) \frac{I_n(pR)}{K_n(pR)} K_{m'+n}(pd) \,. \tag{6.52}$$

This result for the energy can be also obtained from a scalar field theory where TM (TE) modes correspond to Dirichlet (Neumann) boundary conditions [36, 56].

At large separations $d \gg R$, only matrix elements with $m = m' = 0$ for TM modes and $m = m' = 0, \pm 1$ for TE modes contribute to the energy. When the determinant in (6.50) is restricted to these elements, we find for the interaction of two cylinders for large d/R to leading order

$$\mathcal{E}_{TM} = -\hbar c L \frac{1}{8\pi} \frac{1}{d^2 \ln^2(d/R)} \left(1 - \frac{2}{\ln(d/R)} + \cdots \right),$$

$$\mathcal{E}_{TE} = -\hbar c L \frac{7}{5\pi} \frac{R^4}{d^6} \,. \tag{6.53}$$

The asymptotic interaction is dominated by the contribution from TM modes that only vanishes for $R \to 0$ logarithmically.

For arbitrary separations, higher order partial waves have to be considered. The number of partial waves has to be increased with decreasing separation. A numerical evaluation of the determinant and integration has revealed an exponentially fast convergence of the energy in the truncation order for the partial waves, leading to the results shown in Figure 6.7 [56]. It should be noted that the minimum in the curve for the total electromagnetic energy results from the scaling by the PFA estimate of the energy. The total energy is monotonic and the force attractive at all separations.

The interaction between cylinders is very distinct from the Casimir or van der Waals interaction which is reported in literature [55]. Usually, the interaction is proportional to the volumes of the interacting objects, that is, for two spheres of radius R where the Casimir energy $\sim R^6/d^7$. This scaling with volumes also follows from a pairwise summation of two-body forces. However, from the interaction of two parallel plates, one knows that the interaction can scale also with the surface area. These two examples would suggest for two parallel cylinders of length L an interaction energy $\sim LR^4/d^6$ or $\sim LR^2/d^4$. However, the actual results of (6.53) has a much weaker, only logarithmic dependence on the radius. It is interesting to look at the variation of the decay exponent of d for the Casimir energy as a function of the codimension of the object. The exponent is $(-3, -2+\epsilon, -7)$ for codimensions 1 (plates), 2 (cylinders), 3 (spheres), respectively, and hence *not monotonic*. For a codimension of two, the Casimir interaction is typically long-ranged. The physical reason for the unexpected scaling of the cylinder interaction is explained by considering spontaneous charge fluctuations. On a sphere, the positive and negative charges can be separated at most by distances of order $R \ll d$. The retarded van der Waals interactions between the dipoles on the spheres lead to the Casimir–Polder interaction [15]. In the cylinder, fluctuations of charge along the axis of the cylinder can create arbitrarily large positively (or negatively) charged regions. The retarded interaction of these charges (not dipoles) gives the dominant term of the Casimir force. This interpretation is consistent with the difference between the two types of polarizations since for TE modes such charge modulations cannot occur due to the absence of an electric field along the cylinder axis, as illustrated (6.53) and Figure 6.7.

As a second example, the effect of sidewalls on the interaction of two cylinders is considered. The geometry consisting of either one or two plates at a separation H from the two cylinders is shown in Figure 6.8. For this type of geometry, the mean stress tensor has been computed numerically and it has been observed that the force between two one-dimensional structures changes nonmonotonically when H is increased [57, 59]. This many-body effect can be studied by the scattering approach. Instead of studying the interaction of the cylinders and plates via their T-matrices directly, it is more convenient to employ the method of images to describe the effect of the sidewalls [30, 56]. For perfectly conducting sidewalls, their effect on the electromagnetic field can be taken into account by replacing the free space Green's function by a half-space or slab Green's function. This results in an expression for the Casimir energy similar to (6.44) that depends only on the

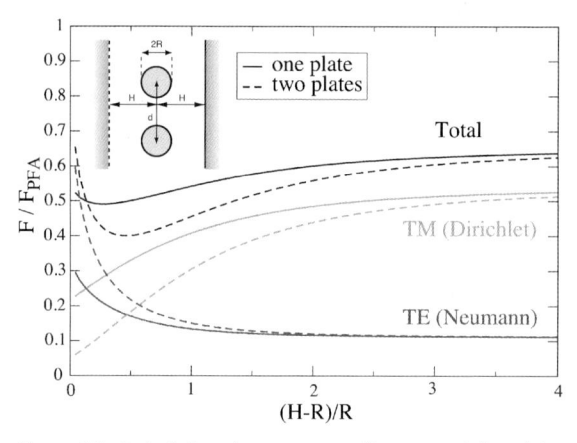

Figure 6.8 Casimir force between two cylinders parallel to one plate or sandwiched between two plates vs. the ratio of sidewall separation to cylinder radius $(H - R)/R$, at fixed distance $d = 4R$ between the cylinders, normalized by the total PFA force per unit length between two isolated cylinders, $F_{PFA} = -\frac{5}{2}(\hbar c \pi^3/1920)\sqrt{R/(d-2R)^7}$. The solid lines refer to the case with one plate, while dashed lines depict the results for two plates. The individual TE and TM contributions to the force are also shown.

T-matrices of the two cylinders and translation matrices that connect the original cylinders and their mirror images. The expression of the energy can be computed again numerically by truncating the partial wave expansion at a sufficiently high order. The resulting Casimir force between two cylinders with one or two sidewalls as a function of the sidewall separation H is shown in Figure 6.8. Two interesting features can be observed. First, the attractive total force varies nonmonotonically with H: Decreasing for small H and then increasing towards the asymptotic limit between two isolated cylinders for large H, as in (6.53). The extremum for the one-sidewall case occurs at $H - R \approx 0.27 R$, and at $H - R \approx 0.46 R$ for the two-sidewall case. Second, the total force between the cylinders for the two-sidewall case in the proximity limit $H \to R$ is larger than for $H/R \to \infty$. As one might expect, the H-dependence for one sidewall is weaker than for two sidewalls, and the effects of the two sidewalls are not additive. Not only is the difference from the $H \to \infty$ force not doubled for two sidewalls compared to one, but the two curves actually intersect.

A simple generic argument for the nonmonotonic sidewall effect has been given in [57]. It arises from a competition between the force from TE and TM polarizations as demonstrated by the results in Figure 6.8. An intuitive perspective for the qualitatively different behavior of the TE and TM force as a function of the sidewall distance is obtained from the method of images. For the TM polarization (corresponding to Dirichlet boundary conditions in a scalar field theory), the Green's function is obtained by subtracting the contribution from the image so that the image sources have *opposite* signs. Any configuration of fluctuating TM charges on one cylinder is thus screened by images, more so as H is decreased, *reducing* the force on the fluctuating charges of the second cylinder. This is similar to the effect of a nearby grounded plate on the force between two opposite electrostatic charges. Since the reduction in force is present for every charge configuration, it is also there for the average over all configurations.

By contrast, the TE polarization (corresponding to Neumann boundary conditions in a scalar field theory) requires image sources of the *same* sign. The total force between fluctuating sources on the cylinders is now larger and increases as the plate separation H is reduced. Note, however, that while for each fluctuating source configuration, the effect of images is additive, this is not the case for the average over all configurations. More precisely, the effect of an image source on the Green's function is not additive because of feedback effects: the image currents change the surface current distribution, which changes the image, and so forth. For example, the net effect of the plate on the Casimir TE force *is not* to double the force as $H \to R$. The increase is in fact larger than two due to the correlated fluctuations.

A similar but weaker nonmonotonic dependence on H of the force between the cylinders is also observed for separations d that are different from the particular choice in Figure 6.8. Also, the force between the cylinders *and the sidewalls* is not monotonic in d but the nonmonotonicity is then smaller since the effect of a cylinder on the force between two bodies is smaller than the effect of an infinite plate.

6.2.4
Spheres

Thus far, geometries with a direction of translational invariance have been considered. In the limit of ideal metal surfaces, this invariance leads to a decoupling of the two polarizations of the electromagnetic field. Any geometry of experimental interest will obviously lack this symmetry beyond some length scale. Hence, it is important to study geometries without this symmetry. *Compact* objects of arbitrary shape obviously do not have an invariant direction. Therefore, the two polarizations are coupled and the matrices in (6.43) assume a more complicated form. A natural choice for a basis are now vector spherical waves for which the translation matrices $\mathbb{U}_{aa',XX'}$ carry an index $X = $ (E or M, l, m) which represents polarization E or M and the order $l \geq 1$, $m = -l, \ldots, l$ of the spherical waves. In contrast to the cylindrical matrices of (6.48), the translation matrix couples E and M polarization and all matrix elements are explicitly known [30].

Here we focus on the simplest case of two compact objects: two perfect metal spheres of equal radius R and center-to-center separation d, see Figure 6.9. The T-matrix of a dielectric sphere is known from the Mie theory for scattering of electromagnetic waves from spherical particles. Due to spherical symmetry, the E and

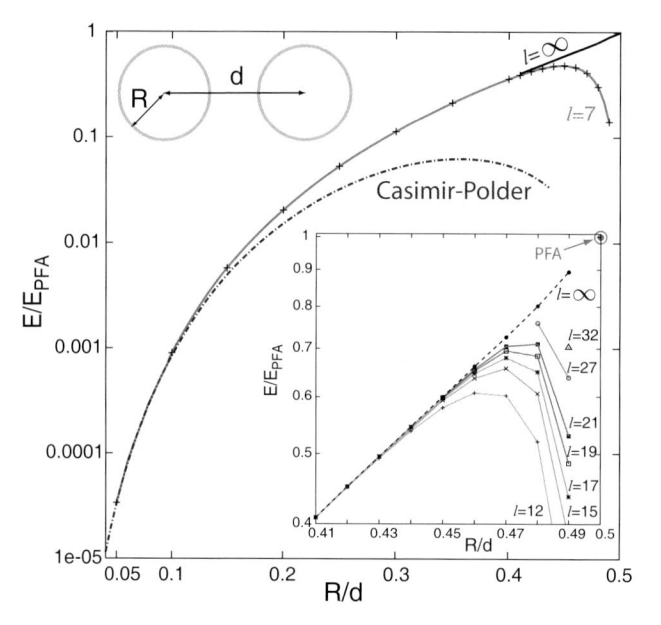

Figure 6.9 Casimir energy of two metal spheres, divided by the PFA estimate $\mathcal{E}_{PFA} = -(\pi^3/1440)\hbar c R/(d - 2R)^2$, which only holds in the limit $R/d \to 1/2$. The label l denotes the multipole order of truncation. The curves $l = \infty$ are obtained by extrapolation. The Casimir–Polder curve is the leading term of (6.56). Inset: Convergence with the truncation order l for partial waves at short separations.

M polarizations for all l, m are decoupled so that the T-matrix is diagonal and the coupling of polarizations only occurs through the translation matrices. After a Wick rotation to imaginary frequency $\omega = i c \kappa$, the matrix elements assume, in the perfect metal limit, the form

$$T_{\text{MM}\,lml'm'} = (-1)^l \frac{\pi}{2} \frac{I_{l+\frac{1}{2}}(\kappa R)}{K_{l+\frac{1}{2}}(\kappa R)} \delta_{ll'} \delta_{mm'} \tag{6.54}$$

$$T_{\text{EE}\,lml'm'} = (-1)^l \frac{\pi}{2} \frac{I_{l+\frac{1}{2}}(\kappa R) + 2\kappa R\, I'_{l+\frac{1}{2}}(\kappa R)}{K_{l+\frac{1}{2}}(\kappa R) + 2\kappa R\, K'_{l+\frac{1}{2}}(\kappa R)} \delta_{ll'} \delta_{mm'}. \tag{6.55}$$

Substitution of these matrix elements together with those of $\mathbb{U}_{\alpha\alpha'}$ from [30] in (6.45) yields the Casimir energy of two spheres. For asymptotically large d, the energy has only contributions from $l = l' = 1$ (dipoles) and one obtains the Casimir–Polder interaction between two polarizable particles [15] where the electric and magnetic dipole polarizabilities of a perfect metal sphere are given by $\alpha_E = R^3$ and $\alpha_M = -R^3/2$. This result can be extended to smaller separations by including higher order multipoles with $l > 1$ that generate higher powers of R/d. One obtains the asymptotic series [32]

$$\mathcal{E}_2 = -\frac{\hbar c}{\pi} \frac{R^6}{d^7} \sum_{n=0}^{\infty} c_n \left(\frac{R}{d}\right)^n, \tag{6.56}$$

where the first eight coefficients are $c_0 = 143/16$, $c_1 = 0$, $c_2 = 7947/160$, $c_3 = 2065/32$, $c_4 = 27\,705\,347/100\,800$, $c_5 = -55\,251/64$, $c_6 = 1\,373\,212\,550\,401/144\,506\,880$, $c_7 = -7\,583\,389/320$. The energy at all separations can be obtained by truncating the matrix $\mathbb{N}$ defined below (6.45) at a finite multipole order l, and by numerically computing the determinant and the integral. The result is shown in Figure 6.9. It provides the force for all separations between the Casimir–Polder limit for $d \gg R$, and the PFA result for $R/d \to 1/2$. At a surface-to-surface distance $4R/3$ ($R/d = 0.3$), the PFA overestimates the energy by a factor of ten. Including up to $l = 32$ partial wave orders and extrapolating based on an exponential convergence in l, the Casimir energy has been determined down to $R/d = 0.49$ [32]. The interaction between a sphere and a plate has been obtained recently and deviations from the PFA have been quantified [30].

6.3
Dependence on Material Properties

In previous sections, we have considered perfectly conducting bodies. For real metals with finite, frequency dependent conductivity or more general dielectric media, Casimir interactions are modified. This is a natural consequence of the fact that dipole and higher multipole polarizabilities depend on the material properties of a body. Thus, the induced fluctuating currents depend not only on shape, but also on material composition. Therefore, for practical applications and experimental tests, it is important to understand the collective effects of shape and material on Casimir interactions. A macroscopic theory that fully accounts for the material dependence of the interaction between two *planar* surfaces was established by Lifshitz [47] in 1956. Until recently, only approximations limited to short separations between bodies or sufficiently diluted media have been available for studying the interaction of dielectric media of arbitrary shapes. The scattering approach described in the previous section has paved the way for studying the material and shape dependence of Casimir forces beyond the case of planar surfaces and without the common approximations in detail. We first provide a simple derivation of the Lifshitz result for two surfaces within the scattering approach. Then, we focus on an example that is of particular interest to the behavior of nanoparticles as they appear, that is, in suspensions where correlations between material and shape effects are important.

6.3.1
Lifshitz Formula

Consider two material half-spaces that are bounded by planar, parallel surfaces with a vacuum gap of width d between them. The material in the two halfspaces can be different and is characterized by the dielectric functions $\epsilon_\alpha(\omega)$ and magnetic permeabilities $\mu_\alpha(\omega)$ where $\alpha = 1, 2$ numbers the halfspaces. A compact derivation of the Casimir–Lifshitz interaction between the two surfaces follows from the scattering formula of (6.44). The T-matrix of a planar dielectric surface is given by the Fresnel coefficients which are usually expressed in a planar wave basis. When we define the two polarizations relative to the surface normal vector, the T-matrix is diagonal in polarization and parallel to the surface in the wave vector $k_\parallel$. The diagonal matrix elements are

$$T_{\alpha,M\,k_\parallel} = \frac{\mu_\alpha(ic\kappa)p - p_\alpha}{\mu_\alpha(ic\kappa)p + p_\alpha},$$

$$T_{\alpha,E\,k_\parallel} = \frac{\epsilon_\alpha(ic\kappa)p - p_\alpha}{\epsilon_\alpha(ic\kappa)p + p_\alpha}, \qquad (6.57)$$

where $p = \sqrt{\kappa^2 + k_\parallel^2}$ and $p_\alpha = \sqrt{\epsilon_\alpha(ic\kappa)\mu_\alpha(ic\kappa)\kappa^2 + k_\parallel^2}$. The translation matrices for translations perpendicular to the surfaces by a distance d are also diagonal in $k_\parallel$ in the planar wave basis and the diagonal elements have the simple form

$$U_{\alpha\alpha',M\,k_\parallel} = U_{\alpha\alpha',E\,k_\parallel} = e^{-pd} \qquad (6.58)$$

for $\alpha \neq \alpha' = 1, 2$, that is, they do not couple E and M polarizations and are identical for the two polarizations. The determinant of (6.44) leads to a product over all $k_\parallel$ which becomes an integral after taking the logarithm. The resulting Casimir–Lifshitz energy has two separate contributions form M and E polarizations (TE and

TM modes, respectively),

$$\mathcal{E} = \frac{\hbar c A}{4\pi^2} \int_0^\infty d\kappa \int_0^\infty k_\| \, dk_\| \ln\left[\left(1 - \frac{\epsilon_1(ic\kappa)p - p_1}{\epsilon_1(ic\kappa)p + p_1}\frac{\epsilon_2(ic\kappa)p - p_2}{\epsilon_2(ic\kappa)p + p_2}e^{-2pd}\right)\right.$$
$$\left.\times \left(1 - \frac{\mu_1(ic\kappa)p - p_1}{\mu_1(ic\kappa)p + p_1}\frac{\mu_2(ic\kappa)p - p_2}{\mu_2(ic\kappa)p + p_2}e^{-2pd}\right)\right], \quad (6.59)$$

where A is the surface area. This result generalizes the Casimir interaction between two perfect metal plates of (6.5) to dielectric materials.

6.3.2
Nanoparticles: Quantum Size Effects

The Lifshitz formula of (6.59), while derived for infinitely extended planar surfaces, is also commonly applied to curved surfaces of particles of finite size within a proximity approximation. This leads to predictions for the interaction that are limited to particles that a very large compared to their separations. To be able to study the interaction of particles of arbitrary sizes and separations, a theory is needed that is a generalization of the Lifshitz formula to bodies of arbitrary shape. Such a general theory provides the scattering formula of (6.44). The challenges in applying this formula consist in the computation of the T-matrix for bodies with general dielectric functions and in the proper modeling of the dielectric response of the bodies. The latter is especially important for nanoparticles for which bulk optical properties are modified by finite-size effects.

Some characteristic effects of the Casimir interaction between nanoparticles will be discussed in this section by studying two spheres with *finite* conductivity in the limit where their radius R is much smaller than their separation d. We assume further that R is large compared to the inverse Fermi wave vector π/k_F of the metal. Since typically π/k_F is of the order of a few Angstrom, this assumption is reasonable even for nanoparticles. To employ (6.44), we need the T-matrix of a sphere with general dielectric function $\epsilon(\omega)$ which generalizes the matrix of (6.54), (6.55). All elements of this matrix are known explicitly, see, for example, [32]. Relevant to the interaction for $d \gg R$ are the dipole matrix elements ($l = l' = 1$) at low frequencies κ. In order to proceed, we need information about the dielectric function on the imaginary frequency axis $\omega = ic\kappa$ for small κ. Theories for the optical properties of small metallic particles [62] suggest a Drude-like response

$$\epsilon(ic\kappa) = 1 + 4\pi \frac{\sigma(ic\kappa)}{c\kappa}, \quad (6.60)$$

where $\sigma(ic\kappa)$ is the conductivity which approaches for $\kappa \to 0$ the dc conductivity σ_{dc}. For bulk metals, $\sigma_{dc} = \omega_p^2 \tau/4\pi$ where $\omega_p = \sqrt{4e^2 k_F^3/3\pi m_e}$ is the plasma frequency with electron charge e and electron mass m_e, and τ is the relaxation time. With decreasing dimension of the particle, $\sigma_{dc}(R)$ is reduced compared to its bulk value due to finite size effects and hence becomes a function of R [62].

In the low frequency limit, with $\epsilon(ic\kappa)$ of (6.60), the T-matrix elements for magnetic and electric dipole scattering ($l = l' = 1$) are diagonal in m and have the series expansion

$$T_{MM\,1m1m} = -\frac{4\pi}{45}\frac{R\sigma_{dc}(R)}{c}(\kappa R)^4 + \cdots \tag{6.61}$$

$$T_{EE\,1m1m} = \frac{2}{3}(\kappa R)^3 - \frac{1}{2\pi}\frac{c}{R\sigma_{dc}(R)}(\kappa R)^4 + \cdots. \tag{6.62}$$

To leading order $\sim \kappa^3$, the electric dipole matrix elements are identical to those of a perfectly conducting sphere and finite conductivity only modifies higher orders. In the magnetic dipole matrix elements, however, the leading term $-(\kappa R)^3/3$ of the perfect conductor result of (6.54) is absent. This is consistent with the observation that the magnetic dipole polarizability is reduced by a factor $\sim (\kappa R)^2[\epsilon(ic\kappa) - 1]$ and $\epsilon(ic\kappa) - 1 \sim \kappa^{-1}$ due to (6.60).

When the matrix elements of (6.61), (6.62) together with the translation matrices $\mathbb{U}_{\alpha\alpha'}$ in spherical coordinates are substituted into (6.45), an expansion for large distance d yields the Casimir energy of two spheres

$$\frac{E}{\hbar c} = -\frac{23}{4\pi}\frac{R^6}{d^7} - \left(\frac{R\sigma_{dc}(R)}{c} - \frac{45}{4\pi^2}\frac{c}{R\sigma_{dc}(R)}\right)\frac{R^7}{d^8} + \cdots. \tag{6.63}$$

The leading term is material independent but different from that of the perfect metal sphere interaction of (6.56) since only the electric polarization contributes to it. At next order, the first and second terms in the parentheses come from magnetic and electric dipole fluctuations, respectively. Notice that the term $\sim 1/d^8$ is absent in the interaction between perfectly conducting spheres, see (6.56). The limit of perfect conductivity, $\sigma_{dc} \to \infty$, cannot be taken in (6.63) since this limit does not commute with the low κ or large d expansion.

In order to estimate the effect of finite conductivity and its dependence on the size of the nanoparticle, we have to employ a theory that can describe the evolution of $\sigma_{dc}(R)$ with the particle size. A theory for the dielectric function of a cubical metallic particle of dimensions $R \gg \pi/k_F$ has been developed within the random phase approximation in the limit of low frequencies $\ll c/R$ [62]. In this theory, it is further assumed that the discreteness of the electronic energy levels, and not the inhomogeneity of the charge distribution, is important. This implies that the particle responds only at the wave vector of the incident field which is a rather common approximation for small particles. From an electron number-conserving relaxation time approximation, the complex dielectric function is obtained which yields the size-dependent dc conductivity for a cubic particle of volume a^3 [62]. It has been shown that the detailed shape of the particle barely matters, and we can set $a = (4\pi/3)^{1/3}R$ which defines the volume equivalent sphere radius R. This yields the estimate

$$\sigma_{dc}(R) = \sigma_{dc}(\infty) \left[1 - \frac{3\pi k_F a + \pi^2}{4(k_F a)^2} - \frac{48\pi}{(k_F a)^3 \Gamma^2} \right.$$

$$\left. \times \text{Re} \sum_{m=1}^{k_F a/\pi} m^2((k_F a/\pi)^2 - m^2) \times \begin{cases} -z_m \tan z_m & m \text{ even} \\ +z_m \cot z_m & m \text{ odd} \end{cases} \right] \quad (6.64)$$

with

$$z_m = \frac{\pi m}{2} \sqrt{1 - \frac{i\Gamma}{m^2}}, \quad (6.65)$$

where $\sigma_{dc}(\infty) = \omega_p^2 \tau / 4\pi$ is the bulk Drude dc conductivity and $\Gamma = (\hbar/\tau\epsilon_F)(k_F a/\pi)^2$ is a linewidth with Fermi energy ϵ_F. The factor in square parentheses multiplying $\sigma_{dc}(\infty)$ describes quantum size effects and leads to a substantial *reduction* of the dc conductivity for nanoscale particles. While the above expression

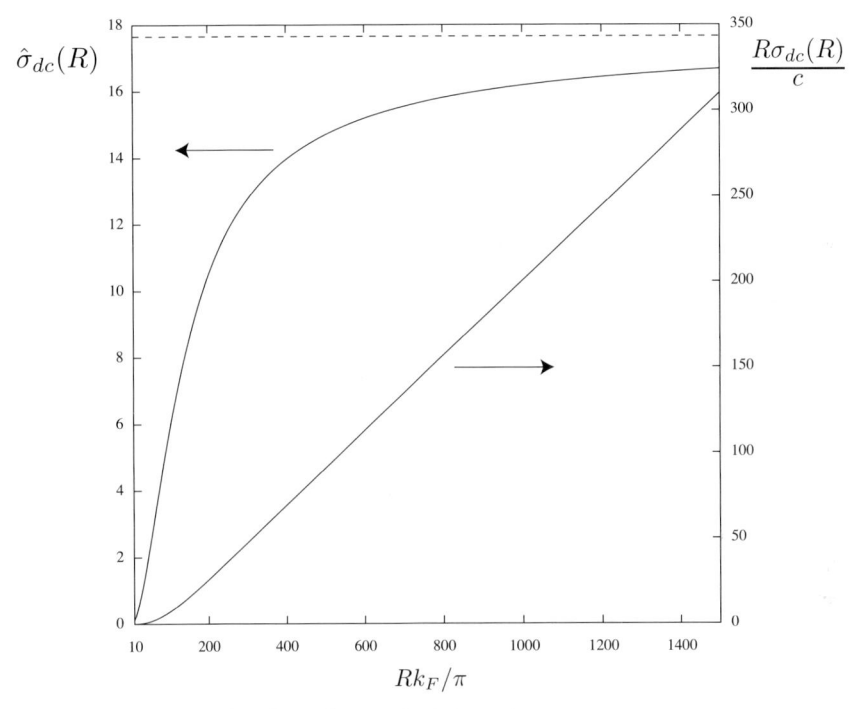

Figure 6.10 Dimensionless dc conductivity $\hat{\sigma}_{dc}(R)$ in units of $e^2/2\hbar a_0$ (with Bohr radius a_0) for a Aluminum sphere with $\epsilon_F = 11.63$ eV, $\pi/k_F = 1.8$ Å and $\tau = 0.8 \cdot 10^{-14}$ sec as function of the radius R, measured in units of π/k_F, see (6.64). The corresponding ratio $R\sigma_{dc}(R)/c$ that determines the Casimir interaction of (6.63) is also shown. The bulk dc conductivity $\hat{\sigma}_{dc}(\infty) = 17.66$ is indicated by a dashed line.

is applicable for $\pi/k_F \ll a$, it suggests that for $\pi/k_F \simeq a$, the particle ceases to conduct, which is consistent with a metal–insulator transition due to the localization of electrons for particles with a size of the order of the mean free path. It is instructive to consider the size dependence of $\sigma_{dc}(R)$ and of the Casimir interaction for a particular choice of material. Following [62], we focus on small Aluminum spheres with $\epsilon_F = 11.63\,\text{eV}$ and $\tau = 0.8 \cdot 10^{-14}\,\text{s}$. These parameters correspond to $\pi/k_F = 1.8\,\text{Å}$ and a plasma wavelength $\lambda_p = 79\,\text{nm}$. It is useful to introduce the dimensionless conductivity $\hat{\sigma}_{dc}(R)$, which is measured in units of $e^2/2\hbar a_0$ with Bohr radius a_0, so that the important quantity of (6.63) can be written as $R\sigma_{dc}(R)/c = (\alpha/2)(R/a_0)\hat{\sigma}_{dc}(R)$ where α is the fine-structure constant. The results following from (6.64) are shown in Figure 6.10. For example, for a sphere of radius $R = 10\,\text{nm}$, the dc conductivity is reduced by a factor ≈ 0.15 compared to the bulk Drude value. If the radius of the sphere is equal to the plasma wavelength λ_p, the reduction factor ≈ 0.8. These results show that shape and material properties are important for the Casimir interaction between nanoparticles. Potential applications include the interaction between dilute suspensions of metallic nanoparticles.

6.4
Casimir Force Driven Nanosystems

We have seen that Casimir forces increase strongly with decreasing distance and hence it can be expected that they are important in devices that are composed of moving elements at short separations. Indeed, a common phenomena seen in nanomechanical devices is stiction due to attractive van der Waals and Casimir forces. This effect imposes a minimum separation between objects in order to prevent them from sticking together. However, one can also make good use of Casimir interactions in nanodevices by employing them to actuate components of small devices without contact [1, 2, 29]. In [29], it has been demonstrated that this can be achieved by coupling two periodically structured parallel surfaces by the zero-point fluctuations of the electromagnetic field between them. We will consider this effect as an example for Casimir force induced nonlinear dynamics, providing a direct application of the results obtained in Section 6.2.2. We have seen that the broken translation symmetry parallel to the surfaces results in a sideways force which has been predicted theoretically [34, 35] and observed experimentally between static surfaces [21]. If at least one of the surfaces is structured *asymmetrically*, there is an additional breaking of reflection symmetry and the surfaces can in principle be set into relative lateral motion in the direction of broken symmetry. The energy for this transport has to be pumped into the system by external driving. This can be realized by setting the surfaces into relative oscillatory motion so that their normal distance is an unbiased periodic function of time. Since the sideways Casimir force decays exponentially with the normal distance (see 6.41), the surfaces experience an asymmetric periodic potential that strongly varies in time.

This scenario resembles so-called ratchet systems [58] that have been studied extensively during the last decade in the context of Brownian particles [41], molecular motors [3] and vortex physics in superconductors [24], to name a few recent examples. Most of the works on ratchets consider an external time-dependent driving force acting on overdamped degrees of freedom to rectify thermal noise. For nanosystems, however, it has been pointed out that inertia terms due to finite mass should not be neglected and, actually, can help the ratchets to perform more efficiently than their overdamped companions [48]. Finite inertia typically induces deterministic chaos in Langevin dynamics. This chaos has been shown to be capable of mimicking the role of noise, and hence to generate directed transport in the absence of external noise [49]. Here, we use this effect in the different context of so-called pulsating (or effectively on–off) ratchets where the strengths of the periodic potential varies in time [58]. We consider weak thermal noise only to test for stability of the inertia induced transport, not as the source of driving[14].

It has been demonstrated that the system described above indeed allows for directed relative motion of the surfaces due to chaotic dynamics caused by the lateral Casimir force [29]. The transport velocity is stable across sizeable intervals of the amplitude and frequency of surface distance oscillations and damping. The velocity scales linear with frequency across these intervals and is almost constant below a critical mean distance beyond which, it drops sharply. The system exhibits multiple current reversals as function of the oscillation amplitude, mean distance and damping. This "Casimir ratchet" allows contact-less transmission of motion which is important since traditional lubrication is not applicable in nanodevices. This actuation mechanism should be compared to other actuation schemes as magnetomotive or capacitive (electrostatic) force transmission. The Casimir effect induced actuation has the advantage of working also for insulators and does not require any electrical contacts and/or external fields. Other applications of zero-point fluctuation induced (van der Waals) interactions to nanodevices have already been experimentally realized in order to construct ultra-low friction bearings from multi-wall carbon nanotubes [23].

In the following, we consider two (on average) parallel metallic surfaces with periodic, uni-axial corrugations (along the y_1-axis) that have distance H, see inset (a) of Figure 6.11. To begin with, we assume that both surfaces are at rest with a relative lateral displacement b. Then the surface profiles can be parametrized as

$$h_1(y_1) = a \sum_{n=1}^{\infty} c_n e^{2\pi i n y_1/\lambda_1} + \text{c.c.}, \tag{6.66}$$

$$h_2(y_1) = a \sum_{n=1}^{\infty} d_n e^{2\pi i n (y_1-b)/\lambda_2} + \text{c.c.}, \tag{6.67}$$

where a is the corrugation amplitude, λ_1, λ_2 are the corrugation wave lengths, and c_n, d_n are Fourier coefficients.

14) In the absence of inertia, finite thermal noise *is* necessary for on–off ratchets to generate directed motion.

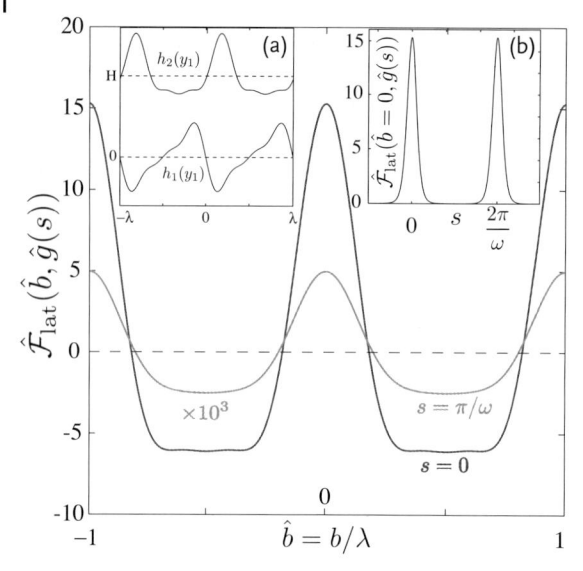

Figure 6.11 The lateral Casimir force acting between the two surfaces as function of the shift $\hat{b}$ at time $s = 0$ and half period $s = \pi/\omega$ (drawn to a larger scale by a factor 10^3) for parameters $\eta = 0.65$, $H_0 = 0.1\lambda$. Insets: (a) Surface profiles at their equilibrium position at $\hat{b} = 0.182$ (b) Periodic variation of the maximum force at $\hat{b} = 0$ with time.

The dependence of the Casimir energy $\mathcal{E}$ on H and b causes macroscopic forces on the surfaces. For a varying separation H, this is the normal Casimir attraction between metallic surfaces modified by the corrugations. Below, we will assume $H = H(t)$ to be a time-dependent distance that is kept at a fixed oscillation by an additional external force from clamping to an oscillator. In such a setup, the surfaces can react freely only to the lateral force component $\mathcal{F}_{\text{lat}}(b, H) = -\partial \mathcal{E}/\partial b$. The results of Section 6.2.2 are readily extended to periodic profiles of arbitrary shapes as described by (6.66). The corrugation lengths have to be commensurate, $\lambda_1/\lambda_2 = p/q$ with integers p, q in order to produce a finite lateral force per surface area. For the purpose of this example, it is sufficient to consider the case $p = 1$. Generalizing the result of (6.37), the lateral (b-dependent) part of the Casimir energy per surface area can then be written as

$$\mathcal{E}(b) = \frac{2\hbar c a^2}{H^5} \sum_{n=1}^{\infty} \left(c_n d_{-nq} e^{-2\pi i n b/\lambda_1} + \text{c.c.} \right) J\left(n \frac{H}{\lambda_1}\right) \tag{6.68}$$

to order a^2. The exact form of the function $J(x) = J_{\text{TM}}(x) + J_{\text{TE}}(x)$ is given by (6.38), (6.39). For the present purpose, it is sufficient to use the simplified expression

$$J(x) \simeq \frac{\pi^2}{120} \left(1 + 2\pi x + \gamma x^2 + 32 x^4\right) e^{-2\pi x} \tag{6.69}$$

with $\gamma = 12.4133$, which is exact for both asymptotically large and small x and approximates the exact results with sufficient accuracy for all x (The maximal deviation from the exact result is $\approx \pm 0.5\%$ around $x = 0.5$). The Casimir potential of (6.68) has two interesting properties which are useful to the construction of a ratchet. First, it decays exponentially with H, and thus can be essentially switched on and off periodically in time by oscillating H. Second, the potential is not only periodic in b, but acquires asymmetry from the surface profiles at small $H \ll \lambda$ and an universal symmetric shape for $H \gg \lambda$ since the effect of higher harmonics of the surface profile is exponentially diminished, as discussed in Section 6.2.2.

The relative surface displacement $b(t)$ can be considered as a classical degree of freedom with inertia. Its equation of motion is described by Langevin dynamics of the form

$$\rho \ddot{b} + \gamma \rho \dot{b} = \mathcal{F}_{\text{lat}}[b, H(t)] + \sqrt{2\gamma \rho T} \xi(t), \qquad (6.70)$$

where ρ is the mass per surface area, γ the friction coefficient, T the intensity (divided by surface area) of the Gaussian noise $\xi(t)$ with zero mean and correlations $\langle \xi(t) \xi(t') \rangle = \delta(t - t')$ so that the Einstein relation is obeyed. This stochastic term describes ambient noise due to effects of temperature and pressure. (Additional contributions from thermally excited photons to the Casimir force can be neglected at surface distances well below the thermal wavelength $\hbar c/(2T)$.) The system is driven by rigid oscillations of one surface so that the distance $H(t) = H_0 g(t)$ oscillates about the mean distance H_0 with $g(t) = 1 - \eta \cos(\Omega t)$. For simplicity, we now consider equal corrugation lengths $\lambda_1 = \lambda_2 \equiv \lambda$. We define the following dimensionless variables: $\hat{b} = b/\lambda$, $s = t/\tau$ for lateral lengths and time with the typical time scale $\tau = (\lambda/a) \sqrt{\rho H_0^5/\hbar c}$ resulting from a balance between inertia and Casimir force. Therefore, velocities will be measured in units of $v_0 = \lambda/\tau$. There are five dimensionless parameters which can be varied independently for fixed surface profiles: the damping $\hat{\gamma} = \tau \gamma$, the angular frequency $\omega = \tau \Omega$, the driving amplitude η, the scaled mean distance H_0/λ and the noise intensity $\hat{T} = (T/\hbar c)(H_0^5/a^2)$. The dimensionless equation of motion for $\hat{b}(s)$ reads

$$\ddot{\hat{b}} + \hat{\gamma} \dot{\hat{b}} = \hat{\mathcal{F}}_{\text{lat}}[\hat{b}, \hat{g}(s)] + \sqrt{2\hat{\gamma} \hat{T}} \hat{\xi}(s) \qquad (6.71)$$

with the Casimir force

$$\hat{\mathcal{F}}_{\text{lat}}(\hat{b}, \hat{g}) = \frac{4\pi}{\hat{g}^5} \sum_{n=1}^{\infty} f_n \cos(2\pi n \hat{b}) J\left(n \hat{g} \frac{H_0}{\lambda}\right), \qquad (6.72)$$

where we have chosen surface profiles with $c_n = i\sqrt{f_n/(2n)}$, $d_n = \sqrt{f_n/(2n)}$ with real coefficients f_n in (6.66), and $\hat{g}(s) = 1 - \eta \cos(\omega s)$.

Directed transport is possible in certain parameter ranges, even in the deterministic case where noise is absent. However, to probe the robustness of transport, we primarily consider the limit of weak noise by choosing $\hat{T} = 10^{-3}$. In fact, it has been shown for underdamped ratchets with time-independent potentials and periodic driving that even an infinitesimal amount of noise can change the rectification from chaotic to stable [48]. To look for similar generic behavior of the pulsating ratchet, we consider a specific geometry consisting of a symmetric and a sawtooth-like surface profile corresponding to three harmonics with $f_1 = 0.0492$, $f_2 = 0.0241$, $f_3 = 0.0059$ and $f_n = 0$ for $n > 3$. Inset (a) of Figure 6.11 shows these profiles in their stable position with $\hat{b} = 0.182$ that minimizes the Casimir energy. The resulting spatial variation of the Casimir force with $\hat{b}$ is plotted in Figure 6.11 for minimal ($s = 0$) and maximal ($s = \pi/\omega$) surface distance with parameters $H_0/\lambda = 0.1$, $\eta = 0.65$. It can be clearly seen that the asymmetry is reduced at a larger distance where the variation of the force becomes more sinusoidal. Inset (b) shows the on–off-like time-dependence of the force amplitude at $\hat{b} = 0$ due to the oscillating surface distance.

The nonlinear equation of motion of (6.71) has to be solved numerically. The trajectory $\hat{b}(s)$ was obtained from a second order Runge–Kutta algorithm. For initial conditions, an equidistant distribution over the interval $[-1, 1]$ for $\hat{b}(0)$ and $\dot{\hat{b}}(0) = 0$ is used. For each set of parameters, 200 different trajectories are calculated from varying initial conditions and noise, each evolving over 4×10^3 periods $2\pi/\omega$ so that transients have decayed. The average velocity $\langle\langle v \rangle\rangle$ involves two different averages of $\dot{\hat{b}}(s)$: The first average is over initial conditions and noise for every time step, then the averaged trajectory is averaged over all discrete times of the numerical solution. For an efficient directed transport, it is not sufficient to have only a finite average $\langle\langle v \rangle\rangle$. To exclude trajectories with a high number of velocity reversals, the fluctuations about the average velocity must be small, that is, the variance $\sigma^2 = \langle\langle v^2 \rangle\rangle - \langle\langle v \rangle\rangle^2$ must be smaller than $\langle\langle v \rangle\rangle^2$.

Naively, one can expect directed motion of the surface profile $h_2(y_1)$ into the positive y_1-direction ($\hat{b} < 0$) since the Casimir force in Figure 6.11 is asymmetric with negative values lasting for a longer time than positive ones. However, the actual behavior is more complicated due to chaotic dynamics. Figure 6.12 shows the dependence of the average velocity and its standard deviation σ on the driving amplitude η and frequency ω for $H_0 = 0.1\lambda$, $\hat{\gamma} = 0.9$. For a fixed frequency, there is an optimal interval of driving amplitudes across which the average velocity is almost constant with $\langle\langle v \rangle\rangle \simeq -\omega/(2\pi)$. Small deviations from the latter value result from noise, as has been checked by studying the dynamics at $\hat{T} = 0$. At higher driving amplitudes, a second narrower interval with maximal $\langle\langle v \rangle\rangle$ is observed which is more strongly reduced and smeared out from its deterministic value $-2 \times \omega/(2\pi)$ by noise. At the plateaus of constant velocity, the standard deviation σ is substantially reduced, rendering transport efficient. Outside the plateaus, velocity reversals occur and σ increases linearly with η. For fixed amplitude η, the average velocity is stable at the value $-\omega/(2\pi)$ over a sizeable frequency range (see inset of Figure 6.12).

In order to understand the observed behavior it is instructive to analyze the dynamics in the three-dimensional extended phase space. Attractors of the long-time dynamics can be identified from Poincaré sections using the period $2\pi/\omega$ of the surface oscillation as stroboscopic time. To obtain a compact section, the trajecto-

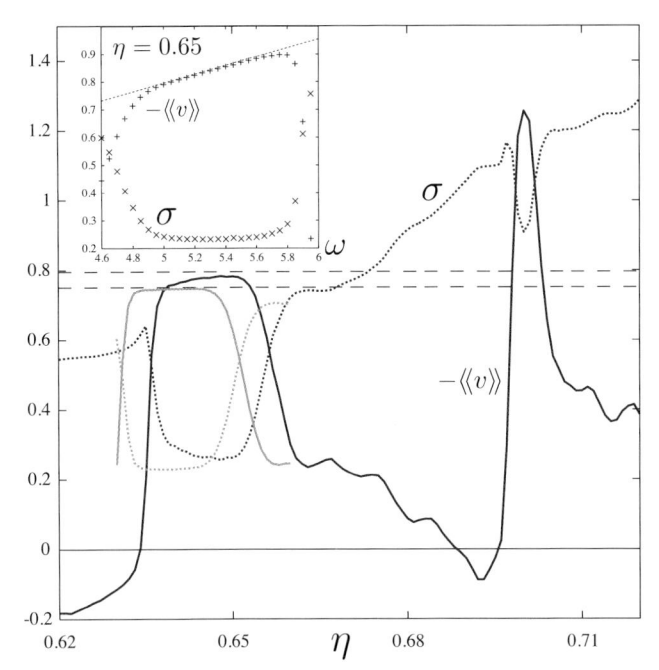

Figure 6.12 Mean $\langle\langle v \rangle\rangle$ and standard deviation σ of the (negative) velocity as function of the driving amplitude η for the frequencies $\omega = 5.0$ and $\omega = 4.72$ (for the latter, only the stable plateau is shown). The parameters are $H_0 = 0.1\lambda$, $\hat{\gamma} = 0.9$, $\hat{T} = 10^{-3}$. Inset: Dependence of the same quantities on frequency for fixed $\eta = 0.65$. Straight dashed lines correspond in both graphs to the velocity $\omega/(2\pi)$.

ry is folded periodically in y_1 on one period of the Casimir potential. From these sections, we can distinguish between periodic and chaotic orbits. As a start, we consider the deterministic limit with $\hat{T} = 0$. The plateaus around $\eta = 0.65$ and $\eta = 0.7$ both result from periodic orbits of period one, corresponding to a single point in the Poincaré section. On the right (downward) edges of the first plateaus, we observe period doubling, that is, a periodic attractor with period two. Upon a further increase of η, chaotic orbits dominate the motion. Therefore, the system exhibits a period-doubling route to chaos with enhanced velocity fluctuations. The findings also basically apply to weak noise ($\hat{T} = 10^{-3}$), but the sharp points of the periodic attractors in the Poincaré sections are smeared out, leading to a decreased $\langle\langle v \rangle\rangle$. The transition from chaotic to periodic dynamics at the beginning of the rising edge of the plateaus is accompanied by a velocity reversal. This is consistent with the observation for non-pulsating potentials that velocity reversals are due to a bifurcation from chaotic to periodic dynamics [49].

The amplitude of the Casimir potential can be tuned by varying the mean distance H_0. From Figure 6.13a, we see that the dynamics show a sharp transition at a critical H_0/λ from efficient transport with large $\langle\langle v \rangle\rangle$ and small σ to chaotic

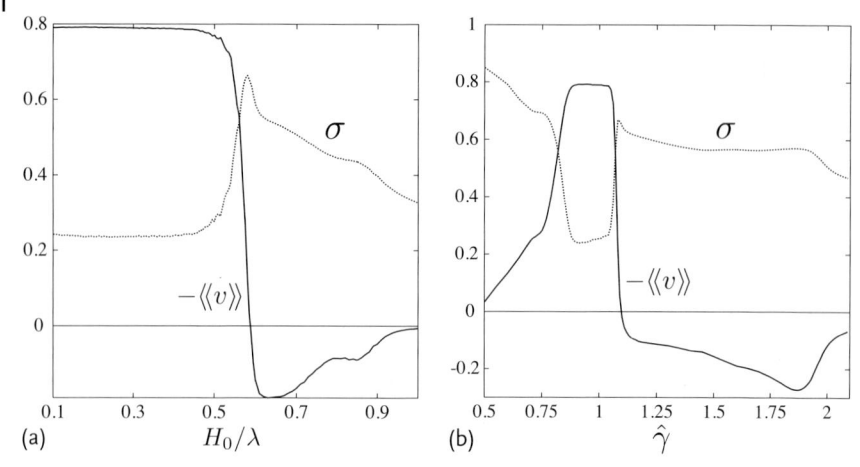

Figure 6.13 Mean $\langle\langle v\rangle\rangle$ and standard deviation σ of the (negative) velocity as a function of (a) the mean plate distance H_0 for $\hat{\gamma} = 0.9$ and (b) damping $\hat{\gamma}$ for $H_0 = 0.1\lambda$. The other parameters are $\eta = 0.65$, $\omega = 5.0$, $\hat{T} = 10^{-3}$.

dynamics with vanishing velocity. The transition is accompanied by a velocity reversal and peaked velocity fluctuations. Interestingly, below the transition $\langle\langle v\rangle\rangle$ is almost constant independently of H_0/λ. The observed transport behavior is also stable against a change of effective damping $\hat{\gamma}$, as shown in Figure 6.13b. Whereas fluctuations increase with decreasing $\hat{\gamma}$, there is a stable plateau of constant average velocity across which fluctuations are diminished. In the deterministic limit, additional plateaus with inverted and doubled average velocity are observed by varying $\hat{\gamma}$ and η. Remnants of a second plateau around $\hat{\gamma} = 1.9$, washed out by noise, can be seen in Figure 6.13b.

It is interesting to estimate typical velocities $v_0 = \lambda/\tau$. With the typical lengths $H_0 = 0.1\,\mu\text{m}$, $a = 10\,\text{nm}$ realized in recent Casimir force measurements [21] and an area mass density of $\rho = 10\,\text{g/m}^2$ for silicon plates with a thickness of a few microns, one obtains $v_0 = \sqrt{\hbar c a^2/\rho H_0^5} \approx 5.5\,\text{mm/s}$. The actual average velocity $v_0\omega/2\pi$ is of the same order for the frequencies studied above. For $\lambda = 1\,\mu\text{m}$, the time scale is $\tau = \lambda/v_0 \approx 10^{-4}\,\text{s}$, leading to driving frequencies and damping rates in the kHz range for the parameters considered here.

The results show that Casimir interactions can offer novel contact-less translational actuation schemes for nanomechanical systems. Similar ratchet-like effects are expected between objects of different shapes as, for example, periodically structured cylinders inducing rotational motion. The use of fluctuation forces also appear promising to move nano-sized objects immersed in a liquid where electrostatic actuation is not possible. Another application is the separation and detection of particles of differing mass adsorbed to the surfaces. For surfaces oscillating at very high frequencies, additional interesting phenomena related to the dynamical Casimir effect occur [37], leading to the emission of photons that could contribute to ratchet-like effects as well.

6.5
Conclusion

Recently, there has been much interest in applying Casimir interactions to the design of nanomechanical devices [9–11, 16, 17]. In such devices as sensors and actuators, attractive Casimir forces can strongly influence their function due to unwanted stiction between small elements at nanoscale separations. However, one can also utilize Casimir interactions in actuators where they can lead to interesting nonlinear dynamics. Recently, repulsive Casimir forces between bodies in a liquid, predicted some decades ago by Lifshitz, Dzyaloshinskii and Pitaevskii for planar surfaces [26], have also been measured between a sphere and a plane [52], suggesting a way to suppress stiction. Thus, it is important to understand the dependence of Casimir forces on shape and material properties beyond common approximations that only apply to weakly curved surfaces. This conclusion is corroborated by the relevance of Casimir interactions to a plethora of phenomena such as wetting, adhesion, friction, and quantum scattering of atoms from surfaces. In this chapter, some characteristic effects of shape and material on Casimir interactions have been presented using the examples of geometries that are typical to nanosystems. Most of the presented results could only be obtained recently by newly developed theoretical tools that have been described here. The important study of correlations between shape and material effects and the additional implications of interacting fields in Casimir effects due to critical fluctuations [39] are largely unexplored. It is expected that the recent progress on the experimental and theoretical side will unveil novel phenomena and provide a better understanding of fluctuation induced interactions with interesting implications for nanosystems.

Acknowledgements

Most of the results presented in this chapter have been obtained in collaboration with Noah Graham, Andreas Hanke, Robert L. Jaffe, Mehran Kardar, Sahand Jamal Rahi, and Antonello Scardicchio. Support from the DFG grant No. EM70/3, the Center for Theoretical Physics at MIT, and the MIT-France Seed Fund is gratefully acknowledged.

References

1 Ashourvan, A., Miri, M., and Golestanian, R. (2007) Noncontact rack and pinion powered by the lateral Casimir force. *Phys. Rev. Lett.*, **98**, 140801.
2 Ashourvan, A., Miri, M., and Golestanian, R. (2007) Rectification of the lateral Casimir force in a vibrating noncontact rack and pinion. *Phys. Rev. E*, **75**, 040103.
3 Astumian, R.D. and Derenyi, I. (1998) Fluctuation driven transport and models of molecular motors and pumps. *Eur. Biophys. J.*, **27**, 474.
4 Balian, R. and Duplantier, B. (1977) Electromagnetic waves near perfect conductors. I. Multiple scattering ex-

pansions. Distribution of modes. *Ann. Phys.*, **104**, 300.
5 Balian, R. and Duplantier, B. (1978) Electromagnetic-waves near perfect conductors. 2. Casimir effect. *Ann. Phys.*, **112**, 165.
6 Bordag, M. (2006) Casimir effect for a sphere and a cylinder in front of a plane and corrections to the proximity force theorem. *Phys. Rev. D*, **73**, 125018.
7 Bordag, M., Mohideen, U., and Mostepanenko, V.M. (2001) New developments in the Casimir effect. *Phys. Rep.*, **353**, 1.
8 Bressi, G., Carugno, G., Onofrio, R., and Ruoso, G. (2002) Measurement of the Casimir force between parallel metallic surfaces. *Phys. Rev. Lett.*, **88**.
9 Buks, E. and Roukes, M.L. (2001) Metastability and the Casimir effect in micromechanical systems. *Europhys. Lett.*, **54**, 220.
10 Buks, E. and Roukes, M.L. (2001) Stiction, adhesion energy, and the Casimir effect in micromechanical systems. *Phys. Rev. B*, **63**, 033402.
11 Buks, E. and Roukes, M.L. (2002) Quantum physics: Casimir force changes sign. *Nature*, **419**, 119.
12 Buscher, R. and Emig, T. (2004) Nonperturbative approach to Casimir interactions in periodic geometries. *Phys. Rev. A*, **69**, 062101.
13 Buscher, R. and Emig, T. (2005) Geometry and spectrum of Casimir forces. *Phys. Rev. Lett.*, **94**, 133901.
14 Casimir, H.B.G. (1948) On the attraction between two perfectly conducting plates. *Kon. Ned. Akad. Wetensch. Proc.*, **51**, 793.
15 Casimir, H.B.G. and Polder, D. (1948) The influence of retardation on the London–van der Waals forces. *Phys. Rev.*, **73**, 360.
16 Chan, H.B., Aksyuk, V.A., Kleiman, R.N., Bishop, D.J., and Capasso, F. (2001) Nonlinear micromechanical Casimir oscillator. *Phys. Rev. Lett.*, **87**, 211801.
17 Chan, H.B., Aksyuk, V.A., Kleiman, R.N., Bishop, D.J., and Capasso, F. (2001) Quantum mechanical actuation of microelectromechanical systems by the Casimir force. *Science*, **291**, 1941.
18 Chan, H.B., Bao, Y., Zou, J., Cirelli, R.A., Klemens, F., Mansfield, W.M., and Pai, C.S. (2008) Measurement of the Casimir force between a gold sphere and a silicon surface with nanoscale trench arrays. *Phys. Rev. Lett.*, **101**, 030401.
19 Chen, F., Klimchitskaya, G.L., Mostepanenko, V.M., and Mohideen, U. (2006) Demonstration of the difference in the Casimir force for samples with different charge-carrier densities. *Phys. Rev. Lett.*, **97**, 170402.
20 Chen, F., Klimchitskaya, G.L., Mostepanenko, V.M., and Mohideen, U. (2007) Control of the Casimir force by the modification of dielectric properties with light. *Phys. Rev. B*, **76**, 035338.
21 Chen, F., Mohideen, U., Klimchitskaya, G.L., and Mostepanenko, V.M. (2002) Demonstration of the lateral Casimir force. *Phys. Rev. Lett.*, **88**, 101801.
22 Chen, F., Mohideen, U., Klimchitskaya, G.L., and Mostepanenko, V.M. (2002) Experimental and theoretical investigation of the lateral Casimir force between corrugated surfaces. *Phys. Rev. A*, **66**, 032113.
23 Cumings, J. and Zettl, A. (2000) Low-friction nanoscale linear bearing realized from multiwall carbon nanotubes. *Science*, **289**, 602.
24 de Souz Silva, C.C., Van de Vondel, J., Morelle, M., and Moshchalkov, V.V. (2006) Controlled multiple reversals of a ratchet effect. *Nature*, **440**, 651. 10.1038/nature0(4595).
25 Decca, R.S., López, D., Fischbach, E., Klimchitskaya, G.L., Krause, D.E., and Mostepanenko, V.M. (2007) Tests of new physics from precise measurements of the Casimir pressure between two gold-coated plates. *Phys. Rev. D*, **75**, 077101.
26 Dzyaloshinskii, I.E., Lifshitz, E.M., and Pitaevskii, L.P. (1961) The general theory of van der waals forces. *Adv. Phys.*, **10**, 165.
27 Ederth, T. (2000) Template-stripped gold surfaces with 0.4-nm rms roughness suitable for force measurements: Application to the Casimir force in the 20–100-nm range. *Phys. Rev. A*, **62**, 062104.
28 Emig, T. (2003) Casimir forces: An exact approach for periodically deformed objects. *Europhys. Lett.*, **62**, 466.
29 Emig, T. (2007) Casimir-force-driven ratchets. *Phys. Rev. Lett.*, **98**, 160801.
30 Emig, T. (2008) Fluctuation-induced quantum interactions between compact objects

and a plane mirror. *J. Stat. Mech.: Theory Exp.*, **4**, P04007.

31 Emig, T., and Buscher, R. (2004) Towards a theory of molecular forces between deformed media. *Nucl. Phys. B*, **696**, 468.

32 Emig, T., Graham, N., Jaffe, R.L., and Kardar, M. (2007) Casimir forces between arbitrary compact objects. *Phys. Rev. Lett.*, **99**, 170403.

33 Emig, T., Graham, N., Jaffe, R.L., and Kardar, M. (2008) Casimir forces between compact objects: The scalar case. *Phys. Rev. D*, **77**, 025005.

34 Emig, T., Hanke, A., Golestanian, R., and Kardar, M. (2001) Probing the strong boundary shape dependence of the Casimir force. *Phys. Rev. Lett.*, **87**, 260402.

35 Emig, T., Hanke, A., Golestanian, R., and Kardar, M. (2003) Normal and lateral Casimir forces between deformed plates. *Phys. Rev. A*, 67(022114).

36 Emig, T., Jaffe, R.L., Kardar, M., and Scardicchio, A. (2006) Casimir interaction between a plate and a cylinder. *Phys. Rev. Lett.*, **96**, 080403.

37 Golestanian, R. and Kardar, M. (1997) Mechanical response of vacuum. *Phys. Rev. Lett.*, **78**, 3421.

38 Golestanian, R. and Kardar, M. (1998) Path-integral approach to the dynamic Casimir effect with fluctuating boundaries. *Phys. Rev. A*, **58**, 1713.

39 Hertlein, C., Helden, L., Gambassi, A., Dietrich, S., and Bechinger, C. (2008) Direct measurement of critical Casimir forces. *Nature*, **451**, 172.

40 Jaekel, M.T. and Reynaud, S. (1991) Casimir force between partially transmitting mirrors. *J. Phys. I*, **1**, 1395.

41 Jülicher, F., Ajdari, A., and Prost, J. (1997) Modeling molecular motors. *Rev. Mod. Phys.*, **69**, 1269.

42 Kenneth, O. and Klich, I. (2008) Casimir forces in a *t*-operator approach. *Phys. Rev. B*, **78**, 014103.

43 Kenneth, O., Klich, I., Mann, A., and Revzen, M. (2002) Repulsive Casimir forces. *Phys. Rev. Lett.*, 89(033001).

44 Krause, D.E., Decca, R.S., López, D., and Fischbach, E. (2007) Experimental investigation of the Casimir force beyond the proximity-force approximation. *Phys. Rev. Lett.*, **98**, 050403.

45 Lambrecht, A., Maia-Neto, P.A., and Reynaud, S. (2006) The Casimir effect within scattering theory. *New. J. Phys.*, **8**, 243.

46 Lamoreaux, S.K. (1997) Demonstration of the Casimir force in the 0.6 to 6 μm range. *Phys. Rev. Lett.*, **78**, 5.

47 Lifshitz, E.M. (1956) The theory of molecular attractive forces between solids. *Sov. Phys. JETP*, **2**, 73.

48 Marchesoni, F., Savel'ev, S., and Nori, F. (2006) Achieving optimal rectification using underdamped rocked ratchets. *Phys. Rev. E*, **73**, 021102.

49 Mateos, J.L. (2000) Chaotic transport and current reversal in deterministic ratchets. *Phys. Rev. Lett.*, **84**, 258.

50 Mohideen, U. and Roy, A. (1998) Precision measurement of the Casimir force from 0.1 to 0.9 μm. *Phys. Rev. Lett.*, **81**, 4549.

51 Munday, J.N. and Capasso, F. (2007) Precision measurement of the Casimir–Lifshitz force in a fluid. *Phys. Rev. A*, **75**, 060102.

52 Munday, J.N., Capasso, F., and Parsegian, V.A. (2009) Measured long-range repulsive Casimir–Lifshitz forces. *Nature*, **457**, 170.

53 Neto, P.A.M., Lambrecht, A., and Reynaud, S. (2005) Casimir effect with rough metallic mirrors. *Phys. Rev. A*, **72**, 012115.

54 Neto, P.A.M., Lambrecht, A., and Reynaud, S. (2005) Roughness correction to the Casimir force: Beyond the proximity force approximation. *Europhys. Lett.*, **69**, 924.

55 Parsegian, V.A. (2005) *Van der Waals Forces*, Cambridge University Press, Cambridge, England.

56 Rahi, S.J., Emig, T., Jaffe, R.L., and Kardar, M. (2008) Casimir forces between cylinders and plates. *Phys. Rev. A*, **78**, 012104.

57 Rahi, S.J., Rodriguez, A.W., Emig, T., Jaffe, R.L., Johnson, S.G., and Kardar, M. (2008) Nonmonotonic effects of parallel sidewalls on Casimir forces between cylinders. *Phys. Rev. A*, **77**, 030101.

58 Reimann, P. (2002) Brownian motors: noisy transport far from equilibrium. *Phys. Rep.*, **361**, 57.

59 Rodriguez, A.W., Ibanescu, M., Iannuzzi, D., Capasso, F., Joannopoulos, J.D., and Johnson, S.G. (2007) Computation and visualization of Casimir forces in arbitrary geometries: Nonmonotonic lateral-wall forces and the failure of proximity-force approximations. *Phys. Rev. Lett.*, **99**, 080401.

60 Roy, A., Lin, C.Y., and Mohideen, U. (1999) Improved precision measurement of the Casimir force. *Phys. Rev. D*, **60**, 111101.

61 Roy, A. and Mohideen, U. (1999) Demonstration of the nontrivial boundary dependence of the Casimir force. *Phys. Rev. Lett.*, **82**, 4380.

62 Wood, D.M. and Ashcroft, N.W. (1982) Quantum size effects in the optical properties of small metallic particles. *Phys. Rev. B*, **25**, 6255.

Part III Nanoelectromechanics

7
The Duffing Oscillator for Nanoelectromechanical Systems
Sequoyah Aldridge

This chapter will explore the Duffing nonlinearity in the context of nanoelectromechanical systems (NEMS). As it turns out, nanoresonators are simple experimental devices for studying this nonlinearity. Both NEMS and MEMS devices will show nonlinear phenomena. These phenomena include the Duffing instability and parametric amplification. The consequences of the Duffing nonlinearity in NEMS devices has not been thoroughly explored to date. As resonators shrink to the nanoscale, the onset of nonlinearity becomes more and more relevant because the dynamic range of the oscillator, defined as the ratio of the kinetic energy at the critical amplitude divided by the thermomechanical energy, scales linearly with dimension [1]. Nanomechanical resonators have recently been the subject of much attention due to the ability to make very high frequency, high quality factor resonators with applications in weak force and small mass detection, frequency stabilization, and possibly quantum computation [2–14].

7.1
Basics of the Duffing Oscillator

A mechanical oscillator with a nonlinear restoring force was first studied by Duffing in 1918 [15]. The equation of motion for the Duffing equation, for a natural resonance frequency Ω_0 and quality factor Q, driven at frequency Ω, has the form

$$M \frac{d^2 Y}{dt^2} + M \frac{\Omega_0}{Q} \frac{dY}{dt} + M \Omega_0^2 Y + K Y^3 = B \cos(\Omega t) + B_{\text{noise}}(t) \,, \qquad (7.1)$$

where Y denotes the displacement amplitude, M denotes the mass of the resonator, B the amplitude of the external driving force, and $B_{\text{noise}}(t)$ the stochastic forcing function due to thermal and external noise [7, 16, 17].

This equation assumes that the beam oscillates in the mode with natural frequency Ω_0, that the displacement amplitude $Y(t)$ is the only relevant degree of freedom, and that the equation of motion includes only the third-order nonlinearity, with strength K. The second-order nonlinearity can be ignored for now be-

---

Nonlinear Dynamics of Nanosystems. Edited by Günter Radons, Benno Rumpf, and Heinz Georg Schuster
Copyright © 2010 WILEY-VCH Verlag GmbH & Co. KGaA, Weinheim
ISBN: 978-3-527-40791-0

cause a second-order nonlinearity only mixes signals to dc and to twice the driving frequency. Only odd orders of nonlinearity can mix signals back to the operating frequency band. We will also disregard fifth order and higher odd terms at the moment.

The parameter K determines the strength of the nonlinearity. Deviations from a linear spring can be either softening or stiffening. Positive K yields a stiffening spring; Negative K yields a softening spring.

The basic principle of the Duffing nonlinearity is the following: For the simple harmonic oscillator with no nonlinearity, the resonance curve is a Lorentzian. The root mean square amplitude of motion of the resonator follows (7.2)

$$Y_{\text{rms}} = \frac{B/M}{\sqrt{2}(\Omega_0^2/Q^2 + (\Omega - \Omega_0^2)^2)^{1/2}} . \tag{7.2}$$

However, for a stiffening nonlinearity, as the amplitude is increased, the resonant frequency must increase. This nonlinear detuning of the resonator distorts the shape of the resonance curve. The peak of the curve is pulled in one direction until, ultimately, the curve will have three values for a given frequency. A good reference that describes how to calculate the new shape of the resonance curve for the Duffing resonator is [16]. Figure 7.1 shows the calculated resonance curve after it has been distorted by the Duffing nonlinearity. When there are three possible values for amplitude at a given frequency, two values are stable and one is metastable. The large amplitude and small amplitude branch are stable, whereas the branch in between is metastable.

Because the resonance curve is multi-valued, hysteresis can occur. Figure 7.2 shows the hysteresis curve of a nanobeam resonator. The amplitude is normalized to the peak maximum expected for the calculated curve. If the frequency of the drive is swept from the left to the right in Figure 7.2, the motion will remain stable in the upper branch of the loop. If the frequency is swept high enough, the amplitude will catastrophically drop to the lower amplitude. On the other hand, if the frequency is swept from *right to left*, the motion will remain stable in the *lower* branch of

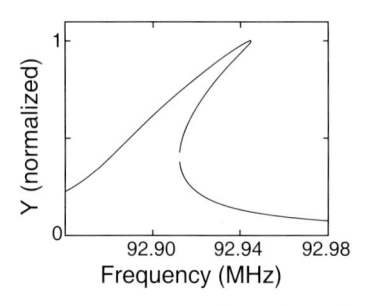

Figure 7.1 This figure shows the amplitude as a function of frequency calculated for a Duffing resonator 7 dB past the critical point. The peak amplitude is normalized to one. At this drive level, three values of amplitude can exist for one value of frequency.

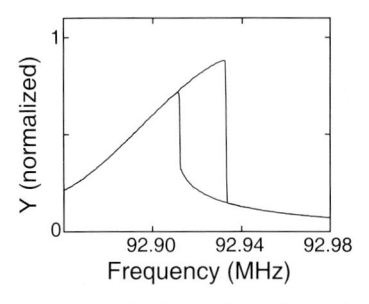

Figure 7.2 This figure shows the amplitude as a function of frequency for a Duffing resonator 7 dB past the critical point. The amplitude is normalized to calculated peak maximum. Hysteresis is observed where three values of amplitude can exist for one frequency.

the loop. If the frequency is swept low enough, the amplitude will catastrophically switch to the upper branch.

Because of the hysteresis, one bit of information can be stored. We can assign the large amplitude state as the one state and the small amplitude state as the zero state. To load the bit into the one state, the frequency must be swept from left to right and then held constant on the upper branch. To load the bit into the zero state, the frequency must be swept from right to left and held constant on the lower branch. Later in the chapter, I will discuss the lifetime of this data bit for a nanoresonator.

For a stiffening nonlinearity, the resonance curve will be pulled upward in frequency. However, it is possible to fabricate devices with a *softening* nonlinearity. In this case, the nonlinearity coefficient K becomes negative. At large amplitudes, for a softening nonlinearity, the peak of the resonance curve is pulled toward negative frequency.

7.2
NEMS Resonators and Their Nonlinear Properties

Using advanced processing techniques, mechanical resonators have been scaled down to sub-micron dimensions. At this size scale, there are three common geometries. They are the doubly-clamped beam, the nanocantilever, and the torsional resonator. These are shown in Figure 7.3.

It is interesting to consider the effect of the geometry of the nanodevice on the nonlinear coefficient. For the doubly-clamped beam system, at large amplitudes, the beam must stretch in order to move. This extra stretching effect causes a stiffening nonlinearity. Therefore, conceptually, a doubly-clamped beam will have a positive value of K.

For the cantilever system, we find a different result. An to consider when thinking about the cantilever system is the simple pendulum. It is easy to show that the

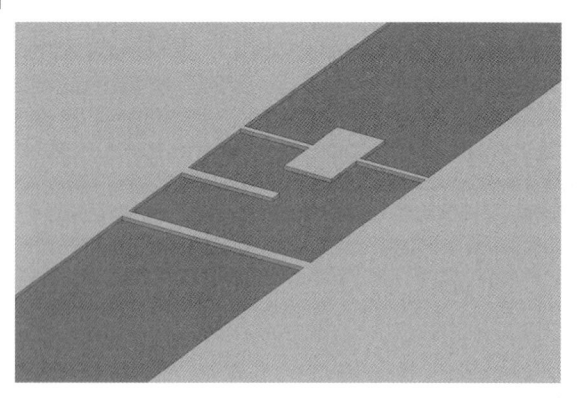

Figure 7.3 This figure shows three typical geometries for nanodevices. They are the doubly-clamped beam, the cantilever, and the torsional resonator.

Figure 7.4 This figure shows a suspended nanowire made from aluminum nitride. The top surface has been covered with titanium then gold. The wire is three microns long.

equation of motion for a driven simple pendulum follows (7.3):

$$Ml\frac{d^2\Theta}{dt^2} + Mg\sin\Theta = B\cos(\Omega t) \,. \tag{7.3}$$

Here, Θ is the angle the pendulum deviates from vertical. Taylor expanding for small Θ gives

$$Ml\frac{d^2\Theta}{dt^2} + Mg(\Theta - \Theta^3/6 + \cdots) = B\cos(\Omega t) \,. \tag{7.4}$$

We find that in this case, the nonlinearity coefficient is in fact negative, leading to a softening spring. As the pendulum swings to large amplitudes, it gains height. This increase in height requires potential energy and saps kinetic energy of the motion. Therefore, the vibrational frequency of the pendulum is decreased at large amplitudes.

Figure 7.5 This figure shows a suspended nanowire viewed from the side.

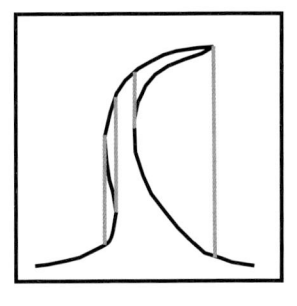

Figure 7.6 This figure illustrates the calculated curve of a resonator that will show dual hysteresis. The hysteretic regions are bounded by the vertical lines.

Similarly, the tip of a cantilever will begin to have motion that is not precisely in and out of plane at large amplitudes. This motion requires extra energy, and causes a frequency downshift. Therefore, cantilevers can have a negative value of K.

A third effect that determines the sign of the nonlinearity is the material nonlinearity. Materials become either stiffer of softer for large strains. At some large deflection, the NEMS device will become affected by the material nonlinearity.

Finally, nanodevices have very large built-in strains. For example, a doubly-clamped beam may have a very large compressional strain in its rest state. The amount of strain will vary to a large degree from device to device. This can lead to a positive or negative nonlinearity coefficient K.

It may be possible to tune the material nonlinearity of a cantilever against the natural softening nonlinearity. Both terms may cancel each other. This would lead to a cantilever with higher linearity and therefore larger dynamic range.

Because a cantilever is free to move at one end, it generally has a higher dynamic range than a doubly-clamped beam. Therefore, a doubly-clamped beam will generally have a higher value of K relative to a cantilever.

I have observed nanodevices that are both softening and stiffening at the same time. A situation where this may occur is when the leading order term yields a softening spring, but higher order terms are stiffening in nature. In this strange case, hysteresis becomes observed on the left side of the peak at some drive power. At some higher drive power, hysteresis will then be observed on the right side of the peak as well. The shape of the calculated resonance peak is an "S" in this case. This is illustrated in Figure 7.6.

The doubly-clamped nanobeam is extraordinarily strong. I have found that drive powers far beyond the critical point do not snap the beam. Instead, they usually survive to the point of melting the metal layer that carries the magnetomotive current.

At large drives, the motion of doubly-clamped beams becomes limited at some point. In other words, a maximum displacement becomes achieved. Higher and higher orders of nonlinearity come in to play. That is the fifth, seventh, ninth term and so on. causing further motion to be impossible. Whether or not the motion of the resonator is chaotic at this point is unknown to the author. Furthermore, to the author's knowledge, no one has demonstrated classical chaos in a nanoresonator. Phenomena in the very large drive regime, say 10 dB above the critical drive, are very complicated. Classical chaos may be observed.

7.3
Transition Dynamics of the Duffing Resonator

This section will discuss the transition dynamics of the Duffing resonator. In other words, how does the resonator move when it undergoes a switch from state to state.

The displacement $Y(t)$ in (7.1) can be written as

$$Y(t) = U_1(t) \cos(\Omega t) + U_2(t) \sin(\Omega t), \qquad (7.5)$$

in terms of the two quadrature amplitudes $U_{1,2}(t)$. For a high Q system driven at frequency Ω near Ω_0, the slowly-varying envelope approximation can be used [16, 18], where the functions $U_{1,2}(t)$ are replaced by their slowly varying averages, $u_{1,2}(t)$, respectively. This type of analysis is commonly used in radio frequency(RF) systems. Using the RF nonmenclature, $u_{1,2}(t)$ are check comma placement the in-phase and quadrature amplitudes of the displacement signal.

In the absence of noise, the average functions $u_{1,2}(t)$ satisfy the equations of motion

$$\left. \begin{aligned} \frac{d^2 u_1}{dt^2} &= \left(\Omega^2 - \Omega_0^2\right) u_1 - \frac{3}{4}\frac{K}{M} u_1 \left(u_1^2 + u_2^2\right) \\ &\quad - \frac{\Omega_0}{Q} \Omega u_2 - \frac{\Omega_0}{Q}\frac{du_1}{dt} - 2\Omega \frac{du_2}{dt} + \frac{B}{M}, \\ \frac{d^2 u_2}{dt^2} &= \left(\Omega^2 - \Omega_0^2\right) u_2 - \frac{3}{4}\frac{K}{M} u_2 \left(u_1^2 + u_2^2\right) \\ &\quad + \frac{\Omega_0}{Q} \Omega u_1 - \frac{\Omega_0}{Q}\frac{du_2}{dt} + 2\Omega \frac{du_1}{dt}. \end{aligned} \right\} \qquad (7.6)$$

This can be found by inserting (7.5) into (7.1) and evaluating. All high frequency terms are ignored, such as terms of $\cos(3\Omega t)$.

One can create a configuration space spanned by the state variables u_1 and u_2. It is interesting to study the dynamic trajectory of the oscillator in this configuration space. Equation 7.6 gives the equation of motion for the state variables.

If the time derivatives of (7.6) are set to zero, then the system must be in a stable, unstable, or metastable state. At is turns out, the stable and metastable states are exactly the branches of the resonance curve discussed earlier. These stable points are called foci.

The Duffing oscillator exhibits one stable state for small drive amplitudes $\mathcal{B}$, while above a critical amplitude $\mathcal{B}_c$ a bifurcation occurs, creating two stable basins of attraction. One basin corresponds to larger displacement amplitudes and is stable for drive frequencies up to an upper critical frequency ν_U ($\nu = \Omega/2\pi$), determined by the drive amplitude $\mathcal{B}$. The other stable basin has smaller displacement amplitude and is stable for frequencies down to a lower critical frequency ν_L, also determined by the drive amplitude. The stable attractors are found by setting all time derivatives in (7.6) to zero and solving for $u_{1,2}$, yielding three equilibrium points. Two of these equilibrium points are stable foci, and the third is a metastable separatrix.

Figure 7.7 shows the numerically-generated flow from initial points near the saddle point. If the initial point is in a given basin, it quickly relaxes to the appropriate focus.

To date, there are many techniques for actuation and detection for NEMS devices [5, 6, 8, 19, 20]. This chapter will neglect the particular choice of transduction and assume the experimenter is given an electrical signal proportional to the motion of the NEMS device. Ultimately, this is the situation normally encountered in practice.

With sufficient measurement bandwidth, it is possible to measure the relaxation of the oscillator to a stable foci. Figure 7.8 shows measured relaxation for a device identical to Figures 7.4 and 7.5. These curves are qualitatively similar to what is shown in Figure 7.7. This reaffirms that the NEMS oscillator is well modeled as a Duffing oscillator.

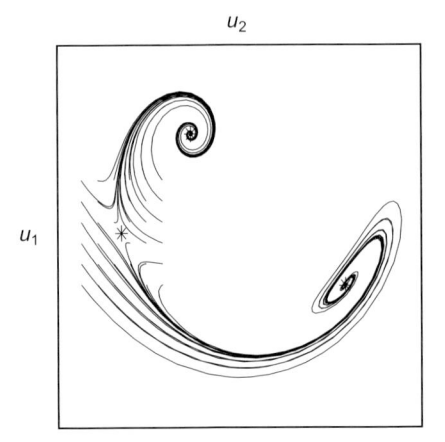

Figure 7.7 Numerically-generated phase-space flow for a drive force 9 dB above the critical point $\mathcal{B}_c$, and drive frequency 40 kHz above $\Omega_0/2\pi$. Flow begins near the separatrix and evolves toward either focus.

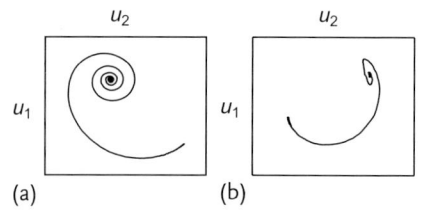

Figure 7.8 (a) Experimental phase space mean trajectory from focus 1 to focus 2 (8000 averages). The resonator was identical to the one pictured in Figures 7.4 and 7.5. (b) Data for phase space mean trajectory from focus 2 to focus 1 (8000 averages).

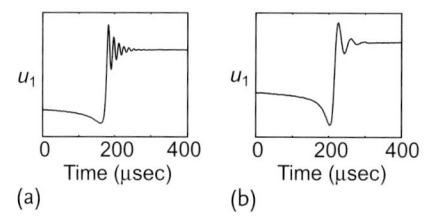

Figure 7.9 (a) and (b) Experimental time traces for the two switching transitions (8000 averages).

If only one state variable, u, is plotted as a function of time, we find the switching curves of Figure 7.9. For a catastrophic switch of the Duffing oscillator, there is ringing that occurs during the transition. This ringing corresponds to the oscillator tracing out a trajectory in the configuration space. We can note that the ringing is not like the ring-down of a linear oscillator. It does not fit to a damped cosine. Instead, the motion is entirely nonlinear in nature and must be calculated by (7.6).

7.4
Energy for "Uphill" Type Transitions

When considering transitions between the two states, there are two STOP possible types of motion, namely, "uphill" and "downhill" motion. The case of downhill motion is simpler. The downhill case is the relaxation case. In this case, a parameter such as frequency or drive amplitude is varied so the state variables can cross the separatrix and relax to the other focus, as described in Section 7.3. The equations of motion for downhill type transitions are (7.6).

The case of "uphill" motion is more complicated. In this case, the oscillator starts at a focus and is perturbed by noise forces. Because the oscillator interacts with the environment, there is a random force that causes stochastic transitions between these stable foci. These noise forces have a small probability to push the oscillator out of the stable focus and into the *other* basin.

We now turn to a discussion of these noise-induced transitions between the stable foci. Thermally-activated escape from a potential landscape with a single basin

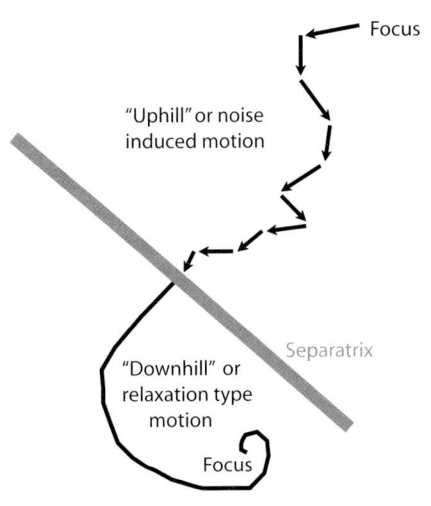

Figure 7.10 Schematic showing "uphill" and "downhill" motion of the oscillator as it switches in the configuration space. For uphill motion, the oscillator is nudged by noise forces into a trajectory that results in a transition.

of attraction is a thoroughly studied problem [21]. The escape rate over a barrier of height E_B is given by $\Gamma = a(Q)\nu_0 \exp(-E_B/k_B T)$, determined predominantly by the Arrhenius factor and less so by the Q-dependent prefactor $a(Q)$. Our system differs from this classic problem: Here, there is a basin of attraction about each of the two foci found on a Poincaré map of the configuration space. Instead of a one-dimensional potential well, there is a quasipotential, with the dynamics governed by the noise energy at each point in the configuration space [22]. The equivalent activation energy, E_A, for transitions between the foci, is found by integrating the minimum available noise energy over the trajectory between the foci.

It is possible to measure the activation energy, E_A, by a histogram measurement technique. By slowly sweeping the drive frequency and recording the precise frequency, a transition occurs and a histogram of frequencies, $h(\nu(t))$, can be generated. Given the sweep rate of the frequency, one can easily calculate the transition rate as a function of frequency. Given the noise power and the transition rate, one can calculate the activation energy, E_A.

Transition histograms were measured by applying a drive signal to the resonator above the critical value, preparing the resonator in one of the two basins of attraction and monitoring the switching transitions to the other basin. Histograms of the switching probability per unit time, $h(t)$, were measured by sweeping the drive frequency $\nu(t) = \Omega(t)/2\pi$ at a constant rate $s = d\nu/dt$, and recording the drive frequency at which a transition occurred. This is a technique that has been extensively used for measuring switching distributions in current-biased Josephson junctions [23].

Transitions were induced by using an external broadband white noise signal, combined with the radiofrequency drive signal using a radio frequency coupler

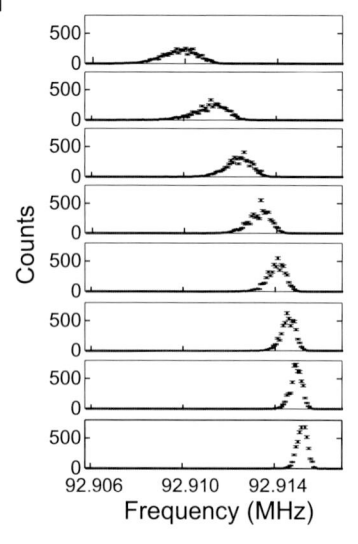

Figure 7.11 Example histograms $h(\nu)$ as a function of noise power, for transitions from focus 1 to 2. The noise power was varied from -127 dBm/Hz to -113 dBm/Hz. Increased noise shifts and broadens the peaks.

in order to generate a signal that included both the drive signal $\mathcal{B}$ and the noise signal $\mathcal{B}_n$. The drive signal itself was produced by a source with very low phase noise; with no additional noise power, transitions were still induced by this remnant phase noise to which the resonator is very sensitive. The thermal noise of the circuit and the mechanical noise associated with the finite resonator Q, were too small to induce measurable transitions in the system. Note that the phase noise injected into the resonator in order to drive the resonator is often much larger than the intrinsic noise of the resonator.

In Figure 7.11, we display a set of histograms $h(\nu(t))$; higher noise powers shift the peak switching frequency and also broaden the distribution. It should be pointed out that the histograms are not Gaussian peaks. Instead they have a particular shape that is *not* symmetric about the center of the peak. Error bars for the points are easily calculated by assuming simple Poisson statistics.

The transition rate $\Gamma(\nu)$ is extracted from the histogram $h(t)$ using $\Gamma(\nu(t)) = (1 - \int_{-\infty}^{t} h(t') \, dt')^{-1} sh(t)$. Some of the transition rate curves are plotted in Figure 7.12. An interesting point about the transition rates is that they have very large variability. The fastest transition rates are on the order of kHz, but the slowest transition rates can have lifetimes of thousands of years if small levels of noise drive the resonator. Of course in practice, it is impossible to wait long enough to see such rare transitions. However, by extrapolating Figure 7.12 for small noise power we can find that such large timescales are possible.

Extracting the quasi-activation energy $E_A(\nu)$, is done by inverting the thermal activation expression $\Gamma(\nu) \equiv \Gamma_0 \exp(-E_A(\nu)/k_B T_{\text{eff}})$, where the effective temperature T_{eff} is proportional to the noise power, and the prefactor Γ_0 is related to the

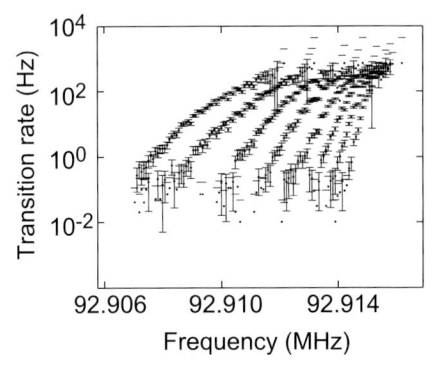

Figure 7.12 Example transition rates $\Gamma(\nu)$ as a function of noise power, for transitions from focus 1 to 2. The noise power was varied from -127 dBm/Hz to -113 dBm/Hz. These are calculated from the histograms in Figure 7.11.

Kramers low-dissipation form [21], $\Gamma_0 \approx \nu_0/Q$. Note that in this technique, the histograms are only logarithmically sensitive to Γ_0, so that a precise determination is difficult. In Figure 7.13, we display the activation energy $E_A(\nu)$ extracted from the histograms, showing the expected decline in the barrier energy as the drive frequency approaches the critical frequency. The distributions shown in Figure 7.12 are seen to collapse onto a single curve $E_A(\nu)$. When measured, these energies reveal, what the author calls, a "butterfly plot". When plotting the energies for both

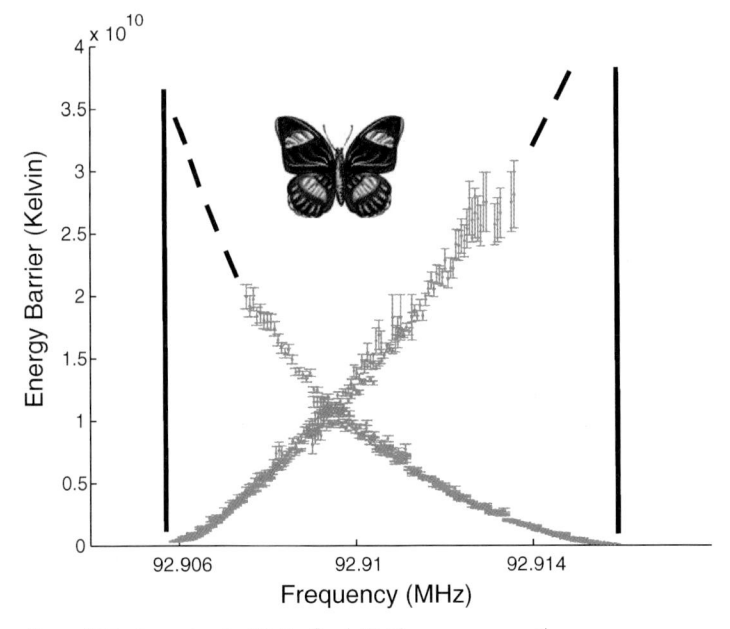

Figure 7.13 Example of a "Butterfly plot". These curves are the quasipotential energy for a fixed drive amplitude.

switching from focus 1 to 2 and from focus 2 to 1, we find two overlapping curves that have the shape of a butterfly. Figure 7.13 shows an example of when the resonator is driven 5 dB past the critical point.

The energy scale for the butterfly plot is enormous. It can be 10^{10} K. Therefore, large noise powers are required to cause transitions at this energy. This is why we find stable lifetimes of thousands of years or more for small noise levels.

A very interesting region of further research is the low noise limit where E_A approaches zero. In this case, the quasienergy barrier is very small and the system becomes very sensitive to remnant noise. If the system is held near the critical point, the butterfly plot energies become small. Therefore, sensor applications are interesting near the critical point [24].

Also, another interesting operating point is the point on the butterfly plot where the two branches cross. At this point, the rates of transition from one to two and from two to one are equal. Therefore, the system will generate a random telegraph signal, switching back and forth at random intervals. Because the lifetimes of the two states are equal, neither focus is preferred.

7.5
Energy Calculation Using a Variational Technique

We calculated the activation energies numerically. The dynamic solutions to (7.1) without noise give the relaxation from the separatrix to one of the foci. During a noise-induced transition, the system is excited from a basin near a focus *towards* the separatrix, which it crosses and then relaxes to the other focus (see Figure 7.10). There is an infinite number of possible trajectories that allow a transition. Given a specific trajectory, it is possible to calculate the contribution of the noise force using (7.1). The total energy transferred to the resonator for a particular trajectory is found by integrating the noise power along that trajectory, thus yielding the effective quasienergy between the foci. The energy transferred is thus an action-like quantity, and the most likely escape trajectory is that which requires the minimum action. The action-like integral S of the system is then $S = \int_{\text{path}} B_n^2(t)\,dt$.

The most likely path $Y_0(t)$ minimizes the integral S. Because the separatrix is a saddle point, the extremal trajectory will most likely travel near the separatrix. The oscillator will naturally evolve from a point near the separatrix to either focus, without contributing to the action-like integral, as this relaxation does not require a noise term. Only when the oscillator is evolving against the dissipative flow field, from a focus toward the separatrix, will it contribute to the action integral.

We used a numerical minimization of the possible trajectories $Y(t)$, using S as a test function to approach the extremum trajectory $Y_0(t)$. Each $Y(t)$ is split into n test points Y_i. Minimization is carried out in the n-dimensional space spanned by the Y_i. Minimum trajectories were calculated for different drive frequencies and amplitudes, yielding the energy barrier as a function of the drive amplitude, as shown in Figure 7.14. We find good agreement (to logarithmic accuracy) between the measured and calculated energy barriers.

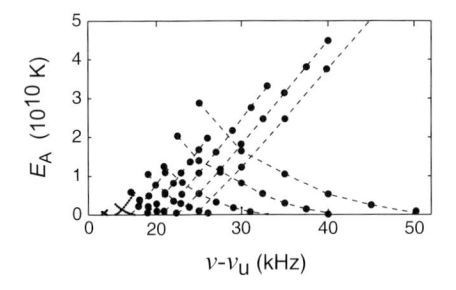

Figure 7.14 Calculated energy curves.

When calculating the energy using this technique, it is easy for the calculation to get trapped in a local minimum in a high-dimensional space. The algorithm must minimize a multi-dimensional function. Often, it is difficult to find the true minimum. Each value of the function corresponds to a different trajectory. Some trajectories contain extra loops in the configuration space. In this case, it is very difficult for the algorithm to converge because of the local minimum.

Furthermore, because of the large number of dimensions that the minimization algorithm must search through, this algorithm takes a very long time to converge. Also, there is never a guarantee that the true global minimum has been reached.

The measurements described in Figure 7.14 were made in the small-to-moderate noise limit, with noise energies much less than the energy barrier. At higher noise powers, the hysteresis due to the nonlinear response can actually be quenched, by rapid noise-induced transitions between the two foci. This quenching is demon-

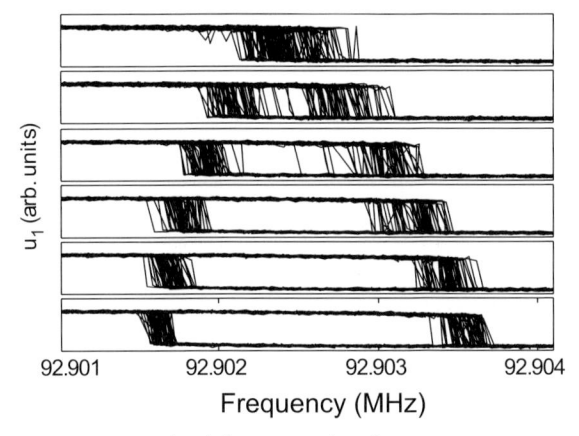

Figure 7.15 Amplitude hysteresis plots, for no noise power (bottom), with the drive amplitude set at −59 dBm, 2 dB above the critical point. The noise power was increased in 2 dB steps for each succeeding frame. At the largest noise power, the hysteresis is quenched.

strated in Figure 7.15: As the noise power is increased, the area of the hysteresis loop grows visibly smaller, until, at the highest noise powers, the switching is no longer hysteretic. In this limit, the oscillator generates random telegraph signals as it makes transitions from one focus to the other. The spectrum of the random telegraph signal is related to the transition rate of the oscillator.

7.6
Frequency Tuning

For magnetomotive detection, it is possible to tune the behavior of the nonlinearity with an external drive current. If a direct current or slowly varying current of large amplitude is applied across the beam, the beam will be bent by the Lorentz force associated with the drive current. This bending changes both the natural frequency and coefficient of nonlinearity for a beam. Furthermore, the frequencies of switching will be shifted due to the drive current. In Figure 7.16, we can see this effect. At positive dc bias, the hysteretic effect of the nonlinearity has disappeared. The frequency shift also occurs at a small drive power. This effect allows for potential circuit applications. An obvious application is to use the beam as a voltage controlled oscillator (VCO).

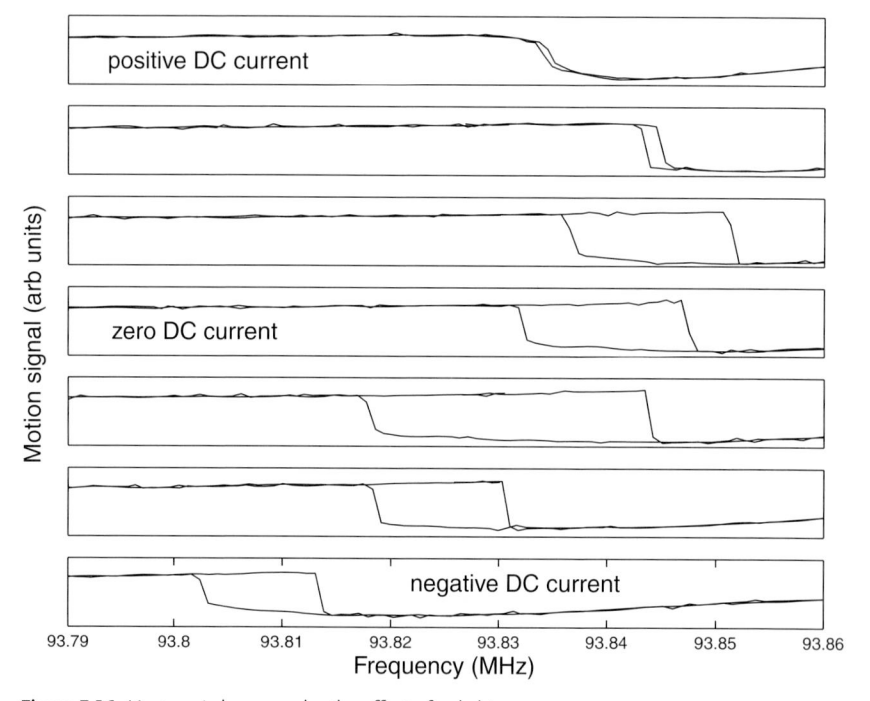

Figure 7.16 Hysteresis loops under the effect of a dc bias current are shown. The extra Lorentz force from the dc bias current bends the beam and shifts the nonlinear hysteresis.

If the frequency is shuttled back and forth, the oscillator will switch and then reset. The frequency at which the switch occurs can be exactly measured. The switch frequency depends on the dc bias current. One can then imagine a "switch and reset amplifier" which operates on this principle. The device performs current to frequency conversion. The bandwidth of the current measurement is related to how fast the oscillator can be switched and reset. For the devices measured, this bandwidth should be in the upper audio frequency range. Also, the performance of the device depends on how much jitter exists in the switching frequency. As was shown earlier, the jitter in the switching frequency is due to noise in the oscillator drive signal.

7.7
Bifurcation Amplifier

We can also engineer a *bifurcation amplifier*. Near the critical point, the derivative of the phase with respect to frequency becomes nearly infinite. In Figure 7.17 we can see the large slope of the phase at the critical point. Therefore, the phase will exhibit a large change for a small change in frequency. This can be used to amplify weak, slowly varying signals that cause a frequency shift. Therefore, a possible use of the Duffing nonlinearity is mass sensing at the critical point. This has recently been discussed in detail in [24]. The result is that the mass sensitivity of such an oscillator may be increased at the expense of detection speed. In practice, nanoresonators

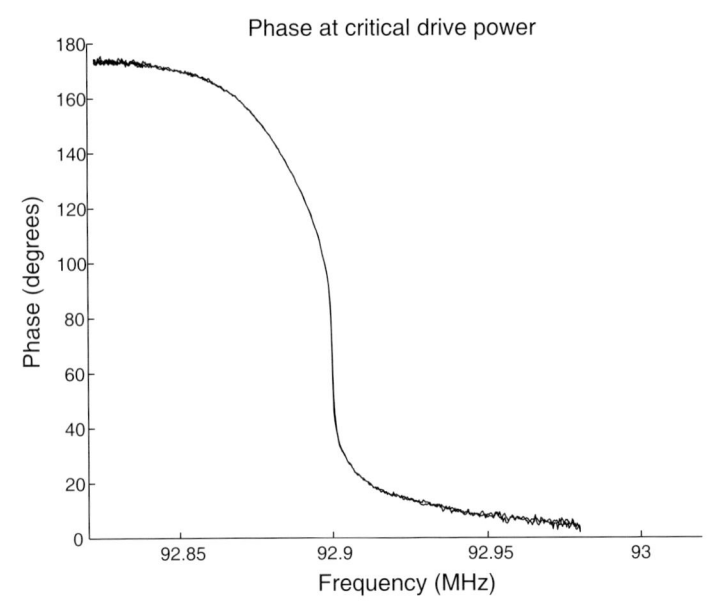

Figure 7.17 Phase with respect to frequency near the critical point. The derivative becomes very large at the critical point.

typically have extra bandwidth to trade for increased mass sensitivity, even when real time operation is required. Therefore, operating a resonator at the critical point may prove to be a promising technology.

7.8 Conclusion

In conclusion, we have studied the properties of the Duffing nonlinearity applied to NEMS devices. We have studied the configuration space trajectories and the transition rates between the bistable states of a nonlinear radiofrequency mechanical resonator. Measurements have been shown to be in good agreement with numerical simulations based on the Duffing oscillator equation of motion.

Acknowledgements

The author would like to thank his former advisor, Andrew Cleland, for his support on this research and authorship on the PRL article that I am revisiting here. Also, the author would like to thank Michael Roukes for his support.

References

1 Postma, H.W.Ch., Kozinsky, I., Husain, A., and Roukes, M.L. (2005) Dynamic range of nanotube- and nanowire-based electromechanical systems. *Appl. Phys. Lett.*, **86**, 223105.

2 Cleland, A.N. and Roukes, M.L. (1996) Fabrication of high frequency nanometer scale mechanical resonators from bulk si crystals. *Appl. Phys. Lett.*, **69**, 2653–2655.

3 Ekinci, K.L., Huang, X.M.H., and Roukes, M.L. (2004) Ultrasensitive nanomechanical mass detection. *Appl. Phys. Lett.*, **84**, 4469–4471.

4 Ilic, B., Czaplewski, D., Craighead, H.G., Neuzil, P., Campagnolo, C., and Batt, C. (2000) Mechanical resonant immunospecific biological detector. *Appl. Phys. Lett.*, **77**, 450–452.

5 Greywall, D.S., Yurke, B., Busch, P.A., Pargellis, A.N., and Willett, R.A. (1994) Evading amplifier noise in nonlinear oscillators. *Phys. Rev. Lett.*, **72**, 2992–2995.

6 Carr, D.W., Evoy, S., Sekaric, L., Craighead, H.G., and Parpia, J.M. (1999) Measurement of mechanical resonance and losses in nanometer scale silicon wires. *Appl. Phys. Lett.*, **75**, 920–922.

7 Cleland, A.N. and Roukes, M.L. (2002) Noise processes in nanomechanical resonators. *J. Appl. Phys.*, **92**, 2758–2769.

8 Knobel, R.G. and Cleland, A.N. (2003) Nanometre-scale displacement sensing using a single electron transistor. *Nature*, **424**, 291.

9 Armour, A.D., Blencowe, M.P., and Schwab, K.C. (2002) Entanglement and decoherence of a micromechanical resonator via coupling to a cooper-pair box. *Phys. Rev. Lett.*, **88**, 148301.

10 Ekinci, K.L., Yang, Y.T., and Roukes, M.L. (2004) Ultimate limits to inertial mass sensing based upon nanoelectromechanical systems. *J. Appl. Phys.*, **95**, 2682–2689.

11 Carr, S.M., Lawrence, W.E., and Wybourne, M.N. (2001) Accessibility of quantum effects in mesomechanical systems. *Phys. Rev. B*, **64**, 220101/1–4.

12 Turner, K.L., Miller, S.A., Hartwell, P.G., MacDonald, N.C., Strogartz, S.H., and Adams, S.G. (1998) Five parametric res-

onances in a microelectromechanical system. *Nature*, **396**, 149–152.

13 Zhang, W., Baskaran, R., and Turner, K.L. (2002) Effect of cubic nonlinearity on auto-parametrically amplified resonant mems mass sensor. *Sens. Actuators A*, **102**, 139–150.

14 Soskin, S.M., Mannella, R., and McClintock, P.V.E. (2003) Zero-dispersion phenomena in oscillatory systems. *Phys. Rep.*, **373**, 247–408.

15 Duffing, G. (1918) *Erzwungene Schwingungen bei veränderlicher Eigenfrequenz und ihre technische Bedeutung.* Friedr. Vieweg & Sohn, Braunschweig.

16 Yurke, B., Greywall, D.S., Pargellis, A.N., and Busch, P.A. (1995) Theory of amplifier-noise evasion in an oscillator employing a nonlinear resonator. *Phys. Rev. A*, **51**, 4211.

17 Nayfeh, A.H. (1979) *Nonlinear oscillations.* Wiley, New York.

18 Dykman, M.I. and Krivoglaz, M.A. (1979) Theory of fluctuational transitions between stable states of a nonlinear oscillator. *Sov. Phys. JETP*, **50** (1), 30.

19 Cleland, A.N., Aldridge, J.S., Driscoll, D.C., and Gossard, A.C. (2002) Nanomechanical displacement sensing using a quantum point contact. *Appl. Phys. Lett.*, **81**, 1699.

20 Bargatin, I., Myers, E.B., Arlett, J., Gudlewski, B., and Roukes, M.L. (2005) Sensitive detection of nanomechanical motion using piezoresistive signal downmixing. *Appl. Phys. Lett.*, **74**, 133109.

21 Kramers, H.A. (1940) Brownian motion in a field of force and the diffusion model of chemical reactions. *Physica*, **7**, 284.

22 Kautz, R.L. (1987) Activation energy for thermally induced escape from a basin of attraction. *Phys. Lett. A*, **125**, 315.

23 Fulton, T.A. and Dunkleberger, L.N. (1974) Lifetime of the zero-voltage state in josephson tunnel junctions. *Phys. Rev. B*, **9**, 4760–4768.

24 Buks, E. and Yurke, B. (2006) Mass detection with a nonlinear resonator. *Phys. Rev. E*, **74**, 046619.

8
Nonlinear Dynamics of Nanomechanical Resonators
Ron Lifshitz and M.C. Cross

8.1
Nonlinearities in NEMS and MEMS Resonators

In the last decade we have witnessed exciting technological advances in the fabrication and control of microelectromechanical and nanoelectromechanical systems (MEMS & NEMS) [16, 19, 26, 54, 55]. Such systems are being developed for a host of nanotechnological applications, such as highly sensitive mass [25, 34, 67], spin [56], and charge detectors [17, 18], as well as for basic research in the mesoscopic physics of phonons [63], and the general study of the behavior of mechanical degrees of freedom at the interface between the quantum and the classical worlds [5, 64]. Surprisingly, MEMS & NEMS have also opened up a whole new experimental window into the study of the nonlinear dynamics of discrete systems in the form of nonlinear micromechanical and nanomechanical oscillators and resonators.

The purpose of this review is to provide an introduction to the nonlinear dynamics of micromechanical and nanomechanical resonators that starts from the basics, but also touches upon some of the advanced topics that are relevant for current experiments with MEMS & NEMS devices. We begin in this section with a general motivation, explaining why nonlinearities are so often observed in NEMS & MEMS devices. In Section 8.2 we describe the dynamics of one of the simplest nonlinear devices, the Duffing resonator, while giving a tutorial in secular perturbation theory as we calculate its response to an external drive. We continue to use the same analytical tools in Section 8.3 to discuss the dynamics of a parametrically-excited Duffing resonator, building up to the description of the dynamics of an array of coupled parametrically-excited Duffing resonators in Section 8.4. We conclude in Section 8.5 by giving an amplitude equation description for the array of coupled Duffing resonators, allowing us to extend our analytic capabilities in predicting and explaining the nature of its dynamics.

Nonlinear Dynamics of Nanosystems. Edited by Günter Radons, Benno Rumpf, and Heinz Georg Schuster
Copyright © 2010 WILEY-VCH Verlag GmbH & Co. KGaA, Weinheim
ISBN: 978-3-527-40791-0

8.1.1
Why Study Nonlinear NEMS and MEMS?

Interest in the nonlinear dynamics of microelectromechanical and nanoelectromechanical systems (MEMS & NEMS) has grown rapidly over the last few years, driven by a combination of practical needs as well as fundamental questions. Nonlinear behavior is readily observed in micro- and nanoscale mechanical devices [1, 2, 9–12, 19, 24, 27, 30, 33, 50, 57, 61, 62, 66, 68, 71, 72]. Consequently, there exists a practical need to understand this behavior in order to avoid it when it is unwanted, and exploit it efficiently when it is wanted. At the same time, advances in the fabrication, transduction, and detection of MEMS & NEMS resonators has opened up an exciting new experimental window into the study of fundamental questions in nonlinear dynamics. Typical nonlinear MEMS & NEMS resonators are characterized by extremely high frequencies, recently going beyond 1 GHz [15, 32, 48], and relatively weak dissipation, with quality factors in the range of 10^2–10^4. For such devices the regime of physical interest is that of steady state motion, as transients tend to disappear before they are detected. This, and the fact that weak dissipation can be treated as a small perturbation, provide a great advantage for quantitative theoretical study. Moreover, the ability to fabricate arrays of tens to thousands of coupled resonators opens new possibilities in the study of nonlinear dynamics of intermediate numbers of degrees of freedom, much larger than one can study in macroscopic or tabletop experiments, yet much smaller than one studies when considering nonlinear aspects of phonon dynamics in a crystal.

The collective response of coupled arrays might be useful for signal enhancement and noise reduction [21, 22], as well as for sophisticated mechanical signal processing applications. Such arrays have already exhibited interesting nonlinear dynamics, ranging from the formation of extended patterns [8, 38], as one commonly observes in analogous continuous systems such as Faraday waves, to that of intrinsically localized modes [39, 58–60]. Thus, nanomechanical resonator arrays are perfect for testing dynamical theories of discrete nonlinear systems with many degrees of freedom. At the same time, the theoretical understanding of such systems may prove useful for future nanotechnological applications.

8.1.2
Origin of Nonlinearity in NEMS and MEMS Resonators

We are used to thinking about mechanical resonators as being simple harmonic oscillators, acted upon by linear elastic forces that obey Hooke's law. This is usually a very good approximation, as most materials can sustain relatively large deformations before their intrinsic stress-strain relation breaks away from a simple linear description. Nevertheless, one commonly encounters nonlinear dynamics in micromechanical and nanomechanical resonators long before the intrinsic nonlinear regime is reached. Most evident are nonlinear effects that enter the equation of motion in the form of a force that is proportional to the cube of the displacement αx^3. These turn a simple harmonic resonator with a linear restoring force into a so-called Duffing resonator. The two main origins of the observed nonlinear effects are illustrated below with the help of two typical examples. These are due to the effect of external potentials that are often nonlinear, and geometric effects that introduce nonlinearities even though the individual forces that are involved are all linear. The Duffing nonlinearity αx^3 can be positive, assisting the linear restoring force, making the resonator stiffer, and increasing its resonance frequency. It can also be negative, working against the linear restoring force, making the resonator softer, and decreasing its resonance frequency. The two examples we give below illustrate how both of these situations can arise in realistic MEMS & NEMS devices.

Additional sources of nonlinearity may be found in experimental realizations of MEMS and NEMS resonators due to practical reasons. These may include nonlinearities in the actuation and in the detection mechanisms that are used for interacting with the resonators. There could also be nonlinearities that result from the manner in which the resonator is clamped by its boundaries to the surrounding material. These all introduce external factors that may contribute to the overall nonlinear behavior of the resonator.

Finally, nonlinearities often appear in the damping mechanisms that accompany every physical resonator. We shall avoid going into the detailed description of the variety of physical processes that govern the damping of a resonator. Suffice it to say that whenever it is reasonable to expand the forces acting on a resonator up to the cube of the displacement x^3, it should correspondingly be reasonable to add to the linear damping, which is proportional to the velocity of the resonator $\dot{x}$, a nonlinear damping term of the form $x^2\dot{x}$, which increases with the amplitude of motion. Such nonlinear damping will be considered in our analysis below.

8.1.3
Nonlinearities Arising from External Potentials

As an example of the effect of an external potential, let us consider a typical situation, discussed for example by Cleland and Roukes [17, 18], and depicted in Figure 8.1, in which a harmonic oscillator is acted upon by an external electrostatic force. This could be implemented by placing a rigid electrically charged base elec-

Figure 8.1 A 43 nanometer thick doubly-clamped platinum nanowire with an external electrode that can be used to tune its natural frequency as well as its nonlinear properties. Adapted with permission from [33].

trode near an oppositely charged NEMS or MEMS resonator. If the equilibrium separation between the resonator and the base electrode in the absence of electric charge is d, the deviation away from this equilibrium position is denoted by X, the effective elastic spring constant of the resonator is K, and the charge q on the resonator is assumed to be constant, then the potential energy of the resonator is given by

$$V(X) = \frac{1}{2}KX^2 - \frac{C}{d+X}. \tag{8.1}$$

In SI units $C = Aq^2/4\pi\epsilon_0$, where A is a numerical factor of order unity that takes into account the finite dimensions of the charged resonator and base electrode. The new equilibrium position X_0 in the presence of charge can be determined by solving the cubic equation

$$\frac{dV}{dX} = KX + \frac{C}{(d+X)^2} = 0. \tag{8.2}$$

If we now expand the potential acting on the resonator in a power series in the deviation $x = X - X_0$ from this new equilibrium, we obtain

$$\begin{aligned}V(x) &\simeq V(X_0) + \frac{1}{2}\left(K - \frac{2C}{(d+X_0)^3}\right)x^2 + \frac{C}{(d+X_0)^4}x^3 - \frac{C}{(d+X_0)^5}x^4 \\ &= V(X_0) + \frac{1}{2}kx^2 + \frac{1}{3}\beta x^3 + \frac{1}{4}\alpha x^4.\end{aligned} \tag{8.3}$$

This gives rise, without any additional driving or damping, to an equation of motion of the form

$$m\ddot{x} + kx + \beta x^2 + \alpha x^3 = 0, \quad \text{with} \quad \beta > 0, \alpha < 0, \tag{8.4}$$

where m is the effective mass of the resonator and k is its new effective spring constant, which is softened by the electrostatic attraction to the base electrode. Note that if $2C/(d+X_0)^3 > K$, the electrostatic force exceeds the elastic restoring force and the resonator is pulled onto the base electrode. β is a positive symmetry breaking quadratic elastic constant that pulls the resonator towards the base electrode regardless of the sign of x, and α is the cubic, or Duffing, elastic constant that, owing to its negative sign, softens the effect of the linear restoring force. It should be sufficient to stop the expansion here, unless the amplitude of the motion is much larger than the size of the resonator, or if by some coincidence the effects of the quadratic and cubic nonlinearities happen to cancel each other out, a situation that will become clearer after reading Section 8.2.3.

8.1.4
Nonlinearities Due to Geometry

As an illustration of how nonlinearities can emerge from linear forces due to geometric effects, consider a doubly-clamped thin elastic beam, which is one of the most commonly encountered NEMS resonators. Because of the clamps at both ends, as the beam deflects in its transverse motion it necessarily stretches. As long as the amplitude of the transverse motion is much smaller than the width of the beam, this effect can be neglected. But with NEMS beams it is often the case that they are extremely thin, and are driven quite strongly, making it common for the amplitude of vibration to exceed the width. Let us consider this effect in some detail by starting with the Euler-Bernoulli equation, which is the commonly used approximate equation of motion for a thin beam [43]. For a transverse displacement $X(z,t)$ from equilibrium, which is much smaller than the length L of the beam, the equation is

$$\rho S \frac{\partial^2 X}{\partial t^2} = -EI \frac{\partial^4 X}{\partial z^4} + T \frac{\partial^2 X}{\partial z^2}, \tag{8.5}$$

where z is the coordinate along the length of the beam ρ is the mass density, S is the area of the cross section of the beam, E is the Young's modulus, I is the moment of inertia, and T the tension in the beam. The latter is composed of its inherent tension T_0 and the additional tension ΔT due to bending that induces an extension ΔL in the length of the beam. Inherent tension results from the fact that in equilibrium in the doubly-clamped configuration, the actual length of the beam may differ from its rest length, being either extended (positive T_0) or compressed (negative T_0). The additional tension ΔT is given by the strain, or relative extension of the beam $\Delta L/L$, multiplied by Young's modulus E and the area of the beam's cross section S. For small displacements, the total length of the beam can be expanded as

$$L + \Delta L = \int_0^L dz \sqrt{1 + \left(\frac{\partial X}{\partial z}\right)^2} \simeq L + \frac{1}{2} \int_0^L dz \left(\frac{\partial X}{\partial z}\right)^2. \tag{8.6}$$

The equation of motion (8.5) then clearly becomes nonlinear

$$\rho S \frac{\partial^2 X}{\partial t^2} = -EI \frac{\partial^4 X}{\partial z^4} + \left[T_0 + \frac{ES}{2L} \int_0^L dz \left(\frac{\partial X}{\partial z}\right)^2\right] \frac{\partial^2 X}{\partial z^2}. \tag{8.7}$$

We can treat this equation perturbatively [49, 69]. We first consider the linear part of the equation, which has the form of (8.5) with T_0 in place of T, separate the variables,

$$X_n(z,t) = x_n(t)\phi_n(z), \tag{8.8}$$

and find its spatial eigenmodes $\phi_n(z)$. For the eigenmodes, we use the convention that the local maximum of the eigenmode $\phi_n(z)$ that is nearest to the center of the beam is scaled to 1. Thus $x_n(t)$ measures the actual deflection of the beam at the point nearest to its center that extends the furthest. Next, we assume that the beam is vibrating predominantly in one of these eigenmodes and use this assumption to evaluate the effective Duffing parameter α_n, multiplying the x_n^3 term in the equation of motion for this mode. Corrections to this approximation will appear only at higher orders of x_n. We multiply (8.7) by the chosen eigenmode $\phi_n(z)$ and integrate over z to get, after some integration by parts, a Duffing equation of motion for the amplitude of the nth mode $x_n(t)$,

$$\ddot{x}_n + \left[\frac{EI}{\rho S} \frac{\int \phi_n''^2 dz}{\int \phi_n^2 dz} + \frac{T_0}{\rho S} \frac{\int \phi_n'^2 dz}{\int \phi_n^2 dz} \right] x_n + \left[\frac{E}{2\rho L} \frac{\left(\int \phi_n'^2 dz \right)^2}{\int \phi_n^2 dz} \right] x_n^3 = 0, \tag{8.9}$$

where primes denote derivatives with respect to z, and all the integrals are from 0 to L. Note that we have obtained a positive Duffing term, indicating a stiffening nonlinearity, as opposed to the softening nonlinearity that we saw in the previous section. Also note that the effective spring constant can be made negative by compressing the equilibrium beam, thus making T_0 large and negative. This may lead to the so-called Euler instability, which is a buckling instability of the beam.

To evaluate the effective Duffing nonlinearity α_n for the nth mode, we introduce a dimensionless parameter $\hat{\alpha}_n$ by rearranging the equation of motion (8.9) to have the form

$$\ddot{x}_n + \omega_n^2 x_n \left[1 + \hat{\alpha}_n \frac{x_n^2}{d^2} \right] = 0, \tag{8.10}$$

where ω_n is the normal frequency of the nth mode, d is the width or diameter of the beam in the direction of the vibration, and x_n is the maximum displacement of the beam near its center. This parameter can then be evaluated regardless of the actual dimension of the beam.

In the limit of small residual tension T_0, the eigenmodes are those dominated by bending given by [43]

$$\phi_n(z) = \frac{1}{a_n} [(\sin k_n L - \sinh k_n L)(\cos k_n z - \cosh k_n z)$$
$$- (\cos k_n L - \cosh k_n L)(\sin k_n z - \sinh k_n z)], \tag{8.11}$$

where a_n is the value of the function in the square brackets at its local maximum that is closest to $z = 0.5$, and the wave vectors k_n are solutions of the transcendental equation $\cos k_n L \cosh k_n L = 1$. The first few values are

$$\{k_n L\} \simeq \{4.7300, 7.8532, 10.9956, 14.1372, 17.2788, 20.4204 \ldots\}, \tag{8.12}$$

and the remaining ones tend towards odd-integer multiples of $\pi/2$ as n increases. Using these eigenfucntions, we can obtain explicit values for the dimensionless Duffing parameters for the different modes by calculating

$$\hat{\alpha}_n = \frac{S d^2}{2I} \frac{\left(\frac{1}{L} \int \phi_n'^2 dz \right)^2}{\frac{1}{L} \int \phi_n''^2 dz} \equiv \frac{S d^2}{2I} \hat{\beta}_n. \tag{8.13}$$

The first few values are

$$\{\hat{\beta}_n\} \simeq \{0.1199, 0.2448, 0.3385, 0.3706, 0.3908, 0.4068, 0.4187, \ldots\}, \tag{8.14}$$

tending to an asymptotic value of 1/2 as $n \to \infty$. For beams with rectangular or circular cross sections, the geometric prefactor evaluates to

$$\frac{Sd^2}{2I} = \begin{cases} 16 & \text{Circular cross section}, \\ 6 & \text{Rectangular cross section}. \end{cases} \quad (8.15)$$

Thus the dimensionless Duffing parameters are of order 1, and therefore the significance of the nonlinear behavior is solely determined by the ratio of the deflection to the width of the beam.

In the limit of large equilibrium tension, the beam essentially behaves as a string with relatively negligible resistance to bending. The eigenmodes are those of a string,

$$\phi_n(z) = \sin\left(\frac{n\pi}{L}z\right), \quad n = 1, 2, 3 \ldots, \quad (8.16)$$

and, if we denote the equilibrium extension of the beam as $\Delta L_0 = LT_0/ES$, the dimensionless Duffing parameters are exactly given by

$$\hat{\alpha}_n = \frac{d^2}{2\Delta L_0} \int \phi_n'^2 \, dz = \frac{(n\pi d)^2}{4L\Delta L_0}. \quad (8.17)$$

In the large tension limit, as in the case of a string, the dimensionless Duffing parameters are proportional to the inverse aspect ratio of the beam d/L times the ratio between its width and the extension from its rest length $d/\Delta L_0$, at least one of which can be a very small parameter. For this reason nonlinear effects are relatively negligible in these systems.

8.2
The Directly-Driven Damped Duffing Resonator

8.2.1
The Scaled Duffing Equation of Motion

Let us begin by considering a single nanomechanical Duffing resonator with linear and nonlinear damping that is driven by an external sinusoidal force. We shall start with the common situation where there is symmetry between x and $-x$, and consider the changes that are introduced by adding symmetry-breaking terms later. Such a resonator is described by the equation of motion

$$m\frac{d^2\tilde{x}}{d\tilde{t}^2} + \Gamma\frac{d\tilde{x}}{d\tilde{t}} + m\omega_0^2\tilde{x} + \tilde{\alpha}\tilde{x}^3 + \tilde{\eta}\tilde{x}^2\frac{d\tilde{x}}{d\tilde{t}} = \tilde{G}\cos\tilde{\omega}\tilde{t}, \quad (8.18)$$

where m is its effective mass, $k = m\omega_0^2$ is its effective spring constant, $\tilde{\alpha}$ is the cubic spring constant or Duffing parameter, Γ is the linear damping rate, and $\tilde{\eta}$ is the coefficient of nonlinear damping – damping that increases with the amplitude of oscillation. We follow the convention that physical parameters that are to be immediately rescaled appear with twiddles, as the first step in dealing with such an equation is to scale away as many unnecessary parameters as possible, leaving only those that are physically significant. This then removes all of the twiddles. We do so by: (1) Measuring time in units of ω_0^{-1} so that the dimensionless time variable is $t = \omega_0 \tilde{t}$. (2) Measuring amplitudes of motion in units of length for which a unit-amplitude oscillation doubles the frequency of the resonator. This is achieved by taking the dimensionless length variable to be $x = \tilde{x}\sqrt{\tilde{\alpha}/m\omega_0^2}$. For the doubly-clamped beam of width or diameter d, discussed in Section 8.1.4, this length is $x = \tilde{x}\sqrt{\tilde{\alpha}_n}/d$. (3) Dividing the equation by an overall factor of $\omega_0^3\sqrt{m^3/\tilde{\alpha}}$. This yields a scaled Duffing equation of the form

$$\ddot{x} + Q^{-1}\dot{x} + x + x^3 + \eta x^2 \dot{x} = G \cos \omega t, \tag{8.19}$$

where dots denote derivatives with respect to the dimensionless time t, all the dimensionless parameters are related to the physical ones by

$$Q^{-1} = \frac{\Gamma}{m\omega_0}, \quad \eta = \frac{\tilde{\eta}\omega_0}{\tilde{\alpha}}, \quad G = \frac{\tilde{G}}{\omega_0^3}\sqrt{\frac{\tilde{\alpha}}{m^3}}, \quad \text{and} \quad \omega = \frac{\tilde{\omega}}{\omega_0}, \tag{8.20}$$

and Q is the quality factor of the resonator.

8.2.2
A Solution Using Secular Perturbation Theory

We proceed to calculate the response of the damped Duffing resonator to an external sinusoidal drive, as given by (8.19), by making use of secular perturbation theory [31, 65]. We do so in the limit of a weak linear damping rate Q^{-1}, which we use to define a small expansion parameter, $Q^{-1} \equiv \epsilon \ll 1$. In most actual applications, Q is at least on the order of 100, making this limit well-justified. We also consider the limit of weak oscillations where it is justified to truncate the expansion of the force acting on the resonator at the third power of x. We do so by requiring that the cubic force x^3 be a factor of ϵ smaller than the linear force, or equivalently, by requiring the deviation from equilibrium x to be on the order of $\sqrt{\epsilon}$. We ensure that the external driving force has the right strength to induce such weak oscillations by having it enter the equation at the same order as all the other physical effects. This, in effect, requires the amplitude of the drive to be $G = \epsilon^{3/2}g$. To see why, recall that for a regular linear resonance, x is proportional to GQ. Q is of order ϵ^{-1} and we want x to be of order $\sqrt{\epsilon}$, and so G must be of order $\epsilon^{3/2}$. Finally, since damping is weak we expect to see a response only close to the resonance frequency. We therefore take the driving frequency to be of the form $\omega = 1 + \epsilon\Omega$. The equation of motion (8.19) thus becomes

$$\ddot{x} + \epsilon\dot{x} + x + x^3 + \eta x^2\dot{x} = \epsilon^{3/2}g\cos(1 + \epsilon\Omega)t. \tag{8.21}$$

This is the equation we shall study using secular perturbation theory, while occasionally comparing the results with the original physical equation (8.18).

With the expectation that the motion of the resonator far from equilibrium will be on the order of $\epsilon^{1/2}$, we try a solution of the form

$$x(t) = \frac{\sqrt{\epsilon}}{2} \left(A(T) e^{it} + \text{c.c.} \right) + \epsilon^{3/2} x_1(t) + \ldots \qquad (8.22)$$

where c.c. denotes complex conjunction.

The lowest order contribution to this solution is based on the solution to the linear equation of motion of a simple harmonic oscillator (SHO) $\ddot{x} + x = 0$, where $T = \epsilon t$ is a slow time variable, allowing the complex amplitude $A(T)$ to vary slowly in time due to the effect of all the other terms in the equation. As we shall immediately see, the slow temporal variation of $A(T)$ also allows us to ensure that the perturbative correction $x_1(t)$ as well as all higher-order corrections to the linear equation do not diverge, as they do if one uses naive perturbation theory. Using the relation

$$\dot{A} = \frac{dA}{dt} = \epsilon \frac{dA}{dT} \equiv \epsilon A', \qquad (8.23)$$

we calculate the time derivatives of the trial solution (8.22)

$$\dot{x} = \frac{\sqrt{\epsilon}}{2} \left([iA + \epsilon A'] e^{it} + \text{c.c.} \right) + \epsilon^{3/2} \dot{x}_1(t) + \ldots \qquad (8.24a)$$

$$\ddot{x} = \frac{\sqrt{\epsilon}}{2} \left([-A + 2i\epsilon A' + \epsilon^2 A''] e^{it} + \text{c.c.} \right) + \epsilon^{3/2} \ddot{x}_1(t) + \ldots \qquad (8.24b)$$

By substituting these expressions back into the equation of motion (8.21) and picking out all terms of order $\epsilon^{3/2}$, we get for the first perturbative correction

$$\ddot{x}_1 + x_1 = \left(-iA' - i\frac{1}{2}A - \frac{3 + i\eta}{8}|A|^2 A + \frac{g}{2} e^{i\Omega T} \right) e^{it} - \frac{1 + i\eta}{8} A^3 e^{3it} + \text{c.c.} \qquad (8.25)$$

The collection of terms proportional to e^{it} on the right-hand side of (8.25), called the secular terms, act like a force that drives the SHO on the left-hand side exactly at its resonance frequency. The sum of all these terms must therefore vanish so that the perturbative correction $x_1(t)$ will not diverge. This requirement is the so-called "solvability condition", giving us an equation for determining the slowly varying amplitude $A(T)$,

$$\frac{dA}{dT} = -\frac{1}{2}A + i\frac{3}{8}|A|^2 A - \frac{\eta}{8}|A|^2 A - i\frac{g}{2} e^{i\Omega T}. \qquad (8.26)$$

This general equation could be used to study many different effects [20]. Here we use it to study the steady-state dynamics of the driven Duffing resonator.

We ignore initial transients and assume that there exists a steady-state solution of the form

$$A(T) = a e^{i\Omega T} \equiv |a| e^{i\varphi} e^{i\Omega T}. \qquad (8.27)$$

With this expression for the slowly varying amplitude $A(T)$, the solution to the original equation of motion (8.21) becomes an oscillation at the drive frequency $\omega = 1 + \epsilon\Omega$,

$$x(t) = \epsilon^{1/2}|a|\cos(\omega t + \phi) + O(\epsilon^{3/2}), \tag{8.28}$$

where we are not interested in the actual correction $x_1(t)$ of order $\epsilon^{3/2}$, but rather in finding the fixed complex amplitude a of the lowest order term. This amplitude a can be any solution of the equation

$$\left[\left(\frac{3}{4}|a|^2 - 2\Omega\right) + i\left(1 + \frac{\eta}{4}|a|^2\right)\right]a = g, \tag{8.29}$$

obtained by substituting the steady-state solution (8.27) into Eq. (8.26) of the secular terms.

The magnitude and phase of the response are then given explicitly by

$$|a|^2 = \frac{g^2}{\left(2\Omega - \frac{3}{4}|a|^2\right)^2 + \left(1 + \frac{1}{4}\eta|a|^2\right)^2} \tag{8.30a}$$

and

$$\tan\phi = \frac{1 + \frac{1}{4}\eta|a|^2}{2\Omega - \frac{3}{4}|a|^2}. \tag{8.30b}$$

By reintroducing the original physical scales, we can obtain the physical solution to the original equations of motion $\tilde{x}(\tilde{t}) \simeq \tilde{x}_0 \cos(\tilde{\omega}\tilde{t} + \phi)$, where $\tilde{x}_0 = |a|\sqrt{\Gamma\omega_0/\tilde{\alpha}}$, and therefore

$$\tilde{x}_0^2 = \frac{\left(\frac{\tilde{G}}{2m\omega_0^2}\right)^2}{\left(\frac{\tilde{\omega}-\omega_0}{\omega_0} - \frac{3}{8}\frac{\tilde{\alpha}}{m\omega_0^2}\tilde{x}_0^2\right)^2 + \left(\frac{1}{2}Q^{-1} + \frac{1}{8}\frac{\tilde{\eta}}{m\omega_0}\tilde{x}_0^2\right)^2} \tag{8.31a}$$

and

$$\tan\phi = \frac{\frac{\Gamma}{2} + \frac{\tilde{\eta}}{8}\tilde{x}_0^2}{m\tilde{\omega} - m\omega_0 - \frac{3\tilde{\alpha}}{8\omega_0}\tilde{x}_0^2}. \tag{8.31b}$$

The scaled response functions (8.30a) are plotted in Figure 8.2 for a drive with a scaled amplitude of $g = 3$, both with and without nonlinear damping. The response without nonlinear damping is shown also in Figure 8.3 for a sequence of increasing drive amplitudes ranging from $g = 0.1$, where the response is essentially linear, to the value of $g = 4$. Note that due to our choice of a positive Duffing nonlinearity, the resonator becomes stiffer and its frequency higher as the amplitude increases. The response amplitude of the driven resonator therefore increases with increasing frequency until it reaches a saddle-node bifurcation and drops abruptly to zero. A negative Duffing parameter would produce a mirror image of this response curve.

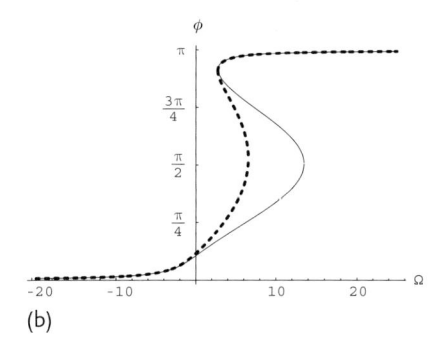

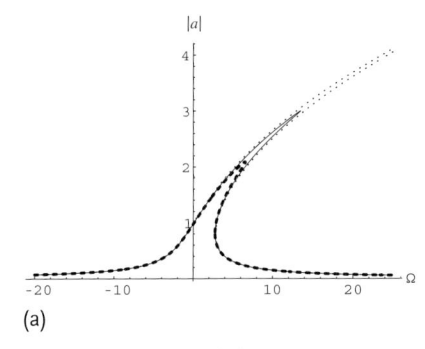

Figure 8.2 Magnitude $|a|$ (a) and phase ϕ (b) of the response of a Duffing resonator as a function of the frequency Ω for a fixed driving amplitude $g = 3$. The thin solid curves show the response without any nonlinear damping ($\eta = 0$). The thick dotted curves show the response with nonlinear damping ($\eta = 0.1$). The thin dotted curve in (a) shows the response without any kind of damping ($Q^{-1} = 0$ and $\eta = 0$ in the original equation (8.19)). The phase in this case is 0 along the whole upper-left branch and π along the whole lower-right branch, and so is not plotted in (b).

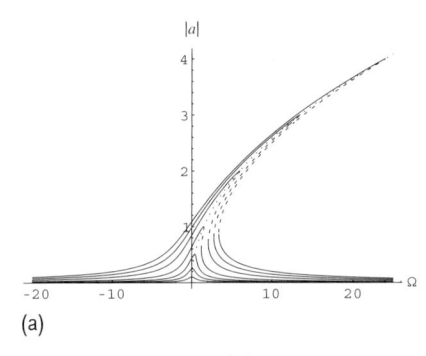

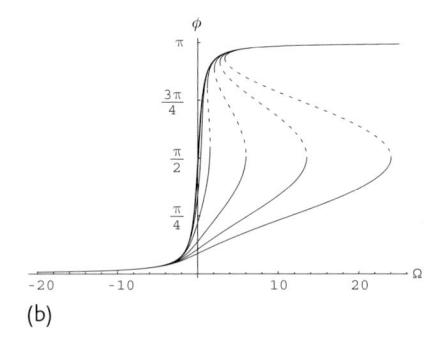

Figure 8.3 Magnitudes $|a|$ (a) and phases ϕ (b) of the response of a Duffing resonator as a function of the frequency Ω for a sequence of increasing values of the drive amplitude $0.1 \leq g \leq 4.0$, without nonlinear damping ($\eta = 0$). Solid curves indicate stable solutions of the response function (8.30a), while dashed curves indicate unstable solutions.

One sees that the magnitude of the response given by (8.30a) formally approaches the Lorentzian response of a linear SHO if we let the nonlinear terms in the original equation of motion tend to zero. Their existence modifies the response function with the appearance of the squared magnitude $|a|^2$ in the denominator on the right-hand side of (8.30a), turning the solution into a cubic polynomial in $|a|^2$. As such there are either one or three real solutions for $|a|^2$, and therefore for $|a|$, as a function of either the drive amplitude g or the driving frequency Ω. We shall analyze the dependence of the magnitude of the response on frequency in some detail, and leave it to the reader to perform such an analysis of the similar dependence on drive amplitude.

In order to analyze the magnitude of the response $|a|$ as a function of driving frequency Ω, we differentiate the response function (8.30a), resulting in

$$\left[\tfrac{3}{64}(9+\eta^2)|a|^4 + \tfrac{1}{4}(\eta - 6\Omega)|a|^2 + \tfrac{1}{4} + \Omega^2\right]d|a|^2$$
$$= \left[\tfrac{3}{4}|a|^4 - 2\Omega|a|^2\right]d\Omega . \tag{8.32}$$

This allows us immediately to find the condition for resonance, where the magnitude of the response is at its peak, by requiring that $d|a|^2/d\Omega = 0$. We find that the resonance frequency $\Omega_{\max}$ depends quadratically on the peak magnitude $|a|_{\max}$, according to

$$\Omega_{\max} = \tfrac{3}{8}|a|^2_{\max}, \tag{8.33a}$$

or in terms of the original variables as

$$\tilde{\omega}_{\max} = \omega_0 + \frac{3}{8}\frac{\alpha}{m\omega_0}(\tilde{x}_0)^2_{\max} . \tag{8.33b}$$

The curve satisfying (8.33a), for which $|a| = \sqrt{8\Omega/3}$, is plotted in Figure 8.3. It forms a square root backbone that connects all the resonance peaks for the different driving amplitudes, which is often seen in typical experiments with nanomechanical resonators. Thus, the peak of the response is pulled further toward higher frequencies as the driving amplitude g is increased, as expected from a stiffening nonlinearity.

When the drive amplitude g is sufficiently strong, we can use Eq. (8.32) to find the two saddle-node bifurcation points, where the number of solutions changes from one to three and then back from three to one. At these points $d\Omega/d|a|^2 = 0$, yielding a quadratic equation in Ω whose solutions are

$$\Omega^{\pm}_{SN} = \tfrac{3}{4}|a|^2 \pm \tfrac{1}{2}\sqrt{\tfrac{3}{16}(3-\eta^2)|a|^4 - \eta|a|^2 - 1} . \tag{8.34}$$

When the two solutions are real, corresponding to the two bifurcation points, a linear stability analysis shows that the upper and lower branches of the response are stable solutions and the middle branch that exists for $\Omega^{-}_{SN} < \Omega < \Omega^{+}_{SN}$ is unstable. When the drive amplitude g is reduced, it approaches a critical value g_c where the two bifurcation points merge into an inflection point. At this point both $d\Omega/d|a|^2 = 0$ and $d^2\Omega/(d|a|^2)^2 = 0$, providing two equations for determining the critical condition for the onset of bistability, or the existence of two stable solution branches,

$$|a|^2_c = \frac{8}{3}\frac{1}{\sqrt{3-\eta}}, \quad \Omega_c = \frac{1}{2\sqrt{3}}\frac{3\sqrt{3}+\eta}{\sqrt{3-\eta}}, \quad g_c^2 = \frac{32}{27}\frac{9+\eta^2}{(\sqrt{3}-\eta)^3} . \tag{8.35}$$

For the case without nonlinear damping, $\eta = 0$, the critical values are $|a|^2_c = (4/3)^{3/2}$ and $\Omega_c = (3/4)^{1/2}$, for which the critical drive amplitude is $g_c = (4/3)^{5/4}$. For $0 < \eta < \sqrt{3}$, the critical driving amplitude g_c that is required for having bistability increases with η, as shown in Figure 8.4. For $\eta > \sqrt{3}$ the discriminant in Eq. (8.34) is always negative, prohibiting the existence of bistability of solutions.

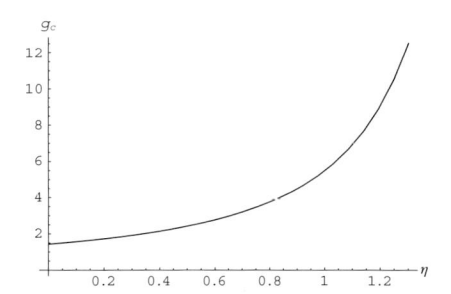

Figure 8.4 Critical driving amplitude g_c for the onset of bistability in the response of the Duffing resonator as a function of nonlinear damping η, as given by Eq. (8.35). Note that $g_c \to (4/3)^{5/4} \simeq 1.43$ as $\eta \to 0$.

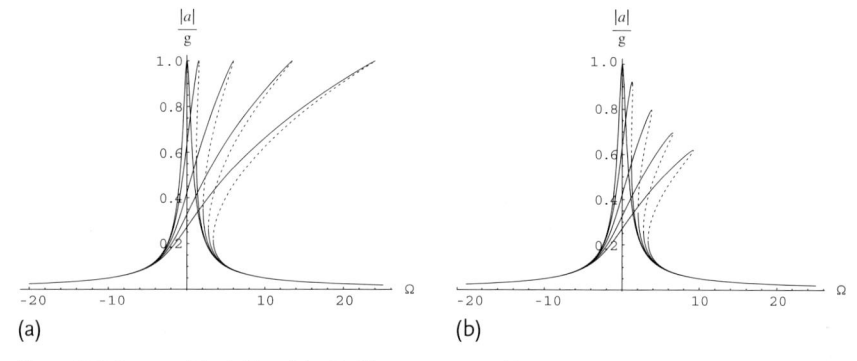

Figure 8.5 Responsivity $|a|/g$ of the Duffing resonator without nonlinear damping (a) and with a small amount of nonlinear damping $\eta = 0.1$ (b), for different values of the driving amplitude g. Viewing the response in this way suggests an experimental scheme by which one could determine the importance of nonlinear damping and extract its magnitude.

Nonlinear damping acts to decrease the magnitude of the response when it is appreciable, that is, when the drive amplitude is large. It gives rise to an effective damping rate for oscillations with magnitude $|a|$ that is given by $1 + \frac{1}{4}\eta|a|^2$, or, in terms of the physical parameters, by $\Gamma + \frac{1}{4}\tilde{\eta}\tilde{x}_0^2$. When viewing the response as it is plotted in Figure 8.3, it is difficult to distinguish between the effects of the two forms of damping. The resonance peaks lie on the same backbone regardless of the existence of a contribution from nonlinear damping. A more useful scheme for seeing the effect of nonlinear damping is to plot the response amplitude scaled by the drive $|a|/g$, often called the responsivity of the resonator, as shown in Figure 8.5. Without nonlinear damping all peaks have the same height of 1. With nonlinear damping, one clearly sees the decrease in the responsivity as the driving amplitude is increased.

The region of bistability that lies between the two saddle-node bifurcations (8.34) in the response of the driven Duffing resonator is the source of a number of interesting dynamical features that are often observed in experiments with MEMS

& NEMS resonators [3, 19, 28, 70]. Most obvious is the existence of hysteresis in quasistatic sweeps of either driving frequency or driving amplitude, which is readily observed in experiments. For example, if we start below resonance and sweep the frequency upwards along one of the constant drive amplitude curves shown in Figure 8.3, the response will gradually increase, climbing up on the curve until it reaches the upper saddle-node bifurcation $\Omega_{SN}^+(g)$. It will then abruptly drop down to the lower stable solution branch and continue toward lower response amplitudes to the right of the resonance. Upon switching the direction of the quasistatic sweep, the response amplitude will gradually increase until it reaches the lower saddle-node bifurcation $\Omega_{SN}^-(g)$, where it will abruptly jump up to the upper stable solution branch. From this point it will gradually follow it downwards towards lower frequencies with diminishing response amplitude.

Another interesting aspect involves basins of attraction. If we fix the values of the driving amplitude and frequency, the driven damped Duffing resonator will deterministically approach one of the two possible solutions, depending on its initial conditions. One can then map the regions of the phase space of initial conditions into the two so-called basins of attraction of the two possible stable solutions, where the unstable solution lies along the separatrix, or border line between the two basins of attraction. These basins of attraction were mapped out in a recent experiment using a suspended platinum nanowire by Kozinsky *et al.* [41]. If one additionally considers the existence of random noise, which is always the case in real systems, then the separatrix becomes fuzzy and it is possible to observe thermally activated switching of the resonator between its two possible solutions. What is in fact observed, for example in an upward frequency scan, is that the resonator can drop to the small amplitude solution before it actually reaches the upper saddle-node bifurcation $\Omega_{SN}^+(g)$. Similar behavior is also observed for the lower bifurcation point. As the noise increases, the observed size of the bistability region effectively shrinks. This was demonstrated with a doubly-clamped nanomechanical resonator made of aluminum nitride in a recent experiment by Aldridge and Clelend [1]. The existence of the saddle-node bifurcation has also been exploited for applications because the response of the resonator at the bifurcation point can change dramatically if one changes the drive frequency, or any of the resonator's physical parameters that can alter the response curve. This idea has been used for signal amplification [10] as well as squeezing of noise [3, 69].

Finally, much effort has been recently invested to push experiments with nanomechanical resonators towards the quantum regime. In this context, it has been shown that the bistability region in the response of the driven damped Duffing resonator offers a novel approach for observing the transition from classical to quantum mechanical behavior as the temperature is lowered [36, 37]. The essential idea is that one can find a regime in frequency and temperature where thermal switching between the two basins of attraction is essentially suppressed when the dynamics is classical, whereas if the resonator has already started entering the quantum regime, quantum dynamics allow it to switch between the two basins. Thus, an observation of switching can be used to ascertain whether or not a Duffing resonator is behaving quantum mechanically.

8.2.3
Addition of Other Nonlinear Terms

It is worth considering the addition of other nonlinear terms that were not included in our original equation of motion (8.18). Without increasing the order of the nonlinearity, we could still add quadratic symmetry breaking terms of the form x^2, $x\dot{x}$, and $\dot{x}^2$ as well as additional cubic damping terms of the form $\dot{x}^3$ and $x\dot{x}^2$. Such terms may appear naturally in actual physical situations, like the examples discussed in Section 8.1.2. For the reader who wishes to skip to the following section on parametrically-driven Duffing resonators, we state at the outset that the addition of such terms does not alter the response curves that we described in the previous section in any fundamental way. They merely conspire to renormalize the effective values of the coefficients used in the original equation of motion. Thus, without any particular model at hand, it is difficult to discern the existence of such terms in the equation.

Consider an equation like (8.18), but with additional terms of the form given above,

$$m\frac{d^2\tilde{x}}{d\tilde{t}^2} + \Gamma\frac{d\tilde{x}}{d\tilde{t}} + m\omega_0^2\tilde{x} + \tilde{\beta}\tilde{x}^2 + \tilde{\mu}\tilde{x}\frac{d\tilde{x}}{d\tilde{t}} + \tilde{\rho}\left(\frac{d\tilde{x}}{d\tilde{t}}\right)^2 + \tilde{\alpha}\tilde{x}^3 + \tilde{\eta}\tilde{x}^2\frac{d\tilde{x}}{d\tilde{t}}$$

$$+ \tilde{\nu}\tilde{x}\left(\frac{d\tilde{x}}{d\tilde{t}}\right)^2 + \tilde{\zeta}\left(\frac{d\tilde{x}}{d\tilde{t}}\right)^3 = \tilde{G}\cos\tilde{\omega}\tilde{t}, \qquad (8.36)$$

and then perform the same scaling as in (8.20) for the additional parameters, producing

$$\beta = \frac{\tilde{\beta}}{\omega_0\sqrt{m\tilde{\alpha}}}, \quad \mu = \frac{\tilde{\mu}}{\sqrt{m\tilde{\alpha}}}, \quad \rho = \frac{\tilde{\rho}\omega_0}{\sqrt{m\tilde{\alpha}}}, \quad \nu = \frac{\tilde{\nu}\omega_0^2}{\tilde{\alpha}}, \quad \zeta = \frac{\tilde{\zeta}\omega_0^3}{\tilde{\alpha}}. \qquad (8.37)$$

After performing the same scaling as before with the small parameter $\epsilon = Q^{-1}$, this yields a scaled equation of motion with all the additional nonlinearities,

$$\ddot{x} + \epsilon\dot{x} + x + \beta x^2 + \mu x\dot{x} + \rho\dot{x}^2 + x^3 + \eta x^2\dot{x} + \nu x\dot{x}^2 + \zeta\dot{x}^3 = \epsilon^{3/2}g\cos\omega t. \qquad (8.38)$$

The important difference between this equation and the one we solved earlier (8.21) is that with a similar scaling of x with $\sqrt{\epsilon}$, we now have terms on the order of ϵ. We therefore need to modify our trial expansion to contain such terms as well, yielding

$$x(t) = \sqrt{\epsilon}x_0(t, T) + \epsilon x_{1/2}(t, T) + \epsilon^{3/2}x_1(t, T) + \ldots, \qquad (8.39)$$

with $x_0 = \frac{1}{2}\left[A(T)e^{it} + \text{c.c.}\right]$ as before.

We begin by collecting all terms on the order of ϵ, arriving at

$$\ddot{x}_{1/2} + x_{1/2} = -\frac{1}{2}(\beta + \rho)|A|^2 - \frac{1}{4}\left[(\beta - \rho + i\mu)A^2 e^{2it} + \text{c.c.}\right]. \qquad (8.40)$$

This equation for the first correction $x_{1/2}(t)$ contains no secular terms, and therefore can be solved immediately to give

$$x_{1/2}(t) = -\tfrac{1}{2}(\beta+\rho)|A|^2 + \tfrac{1}{12}\left[(\beta-\rho+\mathrm{i}\mu)A^2\mathrm{e}^{2\mathrm{i}t} + \text{c.c.}\right]. \tag{8.41}$$

We substitute this solution into the ansatz (8.39) and back into the equation of motion (8.38), and proceed by collecting terms on the order of $\epsilon^{3/2}$. We find a number of additional terms of this order that did not appear earlier on the right-hand side of (8.25) for the correction $x_1(t)$,

$$-2\beta x_0 x_{1/2} - \mu\left(x_0\dot{x}_{1/2} + \dot{x}_0 x_{1/2}\right) - 2\rho\dot{x}_0\dot{x}_{1/2} - \nu x_0\dot{x}_0^2 - \zeta\dot{x}_0^3$$
$$= \left\{\left[\tfrac{5}{12}\beta(\beta+\rho) + \tfrac{1}{6}\rho^2 + \tfrac{1}{24}\mu^2 - \tfrac{1}{8}\nu\right] + \mathrm{i}\left[\tfrac{1}{8}\mu(\beta+\rho) - \tfrac{3}{8}\zeta\right]\right\}|A|^2 A\mathrm{e}^{\mathrm{i}t}$$
$$+ \text{nonsecular terms}. \tag{8.42}$$

After adding the additional secular terms, we obtain a modified equation for the slowly varying amplitude $A(T)$,

$$\frac{\mathrm{d}A}{\mathrm{d}T} = -\tfrac{1}{2}A + \mathrm{i}\tfrac{3}{8}\left(1 - \tfrac{10}{9}\beta(\beta+\rho) - \tfrac{4}{9}\rho^2 - \tfrac{1}{9}\mu^2 + \tfrac{1}{3}\nu\right)|A|^2 A$$
$$-\tfrac{1}{8}(\eta - \mu(\beta+\rho) + 3\zeta)|A|^2 A - \mathrm{i}\tfrac{g}{2}\mathrm{e}^{\mathrm{i}\Omega T}$$
$$\equiv -\tfrac{1}{2}A + \mathrm{i}\tfrac{3}{8}\alpha_{\text{eff}}|A|^2 A - \tfrac{1}{8}\eta_{\text{eff}}|A|^2 A - \mathrm{i}\tfrac{g}{2}\mathrm{e}^{\mathrm{i}\Omega T}. \tag{8.43}$$

We find that the equation is formally identical to the previous result (8.26) before adding the extra nonlinear terms. The response curves and the discussion of the previous section therefore still apply after taking into account all of the quadratic and cubic nonlinear terms. All of these terms combine in a particular way, giving rise to the two effective cubic parameters defined in (8.43). This, in fact, allows one some flexibility in tuning the nonlinearities of a Duffing resonator in real experimental situations. For example, Kozinsky et al. [40] use this flexibility to tune the effective Duffing parameter α_{eff} via an external electrostatic potential, as described in Section 8.1.3 and shown in Figure 8.1. This affects both the quadratic parameter $\tilde{\beta}$ and the cubic parameter $\tilde{\alpha}$ in the physical equation of motion (8.36). Note that due to the different signs of the various contributions to the effective nonlinear parameters, one could actually cause the cubic terms to vanish, altering the response in a fundamental way.

8.3
Parametric Excitation of a Damped Duffing Resonator

Parametric excitation offers an alternative approach for actuating MEMS or NEMS resonators. Instead of applying an external force that acts directly on the resonator, one modulates one or more of its physical parameters as a function of time, which in turn modulates the normal frequency of the resonator. This is what happens on a swing when the up-and-down motion of the center of mass of the swinging child effectively changes the length of the swing, thereby modulating its natural frequency. The most effective way to swing is to move the center of mass up and down twice in every period of oscillation, but one can also swing by moving up and down at slower rates, namely once every n^{th} multiple of half a period, for any integer n.

Let H be the relative amplitude by which the normal frequency is modulated, and ω_P be the frequency of the modulation, often called the pump frequency. One can show [42] that there is a sequence of tongue shaped regions in the $H - \omega_P$ plane where the smallest fluctuations away from the quiescent state of the swing, or any other parametrically-excited resonator [66], are exponentially amplified. This happens when the amplitude of the modulation H is sufficiently strong to overcome the effect of damping, where the threshold for the nth instability tongue scales as $(Q^{-1})^{1/n}$. Above this threshold, the amplitude of the motion grows until it is saturated by nonlinear effects. We shall describe the nature of these oscillations for driving above threshold later, both for the first ($n = 1$) and the second ($n = 2$) instability tongues, but first we shall consider the dynamics when the driving amplitude is just below threshold, as it also offers interesting behavior and a possibility for novel applications such as parametric amplification [4, 12, 57] and noise squeezing [57].

There are a number of actual schemes for the realization of parametric excitation in MEMS & NEMS devices. The simplest and probably most commonly used on the micron scale is to use an external electrode that can induce an external potential. If the external potential is modulated in time it can change the effective spring constant of the resonator [24, 51, 52, 66, 71, 72]. Based on our treatment of this situation in Section 8.1.3, this method is likely to modulate all the coefficients in the potential felt by the resonator, thus also modulating, for example, the Duffing parameter α. Similarly, one may devise configurations in which an external electrode deflects a doubly-clamped beam from its equilibrium, thereby inducing extra tension within the beam itself that can be modulated in time, as described in Section 8.1.4. Alternatively, one may generate motion in the clamps holding a doubly-clamped beam by its ends, thus inducing in it a time-varying tension which is likely to affect the other physical parameters to a lesser extent. An example of this method is shown in Figure 8.6. These methods allow one to modulate the tension in the beam directly and thus modulate its normal frequency. More recently, Masmanidis et al. [45] developed layered piezoelectric NEMS structures whose tension can be fine tuned in doubly-clamped configurations, thus enabling fine control of the normal frequency of the beam with a simple turn of a knob.

Only a minor change is required in our equation of the driven damped Duffing resonator to accommodate this new situation, namely the addition of a modulation of the linear spring constant. Beginning with the scaled form of the Duffing equation (8.19), we obtain

$$\ddot{x} + Q^{-1}\dot{x} + [1 + H\cos\omega_P t]x + x^3 + \eta x^2 \dot{x} = G\cos\left(\omega_D t + \phi_g\right), \quad (8.44)$$

Figure 8.6 A configuration that uses electromotive actuation to perform parametric excitation of a doubly-clamped beam, the central segment of an H-shaped device. A static magnetic field runs normal to the plane of the device. A metallic wire that runs along the external suspended segments of the H-device carries alternating current in opposite directions, thus applying opposite Lorentz forces that induce a time-varying compression of the central segment. This modulates the tension in the central segment, thus varying its normal frequency. This configuration was recently used by Karabalin et al. [35] to demonstrate parametric amplification of a signal running along the central beam through a separate electric circuit. Image courtesy of Michael Roukes.

where the scaling is the same as before, and we shall again use the damping Q^{-1} to define the small expansion parameter ϵ. The term proportional to H on the left hand side is the external drive that modulates the spring constant, giving a term that is proportional to the displacement x as well as to the strength of the drive. This term is the parametric drive.

We first consider the largest excitation effect that occurs when the pump frequency is close to twice the resonant frequency of the resonator. This is the region in the $H - \omega_P$ plane that we termed the first instability tongue. We therefore take the pump frequency to be an amount $\epsilon \Omega_P$ away from twice the resonant frequency, and take the drive amplitude to scale as the damping, that is, we set $H = \epsilon h$. The term on the right hand side is a direct additive drive or signal, with amplitude scaled as in the discussion of the Duffing equation. The frequency of the drive is an amount $\varepsilon \Omega_D$ away from the resonator frequency that has been scaled to 1.

The scaled equation of motion that we now treat in detail is therefore

$$\ddot{x} + \epsilon \dot{x} + (1 + \epsilon h \cos\left[(2 + \epsilon \Omega_P) t\right]) x + x^3 + \eta x^2 \dot{x}$$
$$= \epsilon^{3/2} |g| \cos\left[(1 + \epsilon \omega_D) t + \phi_g\right], \tag{8.45}$$

where we now use $g = |g| e^{i\phi_g}$ to denote a complex drive amplitude.

We follow the same scheme of secular perturbation theory as in Section 8.2.2, using a trial solution in the form of (8.22) and proceeding as before. The new secular term, appearing on the right-hand side of (8.25) and arising from the parametric drive is

$$-\tfrac{1}{4} h A^* e^{i\Omega_P T} e^{it} . \tag{8.46}$$

This gives the equation for the slowly varying amplitude,

$$\frac{dA}{dT} + \frac{1}{2} A - i\frac{h}{4} A^* e^{i\Omega_P T} - i\frac{3}{8} |A|^2 A + \frac{\eta}{8} |A|^2 A = -i\frac{g}{2} e^{i\Omega_D T} . \tag{8.47}$$

8.3.1
Driving Below Threshold: Amplification and Noise Squeezing

We first study the amplitude of the response of a parametrically-pumped Duffing resonator to an external direct drive $g \neq 0$. We will see that the characteristic behavior changes from amplification of an applied signal to oscillations at a critical value of $h = h_c = 2$, even in the absence of a signal. It is therefore convenient to introduce a reduced parametric drive $\bar{h} = h/h_c = h/2$ that plays the role of a bifurcation parameter with a critical value of 1. We begin by assuming that the drive is small enough so that the magnitude of the response remains small and the nonlinear terms in (8.47) can be neglected. This gives the linear equation

$$\frac{dA}{dT} + \frac{1}{2}A - i\frac{\bar{h}}{2}A^* e^{i\Omega_P T} = -i\frac{g}{2} e^{i\Omega_D T}. \tag{8.48}$$

In general, at long times after transients have died out, the solution will take the form

$$A = a' e^{i\Omega_D T} + b' e^{i(\Omega_P - \Omega_D)T}, \tag{8.49}$$

where a' and b' are complex constants.

We first consider the degenerate case where the pump frequency is tuned such that it is always twice the signal frequency. In this case $\Omega_P = 2\Omega_D$, and the long time solution is

$$A = a e^{i\Omega_D T} \tag{8.50}$$

with a a time independent complex amplitude. Substituting this into (8.48) gives

$$(2\Omega_D - i)a - \bar{h} a^* = -g. \tag{8.51}$$

Equation (8.51) is easily solved. If we first look on resonance, $\Omega_D = 0$, we find

$$a = e^{i\pi/4} \left[\frac{\cos(\phi_g + \pi/4)}{(1 - \bar{h})} + i \frac{\sin(\phi_g + \pi/4)}{(1 + \bar{h})} \right] |g|, \tag{8.52}$$

where we remind the reader that $g = |g| e^{i\phi_g}$ so that ϕ_g measures the phase of the signal relative to the pump. Equation (8.52) shows that on resonance and for $\bar{h} \to 1$ (or $h \to h_c = 2$), the strongest *enhancement* of the response occurs for a signal that has a phase $-\pi/4$ relative to the pump. Physically, this means that the maximum of the signal occurs a quarter of a pump cycle *after* a maximum of the pump. (The phase $3\pi/4$ gives the same result: this corresponds to shifting the oscillations by a complete pump period.) The enhancement diverges as $\bar{h} \to 1$, *provided* that the signal amplitude g is small enough that the enhanced response remains within the linear regime. For a fixed signal amplitude g, the response will become large as $\bar{h} \to 1$, so that the nonlinear terms in (8.47) must be retained and the expressions we have derived no longer hold. This situation is discussed in the next section.

On the other hand, there is a weak suppression, by a factor of 2 as $\bar{h} \to 1$, for a signal that has a relative phase $\pi/4$ or $5\pi/4$. The latter pertains to the case of a signal maximum that occurs a quarter of a pump cycle *before* a maximum of the pump. A noise signal on the right-hand side of the equation of motion (8.45) would have both phase components. This leads to the *squeezing* of the noisy displacement driven by this noise, with the response at phase $-\pi/4$ amplified and the response at phase $\pi/4$ quenched.

The full expression for $\Omega_D \neq 0$ for the response amplitude is

$$a = -\left[\frac{2\Omega_D + (i + \bar{h}\,e^{-2i\phi_g})}{4\Omega_D^2 + (1 - \bar{h}^2)}\right] g \,. \tag{8.53}$$

For $\bar{h} \to 1$ the response is large when $\Omega_D \ll 1$, that is, for frequencies much closer to resonance than the original width of the resonator response. In these limits the first term in the numerator may be neglected *unless* $\phi_g \simeq \pi/4$. This then gives

$$|a| = \frac{2\,|g\cos(\phi_g + \pi/4)|}{4\Omega_D^2 + (1 - \bar{h}^2)} \,. \tag{8.54}$$

This is not the same as the expression for a resonant response, since the frequency dependence of the amplitude, not amplitude squared, is Lorentzian. However, estimating a quality factor from the width of the sharp peak would give an *enhanced* quality factor $\propto 1/\sqrt{1 - \bar{h}^2}$, becoming very large as $\bar{h} \to 1$. For the case $\phi_g = \pi/4$ the magnitude of the response is

$$\left|a_{\phi_g = \pi/4}\right| = \frac{\sqrt{4\Omega_D^2 + (1 - \bar{h})^2}}{4\Omega_D^2 + (1 - \bar{h}^2)} |\bar{g}| \,. \tag{8.55}$$

This initially increases as the frequency approaches resonance, but decreases for $\Omega_D \lesssim \sqrt{1 - \bar{h}}$, approaching $|g|/2$ for $\Omega_D \to 0$, $\bar{h} \to 1$.

For the general or nondegenerate case of $\Omega_P \neq 2\Omega_D$, it is straightforward to repeat the calculation with the ansatz (8.49). The result is

$$a' = -\frac{2(\Omega_P - \Omega_D) + i}{4\Omega_D(\Omega_P - \Omega_D) - 2i(\Omega_P - 2\Omega_D) + 1 - \bar{h}^2} g \,. \tag{8.56}$$

Notice that this does *not* reduce to (8.53) for $\Omega_P = 2\Omega_D$, since we miss some of the interference terms in the degenerate case if we base the calculation on $\Omega_P \neq 2\Omega_D$. Also, of course, there is no dependence of the magnitude of the response on the phase of the signal ϕ_g, since for different frequencies the phase difference cannot be defined independent of an arbitrary choice of the origin of time. If the pump frequency is maintained fixed at twice the resonator resonance frequency, corresponding to $\Omega_P = 0$, the expression for the amplitude of the response simplifies to

$$a' = \frac{2\Omega_D - i}{-4\Omega_D^2 + 4i\Omega_D + 1 - \bar{h}^2} g \,. \tag{8.57}$$

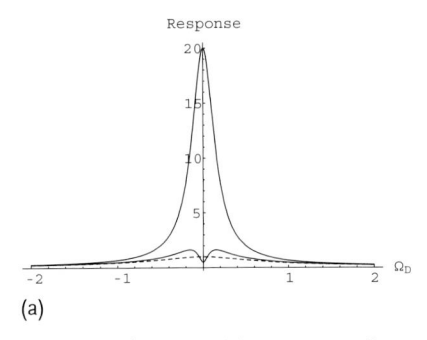

Figure 8.7 Response of the parametrically driven resonator as the signal frequency Ω_D varies for a pump frequency equal to twice the signal frequency (a), and for the pump frequency fixed at the linear resonance frequency (b), given by (8.53) and (8.57), respectively. The dashed curve is the response of the resonator to the same signal without parametric drive. In (a) the upper curve is for the amplified phase $\phi_g = -\pi/4$, and the lower curve for the phase $\phi_g = \pi/4$, giving squeezing on resonance. In both cases the reduced pump amplitude $\bar{h} = h/h_c$ is 0.95.

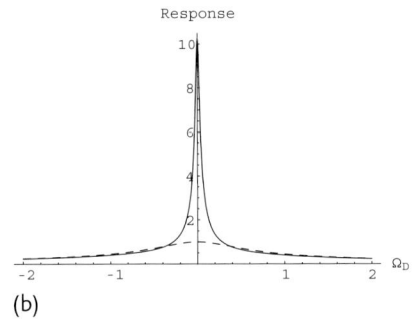

Again, there is an enhanced response for drive frequencies closer to resonance than the width of the original resonator response. In this region $\Omega_D \ll 1$, so that

$$|a'| \simeq |g| \frac{1}{\sqrt{(4\Omega_D)^2 + (1-\bar{h}^2)^2}}. \tag{8.58}$$

This is the usual Lorentzian describing a resonance with a quality factor enhanced by $(1-\bar{h}^2)^{-1}$, as shown in Figure 8.7(b).

For the resonance condition $\Omega_D = \Omega_P = 0$, corresponding to both a pump frequency that is twice the resonance frequency of the device, and to a signal at this resonant frequency, the response amplitude in the linear approximation diverges as the pump amplitude approaches the critical value $h_c = 2$. This is the signature of a linear instability to self sustained oscillations in the absence of any drive. We analyze this *parametric instability* in the next section.

8.3.2
Linear Instability

The divergence of the response as $\bar{h}$ approaches unity from below corresponding to $h \to 2$ suggests a linear instability for $h > 2$, or $QH > 2$ in the original units. We can see this directly from (8.47) by setting $g = 0$ but still ignoring the nonlinear terms, yielding the linear equation

$$\frac{dA}{dT} + \frac{1}{2}A = i\frac{h}{4}A^* e^{i\Omega_P T}. \tag{8.59}$$

We seek a solution of the form

$$A = |a|\, e^{i\phi}\, e^{\sigma T}\, e^{i(\Omega_P/2)T} \tag{8.60}$$

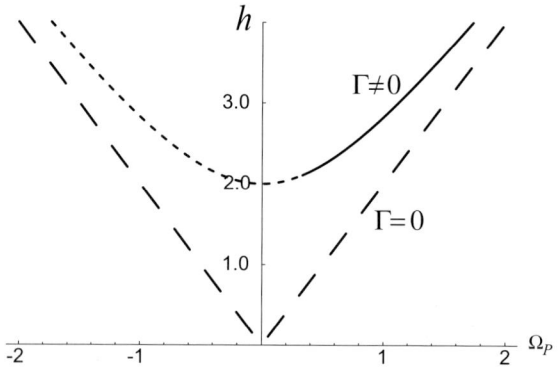

Figure 8.8 The first instability tongue of the parametrically-driven Duffing resonator, the threshold for instability, plotted in the (Ω_P, h) plane. The lower, long-dashed curve shows the threshold without any linear damping ($\Gamma = 0$), which is zero on resonance. The upper curve shows the threshold with linear damping ($\Gamma \neq 0$). The threshold on resonance ($\Omega_P = 0$) is $h = 2$. The solid and short-dashed regions of the upper curve indicate the so-called subcritical and supercritical branches of the instability, respectively, as discussed in Section 8.3.4. On the subcritical branch ($\Omega_P > 4\eta/3$) there will be hysteresis as h is varied, and on the supercritical branch ($\Omega_P < 4\eta/3$) there will not be any hysteresis.

with a real σ giving exponential growth or decay. Substituting into (8.59) gives

$$\sigma = \frac{-1 \pm \sqrt{(h/2)^2 - \Omega_P^2}}{2}, \tag{8.61}$$

$$\phi = \pm \left[\frac{\pi}{4} - \frac{1}{2} \arcsin\left(\frac{2\Omega_P}{h}\right) \right] \tag{8.62}$$

where we take the value of arcsin between 0 and $\pi/2$, and the plus and minus signs in the two equations correspond directly to one another. Note that these expressions apply for $h/2 > \Omega_P$; for $h/2 < \Omega_P$, the value of σ is complex. For pumping at twice the resonance frequency $\Omega_P = 0$, one phase of oscillation $\phi = \frac{\pi}{4}$ has a *reduced damping*, with $\sigma = -(1/2 - h/4)$ for $h < 2$, and an *instability* $\sigma = (h/4 - 1/2) > 0$ signaling exponential growth for $h > 2$. The other phase of oscillation $\phi = -\frac{\pi}{4}$ has an *increased damping*, with $\sigma = -(1/2 + h/4)$. The general condition for instability is

$$h > 2\sqrt{1 + \Omega_P^2}, \tag{8.63}$$

showing an increase of the threshold for nonzero frequency detuning Ω_P, as shown in Figure 8.8. The linear instability that occurs for positive σ gives exponentially growing solutions that eventually saturate due to nonlinearity.

8.3.3
Nonlinear Behavior Near Threshold

Nonlinear effects may also be important below the threshold of the parametric instability in the presence of a periodic signal or noise. As we have seen, in the linear approximation the gain below threshold diverges as $h \to h_c$. This is unphysical, and for a given signal or noise strength there is some h close enough to h_c where nonlinear saturation of the gain will become important. This will give a smooth behavior of the response of the driven system as h passes through h_c into the unstable regime. We first analyze the effects of nonlinearity near the threshold of the instability, and calculate the smooth behavior as h passes through h_c in the presence of an applied signal. In the following section we study the effects of nonlinearity on the self-sustained oscillations above threshold with more generality.

We take h to be close to h_c, and we take the signal to be small. This introduces a second level of "smallness". We have already assumed that the damping and the deviation of the pump frequency from resonance are both small. This means that the critical parametric drive H_c is also small. We now assume that $|H - H_c|$ is small compared with H_c, or, equivalently in scaled units, that $|h - h_c|$ is small compared with h_c. We then introduce the perturbation parameter δ to implement this, that is, we assume that

$$\delta = \frac{h - h_c}{h_c} \ll 1. \tag{8.64}$$

We now use the same type of secular perturbation theory as the method leading to (8.47) to develop the expansion in δ. For simplicity we will develop the theory for the most interesting case of resonant pump and signal frequencies $\Omega_P = \Omega_D = 0$. The critical value of h is then $h_c = 2$, and the solution to (8.47) that becomes marginally stable at this value is

$$A = b\,e^{i\pi/4}, \tag{8.65}$$

with b a real constant.

For h near h_c we make the ansatz for the solution

$$A = \delta^{1/2} b_0(\tau)\,e^{i\pi/4} + \delta^{3/2} b_1(\tau) + \cdots, \tag{8.66}$$

where b_0 is a real function of $\tau = \delta T$. The latter is a new and even slower time scale that determines the time variation of the real amplitude b_0 near threshold. We must also assume that the signal amplitude is very small, that is, $g = \delta^{3/2}\hat{g}$, in total yielding $G = (\epsilon\delta)^{3/2}\hat{g}$. Substituting (8.66) into (8.47) and collecting terms at $O(\delta^{3/2})$ yields

$$\frac{1}{2}(b_1 - b_1^*) = -\frac{\hat{g}}{2}e^{i\pi/4} - \frac{db_0}{d\tau} + \frac{1}{2}b_0 + i\frac{3}{8}b_0^3 - \frac{\eta}{8}b_0^3. \tag{8.67}$$

The left-hand side of this equation is necessarily imaginary, so in order to have a solution for b_1 such that the perturbation expansion is valid, the real part of the right-hand side must be zero. This is the solvability condition for the secular perturbation theory. This gives

$$\frac{db_0}{d\tau} = \frac{1}{2}b_0 - \frac{\eta}{8}b_0^3 - \frac{|\hat{g}|}{2}\cos(\phi_g + \pi/4). \tag{8.68}$$

It is more informative to write this equation in terms of the the variables without the δ scaling. Introducing the "unscaled" amplitude $b = \delta^{1/2} b_0$ and generalizing (8.65) such that

$$A = b\, e^{i\pi/4} + O(\delta^{3/2}), \tag{8.69}$$

we can write the equation as

$$\frac{db}{dT} = \frac{1}{2}\frac{h - h_c}{h_c} b - \frac{\eta}{8} b^3 - \frac{|g|}{2}\cos(\phi_g + \pi/4). \tag{8.70}$$

Equation (8.70) can be used to investigate many phenomena, such as transients above threshold, and how the amplitude of the response to a signal varies as h passes through the instability threshold. The unphysical divergence of the response to a small signal as $h \to h_c$ from below is now eliminated. For example, exactly at threshold $h = h_c$ we have

$$|b| = \left(\frac{4}{\eta}|g\cos(\phi_g + \pi/4)|\right)^{1/3}, \tag{8.71}$$

giving a finite response, but one proportional to $|g|^{1/3}$ rather than to $|g|$. The gain $|b/g|$ scales as $|g|^{-2/3}$ for $h = h_c$, and gets smaller as the signal gets larger, as shown in Figure 8.9. Note that the physical origin of the saturation at the lowest order of perturbation theory is nonlinear damping. Without nonlinear damping the response amplitude (8.71) still diverges. With linear damping that is still small, one would need to go to higher orders of perturbation theory to find a different physical mechanism that can provide this kind of saturation. The response to noise can also be investigated by replacing the $|g|\cos(\phi_g + \pi/4)$ drive by a noise function. Equation (8.70) and the noisy version appear in many contexts of phase transitions and bifurcations, and so solutions are readily found in the literature [20].

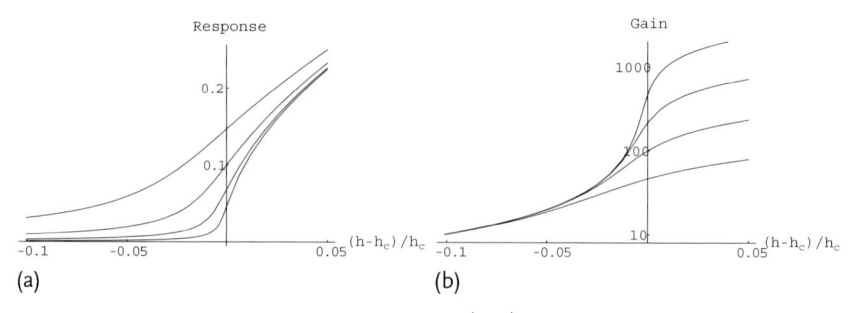

Figure 8.9 Saturation of the response b (a) and gain $|b/g|$ (b) as the parametric drive h passes through the critical value h_c, for four different signal levels g. The signal levels are $\sqrt{\eta/4}$ times $10^{-2.5}$, 10^{-3}, $10^{-3.5}$, and 10^{-4}, increasing upwards for the response figure, and downwards for the gain figure. The response amplitude is also measured in units of $\sqrt{\eta/4}$. The phase of the signal is $\phi_g = -\pi/4$.

8.3.4
Nonlinear Saturation above Threshold

The linear instability leads to exponential growth of the amplitude, regardless of the signal, and results in its saturation. In order to understand this process, we need to return to the full nonlinear treatment of (8.47) with $g = 0$. Ignoring initial transients and assuming that the nonlinear terms in the equation are sufficient to saturate the growth of the instability, we try a steady-state solution of the form

$$A(T) = a e^{i\left(\frac{\Omega_P}{2}\right)T}. \tag{8.72}$$

This amplitude a can be any solution of the equation

$$\left[\left(\frac{3}{4}|a|^2 - \Omega_P\right) + i\left(1 + \frac{\eta}{4}|a|^2\right)\right] a = -\frac{h}{2} a^*, \tag{8.73}$$

obtained by substituting the steady-state solution (8.72) into the equation of the secular terms (8.47). We immediately see that having no response ($a = 0$) is always a possible solution regardless of the excitation frequency Ω_P. Expressing $a = |a| e^{i\phi}$ and taking the magnitude squared of both sides, we obtain the intensity $|a|^2$ of the nontrivial response as all positive roots of the equation

$$\left(\Omega_P - \frac{3}{4}|a|^2\right)^2 + \left(1 + \frac{\eta}{4}|a|^2\right)^2 = \frac{h^2}{4}. \tag{8.74}$$

In addition to the solution $|a| = 0$, we have a quadratic equation for $|a|^2$ and therefore, at most, two additional positive solutions for $|a|$. This has the form of a distorted ellipse in the $(\Omega_P, |a|^2)$ plane and a parabola in the $(|a|^2, h)$ plane. In addition, we obtain for the relative phase of the response

$$\phi = \frac{i}{2} \ln \frac{a^*}{a} = -\frac{1}{2} \arctan \frac{1 + \frac{\eta}{4}|a|^2}{\frac{3}{4}|a|^2 - \Omega_P}. \tag{8.75}$$

In Figure 8.10 we plot the response intensity $|a|^2$ of a Duffing resonator to parametric excitation as a function of the pump frequency Ω_P for a fixed scaled drive amplitude $h = 3$. Solid curves indicate stable solutions, and dashed curves are solutions that are unstable to small perturbations. Thin curves show the response without nonlinear damping ($\eta = 0$), which grows indefinitely with frequency Ω_P and is therefore incompatible with experimental observations [8, 66, 71] as well as the assumptions of our calculation. As we saw for the saturation below threshold, without nonlinear damping and with linear damping being small, one would have to go to higher orders of perturbation theory to search for a physical mechanism that could provide saturation. For large linear damping, or small Q, one sees saturation even without nonlinear damping [47]. Thick curves in Figure 8.10 show the response with finite nonlinear damping ($\eta = 1$). With finite η there is a maximum value for the response $|a|^2_{\max} = 2(h-2)/\eta$, and a maximum frequency

$$\Omega_{SN} = \frac{h}{2}\sqrt{1 + \left(\frac{3}{\eta}\right)^2} - \frac{3}{\eta}, \tag{8.76}$$

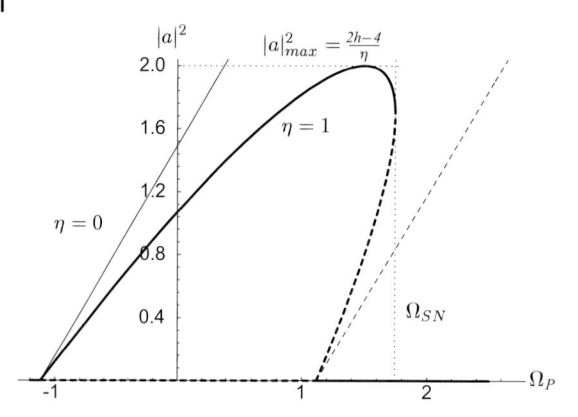

Figure 8.10 Response intensity $|a|^2$ as a function of the pump frequency Ω_P, for fixed amplitude $h = 3$. Solid curves are stable solutions; dashed curves are unstable solutions. Thin curves show the response without nonlinear damping ($\eta = 0$). Thick curves show the response for finite nonlinear damping ($\eta = 1$). Dotted lines indicate the maximal response intensity $|a|^2_{\max}$ and the saddle-node frequency Ω_{SN}.

at a saddle-node bifurcation, where the stable and unstable nontrivial solutions meet. For frequencies above Ω_{SN} the only solution is the trivial one, $a = 0$. These values are indicated by horizontal and vertical dotted lines in Figure 8.10.

The threshold for the instability of the trivial solution is easily verified by setting $a = 0$ in the expression (8.74) for the nontrivial solution, or by inverting the expression (8.63) for the instability that we obtained in the previous section. As seen in Figure 8.10, for a given h the threshold is situated at $\Omega_P = \pm\sqrt{(h/2)^2 - 1}$. This is the same result calculated in the previous section, where we plotted the threshold tongue in Figure 8.8 in the (h, Ω_P) plane. Figure 8.10 is a horizontal cut through that tongue at a constant drive amplitude $h = 3$.

Like the response of a forced Duffing resonator shown in (8.29), the response of a parametrically excited Duffing resonator also exhibits hysteresis in quasistatic frequency scans. If the frequency Ω_P begins at negative values and is increased gradually with a fixed amplitude h, the zero response will become unstable as the lower threshold is crossed at $-\sqrt{(h/2)^2 - 1}$. After this occurs the response will gradually increase along the thick solid curve in Figure 8.10, until Ω_P reaches Ω_{SN} and the response drops abruptly to zero. If the frequency is then decreased gradually, the response will remain zero until Ω_P reaches the upper instability threshold $+\sqrt{(h/2)^2 - 1}$. The response will then jump abruptly to the thick solid curve above, and afterwards gradually decrease to zero along this curve.

Finally, in Figure 8.11 we plot the response intensity $|a|^2$ of the Duffing resonator as a function of drive amplitude h, for fixed frequency Ω_P and finite nonlinear damping $\eta = 1$. This would correspond to performing a vertical cut through the instability tongue Figure 8.8. Again, solid curves indicate stable solutions and dashed curves indicate unstable solutions. Thick curves show the response for $\Omega_P = 1$, and

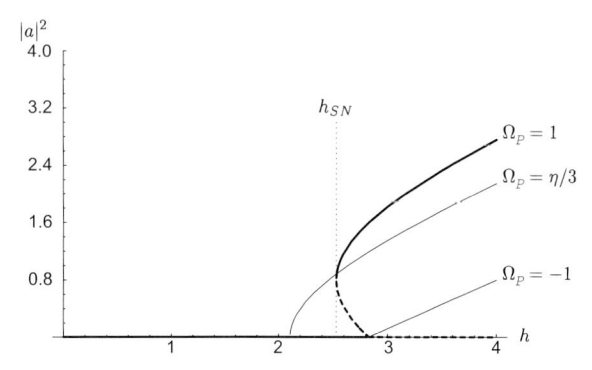

Figure 8.11 Response intensity $|a|^2$ as a function of the parametric drive amplitude h for fixed frequency Ω_P and finite nonlinear damping ($\eta = 1$). Thick curves show the stable (solid curves) and unstable (dashed curves) response for $\Omega_P = 1$. Thin curves show the stable solutions for $\Omega_P = \eta/3$ and $\Omega_P = -1$, and demonstrate that hysteresis as h is varied is expected only for $\Omega_P > \eta/3$.

thin curves show the response for $\Omega_P = \eta/3$ and $\Omega_P = -1$. The intersection of the trivial and the nontrivial solutions, which corresponds to the instability threshold (8.63), occurs at $h = 2\sqrt{\Omega_P^2 + 1}$. For $\Omega_P < \eta/3$, the nontrivial solution for $|a|^2$ grows continuously for h above threshold and is stable. This is a supercritical bifurcation. On the other hand, for $\Omega_P > \eta/3$ the bifurcation is subcritical and the nontrivial solution grows for h below threshold. This solution is unstable until the curve of $|a|^2$ as a function of h turns at a saddle-node bifurcation at

$$h_{SN} = \frac{2 + \frac{2\eta}{3}\Omega_P}{\sqrt{1 + \left(\frac{\eta}{3}\right)^2}}, \tag{8.77}$$

where the solution becomes stable and $|a|^2$ is once more an increasing function of h. For amplitudes $h < h_{SN}$ the only solution is the trivial one $a = 0$. Hysteretic behavior is therefore expected for quasistatic scans of the drive amplitude h only if the fixed frequency $\Omega_P > \eta/3$, as can be inferred from Figure 8.11.

8.3.5
Parametric Excitation at the Second Instability Tongue

We wish to examine the second tongue by looking at the response above threshold and highlighting the main changes from the first tongue. This tongue, it should be noted, is readily accessible in experiments because the pump and the response frequencies are the same. We start with the general equation for a parametrically-driven Duffing resonator (8.44), but with no direct drive ($g = 0$), where the parametric excitation is performed around 1 instead of 2. Correspondingly, the scaling of H with respect to ϵ needs to be changed to $H = h\sqrt{\epsilon}$. The reason for this change is that with the $H = h\epsilon$ scaling, the order $\epsilon^{1/2}$ term in x becomes identically zero. This occurs because the parametric driving term does not contribute to the order $\epsilon^{3/2}$ secular term which we use to find the response. Scaling H in the appropriate manner will introduce a nonsecular correction to x at order ϵ, and this correction will contribute to the order $\epsilon^{3/2}$ secular term and will give us the required response. The equation of motion then becomes

$$\ddot{x} + x = -\frac{h\epsilon^{1/2}}{2}\left(e^{i(t+\Omega_p T)} + \text{c.c.}\right)x - \epsilon \dot{x} - x^3 - \eta x^2 \dot{x}, \qquad (8.78)$$

and we try an expansion of the solution of the form

$$x(t) = \epsilon^{1/2}\tfrac{1}{2}\left(A(T)e^{it} + \text{c.c.}\right) + \epsilon x_{1/2}(t) + \epsilon^{3/2} x_1(t) + \ldots \qquad (8.79)$$

Substituting this expansion into the equation of motion (8.78), we obtain at order $\epsilon^{1/2}$ the linear equation as usual, and at order ϵ

$$\ddot{x}_{1/2} + x_{1/2} = -\frac{h}{4}\left(A\,e^{i\Omega_p T} e^{2it} + A^* e^{i\Omega_p T} + \text{c.c.}\right). \qquad (8.80)$$

As expected, there is no secular term on the right-hand side so we can immediately solve for $x_{1/2}$, yielding

$$x_{1/2}(t) = \frac{h}{4}\left(\frac{A}{3}e^{i\Omega_p T} e^{2it} - A^* e^{i\Omega_p T} + \text{c.c.}\right) + O(\epsilon). \qquad (8.81)$$

Substituting the solution for $x_{1/2}$ into the expansion (8.79), and the expansion back into the equation of motion (8.78), contributes an additional term from the parametric driving which has the form

$$\epsilon^{3/2}\frac{h^2}{8}\left(-\frac{A}{3}e^{i\Omega_p T}e^{2it} + A^* e^{i\Omega_p T} + \text{c.c.}\right)\left(e^{i\Omega_p T}e^{it} + \text{c.c.}\right)$$

$$= \epsilon^{3/2}\frac{h^2}{8}\left(\frac{2}{3}A + A^* e^{i2\Omega_p T}\right)e^{it} + \text{c.c.} + \text{nonsecular terms}. \qquad (8.82)$$

This gives us the required contribution to the equation for the vanishing secular terms. All other terms remain as they were in (8.47), so that the new equation for determining $A(T)$ becomes

$$\frac{dA}{dT} + i\frac{h^2}{8}\left(\frac{2}{3}A + A^* e^{i2\Omega_p T}\right) + \frac{1}{2}A - i\frac{3}{8}|A|^2 A + \frac{\eta}{8}|A|^2 A = 0. \qquad (8.83)$$

Again, ignoring initial transients and assuming that the nonlinear terms in the equation are sufficient to saturate the growth of the instability, we try a steady-state solution, this time of the form

$$A(T) = a\,e^{i\Omega_p T}. \qquad (8.84)$$

The solution to the equation of motion (8.78) is therefore

$$x(t) = \epsilon^{1/2}(a\,e^{i(1+\epsilon\Omega_p)t} + \text{c.c.}) + O(\epsilon), \qquad (8.85)$$

where the correction $x_{1/2}$ of order ϵ is given in (8.81) and, as before, we are not interested in the correction $x_1(t)$ of order $\epsilon^{3/2}$, but rather in the fixed amplitude a of the lowest order term. We substitute the steady-state solution (8.84) into the equation of the secular terms (8.83) and obtain

$$\left[\left(\frac{3}{4}|a|^2 - 2\Omega_P - \frac{h^2}{6}\right) + i\left(1 + \frac{\eta}{4}|a|^2\right)\right] a = \frac{h^2}{4} a^* . \tag{8.86}$$

By taking the magnitude squared of both sides we obtain, in addition to the trivial solution $a = 0$, a nontrivial response given by

$$\left(\frac{3}{4}|a|^2 - 2\Omega_P - \frac{1}{6}h^2\right)^2 + \left(1 + \frac{\eta}{4}|a|^2\right)^2 = \frac{h^4}{16} . \tag{8.87}$$

Figure 8.12 shows the response intensity $|a|^2$ as a function of the frequency Ω_P for a fixed drive amplitude of $h = 3$, producing a horizontal cut through the second instability tongue. The solution looks very similar to the response shown in Figure 8.10 for the first instability tongue, though we should point out two important differences. The first is that the orientation of the ellipse, indicated by the slope of the curves for $\eta = 0$, is different. The slope here is 8/3, whereas for the first instability tongue the slope is 4/3. The second is the change in the scaling of h with ϵ, or the inverse quality factor Q^{-1}. The lowest critical drive amplitude for an instability at the second tongue is again on resonance ($\Omega_P = 0$), and its value is again $h = 2$. This now implies, however, that $H\sqrt{Q} = 2$, or that H scales as the square root of the linear damping rate Γ. This is consistent with the well known result that the minimal amplitude for the instability of the nth tongue scales as $\Gamma^{1/n}$ (for example, see [42], Section 3).

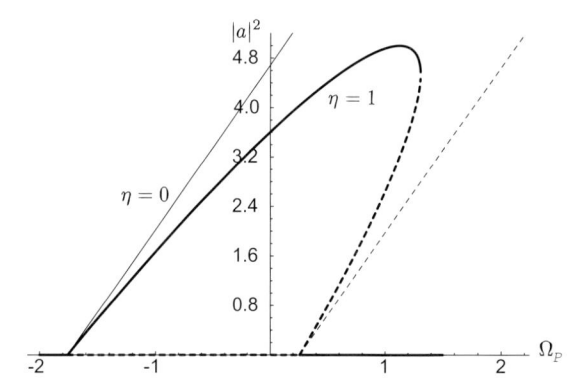

Figure 8.12 Response intensity $|a|^2$ of a parametrically-driven Duffing resonator as a function of the pump frequency Ω_P, for a fixed amplitude $h = 3$ in the second instability tongue. Solid curves are stable solutions and dashed curves are unstable solutions. Thin curves show the response without nonlinear damping ($\eta = 0$). Thick curves show the response for finite nonlinear damping ($\eta = 1$).

8.4
Parametric Excitation of Arrays of Coupled Duffing Resonators

The last two sections of this review describe theoretical work that was motivated directly by the experimental work of Buks and Roukes [8]. They fabricated an array of nonlinear micromechanical doubly-clamped gold beams, and excited them parametrically by modulating the strength of an externally controlled electrostatic coupling between neighboring beams. The Buks and Roukes experiment was modeled by Lifshitz and Cross [44] (henceforth LC) using a set of coupled nonlinear equations of motion. The latter used secular perturbation theory, as we have described so far for a system with just a single degree of freedom, to convert these equations of motion into a set of coupled nonlinear *algebraic* equations for the normal mode amplitudes of the system. This enabled them to obtain exact results for small arrays, but only a qualitative understanding of the dynamics of large arrays. We shall review these results in this section.

In order to obtain analytical results for large arrays, Bromberg, Cross, and Lifshitz [7] (henceforth BCL) studied the same system of equations, approaching it from the continuous limit of infinitely many degrees of freedom. They obtained a description of the slow spatiotemporal dynamics of the array of resonators in terms of an amplitude equation. BCL showed that this amplitude equation could predict the initial mode that develops at the onset of parametric oscillations as the driving amplitude is gradually increased from zero, as well as a sequence of subsequent transitions to other single mode oscillations. We shall review these results in Section 8.5. Kenig, Lifshitz, and Cross [38] have extended the investigation of the amplitude equation to more general questions such as how patterns are selected when many patterns or solutions are simultaneously stable. This extension includes other experimentally relevant questions, such as the response of the system of coupled resonators to time dependent sweeps of the control parameters, rather than quasistatic sweeps like the ones we have been discussing here. Kenig *et al.* [39] have also studied the formation and dynamics of intrinsically-localized modes, or solitons, in the array equations of LC. To this end, they derived a different amplitude equation, which takes the form of a parametrically-driven damped nonlinear Shrödinger equation, also known as a forced complex Ginzburg-Landau equation. We shall not review these last two papers here, but encourage the reader to pursue them independently.

8.4.1
Modeling an Array of Coupled Duffing Resonators

LC modeled the array of coupled nonlinear resonators that was studied by Buks and Roukes using a set of coupled equations of motion (EOM) of the form

$$\ddot{u}_n + u_n + u_n^3 - \tfrac{1}{2}Q^{-1}(\dot{u}_{n+1} - 2\dot{u}_n + \dot{u}_{n-1})$$
$$+ \tfrac{1}{2}(D + H\cos\omega_p t)(u_{n+1} - 2u_n + u_{n-1})$$
$$- \tfrac{1}{2}\eta\left[(u_{n+1} - u_n)^2(\dot{u}_{n+1} - \dot{u}_n) - (u_n - u_{n-1})^2(\dot{u}_n - \dot{u}_{n-1})\right] = 0,$$

(8.88)

where $u_n(t)$ describes the deviation of the nth resonator from its equilibrium, with $n = 1\ldots N$, and fixed boundary conditions $u_0 = u_{N+1} = 0$. Detailed arguments for the choice of terms introduced into the equations of motion are discussed in [44]. The terms include an elastic restoring force with both linear and cubic contributions, whose coefficients are both scaled to 1 as in our discussion of the single degree of freedom. They also include a dc electrostatic nearest neighbor coupling term with a small ac component responsible for the parametric excitation, with coefficients D and H, respectively, and linear as well as cubic nonlinear dissipation terms. Both dissipation terms are assumed to depend on the difference of the displacements of nearest neighbors.

We consider here a slightly simpler and more general model for an array of coupled resonators in order to illustrate the approach. Motivated by the geometry of most experimental NEMS systems, we assume a line of identical resonators although the generalization to two or three dimensions is straightforward. The simplest model is to take the equation of motion of each resonator to be as that in (8.44), with the addition of a coupling term to its two neighbors. A simple choice would be to assume that this coupling does not introduce additional dissipation, which we describe as *reactive* coupling. Elastic and electrostatic coupling might be predominantly of this type. After the usual scaling, the equations of motions would take the form

$$\ddot{u}_n + Q^{-1}\dot{u}_n + u_n^3 + (1 + H\cos\omega_p t)u_n + \eta u_n^2 \dot{u}_n$$
$$+ \tfrac{1}{2} D(u_{n+1} - 2u_n + u_{n-1}) = 0 , \qquad (8.89)$$

where we do not take into account any direct drive for the purposes of the present section.

The equations of motion for particular experimental implementations might have different terms, although we expect all will have linear and nonlinear damping, linear coupling, and parametric drive. For example, to model the experimental setup of Buks and Roukes [8], LC supposed that both linear and nonlinear dissipation terms involved the difference of neighboring displacements, that is, the terms involving $\dot{u}_n$ in our equations of motion (8.89) are replaced with terms involving $u_{n+1} - u_n$ in the equations of motion (8.88) used by LC. This was to describe the physics of electric current damping, with the currents driven by the varying capacitance between neighboring resonators depending on the change in separation and the fixed DC voltage. This effect seemed to be the dominant component of the dissipation in the Buks and Roukes experiments. Similarly, the parametric drive $H\cos\omega_p t$ multiplied $(u_{n+1} - 2u_n + u_{n-1})$ in the equations of LC rather than u_n here, since the voltage between adjacent resonators was the quantity modulated, changing the electrostatic component of the spring constant.

In a more recent implementation [45], the electric current damping has been reduced, and the parametric drive is directly applied to each resonator piezoelectrically, so that the simpler form of (8.89) applies. The method of attack is the same in any case. We will illustrate the approach on the simpler equation, and refer the reader to LC for the more complicated model. An additional complication in a realistic model may be that the coupling is longer range than nearest neighbor.

For example, both electrostatic coupling and elastic coupling through the supports would have longer range components. The general method is the same for these additional effects, and the reader should be able to apply the approach to the model for their particular experimental implementation.

8.4.2
Calculating the Response of an Array

We calculate the response of the array to parametric excitation, again using secular perturbation theory. We suppose Q is large and take $\epsilon = Q^{-1}$ as the small expansion parameter. As in Section 8.3 we take $H = \epsilon h$, but we also take $D = \epsilon d$ so that the width of the frequency band of eigenmodes is also small. This is not quite how LC treated the coupling, but we think the present approach is clearer, and it is equivalent up to the order of the expansion in ϵ that we require. We thank Eyal Kenig for pointing out this simplification.

The equations of motion are now

$$\ddot{u}_n + \epsilon \dot{u}_n + u_n^3 + \left(1 + \epsilon h \cos\left[(2 + \epsilon \Omega_p)t\right]\right) u_n + \eta u_n^2 \dot{u}_n + \tfrac{1}{2}\epsilon d(u_{n+1} - 2u_n + u_{n-1}) = 0, \quad n = 1 \ldots N. \tag{8.90}$$

We expand $u_n(t)$ as a sum of standing wave modes with slowly varying amplitudes. The nature of the standing wave modes will depend on the conditions at the end of the line of resonators. In the experiments of Buks and Roukes there were N mobile beams, with a number of identical immobilized beams at each end. These conditions can be implemented in a nearest neighbor model by taking two additional resonators, u_0 and u_{N+1} and assuming

$$u_0 = u_{N+1} = 0. \tag{8.91}$$

The standing wave modes are then

$$u_n = \sin(nq_m) \quad \text{with} \quad q_m = \frac{m\pi}{N+1}, \quad m = 1\ldots N. \tag{8.92}$$

On the other hand, for a line of N resonators with free ends there is no force from outside the line. For the nearest neighbor model this can be imposed again by taking two additional resonators, but now with the conditions

$$u_0 = u_1 \quad \text{and} \quad u_N = u_{N+1}. \tag{8.93}$$

The standing wave modes are now

$$u_n = \cos\left[\left(n - \tfrac{1}{2}\right)q_m\right] \quad \text{with} \quad q_m = \frac{m\pi}{N}, \quad m = 0\ldots N-1. \tag{8.94}$$

For our illustration we will take (8.91), (8.92). Thus we write

$$u_n(t) = \epsilon^{1/2}\frac{1}{2}\sum_{m=1}^{N}\left(A_m(T)\sin(nq_m)e^{it} + \text{c.c.}\right) + \epsilon^{3/2}u_n^{(1)}(t) + \ldots,$$
$$n = 1\ldots N, \tag{8.95}$$

with q_m as in (8.92).

8.4 Parametric Excitation of Arrays of Coupled Duffing Resonators

We substitute the trial solution (8.95) into the EOM term by term. Up to order $\epsilon^{3/2}$ we have

$$\ddot{u}_n = \epsilon^{1/2} \frac{1}{2} \sum_m \sin(nq_m) \left(\left[-A_m + 2i\epsilon A'_m\right] e^{it} + \text{c.c.}\right) + \epsilon^{3/2} \ddot{u}_n^{(1)}(t) + \ldots , \tag{8.96a}$$

$$\epsilon \dot{u}_n = \epsilon^{3/2} \frac{1}{2} \sum_m \sin(nq_m) \left(iA_m e^{it} + \text{c.c.}\right) + \ldots , \tag{8.96b}$$

$$\frac{1}{2}\epsilon d(u_{n+1} - 2u_n + u_{n-1})$$

$$= -\epsilon^{3/2} \frac{d}{2} \sum_m 2\sin^2\left(\frac{q_m}{2}\right) \sin(nq_m) \left(A_m e^{it} + \text{c.c.}\right) + \ldots \tag{8.96c}$$

$$u_n^3 = \epsilon^{3/2} \frac{1}{8} \sum_{j,k,l} \sin(nq_j) \sin(nq_k) \sin(nq_l)$$

$$\times \left(A_j e^{it} + \text{c.c.}\right) \left(A_k e^{it} + \text{c.c.}\right) \left(A_l e^{it} + \text{c.c.}\right)$$

$$= \epsilon^{3/2} \frac{1}{32} \sum_{j,k,l} \{\sin[n(-q_j + q_k + q_l)] + \sin[n(q_j - q_k + q_l)]$$

$$+ \sin[n(q_j + q_k - q_l)] - \sin[n(q_j + q_k + q_l)]\}$$

$$\times \{A_j A_k A_l e^{3it} + 3A_j A_k A_l^* e^{it} + \text{c.c.}\} , \tag{8.96d}$$

and

$$\eta u_n^2 \dot{u}_n = \epsilon^{3/2} \frac{\eta}{32} \sum_{j,k,l} \{\sin[n(-q_j + q_k + q_l)] + \sin[n(q_j - q_k + q_l)]$$

$$+ \sin[n(q_j + q_k - q_l)] - \sin[n(q_j + q_k + q_l)]\}$$

$$\times \left(A_j e^{it} + \text{c.c.}\right) \left(A_k e^{it} + \text{c.c.}\right) \left(iA_l e^{it} + \text{c.c.}\right). \tag{8.96e}$$

The order $\epsilon^{1/2}$ terms cancel, and at order $\epsilon^{3/2}$ we get N equations of the form

$$\ddot{u}_n^{(1)} + u_n^{(1)} = \sum_m (m\text{th secular term}) \, e^{it} + \text{other terms} , \tag{8.97}$$

where the left-hand sides are uncoupled linear harmonic oscillators, with a frequency unity. On the right-hand sides we have N secular terms which act to drive the oscillators $u_n^{(1)}$ at their resonance frequencies. As we did for all the single resonator examples, here, too, we require that all the secular terms vanish so that the $u_n^{(1)}$ remain finite. Thus, we obtain equations for the slowly varying amplitudes $A_m(T)$. To extract the equation for the mth amplitude $A_m(T)$ we make use of the orthogonality of the modes, multiplying all the terms by $\sin(nq_m)$ and summing over n. We find that the coefficient of the mth secular term, which is required to vanish, is given by

$$-2i\frac{dA_m}{dT} - iA_m + 2d\sin^2\left(\frac{q_m}{2}\right)A_m - \frac{1}{2}hA_m^* e^{i\Omega_p T}$$
$$- \frac{3+i\eta}{16}\sum_{j,k,l} A_j A_k A_l^* \Delta^{(1)}_{jkl;m} = 0, \quad (8.98)$$

where we have used the Δ function introduced by LC, defined in terms of Kronecker deltas as

$$\Delta^{(1)}_{jkl;m} = \delta_{-j+k+l,m} - \delta_{-j+k+l,-m} - \delta_{-j+k+l,2(N+1)-m}$$
$$+ \delta_{j-k+l,m} - \delta_{j-k+l,-m} - \delta_{j-k+l,2(N+1)-m}$$
$$+ \delta_{j+k-l,m} - \delta_{j+k-l,-m} - \delta_{j+k-l,2(N+1)-m}$$
$$- \delta_{j+k+l,m} + \delta_{j+k+l,2(N+1)-m} - \delta_{j+k+l,2(N+1)+m}, \quad (8.99)$$

and have exploited the fact that it is invariant under any permutation of the indices j, k, and l. The function $\Delta^{(2)}_{jkl;m}$, also defined by LC, is not needed for our simplified model. The Δ function ensures the conservation of lattice momentum. In this case, momentum is conserved to within the non-uniqueness of the specification of the normal modes due to the fact that $\sin(nq_m) = \sin(nq_{2k(N+1)\pm m})$ for any integer k. The first Kronecker delta in each line is a condition of direct momentum conservation, and the other two are the so-called umklapp conditions where only lattice momentum is conserved.

As for the single resonator, we again try a steady-state solution, this time of the form

$$A_m(T) = a_m e^{i\left(\frac{\Omega_p}{2}\right)T}, \quad (8.100)$$

so that the solutions to the EOM, after substitution of (8.100) into (8.95), become

$$u_n(t) = \epsilon^{1/2}\frac{1}{2}\sum_m \left(a_m \sin(nq_m) e^{i\left(1+\frac{\epsilon\Omega_p}{2}\right)t} + \text{c.c.}\right) + O(\epsilon^{3/2}), \quad (8.101)$$

where all modes are oscillating at half the parametric excitation frequency.

Substituting the steady state solution (8.100) into the equations (8.98) for the time-varying amplitudes $A_m(T)$, we obtain the equations for the time-independent complex amplitudes, a_m

$$\left[\Omega_p + 2d\sin^2\left(\frac{q_m}{2}\right) - i\right]a_m - \frac{h}{2}a_m^* - \frac{3+i\eta}{16}\sum_{j,k,l} a_j a_k a_l^* \Delta^{(1)}_{jkl;m} = 0.$$

$$(8.102)$$

Note that the first two terms on the left-hand side indicate that the linear resonance frequency is not obtained for $\Omega_p = 0$, but rather for $\Omega_p + 2d\sin^2(q_m/2) = 0$. In terms of the unscaled parameters, this implies that the resonance frequency of the

mth mode is $\omega_m = 1 - D\sin^2(q_m/2)$, which is the same as the expected dispersion relation

$$\omega_m^2 = 1 - 2D\sin^2\left(\frac{q_m}{2}\right) \quad (8.103)$$

to within a correction of $O(\epsilon^2)$.

Equation 8.102 is the main result of the calculation. We have managed to replace N coupled differential equations (8.89) for the resonator coordinates $u_n(t)$ by N coupled algebraic equations (8.102) for the time-independent mode amplitudes a_m. All that remains, in order to obtain the overall collective response of the array as a function of the parameters of the original EOM, is to solve these coupled algebraic equations.

First, one can easily verify that for a single resonator ($N = j = k = l = m = 1$), the general equation (8.102) reduces to the single resonator equation (8.73) that we derived in Section 8.3.4 due to the fact that $\Delta_{111;1} = 4$. Next, one can see that the trivial solution, $a_m = 0$ for all m, always satisfies the equations, though it is not always a stable solution, as we have seen in the case of a single resonator. Finally, one can also verify that a single mode solution exists with $a_m \neq 0$ and $a_j = 0$ for all $j \neq m$ whenever, for any given m, $\Delta^{(1)}_{mmm;j} = 0$ for all $j \neq m$. These single mode solutions have the same type of elliptical shape of the single resonator solution given in (8.74). Note that generically $\Delta^{(1)}_{mmm;m} = 3$, except when umklapp conditions are satisfied.

In general, additional solutions involving more than a single mode exist, but are hard to obtain analytically. LC calculated these multimode solutions for the case of two and three resonators for the model they considered by finding the roots of the coupled algebraic equations numerically. We show some of their results to illustrate the type of behavior that occurs, although the precise details will be slightly different.

8.4.3
The Response of Very Small Arrays and Comparison of Analytics and Numerics

In Figure 8.13 we show the solutions for the response intensity of two resonators as a function of frequency for a particular choice of the equation parameters. Figure 8.13a shows the square of the amplitude of the antisymmetric mode a_2, whereas Figure 8.13b shows the square of the amplitude of the symmetric mode a_1. Solid curves indicate stable solutions and dashed curves indicate unstable solutions. Two elliptical single mode solution branches, similar to the response of the single resonator shown in Figure 8.10 are easily identified. These branches are labeled by S_1 and S_2. LC give the analytical expressions for these two solution branches. In addition, there are two double mode solution branches, labeled D_1 and D_2, involving the simultaneous excitation of both modes. Note that the two branches of double mode solutions intersect at a point where they switch their stability.

With two resonators there are regions in frequency where three stable solutions can exist. If all of the stable solution branches are accessible experimentally then

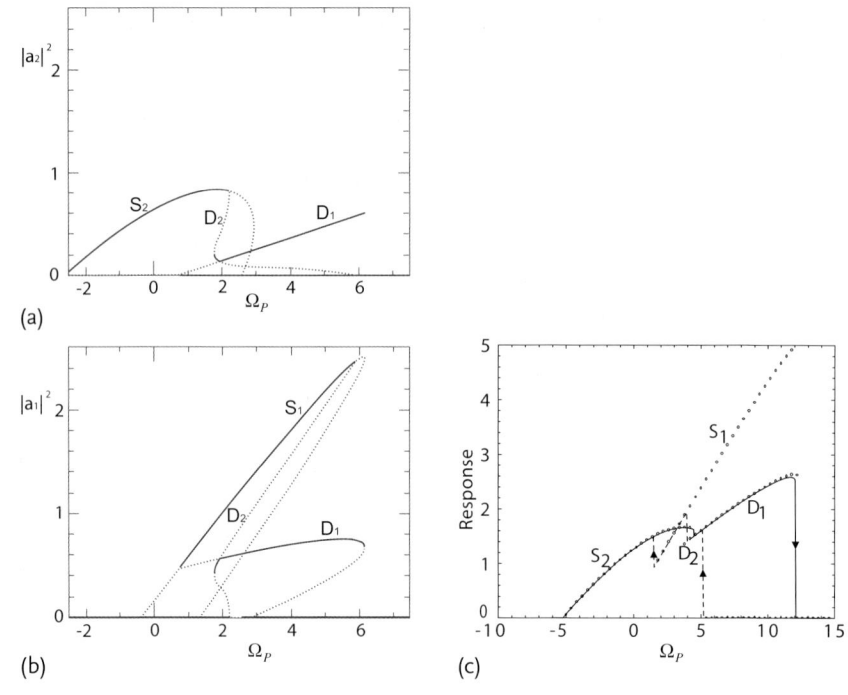

Figure 8.13 Two resonators. (a,b) Response intensity as a function of frequency Ω_P for a particular choice of the equation parameters. (a) shows $|a_2|^2$ and (b) shows $|a_1|^2$. Solid curves indicate stable solutions and dashed curves indicate unstable solutions. The two elliptical single mode solution branches are labeled S_1 and S_2. The two double mode solution branches are labeled D_1 and D_2. (c) Comparison of stable solutions obtained analytically (small circles), with a numerical integration of the equations of motion showing hysteresis in the response (solid curve – frequency swept up; dashed curve – frequency swept down). The averaged response intensity as defined in (8.104) is plotted. Branch labels correspond to those on the left.

the observed effects of hysteresis might be more complex than in the simple case of a single resonator. This is demonstrated in Figure 8.13c, where the analytical solutions are compared with a numerical integration of the differential equations of motion (8.88) for two resonators. The response intensity plotted here is given by the time and space averages of the square of the resonator displacements

$$I = \frac{1}{N} \sum_{n=1}^{N} \langle u_n^2 \rangle, \tag{8.104}$$

where the angular brackets denote time average and $N = 2$. A solid curve shows the response intensity for frequency swept upwards, and a dashed curve shows the response intensity for frequency swept downwards.

Small circles show the analytical response intensity for the stable regions of the four solution branches shown in Figure 8.13. With the analytical solution in the background, one can easily understand the discontinuous jumps and hysteresis effects that are obtained in the numerical solution of the equations of motion. Note that the S_1 branch is missed in the upwards frequency sweep and is only accessed by the system in the downwards frequency sweep. One could trace the whole stable region of the S_1 branch by changing the sweep direction after jumping onto the branch. This would result in climbing all the way up to the end of the S_1 branch and then falling onto the tip of the D_1 branch or to zero. These kinds of changes in the direction of the sweep that occur when one jumps onto a new branch are essential if one wants to trace out as much of the solution as possible. This holds for both real experiments or numerical simulations.

8.4.4
Response of Large Arrays and Numerical Simulation

LC integrated the equations of motion (8.88) numerically for an array of $N = 67$ resonators. The results for the response intensity as a function of the unscaled parametric drive frequency ω_p as given in (8.104) are shown in Figure 8.14. These results must be considered illustrative only, because the structure of the response branches will vary with changes to the model, and will also depend strongly on the chosen equation parameters. First of all, as in the case of a small number of beams, the overall height and width of individual response branches depend on the strength of the drive h and on the nonlinear dissipation coefficient η. Furthermore, if the coupling strength D is increased, for example, such that the width of the frequency response band becomes much larger than N times the width of a single mode response, then very few, if any, multimode solutions exist.

A number of the important features of the response should be highlighted. We concentrate on the solid curve in the figure, which is for frequency swept upwards. First, the response intensity shows features that span a range of frequencies that is large compared with the mode spacing, which is about 0.0006 for the parameters

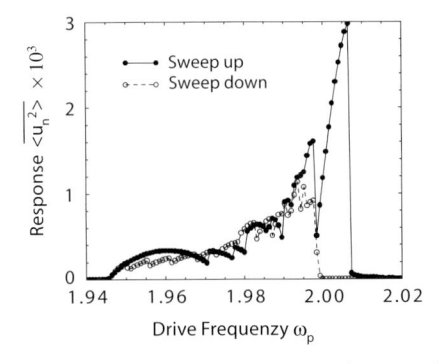

Figure 8.14 Response intensity as a function of the driving frequency ω_p for $N = 67$ parametrically-driven resonators (solid curve – frequency swept up; dashed curve – frequency swept down). The response intensity is defined in (8.104). The response curve was obtained through numerical integration of the equations of motion (8.88).

used. The reason for this is that we skip over many others as we follow a particular solution, as has been seen for the S1 branch in the two resonator case. Second, the variation of the response with frequency shows abrupt jumps as the frequency is raised, particularly on the high frequency side of the features. This happens as we reach saddle-node or other types of bifurcations where we lose the stability of the solution branch, or the branch ends altogether. Third, the response extends to frequencies higher than the band edge for the linear modes, which would give a response only up to $\omega_p = 2.0$. This happens simply due to the positive Duffing parameter which causes frequency pulling to the right. Note that the downwards sweep is able to access additional stable solution branches that were missed in the upwards sweep. There is also no response above $\omega_p = 2.0$ in this case. This is because the zero displacement state is stable for $\omega_p > 2.0$, and the system will remain in this state as the frequency is lowered unless a large enough disturbance kicks it onto another of the solution branches. The hysteresis on reversing the frequency sweep was not examined in any experiment, and it would be interesting to test this prediction of LC in the future.

8.5
Amplitude Equation Description for Large Arrays

We finish this review by describing the approach used by BCL [6, 7] to obtain analytical results for large arrays by approaching them from the continuous limit of infinitely many degrees of freedom. We only summarize the main results of BCL and encourage the reader, who by now has all the required background, to refer to BCL [7] and to Kenig et al. [38] for details of the derivation and for thorough discussions of the results and their experimental consequences. We note that BCL studied the original system of (8.88), where both the parametric excitation and the damping are introduced in terms of the difference variables $u_{n+1} - u_n$. We stick to this model here, and leave it to the reader as an exercise to generalize the BCL derivation for the more general model equations (8.89) that we used in the previous section.

A novel feature of the parametrically-driven instability is that the bifurcation to standing waves switches from supercritical (continuous) to subcritical (discontinuous) at a wave number at or close to the critical one, for which the required parametric driving force is minimum. This changes the form of the amplitude equation that describes the onset of the parametrically-driven waves so that it no longer has the standard "Ginzburg–Landau" form [20]. The central result of BCL is this new scaled amplitude equation (8.112), which is governed by a single control parameter and captures the slow dynamics of the coupled resonators just above the onset of parametric oscillations, including this unusual bifurcation behavior. BCL confirmed the behavior numerically and made suggestions for testing it experimentally. Kenig et al. [38] have extended the investigation of the amplitude equation to include such situations as time-dependent ramps of the drive amplitude, as opposed to the standard quasistatic sweeps of the control parameters. Although our

8.5.1
Amplitude Equations for Counter Propagating Waves

BCL scaled the equations of motion (8.88), as did Lifshitz and Cross [44], without assuming a priori that the coupling D is small. Thus, the scaled equations of motion that they solved were

$$\ddot{u}_n + u_n + u_n^3 - \tfrac{1}{2}\epsilon(\dot{u}_{n+1} - 2\dot{u}_n + \dot{u}_{n-1})$$
$$+ \tfrac{1}{2}\left[D + \epsilon h \cos(2\omega_p t)\right](u_{n+1} - 2u_n + u_{n-1})$$
$$- \tfrac{1}{2}\eta\left[(u_{n+1} - u_n)^2(\dot{u}_{n+1} - \dot{u}_n) - (u_n - u_{n-1})^2(\dot{u}_n - \dot{u}_{n-1})\right] = 0. \tag{8.105}$$

Note the way in which the pump frequency is specified as $2\omega_p$ in the argument of the cosine term, with an explicit factor of two (unlike what we did in Section 8.4), and also without making any assumptions at this point regarding its deviation from twice the resonance. We also remind the reader that this and all other frequencies are measured in terms of the natural frequency of a single resonator, which has been scaled to 1. The first step in treating this system of equations analytically is to introduce a continuous displacement field $u(x, t)$, and slow spatial and temporal scales $X = \epsilon x$ and $T = \epsilon t$. One then tries a solution in terms of a pair of counter-propagating plane waves at half the pump frequency, which is a natural first guess in continuous parametrically-driven systems such as Faraday waves [20]. This yields

$$u(x, t) = \epsilon^{1/2}\left[(A_+(X, T)e^{-iq_p x} + A_-^*(X, T)e^{iq_p x})e^{i\omega_p t} + \text{c.c.}\right]$$
$$+ \epsilon^{3/2} u^{(1)}(x, t, X, T) + \ldots, \tag{8.106}$$

where q_p and ω_p are related through the dispersion relation (8.103)

$$\omega_p^2 = 1 - 2D \sin^2\left(\frac{q_p}{2}\right). \tag{8.107}$$

By substituting this ansatz (8.106) into the equations of motion (8.105) and applying a solvability condition on the terms of order $\epsilon^{3/2}$, BCL obtained a pair of coupled amplitude equations for the counterpropagating wave amplitudes $A_\pm$

$$\frac{\partial A_\pm}{\partial T} \pm v_g \frac{\partial A_\pm}{\partial X} = -\sin^2\left(\frac{q_p}{2}\right) A_\pm \mp i\frac{h}{2\omega_p}\sin^2\left(\frac{q_p}{2}\right) A_\mp$$
$$- \left(4\eta \sin^4\left(\frac{q_p}{2}\right) \mp i\frac{3}{2\omega_p}\right)\left(|A_\pm|^2 + 2|A_\mp|^2\right) A_\pm, \tag{8.108}$$

where the upper signs (lower signs) give the equation for A_+ (A_-) and

$$v_g = \frac{\partial \omega_p}{\partial q_p} = -\frac{D \sin(q_p)}{2\omega_p} \tag{8.109}$$

is the group velocity. This equation is the extension of (8.47) to many coupled resonators, only now the parametric drive couples amplitudes of the two counterpropagating waves A_+ and A_- instead of coupling A and A^*. A detailed derivation of the amplitude equations (8.108) can be found in [6, 7]. We should note that similar equations were previously derived for describing Faraday waves [29, 46].

By linearizing the amplitude equations (8.108) about the zero solution ($A_+ = A_- = 0$), we find that the linear combination of the two amplitudes that first becomes unstable at $h = h_c \equiv 2\omega_p$ is $B \propto (A_+ - iA_-)$. This represents the emergence of a standing wave with a temporal phase of $\pi/4$ relative to the drive. However, the orthogonal linear combination of the amplitudes decays exponentially and does not participate in the dynamics at onset. Thus, just above threshold a single amplitude equation should suffice, describing this standing wave pattern. We describe the derivation of this equation in the next section.

8.5.2
Reduction to a Single Amplitude Equation

Nonlinear dissipation plays an important role in the saturation of the response to parametric excitation, as we saw in Section 8.3.4. Thus, it is natural to try to keep a balance between the strength of this nonlinearity and the amount by which we drive the system above threshold. Assuming that the nonlinear damping is weak, we use it to define a second small parameter $\delta = \sqrt{\eta}$. This particular definition turns out to be useful if we then scale the reduced driving amplitude $(h - h_c)/h_c$ linearly with δ, defining a scaled reduced driving amplitude r by letting $(h - h_c)/h_c \equiv r\delta$. We can then treat the initial linear combination of the two amplitudes in (8.108) that becomes unstable by introducing a second ansatz,

$$\begin{pmatrix} A_+ \\ A_- \end{pmatrix} = \delta^{1/4} \begin{pmatrix} 1 \\ i \end{pmatrix} B(\xi, \tau) + \delta^{3/4} \begin{pmatrix} w^{(1)}(X, T, \xi, \tau) \\ v^{(1)}(X, T, \xi, \tau) \end{pmatrix} + \delta^{5/4} \begin{pmatrix} w^{(2)}(X, T, \xi, \tau) \\ v^{(2)}(X, T, \xi, \tau) \end{pmatrix}, \tag{8.110}$$

where $\xi = \delta^{1/2} X$ and $\tau = \delta T$. Substitution of this ansatz allows one to obtain the correction to the solution at order $\delta^{3/4}$

$$\begin{pmatrix} w^{(1)} \\ v^{(1)} \end{pmatrix} = \frac{1}{2\sin^2(q_p/2)} \left(-v_g \frac{\partial B}{\partial \xi} + i\frac{9}{2\omega_p}|B|^2 B \right) \begin{pmatrix} 1 \\ -i \end{pmatrix}, \tag{8.111}$$

after which a solvability condition applied to the terms of order $\delta^{5/4}$ yields an equation for the field $B(\xi, \tau)$. After scaling, this takes the form

$$\frac{\partial B}{\partial \tau} = rB + \frac{\partial^2 B}{\partial \xi^2} + i\frac{2}{3}\left(4|B|^2 \frac{\partial B}{\partial \xi} + B^2 \frac{\partial B^*}{\partial \xi}\right) - 2|B|^2 B - |B|^4 B . \tag{8.112}$$

This is the BCL amplitude equation. It is governed by a single control parameter, the reduced drive amplitude r, and captures the slow dynamics of the coupled resonators just above the onset of parametric oscillations. The reader is encouraged to consult [7] for a more detailed account of the derivation of the BCL equation. The form of (8.112) is also applicable to the onset of parametrically driven standing waves in continuum systems with weak nonlinear damping, and combines in a single equation a number of effects studied previously [13, 14, 23, 29, 46, 53].

8.5.3
Single Mode Oscillations

Now that this novel amplitude equation has been derived by BCL it can be used to study a variety of dynamical solutions, ranging from simple single mode to more complicated nonlinear extended solutions and, after slight modifications, also for the dynamics of localized solutions. BCL used the amplitude equation to study the stability of single mode steady-state solutions

$$B = b_k e^{-ik\xi}, \qquad (8.113)$$

that is, standing wave solutions that consist of a single sine wave pattern with one of the allowed wave vectors q_m. The wave vector k gives, in some scaled units, the difference between the wave vector q_p determined by the pump frequency through the dispersion relation, and the wave vector $q_m = m\pi/(N+1)$, $m = 1\ldots N$, of the actual mode that is selected by the system.

A number of interesting results are readily evident if we simply substitute the single mode solution (8.113) into the BCL amplitude equation (8.112). From the linear terms in the amplitude equation we find, as expected, that for $r > k^2$ the zero displacement solution is unstable to small perturbations of the form of (8.113). This defines the parabolic neutral stability curve, which is shown as a dashed line in Figure 8.15. The nonlinear gradients and the cubic term take the simple form $2(k-1)|b_k|^2 b_k$. For $k < 1$ these terms immediately act to saturate the growth of the amplitude assisted by the quintic term. Standing waves therefore bifurcate *supercritically* from the zero displacement state. For $k > 1$ the cubic terms act to increase the growth of the amplitude, and saturation is achieved only by the quintic term. Standing waves therefore bifurcate *subcritically* from the zero displacement state. The saturated amplitude $|b_k|$, obtained by setting (8.112) to zero, is given by

$$|b_k|^2 = (k-1) \pm \sqrt{(k-1)^2 + (r-k^2)} \geq 0. \qquad (8.114)$$

In Figure 8.16 we plot $|b_k|^2$ as a function of the reduced driving amplitude r for three different wave number shifts k. The solid (dashed) lines are the stable (unstable) solutions of (8.114). The circles were obtained by numerical integration of the equations of motion (8.105). For each driving amplitude, the Fourier components of the steady state solution were computed to verify that only single modes are found, suggesting that in this regime of parameters only these states are stable.

BCL showed the power of the amplitude equation in predicting the first single mode solution that should appear at onset. In addition it also predicts the sequence of Eckhaus instabilities that switch to other single mode solutions as the reduced

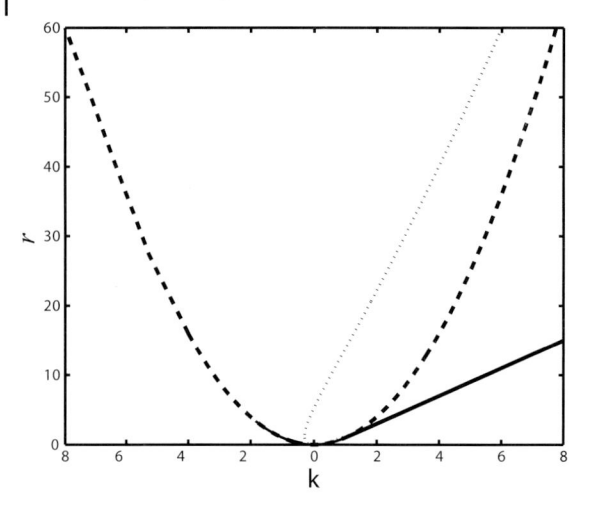

Figure 8.15 Stability boundaries of the single mode solution of (8.112) in the r vs. k plane. Dashed line: neutral stability boundary below which the zero state is stable. Dotted line: stability boundary of the single mode solution (8.113), above which the array experiences an Eckhaus instability and switches to one of the other single mode solutions. For $k > 1$, the bifurcation from zero displacement becomes subcritical and the lower stability boundary is the locus of saddle-node bifurcations (solid line).

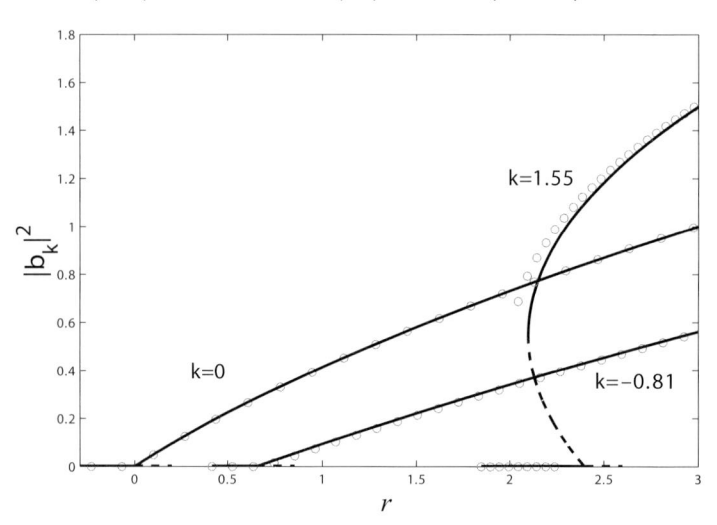

Figure 8.16 Response of the resonator array plotted as a function of reduced amplitude r for three different scaled wave number shifts: $k = 0$ and $k = -0.81$, which bifurcate supercritically, and $k = 1.55$ which bifurcates subcritically and shows clear hysteresis. Solid and dashed lines are the positive and negative square root branches of the calculated response in (8.114). The latter is clearly unstable. Open circles are numerical values obtained by integration of the equations of motion (8.105), with $D = 0.25$, $\omega_p = 0.767445$, $\epsilon = 0.01$, and $\eta = 0.1$.

drive amplitude r is quasistatically increased. Kenig et al. [38] used the amplitude equation for a more general analysis of the question of pattern selection. This question is concerned with predicting which oscillating pattern will be selected, under particular experimental conditions, from among all of the stable steady-state solutions that the array of resonators can choose from. In particular, they have considered experimental situations in which the drive amplitude r is changed abruptly or swept at rates that are faster than typical transient times. In all cases the predictions of the amplitude equations are confirmed with numerical simulations of the original equations of motion (8.105). Experimental confirmation of these predictions is still not available.

Acknowledgments

We wish to thank the students at Tel Aviv University, Yaron Bromberg and Eyal Kenig, who have worked with us on developing and then using the amplitude equation for the treatment of large arrays of parametrically-driven Duffing resonators. We wish to thank our experimental colleagues, Eyal Buks, Rassul Karabalin, Inna Kozinsky, and Michael Roukes, for many fruitful interactions. We also wish to thank Andrew Cleland, Harry Dankowicz, Oleg Kogan, Steve Shaw, and Kimberly Turner for stimulating discussions. This work is funded by the US-Israel Binational Science Foundation (BSF) through Grant No. 2004339, by the US National Science Foundation (NSF) through Grant No. DMR-0314069, by the German-Israeli Foundation (GIF) through Grant No. 981-185.14/2007, and by the Israeli Ministry of Science.

References

1 Aldridge, J.S. and Cleland, A.N. (2005) Noise-enabled precision measurements of a duffing nanomechanical resonator. *Phys. Rev. Lett.*, **94**, 156403.

2 Almog, R., Zaitsev, S., Shtempluck, O., and Buks, E. (2006) High intermodulation gain in a micromechanical duffing resonator. *Appl. Phys. Lett.*, **88**, 213509.

3 Almog, R., Zaitsev, S., Shtempluck, O., and Buks, E. (2007) Noise squeezing in a nanomechanical duffing resonator. *Phys. Rev. Lett.*, **98**, 078103.

4 Baskaran, R. and Turner, K.L. (2003) Mechanical domain coupled mode parametric resonance and amplification in a torsional mode micro electro mechanical oscillator. *J. Micromech. Microeng.*, **13**, 701–707.

5 Blencowe, M.P. (2004) Quantum electromechanical systems. *Phys. Rep.*, **395**, 159–222.

6 Bromberg, Y. (2004) Response of nonlinear systems with many degrees of freedom. Master's thesis, Tel Aviv University.

7 Bromberg, Y., Cross, M.C., and Lifshitz, R. (2006) Response of discrete nonlinear systems with many degrees of freedom. *Phys. Rev. E*, **73**, 016214.

8 Buks, E. and Roukes, M.L. (2002) Electrically tunable collective response in a coupled micromechanical array. *J. MEMS*, **11**, 802–807.

9 Buks, E. and Roukes, M.L. (2001) Metastability and the Casimir effect in micromechanical systems. *Europhys. Lett.*, **54**, 220.

10 Buks, E. and Yurke, B. (2006) Mass detection with a nonlinear nanomechanical resonator. *Phys. Rev. E*, **74**, 046619.

11 Carr, D.W., Evoy, S., Sekaric, L., Craighead, H.G., and Parpia, J.M. (1999) Measurement of mechanical resonance and losses in nanometer scale silicon wires. *Appl. Phys. Lett.*, **75**, 920–922.

12 Carr, D.W., Evoy, S., Sekaric, L., Craighead, H.G., and Parpia, J.M. (2000) Parametric amplification in a torsional microresonator. *Appl. Phys. Lett.*, **77**, 1545–1547.

13 Chen, P. (2002) Nonlinear wave dynamics in Faraday instabilities. *Phys. Rev. E*, **65**, 036308.

14 Chen, P. and Wu, K.-A. (2000) Subcritical bifurcations and nonlinear ballons in Faraday waves. *Phys. Rev. Lett.*, **85**, 3813–3816.

15 Cleland, A.N. and Geller, M.R. (2004) Superconducting qubit storage and entanglement with nanomechanical resonators. *Phys. Rev. Lett.*, **93**, 070501.

16 Cleland, A.N. (2003) *Foundations of Nanomechanics*. Springer, Berlin.

17 Cleland, A.N. and Roukes, M.L. (1998) A nanometer-scale mechanical electrometer. *Nature*, **392**, 160.

18 Cleland, A.N. and Roukes, M.L. (1999) Nanoscale mechanics. In *Proceedings of the 24th International Conference on the Physics of Semiconductors*. World Scientific.

19 Craighead, H.G. (2000) Nanoelectromechanical systems. *Science*, **290**, 1532–1535.

20 Cross, M.C. and Hohenberg, P.C. (1993) Pattern formation outside of equillibrium. *Rev. Mod. Phys.*, **65**, 851–1112.

21 Cross, M.C., Rogers, J.L., Lifshitz, R., and Zumdieck, A. (2006) Synchronization by reactive coupling and nonlinear frequency pulling. *Phys. Rev. E*, **73**, 036205.

22 Cross, M.C., Zumdieck, A., Lifshitz, R., and Rogers, J.L. (2004) Synchronization by nonlinear frequency pulling. *Phys. Rev. Lett.*, **93**, 224101.

23 Deissler, R.J. and Brand, H.R. (1998) Effect of nonlinear gradient terms on breathing localized solutions in the quintic complex Ginzburg–Landau equation. *Phys. Rev. Lett.*, **81**, 3856–3859.

24 DeMartini, B.E., Rhoads, J.F., Turner, K.L., Shaw, S.W., and Moehlis, J. (2007) Linear and nonlinear tuning of parametrically excited mems oscillators. *J. MEMS*, **16**, 310–318.

25 Ekinci, K.L., Huang, X.M.H., and Roukes, M.L. (2004) Ultrasensitive nanoelectromechanical mass detection. *Appl. Phys. Lett.*, **84**, 4469–4471.

26 Ekinci, K.L. and Roukes, M.L. (2005) Nanoelectromechanical systems. *Rev. Sci. Instrum.*, **76**, 061101.

27 Erbe, A., Krommer, H., Kraus, A., Blick, R.H., Corso, G., and Richter, K. (2000) Mechanical mixing in nonlinear nanomechanical resonators. *Appl. Phys. Lett.*, **77**, 3102–3104.

28 Evoy, S., Carr, D.W., Sekaric, L., Olkhovets, A., Parpia, J.M., and Craighead, H.G. (1999) Nanofabrication and electrostatic operation of single-crystal silicon paddle oscillators. *J. Appl. Phys.*, **86**, 6072.

29 Ezerskiǐ, A.B., Rabinovich, M.I., Reutov, V.P., and Starobinets, I.M. (1986) Spatiotemporal chaos in the parametric excitation of capillary ripple. *Zh. Eksp. Teor. Fiz.*, **91**, 2070–2083. [*Sov. Phys. JETP* **64**, 1228 (1986)].

30 Feng, X.L., He, R., Yang, P., and Roukes, M.L. (2007) Very high frequency silicon nanowire electromechanical resonators. *Nano Lett.*, **7**, 1953–1959.

31 Hand, L.N. and Finch, J.D. (1998) *Analytical Mechanics*, chapter 10, Cambridge Univ. Press, Cambridge.

32 Huang, X.M.H., Zorman, C.A., Mehregany, M., and Roukes, M.L. (2003) Nanodevice motion at microwave frequencies. *Nature*, **421**, 496.

33 Husain, A., Hone, J., Postma, H.W.Ch., Huang, X.M.H., Drake, T., Barbic, M., Scherer, A., and Roukes, M.L. (2003) Nanowire-based very-high-frequency electromechanical resonator. *Appl. Phys. Lett.*, **83**, 1240–1242.

34 Ilic, B., Craighead, H.G., Krylov, S., Senaratne, W., Ober, C., and Neuzil, P. (2004) Attogram detection using nanoelectromechanical oscillators. *J. Appl. Phys.*, **95**, 3694–3703.

35 Karabalin, R.B., Feng X.L., and Roukes M.L. (2009) Parametric nanomechani-

cal amplification at very high frequency. *Nano Lett.*, **9**, 3116–3123.

36 Katz, I., Lifshitz, R., Retzker, A., and Straub, R. (2008) Classical to quantum transition of a driven nonlinear nanomechanical resonator. *New J. Phys.*, **10**, 125023.

37 Katz, I., Retzker, A., Straub, R., and Lifshitz, R. (2007) Signatures for a classical to quantum transition of a driven nonlinear nanomechanical resonator. *Phys. Rev. Lett.*, **99**, 040404.

38 Kenig, E., Lifshitz, R., and Cross, M.C. (2007) Pattern selection in parametrically-driven arrays of nonlinear micromechanical or nanomechanical resonators. *Phys. Rev. E*, **79**, 026203.

39 Kenig, E., Malomed, B.A., Cross, M.C., and Lifshitz, R. (2009) Intrinsic localized modes in parametrically-driven arrays of nonlinear resonators. *Phys. Rev. E*, **80**, 046202.

40 Kozinsky, I., Postma, H.W.Ch., Bargatin, I., and Roukes, M.L. (2006) Tuning nonlinearity, dynamic range, and frequency of nanomechanical resonators. *Appl. Phys. Lett.*, **88**, 253101.

41 Kozinsky, I., Postma, H.W.Ch., Kogan, O., Husain, A., and Roukes, M.L. (2007) Basins of attraction of a nonlinear nanomechanical resonator. *Phys. Rev. Lett.*, **99**, 207201.

42 Landau, L.D. and Lifshitz, E.M. (1976) *Mechanics*, Butterworth-Heinemann, Oxford, 3rd edition, §27.

43 Landau, L.D. and Lifshitz, E.M. (1986) *Theory of Elasticity*, Butterworth-Heinemann, Oxford, 3rd edition, §20 & 25.

44 Lifshitz, R. and Cross, M.C. (2003) Response of parametrically driven nonlinear coupled oscillators with application to micromechanical and nanomechanical resonator arrays. *Phys. Rev. B*, **67**, 134302.

45 Masmanidis, S.C., Karabalin, R.B., I. De Vlaminck, Borghs, G., Freeman, M.R., and Roukes, M.L. (2007) Multifunctional nanomechanical systems via tunably-coupled piezoelectric actuation. *Science*, **317**, 780–783.

46 Milner, S.T. (1991) Square patterns and secondary instabilities in driven capillary waves. *J. Fluid Mech.*, **225**, 81–100.

47 Moehlis, J. Private communication.

48 Peng, H.B., Chang, C.W., Aloni, S., Yuzvinsky, T.D., and Zettl, A. (2006) Ultrahigh frequency nanotube resonators. *Phys. Rev. Lett.*, **97**, 087203.

49 Postma, H.W.Ch., Kozinsky, I., Husain, A., and Roukes, M.L. (2005) Dynamic range of nanotube- and nanowire-based electromechanical systems. *Appl. Phys. Lett.*, **86**, 223105.

50 Reichenbach, R.B., Zalalutdinov, M., Aubin, K.L., Rand, R., Houston, B.H., Parpia, J.M., and Craighead, H.G. (2005) Third-order intermodulation in a micromechanical thermal mixer. *J. MEMS*, **14**, 1244–1252.

51 Rhoads, J.F., Shaw, S.W., and Turner, K.L. (2006) The nonlinear response of resonant microbeam systems with purely-parametric electrostatic actuation. *J. Micromech. Microeng.*, **16**, 890–899.

52 Rhoads, J.F., Shaw, S.W., Turner, K.L., Moehlis, J., DeMartini, B.E., and Zhang, W. (2006) Generalized parametric resonance in electrostatically actuated microelectromechanical oscillators. *J. Sound Vib.*, **296**, 797–829.

53 Riecke, H. (1990) Stable wave-number kinks in parametrically excited standing waves. *Europhys. Lett.*, **11**, 213–218.

54 Roukes, M.L. (2001) Nanoelectromechanical systems face the future. *Phys. World*, **14**, 25–31.

55 Roukes, M.L. (2001) Plenty of room indeed. *Sci. Am.*, **285**, 42–49.

56 Rugar, D., Budakian, R., Mamin, H.J., and Chui, B.W. (2004) Single spin detection by magnetic resonance force microscopy. *Nature*, **430**, 329–332.

57 Rugar, D. and Grütter, P. (1991) Mechanical parametric amplification and thermomechanical noise squeezing. *Phys. Rev. Lett.*, **67**, 699.

58 Sato, M., Hubbard, B.E., Sievers, A.J., Ilic, B., and Craighead, H.G. (2004) Optical manipulation of intrinsic localized vibrational energy in cantilever arrays. *Europhys. Lett.*, **66**, 318–323.

59 Sato, M., Hubbard, B.E., Sievers, A.J., Ilic, B., Czaplewski, D.A., and Craighead, H.G. (2003) Observation of locked intrinsic localized vibrational modes in a

micromechanical oscillator array. *Phys. Rev. Lett.*, **90**, 044102.

60 Sato, M., Hubbard, B.E., Sievers, A.J., Ilic, B., Czaplewski, D.A., and Craighead, H.G. (2003) Studies of intrinsic localized vibrational modes in micromechanical oscillator arrays. *Chaos*, **13**, 702–715.

61 Sazonova, V., Yaish, Y., Üstünel, H., Roundy, D., Arias, T.A., and McEuen, P.L. (2004) A tunable carbon nanotube electromechanical oscillator. *Nature*, **431**, 284–287.

62 Scheible, D.V., Erbe, A., Blick, R.H., and Corso, G. (2002) Evidence of a nanomechanical resonator being driven into chaotic response via the ruelle–takens route. *Appl. Phys. Lett.*, **81**, 1884–1886.

63 Schwab, K.C., Henriksen, E.A., Worlock, J.M., and Roukes, M.L. (2000) Measurement of the quantum of thermal conductance. *Nature*, **404**, 974–977.

64 Schwab, K.C. and Roukes, M.L. (2005) Putting mechanics into quantum mechanics. *Phys. Today*, **58** (7), 36–42.

65 Strogatz, S.H. (1994) *Nonlinear dynamics and chaos*, chapter 7, Addison-Wesley, Reading MA.

66 Turner, K.L., Miller, S.A., Hartwell, P.G., MacDonald, N.C., Strogatz, S.H., and Adams, S.G. (1998) Five parametric resonances in a microelectromechanical system. *Nature*, **396**, 149–152.

67 Yang, Y.T., Callegari, C., Feng, X.L., Ekinci, K.L., and Roukes, M.L. (2006) Zeptogram-scale nanomechanical mass sensing. *Nano Lett.*, **6**, 583–586.

68 Yu, M.F., Wagner, G.J., Ruoff, R.S., and Dyer, M.J. (2002) Realization of parametric resonances in a nanowire mechanical system with nanomanipulation inside a scanning electron microscope. *Phys. Rev. B*, **66**, 073406.

69 Yurke, B., Greywall, D.S., Pargellis, A.N., and Busch, P.A. (1995) Theory of amplifier-noise evasion in an oscillator employing a nonlinear resonator. *Phys. Rev. A*, **51**, 4211–4229.

70 Zaitsev, S., Almog, R., Shtempluck, O., and Buks, E. (2005) Nonlinear dynamics in nanomechanical oscillators. *Proceedings of the 2005 International Conference on MEMS,NANO and Smart Systems*, pp. 387–391.

71 Zhang, W., Baskaran, R., and Turner, K. (2003) Tuning the dynamic behavior of parametric resonance in a micromechanical oscillator. *Appl. Phys. Lett.*, **82**, 130–132.

72 Zhang, W., Baskaran, R., and Turner, K.L. (2002) Effect of cubic nonlinearity on auto-parametrically amplified resonant mems mass sensor. *Sens. Actuators A*, **102**, 139–150.

9
Nonlinear Dynamics in Atomic Force Microscopy and Its Control for Nanoparticle Manipulation
Kohei Yamasue and Takashi Hikihara

9.1
Introduction

Science and engineering at the nanoscale is currently one of the most consistently advancing fields [1]. As a tool for directly accessing the nanoscale, scanning probe microscopy (SPM) is now widely accepted [2]. In general, SPM utilizes a probe that interacts with a local area of a sample surface. The probe is precisely located over the area using a three dimensional nanopositioning mechanism. The probe and its positioning technology are the fundamental basis of the SPM, which performs many useful functions such as imaging, profiling, and manipulation of a sample surface on the nanoscale [2, 3]. For example, high resolution imaging of sample topography is achieved by recording this interaction as a function of lateral probe position on the sample surface.

Among the members of the SPM family, atomic force microscopy (AFM), invented by Binning *et al.* [4], plays a particularly important role [5, 6]. Since the AFM detects a *force* interaction between a micromechanical probe and a sample surface, AFM can image insulating samples as well as conducting and semiconducting samples. In this point, AFM is often contrasted with scanning tunneling microscopy (STM) [7], which is also a member of the SPM family. STM cannot image insulating samples because surface imaging by the STM is based on the detection of a tunneling current flowing between the probe and the surface. Another advantage is that AFM can be operated in various environments including vacuum, air, and liquid. These advantages have opened new avenues to SPM applications in biology, polymer science, and organic electronics as well as traditional areas of materials and surface sciences [5].

Of its various operating modes, dynamic mode AFM has been a flagship operating mode for nearly two decades. In the dynamic mode, the probe is oscillated at or near its mechanical resonance. The interaction force is detected as a modulated amplitude or frequency of the oscillation [8–10]. The force sensitivity is much improved by using a probe with a higher quality factor. The important advantage of the dynamic mode is that adhesion of the probe to sample surfaces is avoided by

Nonlinear Dynamics of Nanosystems. Edited by Günter Radons, Benno Rumpf, and Heinz Georg Schuster
Copyright © 2010 WILEY-VCH Verlag GmbH & Co. KGaA, Weinheim
ISBN: 978-3-527-40791-0

oscillating the probe. In addition, the mechanical damage to sample surfaces by a lateral friction during a scan is significantly reduced in the dynamic mode as compared to the traditional, previously developed contact mode. The dynamic mode AFM has thus enabled the high resolution and non-destructive imaging of various samples, including soft samples in liquid [11–16]. In addition, various schemes have been devised for profiling of surface properties [17–19], manipulation of individual atoms and molecules [20–22], and control of surface structures [23].

In this chapter, we focus on two topics related to nonlinear dynamics in dynamic mode AFM. Both are not limited by the phenomenological analysis of nonlinear vibrations in nonlinear systems. The first half of this chapter is devoted to nonlinear probe dynamics and its control in dynamic mode AFM. The probe of dynamic mode AFM is a vibrating micromechanical sensor that detects the force interaction with sample surfaces. This probe-surface interaction is essentially nonlinear, and therefore the nonlinear probe dynamics has recently been extensively studied due to its close relation to imaging characteristics [24–28]. It has been emphasized that the micromechanical probe exposed to an interaction can not be approximated as a harmonic oscillator especially in the AM-AFM (*Amplitude Modulation AFM*), which is a major operating mode in air and liquid [8, 9, 15, 16]. The AM-AFM, in fact, exhibits various nonlinear phenomena including a chaotic oscillation, which nonlinear scientists have focused on for 50 years. Actually, existing and well developed techniques for the analysis of nonlinear systems have been applied to some problems arising in the nonlinear dynamics in the AM-AFM [24–39]. Control strategies for nonlinear systems can be also applied for improving and accelerating the surface imaging. We describe the application of time-delayed feedback control, which is a well known approach in the field of nonlinear dynamics to the stabilization of chaotic oscillations [40]. The efficacy of time-delayed feedback control is successfully demonstrated for cantilever oscillation in the AM-AFM [41–43].

The second topic of this chapter is the manipulation of single atoms and molecules, which is a current challenging topic of nanoscience. Recently, manipulation of single atoms and molecules has been experimentally achieved on surfaces with use of the AFM in lateral as well as vertical processes [20–22]. The lateral processes can transfer atoms and molecules parallel to the surface and vertically between tip of probe and adatom on surface via vibrational excitation of the target-substrate bond. The processes have been also studied theoretically [44–48] and numerically [49]. The last half of this chapter is devoted to a theoretical consideration of van der Waals molecular vibrational predissociation based on a T-shaped model [50–54]. The model was introduced to describe a quantum mechanics that governs the rates of vibrational predissociation of A–B–A triatomic molecules, which are coupled Morse oscillators. Atoms and molecules attached on the material surface are bonded by a van der Waals potential, which is characterized by Morse type atom–atom interaction potential. The dynamics of manipulated particles are described by the fragmentation of the atom B from the coupling of A–A. Here, we introduce a Hamiltonian of the triatomic molecules. The system is comprised of coupled nonlinear oscillators. Assuming that the rotation and bending modes are neglected, it is shown that the eigenfrequency decides the resonance and energy

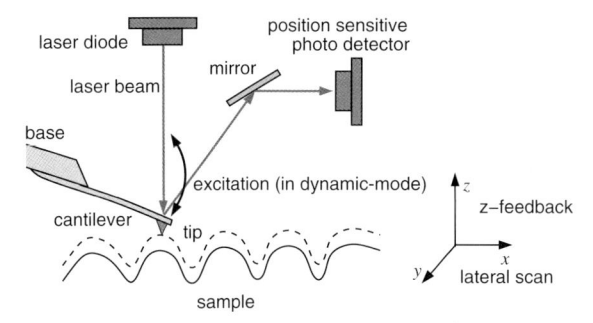

Figure 9.1 Principle of dynamic mode atomic force microscopy with an optical lever type deflection sensor.

exchange. At high energy, the system possibly shows chaotic vibration. It gives the probability of the classical dynamics of atoms and molecules. The fragmentation of atoms from a surface is discussed based on the global phase structure. When the perturbation of energy exceeds the critical value, the dynamics appears the global behavior out of a single potential well. That is, the vibratory dissociation can achieve the manipulation of nanoparticles. This inevitably captures the atom dissociated from the material surface. The results show us possible manipulation methods of nanoparticles in *in situ* conditions using AFM.

9.2
Operation of Dynamic Mode Atomic Force Microscopy

The operation of AFM is briefly introduced in this section. As shown in Figure 9.1, the atomic force microscope in principle consists of several components. These are a microfabricated cantilever that is used as a probe for detecting the interaction force, an actuator for exciting cantilever oscillation (used in the dynamic mode), a sensor for detecting the instantaneous deflection of the cantilever, and a three dimensional positioning mechanism for precisely locating the probe over a surface area. The cantilever has a sharp tip at its free end. The tip of the cantilever is extremely sharp, having a typical radius of approximately 10 nm that is achieved using microfabrication techniques. The positioning device is controlled so that the cantilever tip is placed in close proximity to the sample surface one is going to observe. The tip then feels a force between it and the sample surface. The interaction force is called the *tip–sample interaction*. The tip–sample interaction is on the order of pico- to nanonewtons, but it is sufficient to modify the original cantilever dynamics via deflection or oscillation due to the small dimensions of the cantilever. The cantilever therefore plays a role as a force sensor to detect the tiny tip–sample interaction force. The deflection or oscillation of the cantilever is measured by the optical lever method in the standard device configuration [55].

Assuming that the tip–sample interaction depends on the tip–sample distance, the latter can be regulated at constant value by adjusting the height of the sample surface so that the cantilever dynamics are kept constant. The regulation is typically realized by a PI (proportional-integral) controller and this mechanism is often called *z-feedback*. The surface topography is tracked during a raster scan of the surface. The surface topography is constructed as a three dimensional image from a recorded time series of a signal controlling the z-feedback mechanism. In the contact mode, a cantilever is used without excitation and its deflection is detected for estimating the tip–sample interaction.

The dynamic mode is an improvement introduced immediately after the invention of AFM. The cantilever in the dynamic mode is oscillated at or near its mechanical resonance frequency. Instead of detecting the cantilever deflection, the shift of resonance frequency is detected in this mode, and the amount of the shift depends on the mean tip–sample distance. The dynamic mode has two major operating modes called AM-AFM (Amplitude Modulation AFM) [8] and FM-AFM (Frequency Modulation AFM) [10]. In these modes, the amplitude or frequency modulated by the tip–sample interactions are detected in order to estimate the shift of resonance frequency. In AM-AFM, the cantilever oscillation is excited by applying an external periodic force. The cantilever in FM-AFM is self-excited using an electronic feedback circuit. In both modes, the oscillation is measured using a lock-in amplifier or a RMS-DC (Root-Mean-Square to Direct Current) converter with bandpass filters. The force sensitivity of a cantilever is much improved by increasing the quality factor of the cantilever. Adhesion to surface and destruction of samples are also avoided by using an oscillating cantilever. The height of the sample surface can be precisely adjusted by a positioning device, such as tube scanners.

9.3
Nonlinear Dynamics and Control of Cantilevers

9.3.1
Nonlinear Oscillation and Its Influence on Imaging

Initially, a cantilever subject to a tip–sample interaction was often approximated as a harmonic oscillator. However, there has been growing interest in the nonlinear dynamics of cantilevers and its influence on imaging [24–39]. In particular, the AM-AFM or the tapping-mode AFM has been focused upon due to a strong nonlinearity in an operating range. A great number of experimental, numerical, and theoretical works have been performed in order to clarify the nonlinear dynamics of the AM-AFM. A bistable behavior occurs [56] and the resulting jumping and hysteresis phenomena cause sudden and discontinuous transitions in imaging characteristics [28]. The subharmonics and chaos of the cantilever oscillation have also been predicted numerically and theoretically [30, 31, 34, 37], and they have been demonstrated experimentally [32, 33, 36, 57]. It seems that the chaotic oscillation is experimentally encountered when a soft cantilever is excited with a large amplitude.

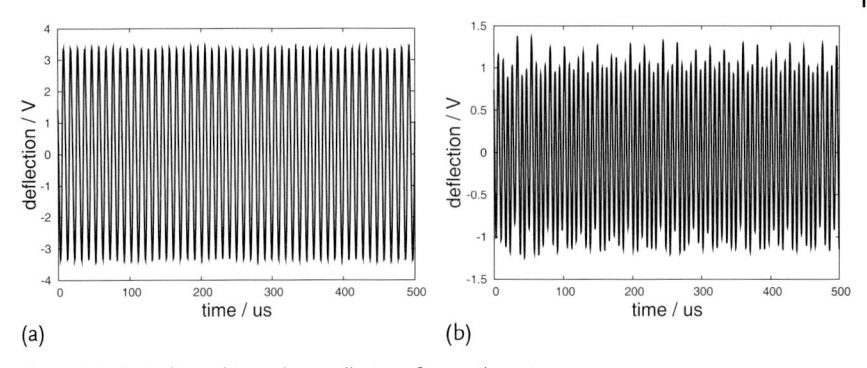

Figure 9.2 Periodic and irregular oscillation of a cantilever in AM-AFM [43]; (a) periodic, (b) irregular and non-periodic.

As an example, Figure 9.2 shows two contrasting oscillation states that have been experimentally observed [43]. The sample was HOPG (Highly Oriented Pyrolytic Graphite) and the imaged area was 500 nm squared. The measurement was performed in air and a magnetically coated cantilever (Agilent, Type I MAC levers C, nominal spring constant and resonance frequency: 0.6 N/m and 75 kHz, respectively) was excited with a large amplitude. A similar result was reported by Hu and Raman in Ref. [36], although they performed experiments in a nitrogen atmosphere in order to eliminate the effect of capillary forces due to the water layer on the surface. When the tip–sample distance was set such that the oscillation amplitude was decreased by 30% compared to the free oscillation amplitude, the oscillation remained periodic. A further decrease of the tip–sample distance, however, made the periodic oscillation unstable and generated an irregular and nonperiodic oscillation. Figure 9.2b shows an oscillation observed when the oscillation amplitude was reduced by 80% compared to the free amplitude. The resolution of images depends on the oscillation state. Figure 9.3 compares an image from the irregular state (Figure 9.3b) to the periodic one (Figure 9.3a). It can be easily seen that the resolution was decreased and the image was much noisier due to the irregular and nonperiodic cantilever oscillations. The z-feedback did not accurately track the surface topography during the raster scan of the surface. The surface of HOPG should be flat except for the large step on the surface. However, Figure 9.3 clearly shows the loss of the original flatness between the steps.

The above experimental results suggest that the resolution of the AM-AFM is reduced and the operating range may be limited due to undesirable irregular and nonperiodic oscillations. In order to overcome this limitation and extend the operating range, the application of control technology is a good candidate for improving the performance of the AM-AFM. In this context, some motivated research groups have already proposed application of control techniques to cantilever oscillation [30, 31, 58]. The authors have also proposed the application of time-delayed feedback control [40], which is a well-known approach that has been used for stabilizing unstable periodic orbits embedded in chaotic attractors [59]. In the following sections we provide an overview of our recent results on this topic.

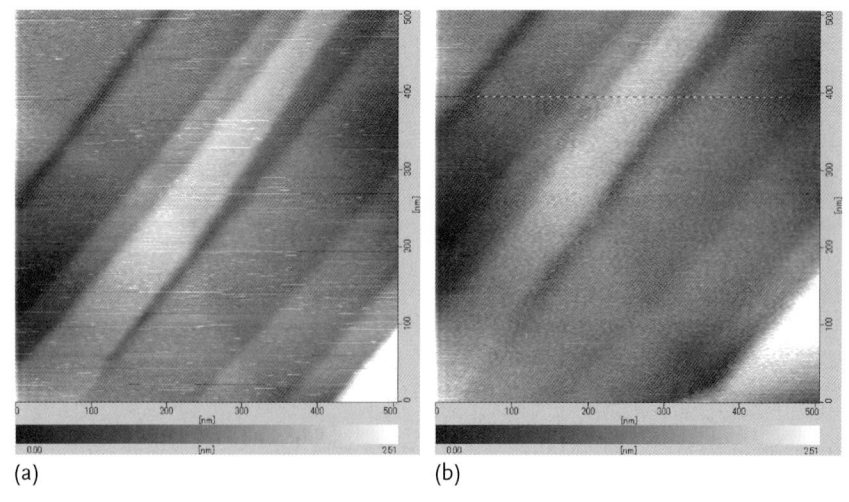

Figure 9.3 Comparison of images by oscillation states; (a) periodic; (b) irregular and non-periodic.

9.3.2
Model of a Cantilever under Tip–Sample Interaction

When the first mode oscillation of a cantilever is considered, the mathematical model of the cantilever is given by

$$\frac{d}{dt}\begin{bmatrix} x \\ y \end{bmatrix} = \begin{bmatrix} y \\ -\omega_0^2 x - \frac{\omega_0}{Q} y + f(x, y, Z) + A \cos \omega t \end{bmatrix} + \boldsymbol{b} u, \tag{9.1}$$

where x and y denote the instantaneous deflection and the velocity of the tip, respectively. A and ω are the amplitude and frequency of the sinusoidal excitation force, respectively. The cantilever has a fundamental resonance frequency at ω_0 and its quality factor is Q. $\boldsymbol{b}$ denotes a two dimensional constant vector describing coupling between the control input and the state variables.

Assuming the tip–sample interaction force is described by the Lennard-Jones potential, $f(x, y, Z)$ is expressed as [31]

$$f(x, y, Z) = -\frac{D\omega_0^2}{(Z+x)^2} + \frac{\sigma^6 D \omega_0^2}{30(Z+x)^8}. \tag{9.2}$$

D denotes a constant related to the Hamaker constant, tip radius, and stiffness of the cantilever. σ denotes the diameter of each molecule organizing the tip and the sample. The first and the second terms denote a long-range attractive force and a short-range repulsive force, respectively. The existence of a chaotic invariant set was proven by applying the Melnikov method to this model [30, 31]. A chaotic cantilever oscillation was subsequently presented numerically based on the same model [34]. It should be mentioned that there are a variety of models that describe

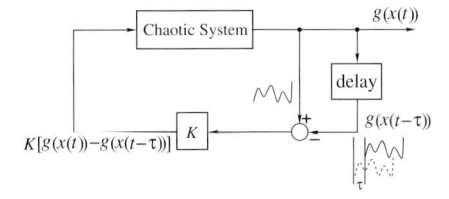

Figure 9.4 Block diagram of time-delayed feedback controlled system.

the cantilever dynamics [25–32, 35, 37–39]. For example, the DMT (Dejarguin–Muller–Toporov) theory has been employed for modeling [27, 28, 32, 35, 39]. An impact oscillator can also be a simple model for the AM-AFM [25, 37]. The effects of a capillary force due to a water layer on the surface also has a strong effect on cantilever dynamics when operating the AM-AFM in air [25, 35]. A neck of water meniscus between the tip and the sample applies a hysteretic force to a cantilever.

9.3.3
Application of Time-Delayed Feedback Control

Time-delayed feedback control was proposed by Pyragas in 1992 [40] and is now well known for its ability to stabilize unstable periodic orbits in chaotic attractors [59]. As shown in Figure 9.4, this continuous control method is a kind of feedback control that exploits a past state or output of a nonlinear system for negative feedback, instead of giving a external reference signal. In his seminal paper, Pyragas showed numerically that the stability of a target unstable periodic orbit can be exclusively changed by choosing an appropriate feedback gain K. This occurs if the time difference τ, namely *the delay time*, between the two outputs is precisely adjusted to the period of the target orbit [40]. Time-delayed feedback control is an invasive control method in this sense and the stability is maintained by the small perturbation ideally converging to the null signal.

The strategy of using an earlier signal has allowed us to readily implement the control method in a real system without identifying the model and parameters of the system. In particular, no complicated time series analysis is needed for reconstruction of underlying dynamics. The simple control law has also enabled applications to chaotic systems operating at a high frequency [60–62]. The number of applications has thus increased since the publication of Pyragas's paper. This area of investigation now includes electronic circuits [60, 61], laser systems [62], magnetoelastic oscillators [63], chemical reactions [64], and gas charge systems [65].

As a novel application to nanosystems, the authors have proposed the stabilization of chaotic cantilever oscillations using time-delayed feedback control [41]. Assuming that the instantaneous velocity of cantilever oscillations are measured as an output of the nonlinear system (9.1), the generation of the control signal $u(t)$ is described by [40]:

$$u(t) = K\left[y(t-\tau) - y(t)\right]. \tag{9.3}$$

This implementation is associated with the cantilever model (9.1) by putting $b = [0\ 1]^T$ into (9.1). The delay time τ is adjusted to the period of the excitation signal in order to stabilize an orbit with a period equal to the period of the excitation signal. The control input then converges to the null signal if the stabilization of a target unstable periodic orbit is completed. The target orbit is an unstable periodic orbit with a period equal to the driving signal. The authors have numerically confirmed control performance in both homoclinic and grazing regimes, and have also presented an application to acceleration of the scanning rate [41, 42]. The invasive control method is significant for dynamic mode AFM because the stabilized orbit should depend on just the pure tip–sample interaction. This is an essential difference from Q-control, which has created controversy concerning the effects of feedback control on measurements [66].

9.3.4
Experimental Setup for Control of Nonlinear Cantilever Dynamics

We have numerically confirmed the possibility of the application of time-delayed feedback control to the AM-AFM in previous studies [41, 42]. The next step is the implementation of a controller to an actual device. This section provides an overview of the circuit implementation of time-delayed feedback control especially designed for cantilevers in the AM-AFM.

9.3.4.1 Circuit Implement of Time-Delayed Feedback Control

Fast feedback electronics is required to achieve time-delayed feedback control of cantilever oscillation because a cantilever in the AM-AFM is excited at high frequencies, ranging from several tens to a few hundreds kilohertz. In addition, the delay time and feedback gain must be flexibly adjusted to appropriate values in order to optimize control performance. In this context, digital facilities can be effectively used for making a flexible controller to examine the control performance.

We made a controller that is schematically illustrated in Figure 9.5 [43]. The controller is composed of a digital delay line that retards the signal and a summing amplifier that generates the error signal between the current and retarded output signal. The digital delay line is constructed using an analog to digital (A/D) converter, first-in-first-out (FIFO) memories, and a digital to analog (D/A) converter. The signal is sequentially stored in the FIFO memories as digital data through the A/D converter and, after a given constant time has elapsed, the digital data are restored as an analog signal through the D/A converter. The signal is sampled at 40 MHz and the resolution is 12 bits. A digital delay line has also been employed in chaos control of a magnetoelastic beam [63] and a gas charge system [65], although the required frequency is much higher in the present case. The instantaneous velocity is estimated by an elemental differentiation circuit using an operational amplifier. The error signal is amplified with the summing amplifier. The amplified error signal, or control signal, is added to a sinusoidal signal for cantilever excitation.

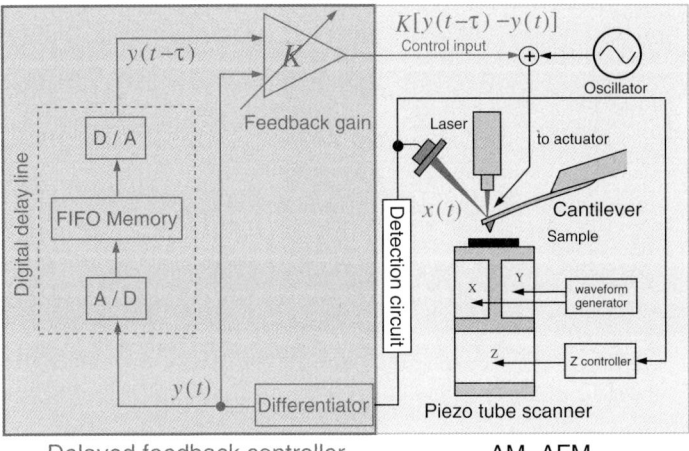

Figure 9.5 System configuration for AM-AFM equipped with time-delayed feedback control [43].

9.3.4.2 Frequency Response of Magnetic Actuators and Deflection Sensors

The numerical results were successful, showing that the time-delayed feedback control has an ability to stabilize the chaotic oscillation of a cantilever, as shown in [41, 42]. In reality, however, there are many factors seriously limiting the control performance.

In our numerical experiment we assumed the ideal conditions for the characteristics of actuator and sensor of the cantilever oscillation. In reality, however, the deflection sensors and dither piezo actuators often used in AFM have their own dynamics. An emphasis should be placed on the frequency characteristics of the actual devices. In the current standard device configuration of dynamic mode AFM, the most critical characteristics for cantilevers are those of the actuators. One can observe many spurious peaks if one actuates the cantilever with a standard piezoelectric actuator. This implies the presence of a large phase delay in the feedback loop and therefore one has to improve the frequency characteristics before applying the controller.

9.3.5
Experimental Demonstration of the Stabilization of Cantilever Oscillations

Our controller is implemented for a commercial AFM (SII, SPA-300/NanoNavi Station), as shown in Figure 9.6. A small solenoid coil is placed beneath a sample stage for excitation of a magnetically coated cantilever (Agilent, Type I MAC levers C, nominal spring constant and resonance frequency: 0.6 N/m and 75 kHz, respectively). A home-built voltage current converter was constructed because the magnetic force generated by the solenoid coil is proportional to the applied current. An additional phase delay of $\pi/2$ therefore arises if the control input generated as a voltage signal is directly applied to the solenoid coil. The home-built voltage-current converter has flat frequency characteristics of amplitude and phase at sufficiently

Figure 9.6 Photograph showing a home-built deflection sensor and the wiring for the magnetic coil beneath the sample holder.

high frequencies. A home-built deflection sensor was also employed to decrease phase delay in the feedback loop. The implemented control system allows us to experimentally investigate the ability of TDFC.

The features of the control method enable us to implement it without identification of the parameters of each cantilever. Parameters such as the spring constant are only given as nominal values and are often quite different from the true values. No analysis on the nonlinear dynamics is needed and only the AM-AFM driving frequency must be known in order to adjust the delay time. The feasibility of high frequency oscillation is also an important advance of time-delayed feedback control [61]. Cantilevers in dynamic mode AFM are typically vibrated around 10 kHz to 300 kHz. From the viewpoint of measurement, it is worth noting that no parameter of a cantilever is modified after control is achieved. The stabilized cantilever oscillation under control depends purely on the tip–sample interaction force in the steady state. This is essentially different from a control method introducing damping to the oscillation, as proposed in [30, 31]. The control method stabilizing the intrinsic orbit of the system should thus be developed from the viewpoint of measurement.

An experimental result is shown in Figure 9.7 [43]. As shown in Figure 9.7a, an irregular and nonperiodic oscillation was observed in close proximity to the sample surface. Figure 9.7b shows the stabilization of the oscillation achieved after control input was activated. It was confirmed that the control input converged to nearly zero volts. These facts suggest that an unstable periodic orbit is successfully stabilized by adding a small perturbation to the excitation signal.

In this section, we have reviewed recent research topics on nonlinear cantilever dynamics and their control using time-delayed feedback. The scheme of controlling chaos was utilized for the stabilization of nonlinear dynamics into the embedded states. The improvement of a limited control performance is currently an ongoing work. Nevertheless, this is the first implementation of the chaos control method to a real device, as far as we know. The stabilization of irregular and nonperiodic

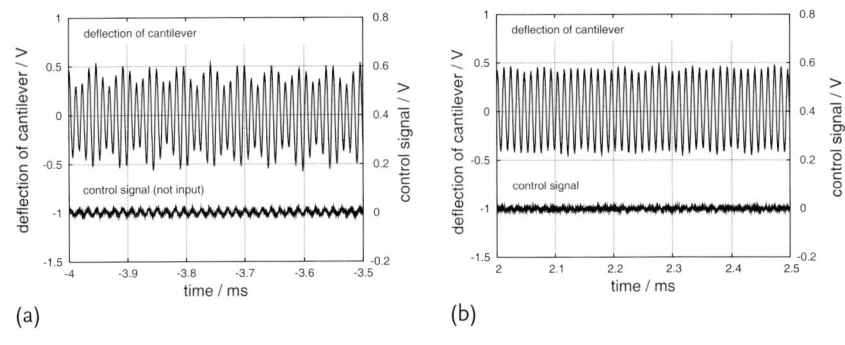

Figure 9.7 Stabilization of irregular oscillations [43]; (a) without control, (b) with control.

oscillations is effective for the z-feedback to accurately track the surface topography, which then improves the image resolution. The stabilized periodic oscillation retains the pure dynamics of the original system. The controlled dynamics should be a probe for detecting the nonlinear force interaction at the nanoscale.

9.4
Manipulation of Single Atoms at Material Surfaces

In this section we go down to the dynamics of single atoms, which can be accessed using dynamic mode AFM. The dynamics of single atoms is formalized based on classical mechanics using the Hamiltonian. We first introduce a model of atomic and molecular alignment for estimating fragmentation of atoms from surface bonds.

9.4.1
Model of Single Atoms and Molecules

We now focus on the dynamics of single atoms and molecules attached to material surfaces at low temperatures and vacuum conditions. As depicted in Figure 9.8, we assume that single atoms or molecules at a surface (B) are bonding to quadriatoms (A). Models of molecular vibrational fragmentation have been discussed for T-shaped structures with van der Waals potentials. At the surface, we assume a pyramid structure of atoms at steady state. In pyramid shape bonding, the rotation and bending modes disappear. Then the DOFs (degrees of freedom) of B are restricted to the direction vertical to the plane in which the rectangle formed by A lies. Hereafter, the system can be modeled by a T-shaped structure with diagonal atoms A and B without rotational dynamics.

The Morse interaction potential limits the distance of the interaction to a short range. In this region, the atoms are called Velet neighbors [67]. The distance between the atoms A is depicted as $\sqrt{2}Q$, and the distance between B and center O of the A-plane as q, which is also on the axis. The distance between A and O then

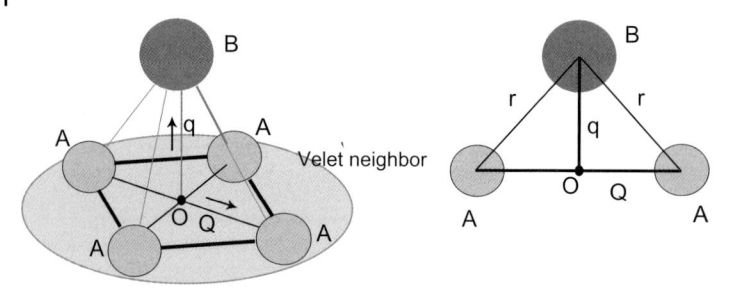

Figure 9.8 Atomic and molecular alignment at material surfaces.

becomes Q, which is perpendicular to the q-axis. Each momentum is given by P and p. The Hamiltonian can be written as

$$H = \frac{P^2}{2m} + \frac{p^2}{2\mu} + V_0(Q, q) + V_1(Q, q), \qquad (9.4)$$

where the angle between the axis of the A-plane and the vector from the center of mass to B is restricted at the rectangle. Then we neglect the kinetics of angular motion. Here m_A and m_B are the masses of A and B, respectively. $m = m_A/2$ is the reduced mass of A and $\mu = 2m_A m_B/(m_A + 2m_B)$ the reduced mass of the whole system. The potentials are given as [53]

$$V_0(Q, q) = W_0(Q) + W_0(q),$$

where

$$W_0(Q) = D_{0A}\left(e^{-2\beta_{0A}(Q-Q_0)} - 2e^{-\beta_{0A}(Q-Q_0)}\right),$$
$$W_0(q) = D_{0B}\left(e^{-2\beta_{0B}(q-q_0)} - 2e^{-\beta_{0B}(q-q_0)}\right),$$

and

$$V_1(Q, q) = W_1(r_+) + W_1(r_-),$$

where

$$W_1(r_\pm) = D_1\left(e^{-2\beta_1(r_\pm - r_{\pm 0})} - 2e^{-\beta_1(r_\pm - r_{\pm 0})}\right).$$

V_1 is the expansion of the van der Waals potential in the Taylor series around the equilibrium point. $r_\pm$ is the distance between B and one of the atoms A. r is a function of Q and q. The equilibrium point is given by r_0. D_{0A}, D_{0B}, and D_1 depict the dissociation energies. β_{0A}, β_{0B}, and β_1 denote the range parameters. The shape of the Morse potential is shown in Figure 9.9.

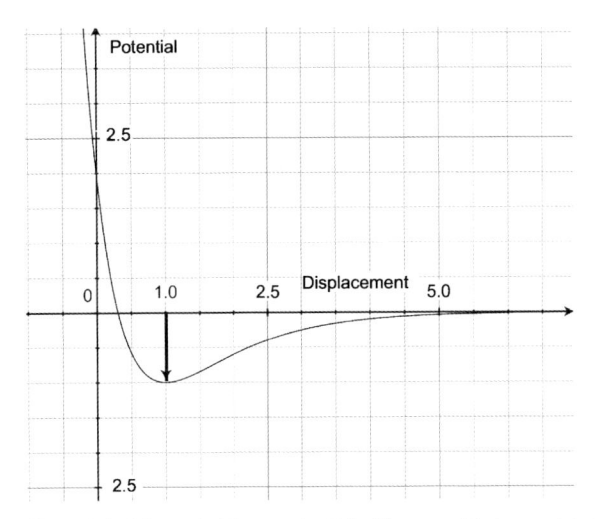

Figure 9.9 Schematic Morse potential with an equilibrium point at 1.0.

Assuming 2-dimensional symmetry for the quadri-atoms A, the potential function around the single atom B possesses an axial symmetric property at steady state. The system then seems to be equivalent to a T-shaped model. It implies that the fragmentation is limited in the direction vertical to the plane A.

Equation 9.4 represents the model system Hamiltonian in coordinates (Q, q, P, p). In classical dynamics, we have the relation

$$\begin{cases} \dot{Q} = \dfrac{\partial H}{\partial P}, \\ \dot{q} = \dfrac{\partial H}{\partial p}, \\ \dot{P} = -\dfrac{\partial H}{\partial Q}, \\ \dot{p} = -\dfrac{\partial H}{\partial q}. \end{cases} \tag{9.5}$$

The system shown in (9.5) is linearized through a Taylor expansion around the equilibrium point for Q and q.

9.4.2
Analysis Based on an Action-Angle Formulation [52]

Under the dissipative or excited state, the perturbed Hamiltonian is given as

$$H = H_0 + \varepsilon H_1, \tag{9.6}$$

where ε is a small parameter. H_1 includes $V_1(Q, q)$. The zero order component of the Hamiltonian is given by

$$H_0 = \frac{P^2}{2m} + \frac{p^2}{2\mu} + W_0(Q) + W_0(q). \tag{9.7}$$

We can rewrite H_0 using an action(I)-angle(θ) form. Based on [52], the relationship becomes

$$H_0 = E_A(I_A) + E_B(I_B). \tag{9.8}$$

I_A and I_B are action variables that are obtained by an action integral. The potential energy of the Morse oscillators E_A and E_B are defined as

$$\begin{cases} E_A(I_A) = \left(I_A + \tfrac{1}{2}\right)\omega_A - \left(I_A + \tfrac{1}{2}\right)^2 \omega_A^2/4D_{0A} - D_{0A}, \\ E_B(I_B) = \left(I_B + \tfrac{1}{2}\right)\omega_B - \left(I_B + \tfrac{1}{2}\right)^2 \omega_B^2/4D_{0B} - D_{0B}, \end{cases}$$

where

$$\omega_A = \left(2D_{0A}\beta_{0A}^2/m\right)^{1/2},$$
$$\omega_B = \left(2D_{0B}\beta_{0B}^2/\mu\right)^{1/2}.$$

D_{0A} and D_{0B} are defined by related Morse potentials. The simple derivatives of E_A and E_B give the zero order frequencies in each motion along Q and q. These are

$$\begin{cases} \dot{\theta}_A = -\omega_A + (2I_A + 1)\omega_A^2/4D_{0A} = \Omega_A, \\ \dot{\theta}_B = -\omega_B + (2I_B + 1)\omega_B^2/4D_{0B} = \Omega_B. \end{cases} \tag{9.9}$$

The energy at which the atom B is separated at infinity depends on D_{0B}. The external energy input for fragmentation is due to the boundary of the trapped motion of H. The phase structure is schematically described in Figure 9.10.

The maximum values of I_A and I_B are obtained from (9.7) [68].

$$\begin{cases} I_{A\max} = -\dfrac{1}{2} + \dfrac{2D_{0A}}{\omega_A}, \\ I_{B\max} = -\dfrac{1}{2} + \dfrac{2D_{0B}}{\omega_B}. \end{cases} \tag{9.10}$$

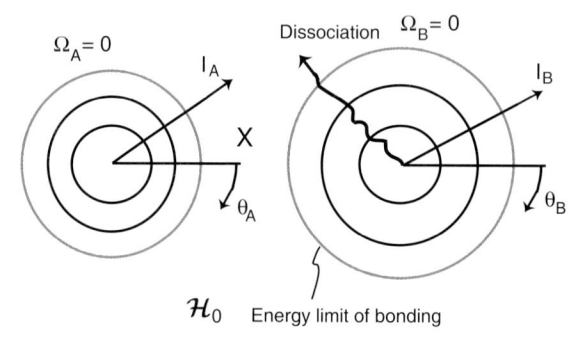

Figure 9.10 Phase portrait of the Hamiltonian.

These are also the limits of the actions for the T-shaped structures in the Hamiltonian. That is, the external energy input to the system can dissociate bonding between atoms A and B.

9.4.3
Dynamics of Single Atoms Induced by Probes

One of the important topics of nanotechnology is the manipulation of single atoms at the material surface. We have already investigated the dynamics of single atoms, which can dissociate from material surfaces. Here we will discuss a mathematical formulation of vibratory fragmentation of single atoms by probes based on a perturbed Hamiltonian system.

The manipulation of single atoms has been achieved by STM [69–71] and AFM [20–22]. The schematic structure is described in Figure 9.11.

The manipulation brings the energy exchange between an atom bonding to surface and probe. The manipulation of atoms is governed by the probability of dissociation of atomic bonds. However, the dynamics and probabilities are not well understood. If the probe is rigid or consists of heavy atoms with strong bonds to the bulk, the dynamics are simply modeled by a T-shaped configuration of atoms and molecules at surface. Then, the vibration of the probe gives us an external energy input to the original Hamiltonian system.

The dynamics of single atoms that are manipulated by a probe are explained in relation to the bonds in a T-shaped structure. The nonlinear resonances under external excitation are related to energy transfer between modes and external vi-

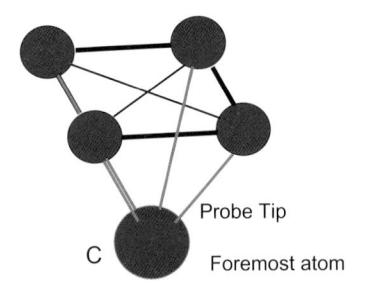

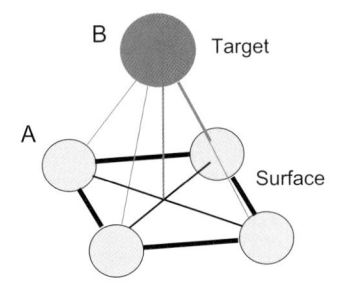

Figure 9.11 Manipulation of single atoms by a probe.

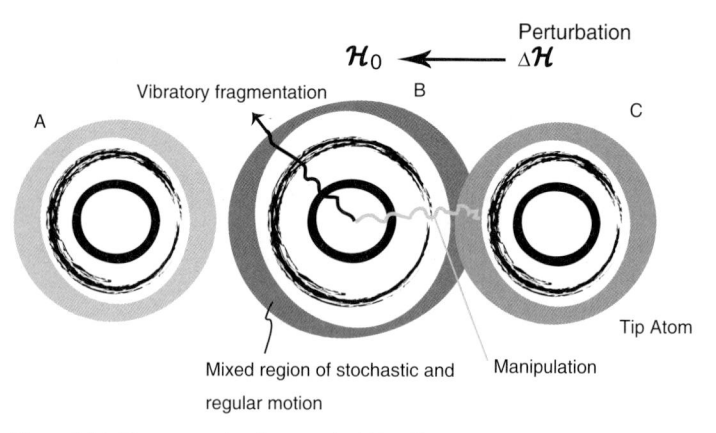

Figure 9.12 Phase portrait of a perturbed Hamiltonian system.

brations. Moreover, the limit of resonance confronts the appearance of instability depending on the nonlinearity. On the other hand, we know that there is a chaotic region around the resonant boundary [51–53]. Under the external Hamiltonian perturbation, ΔH might generate vibratory fragmentation and manipulation of atoms as shown in Figure 9.12. Therefore, the instability of the resonance around the boundary is strongly related to the global phase structure. When the system becomes dissipative, the complexity is due to the intersection of stable and unstable manifolds. Bonding instabilities are also reported between atoms at material surfaces and the foremost tips of the probes [72]. The instability causes the abrupt jump of dynamics. This might be related to the uncertainty in the capture and release of atoms.

In the classical treatment, the probability of fragmentation is defined by the rate of initial conditions from which trajectories escape from the region surrounded by a homoclinic orbit. Consider initial conditions in the space (Q_i, q_j, P_i, p_j), where i and j show the indices of a meshed initial condition space. Their Hamiltonian trajectories are generated by

$$\begin{cases} \dfrac{P_i^2}{2m} + W_0(Q_i) = E_A(I_A), \\ \dfrac{p_j^2}{2\mu} + W_0(q_j) = E_B(I_B). \end{cases} \quad (9.11)$$

These equations possibly show the stochastic region in initial condition space [52]. The dissipation of the coupled Morse oscillators loosens the homoclinic orbit corresponding to the energy limit of bonding. At the same time, homoclinic intersection and folded manifolds appear in the global phase structure. In such a case, the probability is strongly governed by the structure in spite of the uncertainty of the Hamiltonian system [53].

When probe forcing vibrates single atoms, the dynamics can be approximated by (9.5) with dissipation and forcing terms. To achieve vibratory fragmentation, the external energy must satisfy

$$\Delta E_B \geq E_B(I_{Bmax}) - E_B(I_{B0}), \quad (9.12)$$

where I_{B0} is the initial action value. At the same time, the atom B must be captured in the bond of the probe tip. As for the foremost atom (C) of tip, the situation coincides with the energy exchange between A–B and B–C in the bond A–B–C. After the fragmentation of the bond A–B, the atom B generates the new bonding B–C. In this process, the dissipation works for the stabilization of the dynamics. As mentioned above, chaotic dynamics cannot be avoided between release and capture. The manipulation is then governed by the uncertainty of the dynamics.

9.4.4 Control of Manipulation

The certainty of manipulation is completely affected by the nonlinear dynamics between the target atoms and the tips from the viewpoint of classical mechanics. If the cantilever tip is apart from the material surface as in the case of dynamic mode AFM, quantum effects seem to be small. We occasionally encounter the idea that the uncertainty is caused by quantum mechanics and thermal dynamics in the system. However, as dynamic mode AFM does not take place at the material surface, we should be careful to formalize the dynamics.

In the realistic setups, the manipulation process lies in the interaction by multi atoms around the surface and cantilever tip. The simulation show us the possibility of manipulation for several conditions and parameters. Here we show a criteria of energy which can dissociate a single atom bonded to surface atoms. The perturbation to atoms through AFM can change the energy level of target atoms. The possible control of dissociation lies in the tuning resonance conditions between the cantilever tips and the coupling frequencies in atomic bonding. The frequency must be low and in an acoustic mode. Then the interaction by AFM has a possibility to control the dissociation in the system. The Further research on this topic is now underway.

9.5 Concluding Remarks

This chapter discussed two topics: (1) the nonlinear cantilever dynamics in dynamic mode AFM and stabilization of chaotic vibrations for imaging exceeding nonlinear characteristics, and (2) the possibility for manipulating nanoparticles from a material surface through AFM interaction. Both topics relate to the control of AFM cantilevers, but the approaches are completely opposite. The top–down approach restricts the perturbation by nonlinearity between the cantilever tip and material surface at the van der Waals force level. The bottom–up approach gives the perturbation to the bond between atoms through AFM interaction. The top–down approach was begun several years ago from research in nonlinear dynamics, and now the results have been confirmed experimentally and its effectiveness has been demonstrated. The bottom–up theoretical discussion is still far from experimental verification. However, the uncertainty caused by nonlinearities can be controlled through the new well developed control methods and control system designs. We now expect to approach the manipulation of atoms at the limit of classical mechanics with greater control. This control can restrict the nonlinear uncertainty until the appearance of quantum or thermal uncertainty in the system. These are our current ongoing research topics.

Acknowledgments

The authors would like to thank Professor Kazumi Matsushige, Professor Hirofumi Yamada, Professor Kei Kobayashi, Kyoto University for the fruitful discussions and valuable suggestions. The authors would also like to acknowledge Dr. Robert Walter Stark, Dr. Ferdinand Jamitzky, and Dr. Javier Rubio-Sierra, Ludwig-Maximilians-Universität München for their helpful comments and discussions. One of the authors (TH) would like to show his sincere appreciation to Professor Igor Mezić, University of California at Santa Barbara, and Professor Yoshihiko Susuki, Kyoto University for the enthusiastic discussion and comments. This work is partly supported by Grant-in-Aid for Young Scientists (B) No. 20760238, Japan Society for the Promotion of Science (JSPS). The work is also partly supported by the Ministry of Education, Culture, Sports, Science and Technology-Japan (MEXT), the 21st Century COE Program (Grant No. 14213201) and Global COE program.

References

1 Feynman, R.P. (1960) *Eng. Sci.*, **23**, 22
2 Sarid, D. (1994) *Scanning probe microscopy: with applications to electric, magnetic, and atomic forces*, Rev. ed, Oxford University Press, Inc., New York.
3 Binnig, G. and Rohrer, H. (1999) *Rev. Mod. Phys.*, **71**, S324.
4 Binnig, G., Quate, C.F., and Gerber, C. (1986) *Phys. Rev. Lett.*, **56**, 930.
5 García, R. and Pérez, R. (2002) *Surf. Sci. Rep.*, **47**, 197.
6 Giessibl, F.J. (2003) *Rev. Mod. Phys.*, **75**, 949.
7 Binning, G., Rohrer, H., Gerber, C., and Weibel, E. (1982) *Phys. Rev. Lett.*, **49**, 57.
8 Martin, Y., Williams, C.C., and Wickramasinghe, H.K. (1987) *J. Appl. Phys.*, **61**, 4723.
9 Zhong, Q., Inniss, D., Kjoller, K., and Elings, V.B. (1993) *Surf. Sci.*, **290**, L688.
10 Albrecht, T.R., Grütter, P., Horne, D., and Rugar, D. (1991) *J. Appl. Phys.*, **69**, 668.
11 Giessibl, F.J. (1995) *Science*, **267**, 68.
12 Sugawara, Y., Ohta, M., Ueyama, H., and Morita, S. (1995) *Science*, **270**, 1646.
13 Kitamura, S. and Iwatsuki, M. (1995) *Jpn. J. Appl. Phys. 2.*, **34**, L145–L148.
14 Fukuma, T., Kobayashi, K., Matsushige, K., and Yamada, H. (2005) *Appl. Phys. Lett.*, **87**, 034101
15 Hansma, P.K., Cleveland, J.P., Radmacher, M., Walters, D.A., Hillner, P.E., Bezanilla, M., Fritz, M., Vie, D., Hansma, H.G., Prater, C.B., Massie, J., Fukunaga, L., Gurley, J., and Elings, V. (1994) *Appl. Phys. Lett.*, **64**, 1738.
16 Putman, C.A.J., Van der Werf, K.O., De Grooth, B.G., Van Hulst, N.F., and Greve, J. (1994) *Appl. Phys. Lett.*, **64**, 2454.

17 Stern, J.E., Terris, B.D., Mamin, H.J., and Rugar, D. (1988) *Appl. Phys. Lett.*, **53**, 2717–2719.

18 Nonnenmacher, M., O'Boyle, M.P., and Wickramasinghe, H.K. (1991) *Appl. Phys. Lett.*, **58**, 2921–2923.

19 Martin, Y. and Wickramasinghe, H.K. (1987) *Appl. Phys. Lett.*, **50**, 1455.

20 Sugawara, Y., Sano, Y., Suehira, N., and Morita, S. (2002) *Appl. Surf. Sci.*, **188**, 285.

21 Oyabu, N. et al. (2006) *Phy. Rev. Lett.*, **96**, 106101.

22 Sugimoto, Y. et al. (2007) *Phy. Rev. Lett.*, **98**, 106104.

23 Sasaki, N., Watanabe, S., and Tsukada, M. (2002) *Phys. Rev. Lett.*, **88**, 046106.

24 Anczykowski, B., Krüger, D., and Fuchs, H. (1996) *Phys. Rev. B*, **53**, 15485.

25 Berg, J. and Briggs, G.A.D. (1997) *Phys. Rev. B*, **55**, 14899.

26 Sasaki, N., Tsukada, M., Tamura, R., Abe, K., and Sato, N. (1998) *Appl. Phys. A-Mater.*, **66**, S287.

27 Aimé, J.P., Boisgard, R., Nony, L., and Couturier, G. (1999) *Phys. Rev. Lett.*, **82**, 3388.

28 García, R. and San Paulo, A. (2000) *Phys. Rev. B*, **61**, R13381.

29 Rützel, S., Lee, S.I., and Raman, A. (2003) *Proc. R. Soc. Lon. Sea-A.*, **459**, 1925.

30 Ashhab, M., Salapaka, M.V., Dahleh, M., and Mezić, I. (1999) *Automatica*, **35**, 1663.

31 Ashhab, M., Salapaka, M.V., Dahleh, M., and Mezić, I. (1999) *Nonlinear Dynam.*, **20**, 197.

32 Patil, S. and Dharmadhikari, C.V. (2003) *Appl. Surf. Sci*, **217**, 7.

33 Jamitzky, F., Stark, M., Bunk, W., Heckl, W.M., and Stark, R.W. (2006) *Nanotechnology*, **17**, S213.

34 Basso, M., Giarre, L., Dahleh, M., and Mezić, I. (2000) *J. Dyn. Syst.-T. ASME*, **122**, 240.

35 Zitzler, L., Herminghaus, S., and Mugele, F. (2002) *Phys. Rev. B*, **66**, 155436.

36 Hu, S. and Raman, A. (2006) *Phys. Rev. Lett.*, **96**, 036107.

37 van de Water, W. and Molenaar, J. (2000) *Nanotechnology*, **11**, 192.

38 Dankowicz, H. (2006) *Philos. T. R. Soc. A*, **364**, 3505.

39 Hashemi, N., Dankowicz, H., and Paul, M.R. (2008) *J. Appl. Phys.*, **103**, 093512.

40 Pyragas, K. (1992) *Phys. Lett. A*, **170**, 421.

41 Yamasue, K. and Hikihara, T. (2006) *Rev. Sci. Instrum.*, **77**, 053703.

42 Yamasue, K. and Hikihara, T. (2006) In: *2006 International Symposium on Nonlinear Theory and its Applications*, 2006, Institute of Electronics, Information and Communication Engineers, Tokyo, 927.

43 Yamasue, K., Kobayashi, K., Yamada, H., Matsushige, K., and Hikihara, T. (2009) *Phys. Lett. A*, **373**, 3140.

44 Bouju, X., Joachim, C., and Girard, C. (1999) *Phy. Rev. B*, **59**, R7845

45 Lorente, N. and Persson, M. (2001) *R. Soc. Chem. Faraday Discuss.* **117**, 277

46 Kuhnle, A. et al. (2002) *Surf. Sci.*, **499**, 15.

47 Hikihara, T. (2007) In: *2007 International Symposium on Nonlinear Theory and its Applications*, 2007, Institute of Electronics, Information and Communication Engineers, Tokyo, 180.

48 Hikihara, T. (2008) *The 2008 American Control Conference*, 4431.

49 Hikihara, T. and Yamasue, K. (2005) *The Fifth International Symposium on Linear Drives for Industry Applications*, 2005, the Institute of Electrical Engineers of Japan, Tokyo, 29.

50 Beswick, J.A., Delgado-Barrio, G., and Jortner, J. (1979) *J. Chem. Phys.*, **70**, 3895.

51 Hedges, R.M. and Reinhardt, W.P. (1983) *J. Chem. Phys.*, **78**, 3964.

52 Gray, S.K., Rice, S.A., and Noid, D.W. (1986) *J. Chem. Phys.*, **84**, 3745.

53 Toda, M. (1995) *Phy. Rev. Lett.*, **74**, p. 2670.

54 Wolf, R.J. and Hase, W.L. (2006) *J. Chem. Phys.*, **73**, 3779.

55 Meyer, G. and Amer, M. (1988) *Appl. Phys. Lett.*, **53**, 1045.

56 Gleyzes, P., Kuo, P.K., and Boccara, A.C. (1991) *Appl. Phys. Lett.*, **58**, 2989.

57 Burnham, N.A., Kulik, A.J., Germaud, G., and Briggs, G.A.D. (1995) *Phys. Rev. Lett.*, **74**, 5092.

58 Fang, Y., Dawson, D., Feemster, M., and Jalili, N. (2002) In: *Proceedings of 2002 ASME International Mechanical Engineering Congress and Exposition*, 2002, ASME, New Orleans, Louisiana, 33539.

59 Pyragas, K. (2006) *Philos. T. R. Soc. A*, **364**, 2309.

60 Pyragas, K. and Tamaševičius, A. (1993) *Phys. Lett. A*, **180**, 99.
61 Socolar, J.E.S., Sukow, D.W., and Gauthier, D.J. (1994) *Phys. Rev. E*, **50**, 3245.
62 Bielawski, S., Derozier, D., and Glorieux, P. (1994) *Phys. Rev. E*, **49**, R971.
63 Hikihara, T. and Kawagoshi, T. (1996) *Phys. Lett. A*, 211, 29.
64 Parmananda, P., Madrigal, R., Rivera, M. Nyikos, L., Kiss, I.Z., and Gáspár, V. (1999) *Phys. Rev. E*, **59**, 5266.
65 Pierre, Th., Bonhomme, G., and Atipo, A. (1996) *Phys. Rev. Lett.*, **76**, 2290.
66 Tamayo, J. (2005) *J. Appl. Phys.*, **97**, 044903.
67 Allen, M.O. (2004) In: *Computational Soft Matter: From Synthetic Polymers to Proteins, Lecture Notes, NIC Series*, **23**, 1.
68 Chirikov, B.V. (1979) *Phy. Rep.*, **52**, 263.
69 Eigler, D.M. and Schweizer, E.K. (1990) *Nature*, **344**, 524.
70 Bartels, L., Meyer, G., and Rieder, K.-H. (1997) *Phy. Rev. Lett.*, **79**, 697.
71 Stroscio, J.A. and Celotta, R.J. (2004) *Science*, **306**, 242.
72 Caciuc, V., Hölscher, H., and Blügel, S. (2005) *Phy. Rev. B*, **72**, 035423.
73 Kim, B., Putkaradze, V., and Hikihara, T. (2009) *Phy. Rev. Lett.*, **102**, 215502.

…
Part IV Nanoelectronics

10
Classical Correlations and Quantum Interference in Ballistic Conductors

Daniel Waltner and Klaus Richter

10.1
Introduction: Quantum Transport through Chaotic Conductors

Among the most important properties characterizing an electronic nanosystem is its electrical conductance behavior. Therefore, gaining knowledge on charge transport mechanisms, in particular when shrinking conductors from macroscopic sizes down to molecular-sized wires or atomic point contacts, has been the focus of experimental and theoretical research throughout the last decade. Such a reduction in size and spatial dimensionality goes along with a crossover from charge flow in the macroscopic bulk, well described by Ohm's law, to distinct quantum effects in the limit of microscopic or atomistic wires. Nanoconductors in the crossover regime, often referred to as mesoscopic, frequently exhibit a coexistence of both classical remnants of bulk features combined with signatures from wave interference. Such quantum effects usually require low temperatures where coherence of the electronic wave functions is retained up to micron scales. This has lead to the observation of various quantum interference phenomena, for example, quantized steps in the point contact conductance or the Aharonov–Bohm effect and universal conductance fluctuations.

Nonlinear effects can enter into transport through mesoscopic or nanosystems in two ways. First, as nonlinear I–V characteristics and charge flow far from equilibrium for large enough voltages. Second, in the limit of linear response to an applied electric field, the intrinsic nonlinear classical dynamics of the unperturbed conductor can govern its transport properties. In this chapter, we focus on the latter case which is particularly interesting for mesoscopic conductors because the nonlinear charge carrier dynamics can influence both the classical and, in a more subtle way, the quantum transport phenomena.

Initially, disordered metals with underlying diffusive charge carrier motion in mesoscopic matter were focused on. However here, we address ballistic nano- or mesoscopic conductors where impurity scattering is suppressed. The most prominent ballistic systems are nanostructures built from high-mobility semiconductor heterostructures where electrons are confined to two-dimensional, billiard-type

Nonlinear Dynamics of Nanosystems. Edited by Günter Radons, Benno Rumpf, and Heinz Georg Schuster
Copyright © 2010 WILEY-VCH Verlag GmbH & Co. KGaA, Weinheim
ISBN: 978-3-527-40791-0

| t = 0 | t = 1 | t = 2 |
| t = 3 | t = 4 | t = 25 |

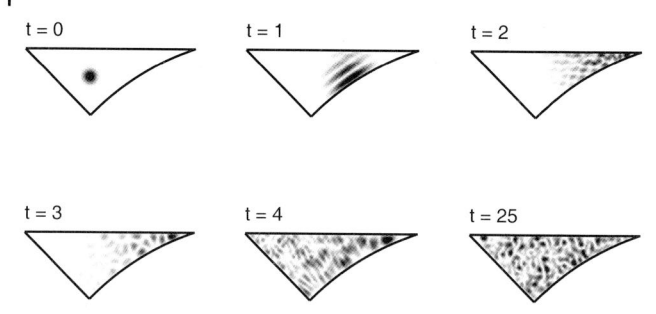

Figure 10.1 Quantum mechanical wave packet launched into a mesoscopic cavity with the geometry of a "desymmetrized diamond billiard". The wave packet evolution is monitored at times $t = 1, 2, 3, 4$, and 25, in units of the average time between collisions with the walls of a corresponding classical particle. (Courtesy of A. Goussev).

cavities of controllable geometry. Though, such systems are also realized as atom optics billiards or through wave scattering as optical, microwave, or acoustic mesoscopic resonators [1].

The quantum dynamics of a single particle in such a geometry is illustrated in Figure 10.1 showing snapshots of a wave packet after multiples of the average classical time between bounces off the billiard walls. The quantum evolution in such a mesoscopic geometry, with corresponding chaotic classical dynamics, is characterized by two main features: The rapid transition from wave packet motion roughly following the path of a classical particle to random wave interference at larger times and, second, the emergence of wave functions of complex morphology with wave lengths much shorter than the system size. The latter can be used to further specify mesoscopic matter, that is, quantum coherent systems where the smallest (quantum) length scale, the de Broglie wave length or Fermi wave length λ_F in electronic conductors, is much smaller than the system size $\mathcal{L}$. Therefore, $1/(k_F \mathcal{L})$ is a small parameter in terms of the Fermi momentum $k_F = 2\pi/\lambda_F$, though not fully negligible as in the case of macroscopic systems.

Such ballistic mesoscopic systems are ideal tools for studying the connection between (chaotic) classical dynamics and wave interference. Presumably, semiclassical techniques provide this link in the most direct way. Modern semiclassical theory is based on trace formulas, sums over Fourier-type components associated with classical trajectories. Analogous to the famous Gutzwiller trace formula for the density of states [2], corresponding expressions for quantum transport in the linear response regime exist. There, semiclassical expressions for the conductance have been obtained within the framework of the Landauer–Büttiker approach, relating conductance to quantum transmission in nanostructures.

Following the early pioneering semiclassical work by Miller [3] for molecular reactions and later by Blümel and Smilansky [4] for quantum chaotic scattering, major advances were made in the context of mesoscopic conductance in the early nineties by Baranger, Jalabert and Stone [5, 6]. All these semiclassical approaches were based on and limited by the so-called diagonal approximation. While most of the features of experimental and numerical magneto-conductance profiles could be well explained qualitatively on the level of the diagonal approximation, it was not current conserving and thus failed to give correct quantitative predictions for the quantum transmission. This was solved about ten years later when an approach was devised to account for off-diagonal contributions to the semiclassical conductance [7], thereby achieving unitarity reflected in (average) current conservation, and furthermore, agreement with existing predictions from random matrix theory (RMT). The applicability of RMT to closed chaotic mesoscopic systems was conjectured after numerical simulations in [8]. Therefore, it was also expected to be applicable to open systems.

Semiclassical ballistic transport on the level of the diagonal approximation was reviewed in detail in [9–11]. In this chapter, we focus on the recent progress beyond the diagonal approximation. This serves as a model case which illustrates how chaotic nonlinear dynamics can govern quantum properties at nanoscales.

10.2
Semiclassical Limit of the Landauer Transport Approach

The Landauer formalism [12], providing a link between the quantum transmission and conductance, has proved to be an appropriate framework to address phase-coherent transport through nanosystems. Consider a sample attached to two leads of width W_1 and W_2 that support, respectively, N_1 and N_2 current carrying transverse modes at (Fermi-)energy E_F. For such a two-terminal setup, the conductance reads at very low temperatures [12]

$$G(E_F) = g_s \frac{e^2}{h} T(E_F) = g_s \frac{e^2}{h} \sum_{m=1}^{N_1} \sum_{n=1}^{N_2} |t_{nm}(E_F)|^2 , \qquad (10.1)$$

with $N_1 = W_1\sqrt{2m E_F}/(\hbar\pi)$ and an analogous relation for N_2. Here, $g_s = 2$ accounts for spin degeneracy, and the $t_{nm}(E)$ are transmission amplitudes between incoming channels m and outgoing channels n in the leads at energy E. They can be expressed in terms of the projections of the Green function of the scattering region onto the transverse modes $\phi_n(y')$ and $\phi_m(y)$ in the two leads [13]

$$t_{nm}(E) = -i\hbar(v_n v_m)^{1/2} \int dy' \int dy \, \phi_n^*(y')\phi_m(y) G(x', y', x, y; E) . \qquad (10.2)$$

Here x and x', respectively, denote the direction along the leads and v_n, v_m, the corresponding longitudinal velocities. The integrals in (10.2) are taken over the cross sections of the (straight) leads at the entrance and the exit.

Figure 10.2 shows the quantum transmission, numerically obtained from (10.2), for a "graphene billiard" [14], fabricated by cutting a cavity out of a two-dimensional graphene flake, a monoatomic layer of carbon atoms arranged in a honeycomb lattice. Two major quantum features are visible: (i) distinct "ballistic conductance fluctuations" as a function of energy. (ii) When subject to an additional perpendicular

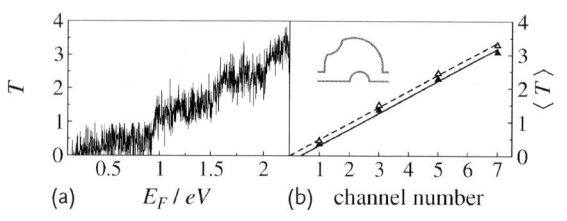

Figure 10.2 Total quantum transmission (as a function of energy E_F, in (b) in units of the channel number in the leads of width $W_1 = W_2 = W$) for transport through a phase-coherent graphene-based quantum dot, see inset. The fluctuating line in (a) is the full quantum transmission at zero magnetic field. In (b), the straight solid and dashed line denote the averaged transmission at zero magnetic field and a magnetic field corresponding to a flux $\phi = 1.6\phi_0$ with the flux quantum $\phi_0 = hc/e$. The difference marks the weak localization correction (from [14]).

magnetic field B with magnetic flux ϕ, the average transmission (straight dashed line in Figure 10.2b) shows a small positive offset compared to the average transmission for $B = 0$ (solid line). This reduction of the average conductance at zero magnetic field reflects a weak localization effect [15]. Its origin is non-classical and due to wave interference.

Here, we focus on this ballistic weak localization effect and present its semiclassical derivation for conductors with classically chaotic analogue. The semiclassical approximation enters in two steps. First, we replace $G(x', y', x, y; E)$ in (10.2) by the semiclassical Green function (in two dimensions) [2]:

$$G^{sc}(\mathbf{r}', \mathbf{r}; E) = \frac{1}{i\hbar(2\,i\pi\hbar)^{1/2}} \sum_t D_t(\mathbf{r}', \mathbf{r}) \exp\left(\frac{i}{\hbar} S_t - i\eta_t \frac{\pi}{2}\right). \tag{10.3}$$

It is given as a sum over contributions from all classical trajectories t connecting the two fixed points $\mathbf{r}$ and $\mathbf{r}'$ at energy E. In (10.3),

$$S_t(\mathbf{r}', \mathbf{r}; E) = \int_{\mathcal{C}_t} \vec{p} \cdot d\vec{q} \tag{10.4}$$

is the classical action along a path $\mathcal{C}_t$ between $\mathbf{r}$ and $\mathbf{r}'$ and governs the accumulated phase.

Second, we evaluate the projection integrals in (10.2) for isolated trajectories within the stationary-phase approximation. For leads with hard-wall boundaries, the mode wave functions are sinusoidal, $\phi_m(y) = \sqrt{2/W_1} \sin(m\pi y/W_1)$. Hence, the stationary-phase condition for the y integral requires [9]

$$\left(\frac{\partial S}{\partial y}\right)_{y'} = -p_y \equiv -\frac{\overline{m}\hbar\pi}{W_1}, \tag{10.5}$$

with $\overline{m} = \pm m$. The stationary-phase solution of the y' integral yields a corresponding "quantization" condition for the transverse momentum $p_{y'}$. Thus, only those paths which enter into the cavity at (x, y) with a fixed angle $\sin\theta = \pm m\pi/k\,W_1$ and exit the cavity at (x', y') with angle $\sin\theta' = \pm n\pi/k\,W_2$ contribute to $t_{nm}(E)$

with $k = \sqrt{2mE}/\hbar$. There is an intuitive explanation: the trajectories are those whose transverse wave vectors on entrance and exit match the wave vectors of the modes in the leads. One then obtains for the semiclassical transmission amplitudes

$$t_{nm}(k) = -\frac{\sqrt{2\pi i\hbar}}{2\sqrt{W_1 W_2}} \sum_{t(\overline{n},\overline{m})} \text{sgn}(\overline{n})\text{sgn}(\overline{m}) \sqrt{A_t} \exp\left[\frac{i}{\hbar}\tilde{S}_t(\overline{n},\overline{m};k) - i\frac{\pi}{2}\tilde{\mu}_t\right]. \tag{10.6}$$

Here, the reduced actions are

$$\tilde{S}_t(\overline{n},\overline{m};k) = S_t(k) + \hbar k y \sin\theta - \hbar k y' \sin\theta', \tag{10.7}$$

which can be considered as Legendre transforms of the original action functional. The phases $\tilde{\mu}_t$ contain both the usual Morse indices and additional phases arising from the y, y' integrations. The prefactors are $A_t = |(\partial y/\partial\theta')_\theta|/(\hbar k |\cos\theta'|)$. The resulting semiclassical expression for the transmission and thereby the conductance (see (10.1)) in chaotic cavities involves contributions from pairs of trajectories t, t'. It reads [5, 6, 9]

$$T(k) = \sum_{m=1}^{N_1}\sum_{n=1}^{N_2} |t_{nm}(k)|^2 = \frac{\pi\hbar}{2 W_1 W_2} \sum_{m=1}^{N_1}\sum_{n=1}^{N_2} \sum_{t,t'} F_{n,m}^{t,t'}(k) \tag{10.8}$$

with

$$F_{n,m}^{t,t'}(k) \equiv \sqrt{A_t A_{t'}} \exp\left[\frac{i}{\hbar}(\tilde{S}_t - \tilde{S}_{t'}) - i\mu_{t,t'}\frac{\pi}{2}\right], \tag{10.9}$$

where the phase $\mu_{t,t'}$ accounts for the differences of the phases $\tilde{\mu}_t$ and the sign factors in (10.6).

In the next two sections, we show how one can calculate quantum corrections to the transmission beyond the diagonal approximation with purely semiclassical methods. The results obtained for chaotic conductors are consistent with RMT-predictions.

10.3
Quantum Transmission: Configuration Space Approach

In evaluating the off-diagonal (interference) contributions to the quantum transmission, we present two approaches. The first approach is based on the analysis of off-diagonal pairs of trajectories and their self-intersections in configuration space. This approach is more illustrative, though less general than the second phase space approach outlined in Section 10.4.

We start from the semiclassical expression for $|t_{nm}(k)|^2$ obtained by squaring the semiclassical transmission amplitudes (10.6), see also (10.8) and (10.9):

$$|t_{nm}(k)|^2 = \frac{\pi\hbar}{2 W_1 W_2} \sum_{t,t'} \sqrt{A_t A_{t'}} \exp\left[\frac{i}{\hbar}(\tilde{S}_t - \tilde{S}_{t'}) - i\mu_{t,t'}\frac{\pi}{2}\right]. \tag{10.10}$$

Due to the action difference of the trajectories t and t' in the exponential in (10.10), this expression is a rapidly oscillating function of k or the energy in the semiclassical limit of large ratios $(S_t - S_{t'})/\hbar$. In the following, we wish to identify those contributions to (10.10) which survive an average over a classically small, but quantum mechanically large k-window Δk. Such contributions must come from very similar trajectories. Afterwards, we wish to evaluate their contributions to $|t_{nm}(k)|^2$ using basic principles of chaotic dynamics. For our calculation, we will need hyperbolicity and ergodicity. Hyperbolicity implies the possible exponential separation of neighboring trajectories for long times with the distance growing proportional to $e^{\lambda T}$, with the Lyapunov exponent λ and the time T of the trajectory. The second principle, ergodicity, means the equidistribution of long trajectories on the energy surface at the energy E of the trajectory.

After calculating the diagonal contribution in the first subsection, we will determine the simplest nondiagonal contribution. The behavior of the latter as a function of a magnetic field and a finite Ehrenfest time is then studied in the third and fourth subsection, respectively.

10.3.1
Diagonal Contribution

The first "diagonal" (D) contribution to (10.10) originates from identical trajectories $t = t'$, meaning $S_t = S_{t'}$. It gives

$$|t_{nm}(k)|^2_{\mathrm{D}} = \frac{\pi \hbar}{2 W_1 W_2} \sum_t A_t . \tag{10.11}$$

The remaining sum over classical trajectories in (10.11) can be calculated using a classical sum rule [7] that can be derived using ergodicity, see for example [16]. It yields

$$\sum_t A_t = \frac{4 W_1 W_2}{\Sigma(E)} \int_0^\infty dT\, \rho(T) , \tag{10.12}$$

where $\Sigma(E)$ denotes the phase space volume of the system at energy E, and $\rho(T)$ is the classical probability to find a particle still inside an open system after a time T. For longer times, the latter decays for a chaotic system exponentially $\rho(T) \sim e^{-T/\tau_D}$, with the dwell time $\tau_D = \frac{\Sigma(E)}{2\pi\hbar(N_1 + N_2)}$. This exponential decay can be easily understood based on the equidistribution of trajectories: the number ΔN of particles leaving the system during ΔT is given by the overall number of particles N times the ratio of the phase space volume from which the particles leave during ΔT and the whole phase space volume of the system. The differential equation for N, obtained in the case of infinitesimal ΔT, obviously has an exponential solution.

By inserting (10.12) into (10.11), we obtain

$$|t_{nm}(k)|^2_{\mathrm{D}} = \frac{1}{N_1 + N_2} . \tag{10.13}$$

Its derivation required ergodicity valid only for long trajectories. We will assume that the classical dwell time is large enough in order to have a statistically relevant number of long trajectories left after time t. The result in (10.13) allows for a very simple interpretation. It is just the probability of reaching one of the $N_1 + N_2$ channels if it is equally likely to reach each of the channels.

10.3.2
Nondiagonal Contribution

In the following, we will calculate the contributions from pairs of different trajectories, however, with similar actions. For a long time, it was not clear how these orbits could look like. There are many orbit pairs in a chaotic system having accidentally equal or nearly equal actions. However, in order to describe universal features of a chaotic system after energy averaging, one has to find orbits that are correlated in a systematic way. These orbits were first identified and analyzed in 2000 in the context of spectral statistics [17, 18]. There, *periodic* orbits were studied to compute correlations between energy eigenvalues of quantum systems with a classically chaotic counterpart. Based on *open*, lead-connecting trajectories, this approach was generalized to the conductance we study here in [7]. Still, the underlying mechanism of forming pairs of classically correlated orbits is the same both cases.

In Figure 10.3, we show an example representative of such a correlated (periodic) orbit pair in the chaotic hyperbola billiard. The two partner orbits are topologically the same up to the region marked by the circle where one orbit exhibits a self-crossing (Figure 10.3a), while the other exhibits an 'avoided' crossing (Figure 10.3b). Usually, such trajectory pairs are drawn schematically, as shown in Figure 10.4.

One considers very long orbits with self-crossings characterized by crossing angles $\epsilon \ll \pi$. In [17, 18], it was shown that for each orbit, a partner orbit starting and

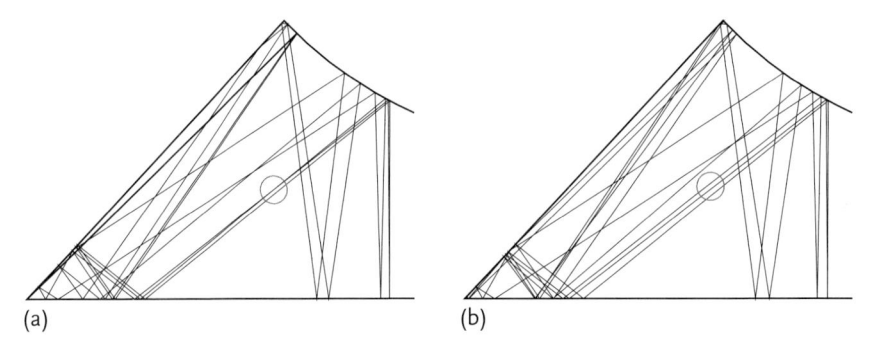

Figure 10.3 Pair of two periodic orbits in the hyperbola billiard essentially differing from each other in the region marked by the circle, where the left orbit (a) exhibits a self-crossing while the right partner orbit (b) shows an "avoided crossing". (Courtesy of M. Sieber).

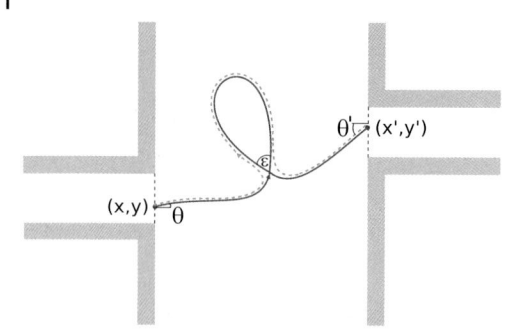

Figure 10.4 Schematic drawing of a pair of orbits yielding the first nondiagonal contribution to the transmission considered in [7]. One of the orbits crosses itself under an angle ϵ, the other one possesses an "avoided crossing". Except for the crossing region, both orbits are almost identical.

ending (exponentially) close to the first one exists. It follows the first orbit until the crossing, avoids it, then traverses the loop in a reversed direction and avoids the crossing again.

In order to quantify the contribution of these trajectory pairs to (10.10), we need two inputs: an expression for the action difference and for the density quantifying how often an orbit of time T exhibits a self-intersection, and both quantities expressed as a function of the parameter ϵ. The formula for the action difference ΔS can be derived by linearizing the dynamics of the orbit without crossing around the reference orbit with crossing. This yields, in the limit $\epsilon \ll \pi$, [17, 18],

$$\Delta S = \frac{p^2 \epsilon^2}{2m\lambda}. \tag{10.14}$$

At this point, we can justify our assumption of small crossing angles ϵ. In the limit $\hbar \to 0$, we expect important contributions to (10.10) only from orbit pairs with small action differences, that is, small crossing angles, as seen in (10.14).

Before deriving the number of self-crossings, $P(\epsilon, T) d\epsilon$, in the range between ϵ and $\epsilon + d\epsilon$ of an orbit of time T, we give rough arguments how this expression depends on ϵ and T for trajectories in billiards. There, each orbit is composed of a chain of N chords connecting the reflection points. Following an orbit, the first two chords cannot intersect, the third chord can cross with up to one, the fourth chord with up to two segments, and so on. Hence, the overall number of self-crossings will be proportional to $\sum_{n=3}^{N}(n-2) \propto N^2$, to leading order in N, that is, proportional to T^2.

The crossing angle dependence of $P(\epsilon, T)$ can be estimated for small ϵ as follows. Given a trajectory chord of length L, a second chord, tilted by an angle ϵ with respect to the first one, will cross it inside the billiard (with area of order L^2) only if the distance between the reflection points of the two chords at the boundary is smaller than $L \sin \epsilon$. The triangle formed in the latter case includes a fraction $\sin \epsilon$ of the entire billiard size. From this rough estimation, we expect $P(\epsilon, T) \propto T^2 \sin \epsilon$.

10.3 Quantum Transmission: Configuration Space Approach

More rigorously, the quantity $P(\epsilon, T)\, d\epsilon$, can be expressed for an arbitrary orbit γ as [17, 18]

$$P(\epsilon, T)\, d\epsilon = \left\langle \int_{T_{\min}(\epsilon)}^{T-T_{\min}(\epsilon)} dt_l \int_{T_{\min}(\epsilon)/2}^{T-t_l-T_{\min}(\epsilon)/2} dt_s\, |J|\, \delta(\mathbf{x}(t_s) - \mathbf{x}(t_s + t_l)) \right.$$
$$\left. \times \delta(\epsilon - \alpha(t_s, t_s + t_l)) \right\rangle d\epsilon \quad (10.15)$$

with the average $\langle \ldots \rangle$ taken over different initial conditions $(\mathbf{x}_0, \mathbf{p}_0)$. The time of the closed loop of the trajectory is denoted by t_l and the time before the loop by t_s. $\alpha(t_s, t_s + t_l)$ denotes the absolute value of the angle between the velocities $\mathbf{v}(t_s)$ and $\mathbf{v}(t_s + t_l)$. $|J|$ is the Jacobian for the transformation from the argument of the first delta function to t_l and t_s, ensuring that $P(\epsilon, T)\, d\epsilon$ yields a 1 for each crossing of γ. With the absolute value of the velocity, v, it can be expressed as

$$|J| = |\mathbf{v}(t_s) \times \mathbf{v}(t_s + t_l)| = v^2 \sin\alpha(t_s, t_s + t_l). \quad (10.16)$$

Starting from the formal expression (10.15), the derivation [17, 18] of the formula for $P(\epsilon, T)$ for a chaotic system is instructive. It shows how information can be extracted from the basic principles of chaotic dynamics beyond the diagonal approximation, and so we will present it here in detail. Hyperbolicity will yield a justification for the minimal time $T_{\min}(\epsilon)$, already introduced in (10.15); we will come back to that point later after.

To proceed, we interchange the phase space integral of the average with the time integrals, substitute $(\mathbf{x}(t_s), \mathbf{p}(t_s)) \mapsto (\mathbf{x}_0, \mathbf{p}_0)$ in (10.15) and obtain

$$P(\epsilon, T) = 2m \int_{T_{\min}(\epsilon)}^{T-T_{\min}(\epsilon)} dt_l\, v^2 \sin\epsilon\, p_E(\epsilon, t_l)(T - t_l - T_{\min}(\epsilon)), \quad (10.17)$$

with the averaged classical return probability density

$$p_E(\epsilon, t_l) = \frac{1}{2m} \langle \delta(\mathbf{x}_0 - \mathbf{x}(t_l))\, \delta(\epsilon - |\angle(\mathbf{v}_0, \mathbf{v}(t_l))|) \rangle. \quad (10.18)$$

This yields the probability density that a particle possessing the energy E returns after the time t_l to its starting point with the angle $|\angle(\mathbf{v}_0, \mathbf{v}(t_l))| = \epsilon$. For long times, this can be replaced by $1/\Sigma(E)$, assuming ergodicity. Then, we obtain

$$P(\epsilon, T) = 2m \int_{T_{\min}(\epsilon)}^{T-T_{\min}(\epsilon)} dt_l\, v^2 \sin\epsilon\, \frac{1}{\Sigma(E)}(T - t_l - T_{\min}(\epsilon))$$
$$= \frac{mv^2}{\Sigma(E)} \sin\epsilon\, (T - 2T_{\min}(\epsilon))^2. \quad (10.19)$$

Now, we return to our assumption of hyperbolicity and explain the cutoff time $T_{\min}(\epsilon)$, introduced in the equations above. To this end, we consider two classical paths leaving their crossing with a small angle ϵ. The initial deviation of their velocities is $\delta v_i = \epsilon v$. In order to form a closed loop, the deviation of the velocities

δv_f, when both paths have traversed half of the closed loop, has to be given by $\delta v_f = cv$ with c of the order unity. Then, for the minimal time $T_{\min}(\epsilon)$ to form a closed loop, due to the exponential divergence of neighboring orbits, we reach

$$c = \epsilon \, e^{(\lambda T_{\min}(\epsilon))/2}, \tag{10.20}$$

implying

$$T_{\min}(\epsilon) = \frac{2}{\lambda} \ln\left(\frac{c}{\epsilon}\right). \tag{10.21}$$

An argument similar to the one used here for the closed loop can also be applied to the other two parts of the trajectory, leaving the crossing with an angle ϵ towards the opening of the conductor. Suppose t_s has a length between 0 and $T_{\min}(\epsilon)/2$, then both parts have to be so close together that they must leave both through the same lead. As we are interested in the transmission, we also have to exclude that case[15]. A similar argument holds for the case where the last part of the orbit has a length between 0 and $T_{\min}(\epsilon)/2$. In this case, the orbit has to come very close to the opening before the crossing and leave before it could have crossed. Accounting for all these restrictions yields the integration limits in (10.15).

Now, we are prepared to calculate the contribution of the considered trajectory pairs to the transmission. Therefore, we keep one sum over trajectories in (10.10). This sum will be performed using the same classical sum rule as in diagonal approximation. We can replace the other by a sum over all the partner trajectories of one trajectory, which can be calculated using $P(\epsilon, T)$. There is, however, one subtlety concerning the survival probability $\rho(T)$ in the sum rule. Once again, if the crossing happens near the opening, both parts of the orbit act in a correlated way; $\rho(T)$ is changed in the case of the trajectory pairs considered here for a similar reason. Since we know that the two parts of the orbit leaving the crossing on each side are very close to each other, the orbit can either leave the cavity during the first stretch, that is, during the first time it traverses the crossing region, or it cannot leave at any point. This implies that we have to change the survival probability from $\rho(T)$ to $\rho(T - T_{\min}(\epsilon))$[16]. Then, we arrive at the loop (L) contribution

$$\begin{aligned}
|t_{nm}(k)|^2_L &= \frac{\pi \hbar}{2 W_1 W_2} \sum_t \sum_P A_t 2 \,\text{Re}\, \exp\left(i \frac{p^2 \epsilon_p^2}{2m\lambda\hbar}\right) \\
&= \frac{4\pi\hbar}{\Sigma(E)} \int_0^\pi d\epsilon \int_{2T_{\min}(\epsilon)}^\infty dT \, e^{-(T - T_{\min}(\epsilon))/\tau_D} P(\epsilon, T) \cos\left(\frac{p^2 \epsilon^2}{2m\lambda\hbar}\right) \\
&= \frac{8\pi\hbar m v^2 \tau_D^3}{\Sigma(E)^2} \int_0^\pi d\epsilon \, e^{-T_{\min}(\epsilon)/\tau_D} \sin\epsilon \cos\left(\frac{p^2\epsilon^2}{2m\lambda\hbar}\right)
\end{aligned} \tag{10.22}$$

---

15) If we would calculate the reflection instead of the transmission, the effect of short legs, referred to as coherent backscattering, must be taken into account.

16) This effect together with the requirement of a finite length of the orbit parts leaving towards the opening was originally not taken into account in [7]. In this calculation, the contributions from these two effects cancel each other. They will only be important when considering more complicated diagrams as in the next section.

with the sum over the partner trajectories P in the first line. As the important contributions require very small action differences, that is, very similar trajectories, and as the prefactor A_t is not as sensitive as the actions to small changes of the trajectories, we can neglect differences between t and t' in the prefactor. In the second line, we applied the classical sum rule with the modification explained prior to (10.22) and used $P(\epsilon, T)$ to evaluate the sum over P. After performing the simple time integral in the third line, we can do the ϵ-integration, as for example in [19], by taking into account that the important contributions come from very small ϵ, yielding

$$\begin{aligned}|t_{nm}(k)|^2_L &= \frac{8\pi\hbar mv^2\tau_D^3}{\Sigma(E)^2}\int_0^\pi d\epsilon(\epsilon/c)^{\frac{2}{\lambda\tau_D}}\sin\epsilon\cos\left(\frac{p^2\epsilon^2}{2m\lambda\hbar}\right)\\ &= \frac{8\pi\hbar mv^2\tau_D^3}{\Sigma(E)^2}\int dz\frac{m\lambda\hbar}{p^2}\left(\frac{1}{c}\right)^{\left(\frac{2}{\lambda\tau_D}\right)}\left(\frac{2m\lambda\hbar z}{p^2}\right)^{\frac{1}{\lambda\tau_D}}\cos z\\ &= -\frac{8\pi\hbar mv^2\tau_D^2}{\Sigma(E)^2}\int dz\frac{m\hbar}{p^2}\left(\frac{1}{c}\right)^{\left(\frac{2}{\lambda\tau_D}\right)}\left(\frac{2m\lambda\hbar z}{p^2}\right)^{\frac{1}{\lambda\tau_D}}\frac{\sin z}{z}.\end{aligned}$$
(10.23)

In the first line, we rewrote $e^{-T_{\min}(\epsilon)/\tau_D}$ as $(\epsilon/c)^{\frac{2}{\lambda\tau_D}}$, and in the second line, we approximated $\sin\epsilon \approx \epsilon$ and substituted $z = p^2\epsilon^2/(2m\lambda\hbar)$. Then, we perform a partial integration with respect to z, neglecting rapidly oscillating terms that are canceled by the k-average, introduced after (10.10). Eventually, we perform the z-integral by pushing the upper limit to infinity, that is, $\hbar \to 0$ and taking into account our assumption of large dwell times, that is, $\lambda\tau_D \to \infty$. Additionally, we assume $(2m\lambda\hbar/p^2)^{\frac{1}{\lambda\tau_D}} \approx 1$; we will return to the last point soon.

Finally, we arrive at the leading nondiagonal contribution to the quantum transmission [7],

$$|t_{nm}(k)|^2_L = -\frac{1}{(N_1+N_2)^2}.$$
(10.24)

10.3.3
Magnetic Field Dependence of the Nondiagonal Contribution

Until now, we assumed time reversal symmetry. If this symmetry is destroyed, for example, by applying a strong magnetic field, the latter contribution will vanish because the closed loop has to be traversed in different directions by the trajectory and its partner. Here, we study the transition region between zero and the finite magnetic field. In particular, we assume a homogeneous magnetic field B_z perpendicular to the sample that is assumed weak enough not to change the classical trajectories, but only the actions in the exponents. Since the closed loop is traversed in different directions by the two trajectories, we obtain an additional phase difference $(4\pi AB_z/\phi_0)$ between the two trajectories with the enclosed area A of the loop and the flux quantum $\phi_0 = (hc/e)$. Additionally, we need the distribution of enclosed areas for a trajectory with a closed loop of time T in chaotic systems, given by

$$P(A, T) = \frac{1}{\sqrt{2\pi T \beta}} \exp\left(-\frac{A^2}{2T\beta}\right) \tag{10.25}$$

with a system specific parameter β. A derivation of this formula can be found, for example, in [20]. Incorporating the phase difference and the area distribution in a modified $P(\epsilon, T)$ yields

$$\begin{aligned} P_B(\epsilon, T) &= \frac{2mv^2}{\Sigma(E)} \sin \epsilon \int_{T_{\min}(\epsilon)}^{T-T_{\min}(\epsilon)} dt_l \, (T - t_l - T_{\min}(\epsilon)) \\ &\quad \times \int_{-\infty}^{\infty} dA \, P(A, t_l - T_{\min}(\epsilon)) \cos \frac{4\pi A B_z}{\Phi_0} \\ &= \frac{2mv^2}{\Sigma(E)} \sin \epsilon \int_{T_{\min}(\epsilon)}^{T-T_{\min}(\epsilon)} dt_l \, (T - t_l - T_{\min}(\epsilon)) \, e^{-(t_l - T_{\min}(\epsilon))/t_B} \end{aligned} \tag{10.26}$$

with $t_B = \frac{\phi_0^2}{8\pi^2 \beta B_z^2}$. In the first line, we employed that paths leaving the crossing to form a closed loop enclose a negligible flux as long as they are correlated; for a more detailed analysis see Appendix D of [21]. Performing the T- and ϵ-integrals similar to the case without magnetic field yields [7]

$$|t_{nm}(k, B_z)|_L^2 = -\frac{1}{(N_1 + N_2)^2} \frac{1}{1 + \tau_D/t_B} . \tag{10.27}$$

We obtain an inverted Lorentzian with minimum at zero magnetic field, implying that the transmission through our sample increases with increasing magnetic field. This weak localization phenomenon, a precursor of strong localization, is visible as the reduction of the average quantum transmission in Figure 10.2.

10.3.4
Ehrenfest Time Dependence of the Nondiagonal Contribution

This semiclassical approach can also be applied in order to calculate the Ehrenfest time dependence of the transmission. The Ehrenfest time $\tau_E \equiv (1/\lambda) \ln(E/(\lambda \hbar))$, more generally a time proportional to $\ln \hbar$ [22], is the time a wave packet needs to reach a size such that it can no longer be described by a single classical particle. Therefore, the Ehrenfest time separates the evolution of wave packets, essentially following the classical dynamics (for example, up to a few bounces of the wave packet in Figure 10.1) from longer time scales dominated by wave interference (last panel in Figure 10.1). Based on field theoretical methods, Aleiner and Larkin showed that a minimal time is required for quantum effects in the transmission to appear, the Ehrenfest time [23]. We will now apply our semiclassical methods to determine the Ehrenfest time dependence of the transmission following the pioneering work [24] that has later been extended to the reflection [21, 25, 26], including a distinction between different Ehrenfest times. To this end, we directly start from (10.22). However, we have to be careful when evaluating the ϵ-integral. In our former calculation, we assumed $\mathrm{e}^{-\tau_\mathrm{E}/\tau_\mathrm{D}} = (\lambda\hbar/E)^{\frac{1}{\lambda\tau_\mathrm{D}}} \approx 1$, requiring $\tau_\mathrm{E} \ll \tau_\mathrm{D}$. However, by lifting this strong restriction for τ_D, but still keeping it large enough to fulfill our assumption of chaotic dynamics, we obtain

$$|t_{nm}(k,\tau_\mathrm{E})|^2_\mathrm{L} = -\frac{1}{(N_1+N_2)^2}\, \mathrm{e}^{-\tau_\mathrm{E}/\tau_\mathrm{D}}\,, \tag{10.28}$$

that is, an exponential suppression of the nondiagonal contribution due to the Ehrenfest time. This dependence has been confirmed in numerical simulations for quantum maps.

Following our introduction of semiclassical methods for the evaluation of nondiagonal contributions in configuration space, we will now turn to the generalization to phase space. This also provides an elegant way to compute higher-order corrections to the leading weak localization contribution presented above.

10.4
Quantum Transmission: Phase Space Approach

The above configuration space treatment, based on self-crossings, is restricted to systems with two degrees of freedom. More generally, for higher-dimensional dynamical systems, one cannot assume to find a one-to-one correspondence between partner orbits and crossings of an orbit [27, 28]. In order to overcome these difficulties, a phase space approach was developed for calculating the spectral form factor in spectral statistics in [27–29] involving periodic orbits. The next challenge was the generalization of this theory to trajectory pairs differing from one other at several places, solved again first for the spectral form factor [30] and generalized to the transport situation considered here in [31, 32], which serves as the basis of the following discussion.

In this section, we first explain the phase space approach and use it afterwards in the way developed in [30, 31] for the calculation of the quantum transmission also involving higher-order semiclassical diagrams.

10.4.1
Phase Space Approach

Contrary to the last section, we must initially replace the role of the reference orbit. Whereas before, we used the crossing orbit as a reference and then calculated the action difference and the crossing angle distribution in terms of the crossing angle ϵ, we now consider the orbit without the crossing that is close to itself in the region where the partner orbit crossed itself in the last section. Inside the region, where the orbit is close to itself – we will refer to it as encounter region and to the parts of the orbit inside as encounter stretches that are connected by so called links – we place a so called Poincaré surface of section $\mathcal{P}$ with its origin at $\tilde{x} = (\mathbf{x}, \mathbf{p})$. The section consists of all points $\tilde{x} + \delta\tilde{y} = (\mathbf{x} + \delta\mathbf{x}, \mathbf{p} + \delta\mathbf{p})$ in the same energy shell as the reference point with the $\delta\mathbf{x}$ perpendicular to the momentum $\mathbf{p}$ of the trajectory. For the two-dimensional systems considered here, $\mathcal{P}$ is a two-dimensional surface where every vector $\delta\tilde{y}$ can be expressed in terms of the stable direction $e_s(\tilde{x})$ and the unstable one $e_u(\tilde{x})$ [33]

$$\delta\tilde{y} = se_s(\tilde{x}) + ue_u(\tilde{x}). \tag{10.29}$$

The stable and unstable expressions refer to the following. We consider two orbits, one starting at $\tilde{x}$ and the other one at $\tilde{x} + \delta\tilde{y}_0$, then the difference between the stable coordinates will decrease exponentially for positive times and exponentially increase for negative time in the limit of long time T, unstable coordinates have opposite behavior. The functional form of the exponentials can be determined as $e^{\lambda T}$ and $e^{-\lambda T}$. Now, we can come back to the trajectory with the encounter region where we have put our Poincaré surface of section. The trajectory considered in the last section will pierce through the Poincaré section twice: we will consider one of the points as the origin of the section and the other piercing will take place at the distance (s, u). The coordinates of the piercing points of the partner trajectory are determined in the following way. The unstable coordinate of the partner trajectory has to be the same as the one of that part of the first trajectory that the second will follow for positive times. The stable coordinate is determined by the same requirement for negative times.

After this introduction into the determination of the (s, u)-coordinates, we are now ready to treat trajectory pairs that differ in encounters of arbitrary complexity. Following [30, 31] and using the notation introduced there we will allow the two trajectories to differ in several encounters involving an arbitrary number of stretches. In order to organize this structure, we introduce a vector $\vec{v} = (v_2, v_3, \ldots)$ with the component v_l determining the number of encounters with l stretches involved. The overall number of encounters during an orbit will be denoted by $V = \sum_{l=2}^{\infty} v_l$, the overall number of encounter stretches by $L = \sum_{l=2}^{\infty} lv_l$. In an encounter of l stretches, we will get $l - 1$ (s, u)-coordinates.

Now, we can proceed by replacing the former expressions for the minimal loop time, the action difference and the crossing angle distribution depending on the former small parameter ϵ by the corresponding expressions depending on the new small parameters (s, u).

We start with the minimal loop time, also known as the duration of the encounter. Shifting the Poincaré surface of section through our encounter, the stable components will asymptotically decrease and the unstable ones will increase for increasing time. We then claim that both components have to be smaller than a classical constant c. Its exact value will again, as in the last section, be unimportant for our final results. We finally obtain the encounter duration t_{enc} as the sum of the times t_u, that the trajectory needs from $\mathcal{P}$ till the point where the first unstable component reaches c, and the time t_s, that the trajectory needs from $\mathcal{P}$ till the point where the last stable component falls below c. Thus, we get

$$t_{\text{enc}} = t_s + t_u = \frac{1}{\lambda} \ln \frac{c^2}{\max_i \{|s_i|\} \max_j \{|u_j|\}}. \tag{10.30}$$

Now, we address the action difference between the two trajectories. By expressing the actions of the paired trajectories as the line integral of the momentum along the trajectory, we can expand [27, 28] one action around the other and express the result in terms of the (s, u)-coordinates. For an l-encounter, this yields[17]

$$\Delta S = \sum_{j=1}^{l-1} s_j u_j . \tag{10.31}$$

The action difference of a trajectory pair is then obtained by adding the differences resulting from all encounters.

Finally we come to the crossing angle distribution that will be replaced by a weight function for the stable and unstable coordinates for a trajectory of time T. We first notice that the uniformity of the trajectory distribution implies, in terms of our coordinates in $\mathcal{P}$, that a trajectory pierces through the section with the coordinates $(u, u + du)$ and $(s, s + ds)$ within the time interval $(t, t + dt)$ with the probability $1/\Sigma(E) ds du dt$. In general, we obtain for an l-encounter $(1/\Sigma(E))^{l-1} ds^{l-1} du^{l-1} dt^{l-1}$. By integrating the product of the latter quantities for all encounters over all possible durations of the $L-V$ intra-encounter links in a way that their durations are positive, yields our weight function for a fixed position of $\mathcal{P}$. To take into account all possible positions of $\mathcal{P}$, we also integrate over all possible positions, where it can be placed and divide by t_{enc} to avoid overcounting of equivalent positions. By taking all link times positive, we obtain for the weight function for an orbit of time T

$$\begin{aligned} w_T(\mathbf{s}, \mathbf{u}) &= \frac{1}{(\Sigma(E))^{L-V} \prod_{\alpha=1}^{V} t_{\text{enc}}^{\alpha}} \\ &\times \int_0^\infty dt_1 \ldots dt_L \, \Theta\left(T - \sum_{\alpha=1}^{V} l_\alpha t_{\text{enc}}^\alpha - \sum_{\alpha=1}^{L} t_\alpha\right) \\ &= \frac{\left(T - \sum_{\alpha=1}^{V} l_\alpha t_{\text{enc}}^\alpha\right)^L}{L! \, (\Sigma(E))^{L-V} \prod_{\alpha=1}^{V} t_{\text{enc}}^\alpha} . \end{aligned} \tag{10.32}$$

One additional problem arises when treating trajectory pairs differing not only in one 2-encounter, as in the last section. One can construct for one $\vec{v}$ different trajectory pairs, varying for example in the relative orientation, in which the encounter stretches are traversed. We will count this number by a function $N(\vec{v})$ and briefly describe later how it can be calculated.

10.4.2
Calculation of the Full Transmission

After the introduction into the phase space approach, we are now ready to calculate the transmission. Taking the weight function, the action difference and the number

---

[17] Strictly speaking [30], the (s, u) coordinates used here and in the following calculation are for encounters involving more than two stretches *not* the same as the described before, but related to them via a linear and volume preserving transformation.

of structures, we can transform the nondiagonal (ND) part of (10.10) into

$$|t_{nm}(k)|^2_{\text{ND}} = \frac{\pi\hbar}{2W_1 W_2} \sum_t \sum_{\vec{v}} A_t N(\vec{v})$$

$$\times \left\langle \int_{-c}^{c} \cdots \int_{-c}^{c} d^{L-V}u\, d^{L-V}s \exp\left(\frac{i}{\hbar}\Delta S\right) w_T(\mathbf{s}, \mathbf{u}) \right\rangle_{\Delta k} \quad (10.33)$$

with the average over a small k-window denoted by $\langle \ldots \rangle_{\Delta k}$. By inserting the formulas for the action difference, the weight function and using the classical sum rule with the modification of the survival probability, discussed in the last section, we can then transform the integral with respect to the length of the trajectory into one over the last link and obtain

$$|t_{nm}(k)|^2_{\text{ND}} = \frac{2\pi\hbar}{\Sigma(E)} \sum_{\vec{v}} N(\vec{v}) \left(\prod_{i=1}^{L+1} \int_0^{\infty} dt_i\, \exp\left(-\frac{t_i}{\tau_D}\right)\right)$$

$$\times \left\langle \int_{-c}^{c} \cdots \int_{-c}^{c} \frac{d^{L-V}u\, d^{L-V}s}{(\Sigma(E))^{L-V}} \right.$$

$$\left. \times \prod_{\alpha=1}^{V} \frac{\exp\left(-\frac{t^{\alpha}_{\text{enc}}}{\tau_D} + \frac{i}{\hbar}\sum_{j=1}^{l_{\alpha}-1} s_{\alpha j} u_{\alpha j}\right)}{t^{\alpha}_{\text{enc}}} \right\rangle_{\Delta k}$$

$$= \frac{1}{N_1 + N_2} \sum_{n=1}^{\infty} \left(\frac{1}{N_1 + N_2}\right)^n \sum_{\vec{v}}^{L-V=n} (-1)^V N(\vec{v}) \quad (10.34)$$

with the $L+1$ link times t_i. The (s, u)-integrals are calculated using the rule [30], that after expanding the exponential $e^{t_{\text{enc}}/\tau_D}$ into a Taylor series, only the t_{enc}-independent term contributes and yields in leading order in $\hbar$

$$\left\langle \frac{1}{\Sigma(E)} \int_{-c}^{c} \int_{-c}^{c} ds\, du \exp\left(\frac{isu}{\hbar}\right) \right\rangle_{\Delta k} \sim \frac{1}{T_H} \quad (10.35)$$

with the so called Heisenberg time $T_H = \Sigma(E)/(2\pi\hbar)$. For the sum with respect to $\vec{v}$, one can derive recursion relations, yielding [30]

$$\sum_{\vec{v}}^{L-V=n} (-1)^V N(\vec{v}) = \left(1 - \frac{2}{\beta}\right)^n \quad (10.36)$$

with $\beta = 1$ and $\beta = 2$ for the case with and without time reversal symmetry, respectively. Relations of this kind are derived by describing our trajectories by permutations expressing the connections inside and between encounters and considering the effect of shrinking one link in an arbitrarily complicated structure to zero.

We then obtain for $T(E_F)$, given in (10.1), in the case with time reversal symmetry [31, 32]

$$T(E_F)^{\beta=1} \approx \frac{N_1 N_2}{N_1 + N_2} + \frac{N_1 N_2}{N_1 + N_2} \sum_{n=1}^{\infty} \left(\frac{-1}{N_1 + N_2}\right)^n = \frac{N_1 N_2}{N_1 + N_2 + 1}$$

$$(10.37)$$

and in the case without time reversal symmetry

$$T(E_F)^{\beta=2} \approx \frac{N_1 N_2}{N_1 + N_2}, \qquad (10.38)$$

which agrees with the diagonal contribution already obtained in the previous section. Both results are in agreement with RMT predictions [34].

10.5
Semiclassical Research Paths: Present and Future

In the above sections, we outlined the recently developed semiclassical techniques for treating off-diagonal contributions for a paradigmatic example of coherent transport, weak localization. Nonlinear (hyperbolic) dynamics, as a prerequisite for the formation of correlated orbit pairs together with classical ergodicity, are at the core of the universality in the interference contribution to the ballistic conductance. Recently, this semiclassical approach to quantum transport has been extended along various paths, thereby gaining a more and more closed semiclassical framework of mesoscopic quantum effects.

In the presence of spin-orbit interaction, spin relaxation in confined ballistic systems and weak antilocalization, the *enhancement* of the conductance for systems obeying time-reversal symmetry (i.e. at zero magnetic field), has been semiclassically predicted [35] and generalized to other types of spin-orbit interaction [36] and higher-order contributions [37].

Ballistic conductance fluctuations, the analogue of the universal conductance fluctuations in the diffusive case, require the computation of semiclassical diagrams involving four trajectories. Again, RMT predictions could be confirmed [26]. Accordingly, RMT predictions for shot noise in Landauer transport agree with recent semiclassical results [38, 39]. This could also be shown for higher moments of the conductance in [40]. Even strong localization effects (in certain systems, i.e. chains of chaotic cavities) could be derived by making use of semiclassical loop-contributions [41].

Finally, beyond RMT, Ehrenfest time effects on transport properties have recently been in the spotlight. As a result, ballistic conductance fluctuations turn out to be, to leading order, τ_E-independent [25], contrary to weak localization which is suppressed with $e^{-\tau_E/\tau_D}$ [24], as we have shown. Recently these results were extended to the ac-conductance in [42] and to a semiclassical calculation of the Andreev gap in [43].

In addition to universal spectral statistics of closed systems and transport through open systems, the semiclassical techniques have been recently generalized to decay and photo fragmentation of complex systems [44, 45].

To summarize, by now a newly developed semiclassical machinery to compute systematically quantum coherence effects for quantities being composed of products of Green functions surely exists. This opens up various possible future research directions. With respect to system classes, mesoscopic transport theory has been predominantly focused on electron transport, leaving the conductance of ballistic hole systems aside, which experimentally plays an important role and for which a semiclassical theory remains to be developed. The situation is similar for a novel class of ballistic conductors: graphene based nanostructures, see Figure 10.2. More generally, a proper treatment of interaction effects in mesoscopic physics would probably represent the major challenge to future semiclassical theory.

To conclude, at mesoscopic scales where linear quantum evolution meets nonlinear classical dynamics, interesting phenomena will surely emerge in view of increasingly controllable future experiments.

References

1 Stöckmann, H.-J. (1999) *Quantum chaos-an introduction*, Cambridge Univ. Press, Cambridge.
2 Gutzwiller, M.C. (1990) *Chaos in Classical and Quantum Mechanics*, Springer, New York.
3 Miller, W.H. (1974) *Adv. Chem. Phys.*, **25**, 69.
4 Blümel, R. and Smilansky, U. (1988) *Phys. Rev. Lett.*, **60**, 477.
5 Jalabert, R.A., Baranger, H.U., and Stone, A.D. (1990) *Phys. Rev. Lett.*, **65**, 2442.
6 Baranger, H.U., Jalabert, R.A., and Stone, A.D. (1993) *Phys. Rev. Lett.*, **70**, 3876.
7 Richter, K. and Sieber, M. (2002) *Phys. Rev. Lett.*, **89**, 206801.
8 Bohigas, O., Giannoni, M.J., and Schmit, C. (1984) *Phys. Rev. Lett.*, **52**, 1.
9 Baranger, H.U., Jalabert, R.A., and Stone, A.D. (1993) *Chaos*, **3**, 665.
10 Richter, K. (2000) *Semiclassical Theory of Mesoscopic Quantum Systems*, Springer, Berlin.
11 Jalabert, R.A. (2000) The semiclassical tool in mesoscopic physics, in *Proceedings of the International School of Physics Enrico Fermi, New Directions in Quantum Chaos*, Course CXLIII, (eds G. Casati, I. Guarneri, and U. Smilansky), IOS Press, Amsterdam.
12 Datta, S. (1995) *Electronic Transport in Mesoscopic Systems*, Cambridge University Press, Cambridge.
13 Fisher, D.S. and Lee, P.A. (1981) *Phys. Rev. B*, **23**, 6851.
14 Wurm, J., Rycerz, A., Adagideli, I., Wimmer, M., Richter, K. and Baranger, H.U. (2009) *Phys. Rev. Lett.*, **102**, 056806.
15 Chakravarty, S. and Schmid, A. (1986) *Phys. Rep.*, **140**, 193.
16 Sieber, M. (1999) *J. Phys. A*, **32**, 7679.
17 Sieber, M. and Richter, K. (2001) *Phys. Scr.*, **T90**, 128.
18 Sieber, M. (2002) *J. Phys. A*, **35**, L613.
19 Lassl, A. (2003) *Semiklassik jenseits der Diagonalnäherung: Anwendung auf ballistische mesoskopische Systeme*, diploma thesis, Universität Regensburg.
20 Jensen, R.V. (1991) *Chaos*, **1**, 101.
21 Jacquod, Ph. and Whitney, R.S. (2006) *Phys. Rev. B*, **73**, 195115.
22 Chirikov, B.V., Izrailev, F.M., and Shepelyansky, D.L. (1981) *Sov. Sci. Rev. Sect. C*, **2**, 209.
23 Aleiner, I.L. and Larkin, A.I. (1996) *Phys. Rev. B*, **54**, 14423.
24 Adagideli, I. (2003) *Phys. Rev. B*, **68**, 233308.
25 Rahav, S. and Brouwer, P.W. (2006) *Phys. Rev. Lett.*, **96**, 196804.
26 Brouwer, P.W. and Rahav, S. (2006) *Phys. Rev. B*, **74**, 075322.
27 Turek, M. and Richter, K. (2003) *J. Phys. A*, **36**, L455.
28 Turek, M. (2004) *Semiclassics beyond the diagonal approximation*, PhD thesis, Universität Regensburg.
29 Spehner, D. (2003) *J. Phys. A*, **36**, 7269.
30 Müller, S., Heusler, S., Braun, P., Haake, F., and Altland, A. (2004) *Phys. Rev. Lett.*, **93**, 014103.
31 Heusler, S., Müller, S., Braun, P., and Haake, F. (2006) *Phys. Rev. Lett.*, **96**, 066804.
32 Müller, S., Heusler, S., Braun, P., and Haake, F. (2007) *New J. Phys.*, **9**, 12.

33 Gaspard, P. (1998) *Chaos, Scattering and Classical Mechanics*, Cambridge University Press, Cambridge.
34 Beenakker, C.W.J. (1997) *Rev. Mod. Phys.*, **69**, 731.
35 Zaitsev, O., Frustaglia, D., and Richter, K. (2005) *Phys. Rev. Lett.*, **94**, 026809.
36 Zaitsev, O., Frustaglia, D., and Richter, K. (2005) *Phys. Rev. B*, **72**, 155325.
37 Bolte, J. and Waltner, D. (2007) *Phys. Rev. B*, **76**, 075330.
38 Braun, P., Heusler, S., Müller, S., and Haake, F. (2006) *J. Phys. A*, **39**, L159.
39 Whitney, R.S. and Jacquod, Ph. (2006) *Phys. Rev. Lett.*, **96**, 206804.
40 Berkolaiko, G., Harrison, J.M., and Novaes M. (2008) *J. Phys. A*, **41**, 365102.
41 Brouwer, P.W. and Altland, A. (2008) *Phys. Rev. B*, **78**, 075304.
42 Petitjean, C., Waltner, D., Kuipers, J., Adagideli, I., and Richter, K. (2009) *Phys. Rev. B*, **80**, 115310.
43 Kuipers, J., Waltner, D., Petitjean, C., Berkolaiko, G., and Richter, K. (2009) preprint, arXiv:0907.2660.
44 Waltner, D., Gutiérrez, M., Goussev, A., and Richter, K. (2008) *Phys. Rev. Lett.*, **101**, 174101.
45 Gutiérrez, M., Waltner, D., Kuipers, J., and Richter, K. (2009) *Phys. Rev. E*, **79**, 046212.

11
Nonlinear Response of Driven Mesoscopic Conductors
Franz J. Kaiser and Sigmund Kohler

11.1
Introduction

Time-dependent systems represent a frequently studied class of models in the field of nonlinear dynamics. Of foremost interest is the interplay of an external time-dependent ac driving force with the nonlinearities of the system, which generally leads to frequency mixing and is visible in the dc response of the system. Tight-binding models are systems with intrinsic nonlinearities and when coupled to electron reservoirs describe mesoscopic conductors, such as coherently coupled quantum dots [1–4] and molecular wires [5–8]. In experiments, the physical realization of time-dependent fields acting on these systems range from oscillating gate voltages to microwaves and infrared lasers. Excitations with relatively small amplitudes can already induce electron excitations that lead to an enhanced transport via photon-assisted tunneling [3, 9–14]. This phenomenon is typically associated with reduced shot noise and can be described as the linear response of the conductor [14, 15].

When increasing the driving strength such that one leaves the linear regime, two interesting effects have been predicted. First, in spatially asymmetric systems, the nonlinear response of the conductor can comprise pump currents, that is, dc currents even in the absence of any net source-drain voltage [16–24]. In the presence of static magnetic fields, the pump mechanism may favor electrons with a particular spin, so that one observes spin pumping [25–29] and spin filtering [30]. In the non-adiabatic regime, it is possible to tune the pump into regimes with clearly sub-Poissonian noise characteristics [31, 32]. A second phenomenon in driven conductors that is, in a sense, opposite to pumping is current suppression by proper time-dependent fields even in the presence of an applied source-drain voltage [33, 34]. At the same time, the zero frequency noise may be suppressed as well [35–37].

When the electron–electron interaction sets the dominating energy scale of the problem, the transport through quantum dots is governed by resonant tunneling. Then, for a given source-drain and gate voltage, only dot states with particular electron numbers play a role, while others suffer from Coulomb blockade [38]. Con-

Nonlinear Dynamics of Nanosystems. Edited by Günter Radons, Benno Rumpf, and Heinz Georg Schuster
Copyright © 2010 WILEY-VCH Verlag GmbH & Co. KGaA, Weinheim
ISBN: 978-3-527-40791-0

sequently, the current-voltage characteristics show a step-like behavior, and the current as a function of both a bias and a gate voltage exhibits the characteristic Coulomb diamond structure. The main aim of the present work is to study how Coulomb blockade and time-dependent fields influence transport.

A proper theoretical tool for the description of transport in the Coulomb blockade regime is provided by master equations for the reduced density operator of the wire [39–42]. For time-dependent conductors, this enables a rather efficient treatment of transport problems after decomposing the wire density operator into a Floquet basis. Then it is possible to study the response of relatively large driven conductors [14], and to also include electron–electron interactions [43, 44] and electron-phonon interactions [45]. We nevertheless restrict ourselves to two-level systems here because their behavior is sufficiently elementary for a qualitative understanding. For the computation of current fluctuations, one can employ a generalized master equation that resolves the number of the transported electrons. This degree of freedom is traced out after introducing a counting variable [46]. For various static transport problems, this approach has been followed by several groups [47–53]. A Floquet master equation approach for the description of shot noise in driven conductors has recently been developed [54].

In Section 11.2 a model for the description of quantum transport and the central quantities of interest are introduced. We then review in Section 11.3 the Floquet master equation approach derived in [54]. In Section 11.4, we investigate a double quantum dot in the Coulomb blockade regime under the influence of ac driving with this formalism.

11.2
Wire-Lead Model and Current Noise

A frequently used model for nanoscale conductors such as molecular wires or coupled quantum dots is sketched in Figure 11.1. It is described by the time-dependent Hamiltonian

$$H(t) = H_{\text{wire}}(t) + H_{\text{leads}} + H_{\text{contacts}} , \qquad (11.1)$$

where the different terms correspond to the central conductor ("wire"), electron reservoirs ("leads"), and the wire-lead couplings ("contacts"), respectively. We focus on the regime of coherent quantum transport where the main physics at work occurs on the wire itself. In doing so, we neglect other possible influences originating from driving-induced hot electrons in the leads and dissipation on the wire. Then, the wire Hamiltonian reads, in a tight-binding approximation with N orbitals $|n\rangle$,

$$H_{\text{wire}}(t) = \sum_{n,n',s,s'} H_{nn'}(t) c_{ns}^\dagger c_{n's'} + H_{\text{interaction}} . \qquad (11.2)$$

For a molecular wire, this constitutes the Hückel description where each site corresponds to one atom. The fermion operators c_{ns}, $c_{ns}^\dagger$ annihilate and create, respectively, an electron with spin $s = \uparrow, \downarrow$ in the orbital $|n\rangle$. The influence of an applied

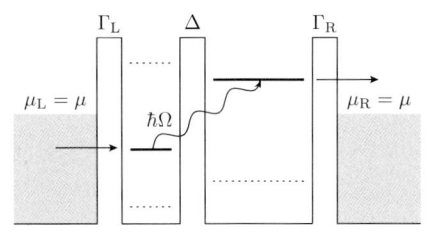

Figure 11.1 Level structure of a double quantum dot with $N = 2$ orbitals. The terminating sites are coupled to leads with chemical potentials μ_L and $\mu_R = \mu_L + eV$, respectively.

ac field or an oscillating gate voltage with frequency $\Omega = 2\pi/T$ results in a periodic time dependence of the wire Hamiltonian, namely $H_{nn'}(t+T) = H_{nn'}(t)$.

For the interaction Hamiltonian, we assume a capacitor model, so that

$$H_{\text{interaction}} = \frac{U}{2} N_{\text{wire}}(N_{\text{wire}} - 1) , \qquad (11.3)$$

where $N_{\text{wire}} = \sum_{ns} c_{ns}^\dagger c_{ns}$ describes the number of electrons on the wire. Below we focus on the limit of strong interaction, $U \to \infty$, which finally means that the Coulomb repulsion is so strong that only states with zero or one excess electron play a role.

The leads are modeled by ideal electron gases,

$$H_{\text{leads}} = \sum_{q,s} \epsilon_q \left(c_{Lqs}^\dagger c_{Lqs} + c_{Rqs}^\dagger c_{Rqs} \right) , \qquad (11.4)$$

where $c_{Lq}^\dagger$ ($c_{Rq}^\dagger$) creates an electron in the state $|Lq\rangle$ ($|Rq\rangle$) in the left (right) lead. The wire-lead tunneling Hamiltonian

$$H_{\text{contacts}} = \sum_{q,s} \left(V_{Lqs} c_{Lqs}^\dagger c_{1s} + V_{Rqs} c_{Rqs}^\dagger c_{Ns} \right) + \text{h.c.} \qquad (11.5)$$

establishes the contact between the sites $|1\rangle$, $|N\rangle$, and the respective lead. This tunneling coupling is described by the spectral density

$$\Gamma_\ell(\epsilon) = 2\pi \sum_q |V_{\ell q}|^2 \delta(\epsilon - \epsilon_q) \qquad (11.6)$$

of lead $\ell = L, R$. In the following, we restrict ourselves to the wide band limit in which the spectral density is assumed to be energy independent, $\Gamma_\ell(\epsilon) \to \Gamma_\ell$.

To fully specify the dynamics, we choose as an initial condition for the left/right lead a grand canonical electron ensemble at temperature T and electrochemical potential $\mu_{L/R}$. Thus, the initial density matrix reads

$$\rho_0 \propto e^{-(H_{\text{leads}} - \mu_L N_L - \mu_R N_R)/k_B T} , \qquad (11.7)$$

where $N_\ell = \sum_{qs} c^\dagger_{\ell qs} c_{\ell qs}$ is the number of electrons in lead ℓ and $k_B T$ denotes the Boltzmann constant times temperature. An applied voltage V maps to a chemical potential difference $\mu_R - \mu_L = eV$ with $-e$ being the electron charge. Then, at initial time t_0, the only nontrivial expectation values of the wire operators read $\langle c^\dagger_{\ell' q' s'} c_{\ell q s} \rangle = f_\ell(\epsilon_q) \delta_{\ell \ell'} \delta_{qq'} \delta_{ss'}$, where $f_\ell(\epsilon) = (1 + \exp[(\epsilon - \mu_\ell)/k_B T])^{-1}$ denotes the Fermi function.

11.2.1
Charge, Current, and Current Fluctuations

To avoid the explicit appearance of commutators in the definition of correlation functions, we perform the derivation of the central transport quantities in the Heisenberg picture. As a starting point we choose the operator

$$Q_\ell(t) = e N_\ell(t) - e N_\ell(t_0), \quad (11.8)$$

which describes the charge accumulated in lead ℓ with respect to the initial state. Due to total charge conservation, Q_ℓ equals the net charge transmitted across the contact ℓ, and its time derivative defines the corresponding current

$$I_\ell(t) = \frac{d}{dt} Q_\ell(t). \quad (11.9)$$

The current noise is described by the symmetrized correlation function

$$S_\ell(t, t') = \frac{1}{2} \langle [\Delta I_\ell(t), \Delta I_\ell(t')]_+ \rangle \quad (11.10)$$

of the current fluctuation operator $\Delta I_\ell(t) = I_\ell(t) - \langle I_\ell(t) \rangle$, where the anticommutator $[A, B]_+ = AB + BA$ ensures hermiticity. At long times, $S_\ell(t, t') = S_\ell(t + T, t' + T)$ shares the time periodicity of the driving [55]. Therefore, it is possible to characterize the noise level by the zero frequency component of $S_\ell(t, t-\tau)$ averaged over the driving period,

$$\bar{S}_\ell = \frac{1}{T} \int_0^T dt \int_{-\infty}^\infty d\tau \, S_\ell(t, t - \tau). \quad (11.11)$$

Moreover, for two-terminal devices $\bar{S}_\ell$ is independent of the contact ℓ, that is, $\bar{S}_L = \bar{S}_R \equiv \bar{S}$.

The evaluation of the zero frequency noise $\bar{S}$ directly from the definition given in (11.11) can be tedious due to the explicit appearance of both t and $t - \tau$. This inconvenience can be circumvented by employing the relation

$$\frac{d}{dt} (\langle Q_\ell^2(t) \rangle - \langle Q_\ell(t) \rangle^2) = 2 \int_0^\infty d\tau \, S_\ell(t, t - \tau), \quad (11.12)$$

which follows from the integral representation of (11.8) and (11.9), $Q_\ell(t) = \int_{t_0}^t dt' I_\ell(t')$, in the limit $t_0 \to -\infty$. By averaging (11.12) over the driving period and using $S(t, t - \tau) = S(t - \tau, t)$, we obtain

$$\bar{S} = \left\langle \frac{d}{dt} \langle \Delta Q_\ell^2(t) \rangle \right\rangle_t, \quad (11.13)$$

where $\Delta Q_\ell = Q_\ell - \langle Q_\ell \rangle$ denotes the charge fluctuation operator and $\langle \ldots \rangle_t$ the time average. The fact that the time average can be evaluated from the limit $\bar{S} = \lim_{t_0 \to -\infty} \langle \Delta Q_\ell^2(t) \rangle / (t - t_0) > 0$ allows the interpretation of the zero frequency noise as the "charge diffusion coefficient". As a dimensionless measure for the *relative* noise strength, we employ the Fano factor [56]

$$F = \frac{\bar{S}}{e|\bar{I}|}, \tag{11.14}$$

which can provide information about the nature of the transport mechanism [57, 58]. Here, $\bar{I}$ denotes the time average of the current expectation value $\langle I_\ell(t) \rangle$. Historically, the zero frequency noise (11.11) contains a factor of 2, that is, $\bar{S}' = 2\bar{S}$, resulting from a different definition of the Fourier transform. In this case the Fano factor is defined as $F = \bar{S}'/2e|\bar{I}|$.

11.2.2
Full Counting Statistics

A more complete picture of the current fluctuations beyond second order correlations is provided by the full counting statistics. It is determined by the moment generating function

$$\phi(\chi, t) = \langle e^{i\chi N_L} \rangle_t, \tag{11.15}$$

and allows the direct computation of the kth moment of the charge in the left lead via the relation

$$\langle Q_L^k(t) \rangle = e^k \frac{\partial^k}{\partial (i\chi)^k} \phi(\chi, t) \Big|_{\chi=0}. \tag{11.16}$$

Subtracting from the moments the trivial contributions that depend on a shift of the initial values, one obtains the cumulants. They are defined and generated via the cumulant generating function $\ln \phi(\chi, t)$ which replaces ϕ in (11.16) [59], so that the kth cumulant reads

$$C_k = e^k \frac{\partial^k}{\partial (i\chi)^k} \ln \phi(\chi, t) \Big|_{\chi=0}. \tag{11.17}$$

In a continuum limit for the leads, both the moments and the cumulants diverge as a function of time, and one focuses on the rates at which these quantities change in the long time limit. This establishes the relations between the first two cumulants and $I(t)$ and $S(t)$, namely

$$I(t) = -i e \frac{\partial}{\partial \chi} \dot{C}(\chi, t) \Big|_{\chi=0}, \tag{11.18}$$

$$S(t) = -e^2 \frac{\partial^2}{\partial \chi^2} \dot{C}(\chi, t) \Big|_{\chi=0}. \tag{11.19}$$

For driven systems, these quantities are time-dependent even in the asymptotic limit and thus, we characterize the transport by the corresponding averages over one driving period. Then expressions (11.18) and (11.19) become identical to the previously defined time averages $\bar{I}$ and $\bar{S}$, respectively. Herein we restrict ourselves to the computation of the first and the second cumulant, despite the fact that higher order cumulants can also be measured [60, 61].

11.3
Master Equation Approach

In the presence of electron–electron interactions, an exact treatment of the electron transport within a scattering theory is no longer possible, and a master equation formalism can be an appropriate tool for the computation of currents [12, 14, 39, 44, 45]. Recently, master equations have been established for the computation of current noise of various static conductors as well [46–53]. In the following, we develop such an approach for the case of periodically time-dependent conductors.

11.3.1
Perturbation Theory and Reduced Density Operator

We start our derivation of a master equation formalism from the Liouville–von Neumann equation $i\hbar \dot{\mathcal{R}}(t) = [H(t), \mathcal{R}(t)]$ for the total density operator $\mathcal{R}(t)$. By standard techniques we obtain the exact equation of motion

$$\frac{d}{dt}\widetilde{\mathcal{R}}(t) = -\frac{i}{\hbar}\left[\widetilde{H}_{\text{contacts}}(t), \mathcal{R}(0)\right]$$
$$- \frac{1}{\hbar^2}\int_0^\infty d\tau \left[H_{\text{contacts}}, \left[\widetilde{H}_{\text{contacts}}(t-\tau, t), \mathcal{R}(t)\right]\right], \quad (11.20)$$

where the tilde denotes the interaction picture with respect to the lead and the wire Hamiltonian, $\tilde{X}(t, t') = U_0^\dagger(t, t') X U_0(t, t')$, and U_0 is the propagator without the coupling. Below we will employ Floquet theory in order to obtain explicit expressions for these operators.

As already discussed above, the moment generating function $\phi(\chi) = \langle \exp(i\chi N_\text{L}) \rangle$ contains the full information about the counting statistics. For its explicit computation, we define in the Hilbert space of the wire the operator

$$\mathcal{F}(\chi, t) = \text{tr}_{\text{leads}}\left\{e^{i\chi N_\text{L}}\mathcal{R}(t)\right\}, \quad (11.21)$$

whose limit $\chi \to 0$ is obviously the reduced density operator of the wire, $\mathcal{F}(0, t) = \rho(t)$. After tracing out the wire degrees of freedom, $\mathcal{F}$ becomes the moment generating function $\phi(\chi) = \text{tr}_{\text{wire}}\mathcal{F}$. It will prove convenient to decompose $\mathcal{F}$ into a Taylor series,

$$\mathcal{F} = \rho + \sum_{k=1}^\infty \frac{(i\chi)^k}{k!}\mathcal{F}_k, \quad (11.22)$$

where the coefficients $\mathcal{F}_k = \text{tr}_{\text{leads}}(N_L^k \mathcal{R})$ provide direct access to the moments $\langle N_L^k \rangle = \text{tr}_{\text{wire}} \mathcal{F}_k$.

Our strategy is now to derive an equation of motion for the coefficients $\mathcal{F}_k$ from the master equation for the full density operator (11.20). For that purpose, we transform the master equation for $\tilde{\mathcal{R}}$ back to the Schrödinger picture and multiply it from the left by the operator $\exp(i\chi N_L)$. By tracing out the lead degrees of freedom and using the commutation relations $[N_L, V] = V$ and $[N_L, V^\dagger] = -V^\dagger$, where $V = \sum_{q,s} V_{Lqs} c_{Lqs}^\dagger c_{1s}$ we obtain

$$\frac{d}{dt}\mathcal{F}(\chi, t) = \{\mathcal{L} + (e^{i\chi} - 1)\mathcal{J}_+ + (e^{-i\chi} - 1)\mathcal{J}_-\}\mathcal{F}(\chi, t). \quad (11.23)$$

In order to achieve this compact notation, we have defined the superoperators $\mathcal{J}_\pm$ and the time-dependent Liouville operator

$$\mathcal{L}(t)X = -\frac{i}{\hbar}[H_{\text{wire}}(t), X]$$

$$+ \frac{\Gamma_L}{2\pi} \int_0^\infty d\tau \int d\epsilon \left[e^{i\epsilon\tau} \left(-c_1 \tilde{c}_1^\dagger X f_L(\epsilon) + \tilde{c}_1^\dagger X c_1 f_L(\epsilon) \right. \right.$$

$$\left. - X \tilde{c}_1^\dagger c_1 \bar{f}_L(\epsilon) + c_1 X \tilde{c}_1^\dagger \bar{f}_L(\epsilon) \right)$$

$$+ e^{-i\epsilon\tau} \left(-X \tilde{c}_1 c_1^\dagger f_L(\epsilon) + c_1^\dagger X \tilde{c}_1 f_L(\epsilon) \right.$$

$$\left. \left. - c_1^\dagger \tilde{c}_1 X \bar{f}_L(\epsilon) + \tilde{c}_1 X c_1^\dagger \bar{f}_L(\epsilon) \right) \right]$$

$$+ \text{same terms with the replacement } 1, L \to N, R, \quad (11.24)$$

which also determines the time evolution of the reduced density operator, $\dot{\rho} = \mathcal{L}(t)\rho$. The tilde denotes the interaction picture operator $\tilde{c} = \tilde{c}(t, t-\tau)$ and f_ℓ the Fermi function of lead ℓ, while $\bar{f}_\ell = 1 - f_\ell$. The current operators

$$\mathcal{J}_+(t)X = \frac{\Gamma_L}{2\pi} \int_0^\infty d\tau \int d\epsilon \left(e^{i\epsilon\tau} \tilde{c}_1^\dagger X c_1 + e^{-i\epsilon\tau} c_1^\dagger X \tilde{c}_1 \right) f_L(\epsilon), \quad (11.25)$$

$$\mathcal{J}_-(t)X = \frac{\Gamma_L}{2\pi} \int_0^\infty d\tau \int d\epsilon \left(e^{i\epsilon\tau} c_1 X \tilde{c}_1^\dagger + e^{-i\epsilon\tau} \tilde{c}_1 X c_1^\dagger \right) \bar{f}_L(\epsilon), \quad (11.26)$$

describe the tunneling of an electron from the left lead to the wire and the opposite process, respectively. Note that these superoperators still contain a nontrivial time dependence stemming from the interaction picture representation of the creation and annihilation operators of wire electrons.

11.3.2
Computation of Moments and Cumulants

For the computation of the current given in (11.18) and the zero frequency noise given in (11.19), we generalize the approach of [49] to the time-dependent case. Since we restrict the noise characterization to the Fano factor, it is sufficient to compute the long time behavior of the first and the second moment of the electron number in the left lead. This information is fully contained in the time derivative of the operator $\mathcal{F}$ up to second order in χ, for which we obtain the hierarchy

$$\dot{\rho} = \mathcal{L}(t)\rho ,\tag{11.27}$$

$$\dot{\mathcal{F}}_1 = \mathcal{L}(t)\mathcal{F}_1 + (\mathcal{J}_+(t) - \mathcal{J}_-(t))\,\rho ,\tag{11.28}$$

$$\dot{\mathcal{F}}_2 = \mathcal{L}(t)\mathcal{F}_2 + 2\,(\mathcal{J}_+(t) - \mathcal{J}_-(t))\,\mathcal{F}_1 + (\mathcal{J}_+(t) + \mathcal{J}_-(t))\,\rho \tag{11.29}$$

by Taylor expansion of the equation of motion (11.23). The first equation determines the time evolution of the reduced density operator, which in the long time limit becomes the stationary solution $\rho_0(t)$. Note that for a driven system it generally is time dependent. Replacing ρ by ρ_0 in (11.28) and using the fact that $\mathrm{tr}_{\mathrm{wire}}\mathcal{L}X = 0$ for any operator X, we obtain the stationary current

$$I(t) = -e\,\mathrm{tr}_{\mathrm{wire}}\dot{\mathcal{F}}_1 = -e\,\mathrm{tr}_{\mathrm{wire}}(\mathcal{J}_+ - \mathcal{J}_-)\rho_0(t) .\tag{11.30}$$

The dc current follows simply by averaging over one driving period and the result is the current formula of [62].

The computation of $\mathcal{F}_1(t)$ is hindered by the fact that the inverse of a Liouvillian generally does not exist. For static systems this is obvious from the fact that the stationary solution fulfills $\mathcal{L}\rho_0 = 0$, which implies that $\mathcal{L}$ is singular. This unfortunately also complicates the computation of the second cumulant, and we proceed in the following way. We start from (11.12) which relates the zero frequency noise to the charge fluctuation in the leads, and write the time derivative of the first and the second moment of the electron number in the left lead by the operators $\dot{\mathcal{F}}_{1,2}$. From the equations of motion (11.28) and (11.29), we then find $S = e^2\mathrm{tr}_{\mathrm{wire}}\{2(\mathcal{J}_+ - \mathcal{J}_- - I)\mathcal{F}_1 + (\mathcal{J}_+ + \mathcal{J}_-)\rho\}$, where we again used the relation $\mathrm{tr}_{\mathrm{wire}}\mathcal{L}X = 0$. An important observation is now that the first part of this expression vanishes for $\mathcal{F}_1 \propto \rho_0$, which can easily be demonstrated by inserting the current expectation value of (11.30). Since $\rho_0\mathrm{tr}_{\mathrm{wire}}$ acts as a projector onto the stationary solution ρ_0, we can define the "perpendicular" part

$$\mathcal{F}_{1\perp} = \mathcal{F}_1 - \rho_0\mathrm{tr}_{\mathrm{wire}}\mathcal{F}_1 ,\tag{11.31}$$

which fulfills the relation $\mathrm{tr}_{\mathrm{wire}}\mathcal{F}_{1\perp} = 0$ and obeys the equation of motion

$$\dot{\mathcal{F}}_{1\perp} = \mathcal{L}(t)\mathcal{F}_{1\perp} + (\mathcal{J}_+(t) - \mathcal{J}_-(t) - I(t))\,\rho_0(t) .\tag{11.32}$$

We will see below that in contrast to $\mathcal{F}_1$, the long time limit of the traceless $\mathcal{F}_{1\perp}$ can be computed directly from the equation of motion (11.32). Upon inserting (11.31) into the equation of motion (11.29), we finally obtain the expression for the still time-dependent "charge diffusion coefficient"

$$S(t) = e^2\mathrm{tr}_{\mathrm{wire}}\{2(\mathcal{J}_+ - \mathcal{J}_-)\mathcal{F}_{1\perp} + (\mathcal{J}_+ + \mathcal{J}_-)\rho_0\} ,\tag{11.33}$$

whose time average finally provides the Fano factor $F = \bar{S}/e\bar{I}$.

11.3.3
Floquet Decomposition

The remaining task is now to compute the stationary solutions $\rho_0(t)$ and $\mathcal{F}_{1\perp}(t)$ from the time-dependent equations of motion (11.27) and (11.28). As for the computation of the dc current in our previous work [62], we solve this problem within a Floquet treatment of the isolated wire, which provides a convenient representation of the electron creation and annihilation operators.

11.3.3.1 Fermionic Floquet Operators

In the driven wire Hamiltonian (11.2), the single particle contribution commutes with the interaction term, and thus these two Hamiltonians possess a complete set of common many particle eigenstates. Here we start by diagonalizing the first part of the Hamiltonian, which describes the single particle dynamics determined by the time-periodic matrix elements $H_{nn'}(t)$. According to the Floquet theorem, the corresponding single particle Schrödinger equation possesses a complete solution of the form

$$|\Psi_\alpha(t)\rangle = e^{-i\epsilon_\alpha t/\hbar}|\varphi_\alpha(t)\rangle, \tag{11.34}$$

with the quasienergies ϵ_α and the $\mathcal{T}$-periodic Floquet states

$$|\varphi_\alpha(t)\rangle = \sum_k e^{-ik\Omega t}|\varphi_{\alpha,k}\rangle. \tag{11.35}$$

The Floquet states and the quasienergies are obtained by solving the eigenvalue problem

$$\left(\sum_{n,n'} |n\rangle H_{nn'}(t)\langle n'| - i\hbar\frac{d}{dt}\right)|\varphi_\alpha(t)\rangle = \epsilon_\alpha|\varphi_\alpha(t)\rangle, \tag{11.36}$$

whose solution allows one to construct via Slater determinants many particle Floquet states. In analogy to the quasimomenta in Bloch theory for spatially periodic potentials, the quasienergies ϵ_α come in classes

$$\epsilon_{\alpha,k} = \epsilon_\alpha + k\hbar\Omega, \quad k \in \mathbb{Z} \tag{11.37}$$

of which all members represent the same physical solution of the Schrödinger equation. Thus, we can restrict ourselves to states within one Brillouin zone as, for example, $0 \leq \epsilon_\alpha < \hbar\Omega$.

For the numerical computation of the operators ρ_0 and $\mathcal{F}_{1\perp}$, it is essential to have an explicit expression for the interaction picture representation of the wire operators. It can be obtained from the fermionic Floquet creation and annihilation operators [14] defined via the transformation

$$c_{\alpha s}(t) = \sum_n \langle\varphi_\alpha(t)|n\rangle c_{ns}. \tag{11.38}$$

The inverse transformation

$$c_{ns} = \sum_{\alpha} \langle n|\varphi_\alpha(t)\rangle c_{\alpha s}(t) \qquad (11.39)$$

follows from the mutual orthogonality and the completeness of the Floquet states at equal times [63]. Note that the right-hand side of (11.39) becomes time independent after the summation. The Floquet annihilation operator (11.38) has the interaction picture representation

$$\tilde{c}_{\alpha s}(t, t') = U_0^\dagger(t, t') c_{\alpha s}(t) U_0(t, t') \qquad (11.40)$$

$$= e^{-i(\epsilon_\alpha + U N_{\text{wire}})(t-t')/\hbar} c_{\alpha s}(t'), \qquad (11.41)$$

with the important feature that the time difference $t - t'$ enters only via the exponential prefactor. This allows us to evaluate the τ-integration of the master equation (11.27) after a Floquet decomposition. Relation (11.41) can easily be shown by computing the time derivative with respect to t, which by use of the Floquet equation (11.36) becomes

$$\frac{d}{dt}\tilde{c}_{\alpha s}(t, t') = -\frac{i}{\hbar}(\epsilon_\alpha + U N_{\text{wire}})\tilde{c}_{\alpha s}(t, t'). \qquad (11.42)$$

From this equation on the initial condition $\tilde{c}_\alpha(t', t') = c_\alpha(t')$ follows relation (11.41). Note that the time evolution induced by $\mathcal{H}_{\text{wire}}(t)$ conserves the number of electrons on the wire.

11.3.3.2 Master Equation and Current Formula

In order to make use of the Floquet ansatz, we decompose the master equation (11.27) and the current formula (11.30) into the Floquet basis derived in the last subsection. For that purpose, we use the fact that we are finally interested in the current at asymptotically large times in the limit of large interaction U. The latter has the consequence that only wire states with at most one excess electron play a role, so that the stationary density operator $\rho_0(t)$ can be decomposed into the $2N + 1$ dimensional basis $\{|0\rangle, c_{\alpha s}^\dagger(t)|0\rangle\}$, where $|0\rangle$ denotes the wire state in the absence of excess electrons and $s = \uparrow, \downarrow$. Moreover, it can be shown that at large times the density operator becomes diagonal in the electron number N_{wire}, so that a proper ansatz reads

$$\rho_0(t) = |0\rangle \rho_{00}(t)\langle 0| + \sum_{\alpha,\beta,s,s'} c_{\alpha s}^\dagger |0\rangle \rho_{\alpha s, \beta s'}(t)\langle 0| c_{\beta s'}. \qquad (11.43)$$

Note that we keep terms with $\alpha \neq \beta$, which means that we work beyond a rotating wave approximation. Indeed, in a nonequilibrium situation the off-diagonal density matrix elements $\rho_{\alpha\beta}$ will not vanish, and neglecting them might lead to artefacts [14, 64].

By inserting the decomposition shown in (11.43) into the master equation (11.27), we obtain an equation of motion for the matrix elements $\rho_{\alpha s,\beta s'} = \langle 0|c_{\alpha s}\rho_{\text{wire}}c^\dagger_{\beta s'}|0\rangle$. We evaluate the trace over the lead states and compute the matrix element $\langle 0|c_{\alpha s}(t)\ldots c^\dagger_{\beta s'}(t)|0\rangle$. Thereby, we neglect the two particle terms which are of the structure $c^\dagger_{\alpha s}c^\dagger_{\beta s'}|0\rangle\langle 0|c_{\beta s}c_{\alpha s}$. Formally, these terms drop out in the limit of strong Coulomb repulsion because they are accompanied by a rapidly oscillating phase factor $\exp(-iUN_{\text{wire}}\tau/\hbar)$. Then the τ-integration results in a factor $f_L(\epsilon_{\alpha,k} + U)$ which vanishes in the limit of large U. Since the total Hamiltonian of (11.1) is diagonal in the spin index s, we find that the density matrix elements $\rho_{\alpha s,\beta s'}$ are spin-independent as well, so that after a transient stage

$$\rho_{\alpha\uparrow,\beta\uparrow}(t) = \rho_{\alpha\downarrow,\beta\downarrow}(t) \equiv \rho_{\alpha\beta}(t) \tag{11.44}$$

and $\rho_{\alpha\uparrow,\beta\downarrow} = 0$. Moreover, the stationary density operator (11.43) obeys the time periodicity of the driving field [14], and thus can be decomposed into the Fourier series

$$\rho_{\alpha\beta}(t) = \sum_k e^{-ik\Omega t}\rho_{\alpha\beta,k} \tag{11.45}$$

and $\rho_{00}(t)$ accordingly. After some algebra, we arrive at a set of N^2 coupled equations of motion for $\rho_{\alpha\beta}(t)$, which in Fourier representation read

$$i(\epsilon_\alpha - \epsilon_\beta - k\hbar\Omega)\rho_{\alpha\beta,k}$$
$$= \frac{\Gamma_L}{2}\sum_{k',k''}\langle\varphi_{\alpha,k'+k''}|1\rangle\langle 1|\varphi_{\beta,k+k''}\rangle\rho_{00,k'}\left(f_L(\epsilon_{\alpha,k'+k''}) + f_L(\epsilon_{\beta,k+k''})\right)$$
$$- \frac{\Gamma_L}{2}\sum_{\alpha',k',k''}\langle\varphi_{\alpha,k'+k''}|1\rangle\langle 1|\varphi_{\alpha',k+k''}\rangle\rho_{\alpha'\beta,k'}\,\bar{f}(\epsilon_{\alpha',k+k''})$$
$$- \frac{\Gamma_L}{2}\sum_{\beta',k',k''}\langle\varphi_{\beta',k'+k''}|1\rangle\langle 1|\varphi_{\beta,k+k''}\rangle\rho_{\alpha\beta',k'}\,\bar{f}(\epsilon_{\beta',k'+k''})$$
$$+ \text{same terms with the replacement } 1, L \to N, R\ . \tag{11.46}$$

In order to solve these equations we have to eliminate $\rho_{00,k}$, which is most conveniently done by inserting the Fourier representation of the normalization condition

$$\text{tr}_{\text{wire}}\rho_0(t) = \rho_{00}(t) + 2\sum_\alpha \rho_{\alpha\alpha}(t) = 1\ . \tag{11.47}$$

In order to obtain an expression for the current that is consistent with the restriction to one excess electron, we compute the expectation values in the current formula of (11.30) with the reduced density operator given in (11.43), and insert the Floquet representation of the wire operators from (11.39). Performing an average over the driving period, we obtain for the dc current the expression [62]

$$\bar{I} = \frac{2e\Gamma_L}{\hbar} \operatorname{Re} \sum_{\alpha,k} \left(\sum_{\beta,k'} \langle \varphi_{\beta,k'+k} | 1 \rangle \langle 1 | \varphi_{\alpha,k} \rangle \rho_{\alpha\beta,k'} \bar{f}_L(\epsilon_\alpha + k\hbar\Omega) \right.$$
$$\left. - \sum_{k'} \langle \varphi_{\alpha,k'+k} | 1 \rangle \langle 1 | \varphi_{\alpha,k} \rangle \rho_{00,k'} \bar{f}_L(\epsilon_\alpha + k\hbar\Omega) \right). \quad (11.48)$$

Physically, the second contribution of the current formula (11.48) describes the tunneling of an electron from the left lead to the wire, and thus is proportional to $\rho_{00} f_L$, which denotes the probability that a lead state is occupied while the wire is empty. The first terms corresponds to the reversed process, namely the tunneling on an electron from site $|1\rangle$ to the left lead.

The decomposition of the equation of motion (11.32) for the long time limit of $\mathcal{F}_{1\perp}$ and the subsequent computation of the $\bar{S}$ from (11.33) proceed along the same lines. The only difference is that the current operators $\mathcal{J}_\pm$ yield an inhomogeneity, and that the r.h.s. of the trace condition shown in (11.47) is

$$\operatorname{tr}_{\text{wire}} \mathcal{F}_{1\perp} = (\mathcal{F}_{1\perp})_{00} + 2 \sum_\alpha (\mathcal{F}_{1\perp})_{\alpha\alpha} = 0. \quad (11.49)$$

11.3.4
Spinless Electrons

A particular consequence of strong Coulomb repulsion is the mutual blocking of different spin channels. This also motivates us to compare to the case of spinless electrons, which is physically realized by spin polarization. For spinless electrons, we drop all spin indices in the Hamiltonian given in (11.1). Physically, this limit is realized by a sufficiently strong magnetic field that polarizes all electrons contributing to the transport. By the same calculation as that shown above, we can also obtain the expression given in (11.48) for the current, but without the prefactor 2. The factor 2 is also no longer present in the normalization condition (11.47), which now reads

$$\operatorname{tr}_{\text{wire}} \rho_0(t) = \rho_{00}(t) + \sum_\alpha \rho_{\alpha\alpha}(t) = 1, \quad (11.50)$$

and accordingly in the equation of motion for $\mathcal{F}_{1\perp}$.

11.4
Transport under Multi-Photon Emission and Absorption

As an application of the master equation approach derived in the last section, we study the transport through a double quantum dot in which multi-photon transitions entail significant consequences for the dc current. In particular, we focus on the creation of a dc current in the absence of a transport voltage and on the current suppression by the purely coherent influence of an oscillating field.

11.4.1
Electron Pumping

In experiments with double quantum dots [2, 3, 65], it has been demonstrated that resonant excitations may induce pump currents that are much larger than those predicted in the limit of an adiabatically slow driving [66]. This effect requires internal asymmetries of the quantum dots which can be induced by gate voltages, such that the relevant level of one dot has a different on site energy than the other one (see Figure 11.1). Corresponding theoretical studies explained this behavior by the fact that the driving bridges the energy difference, and in this way opens an inelastic transport channel. This holds both for the case of strong Coulomb repulsion [21, 44, 62] and for non-interacting electrons [67]. For the latter case, which can be treated within a Floquet scattering approach [14], the investigation of the corresponding shot noise confirmed this picture. As expected for an open channel, the electron flow through this system is sub-Poissonian, which is reflected by the observed small Fano factor [67].

Figure 11.2a shows the behavior of the current in the limit of strong Coulomb repulsion. It demonstrates that whenever the driving field is in resonance with the energy difference, an electron residing in the left quantum dot can absorb a photon which leads to a transition to the higher levels in the right quantum dot. Since this level lies above the Fermi level of the right lead, the electron can leave the central system and will be replaced by an electron that tunnels from the left lead to the left dot. For off-resonant driving, photon absorption is not very efficient, such that only a few electrons can tunnel from the left to the right quantum dot. Then only a small tunnel current emerges. The Fano factor shown in Figure 11.2b

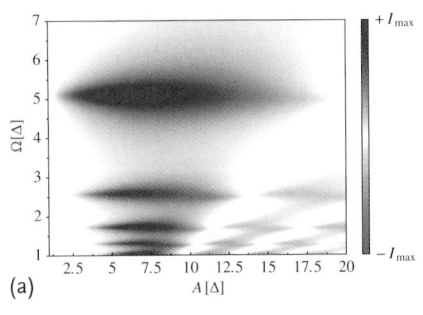

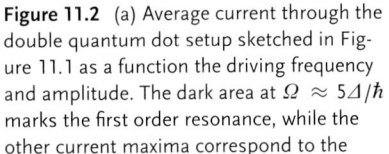

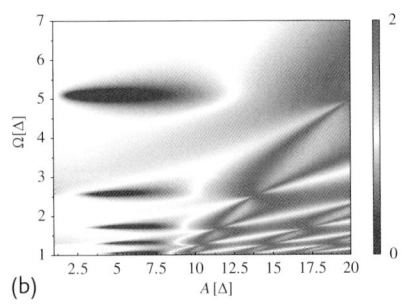

Figure 11.2 (a) Average current through the double quantum dot setup sketched in Figure 11.1 as a function the driving frequency and amplitude. The dark area at $\Omega \approx 5\Delta/\hbar$ marks the first order resonance, while the other current maxima correspond to the nonlinear response. (b) Corresponding Fano factor. White marks Poissonian noise ($F \approx 1$). At the first order resonances, the noise is sub-Poissonian (elliptically shaped dark areas), while it becomes super-Poissonian with an increasing driving amplitude.

is much smaller than unity at the resonances. This reveals that the transport is quite regular.

The current as a function of the driving amplitude exhibits an interesting non-monotonic behavior. This can be attributed to the fact that in strongly driven systems, the response is governed by an effective Hamiltonian whose matrix elements are dressed by Bessel functions $J_\nu(A/\hbar\Omega)$, where ν reflects the order of the resonance. For noninteracting electrons, this behavior has even been predicted in analytical calculations [67], while here our numerical results provide a strong indication for a similar behavior.

11.4.2
Coherent Current Suppression

A transport effect opposite to pumping is the suppression of a dc current by an ac field with a proper ratio between driving amplitude and frequency [33]. The physical cause of this effect is the so-called coherent destruction of tunneling [68, 69] between the two quantum dots, which turns out to be stable in the presence of Coulomb repulsion [70]. This phenomenon can be quantitatively understood in terms of a driven two level system in the absence of external leads. We consider a single particle in a driven two level system described by the Hamiltonian

$$H_{\text{TLS}}(t) = -\frac{\Delta}{2}\sigma_x + A\sigma_z \cos(\Omega t) , \qquad (11.51)$$

where $\sigma_{x,z}$ denotes the usual Pauli matrices. If the energy $\hbar\Omega$ of the quanta of the driving field is much larger than the tunnel matrix element Δ, one can transform the Hamiltonian to an interaction picture with respect to the time-dependent part of the Hamiltonian,

$$H_{\text{TLS}}(t) \rightarrow \widetilde{H}_{\text{TLS}}(t) = -\frac{\Delta}{2} U_0^\dagger(t) \sigma_x U_0(t) , \qquad (11.52)$$

where $U_0(t) = \exp\{-(iA/\hbar\Omega)\sigma_z \sin\Omega t\}$. Since we assumed that the driving is much faster than the tunneling, we can separate time scales, and thus replace the interaction-picture Hamiltonian by its time average

$$\bar{H}_{\text{TLS,eff}} = -\frac{\Delta_{\text{eff}}}{2}\sigma_x , \qquad (11.53)$$

with the tunnel matrix element renormalized according to

$$\Delta \longrightarrow \Delta_{\text{eff}} = J_0(A/\hbar\Omega)\Delta , \qquad (11.54)$$

where J_0 denotes the zeroth order Bessel function of the first kind. If the ratio $A/\hbar\Omega$ equals a zero of the Bessel function J_0 (i.e., for the values 2.405.., 5.520.., 8.654.., ...), the effective tunnel matrix element vanishes and the tunneling is brought to a standstill.

When going beyond this approximation, the effective tunnel matrix element no longer vanishes exactly, but it still becomes very small. Then in the presence of two

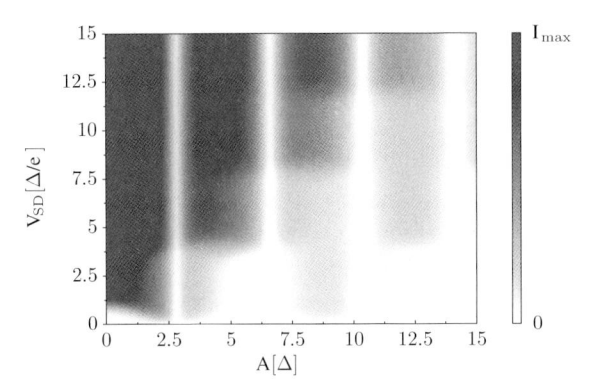

Figure 11.3 Average current through an unbiased two level system as a function the driving amplitude and source-drain voltage. The driving frequency is $\Omega = 5\Delta/\hbar$.

leads with a source-drain voltage, the electron transport acquires a "bottleneck" between the two quantum dots. This situation corresponds to a quantum point contact for which the transmission of an electron is a "rare event". Consequently, the electron transport obeys Poisson statistics, for which the Fano factor is $F = 1$. This has been demonstrated within Floquet scattering theory for noninteracting electrons at low temperature [36].

In the limit of strong Coulomb repulsion, the average current and the corresponding Fano factor are shown in Figure 11.3. For very small driving amplitudes the system is practically undriven, and we find a dc current when the source-drain voltage is so large that the eigenenergies of the two level system lie within the voltage window. Then electrons from the left lead can tunnel into the double dot and proceed further to an empty state in the right lead. Since the transport channel is blocked while an electron resides in the dots, the electrons are anti-bunched. This is reflected by sub-Poissonian noise, with $F \approx 1/2$.

With an increasing driving amplitude, the renormalization of the tunnel matrix element according to (11.54) becomes relevant. When the ratio $A/\hbar\Omega$ matches a zero of the Bessel function, the current is suppressed while the Fano factor increases. Comparing with the case of noninteracting electrons, we find that Coulomb repulsion generally reduces the current. This can be attributed to the fact that electrons approaching the central region from the left-hand side are repelled even by electrons that have already tunneled to the right dot. The Poissonian "rare event" character of the transport process is nevertheless maintained. Thus, we observe $F \approx 1$ for driving parameters that lead to suppression of tunneling, while outside these regions resonant tunneling dominates and the Fano factor assumes values close to $F = 1/2$.

We also find that the current exhibits a step-like behavior as a function of the source-drain voltage, even when the voltage is already quite large. This can be attributed to the photon absorption during the transport. In the current formula of (11.48), photon absorption is manifest in the energy shifts by multiples of $\hbar\Omega$

in the arguments of the Fermi functions. Thus, a particular sideband of a Floquet state $|\phi_\alpha\rangle$ may or may not contribute to the transport, depending on whether its energy $\epsilon_\alpha + k\hbar\Omega$ lies inside or outside the voltage window. When the voltage window increases, more sidebands contribute to the transport. The observed step size of $\hbar\Omega/e$ confirms this scenario.

11.5
Conclusions

We have studied the influence of strong Coulomb repulsion on the transport through a driven double quantum dot, focusing on electron pumping and coherent current suppression. While the former denotes the emergence of a dc current due to a pure ac driving in an asymmetric system, the latter means that an ac driving can substantially reduce the dc current that stems from an applied source-drain voltage. Both effects represent particular nonlinear responses of a coherent conductor to an oscillating force. A proper theoretical tool for the computation of such a nonlinear response is Floquet theory. In the present case we combined a Floquet theory for the isolated conductor with a master equation approach that describes the electron transport in the regime of strong Coulomb repulsion.

Our main quantities of interest were the average current and the Fano factor which provides information about the nature of the transport process. We also revealed that in this limit pumping stems mainly from resonant excitations of electrons from the lower to the higher levels of the double dot. Moreover, it turned out that multi-photon transitions are almost indistinguishable from single photon transitions, since they lead to very similar peaks of the dc current, while shot noise is reduced. Coherent current suppression represents a more involved type of response and can be understood in terms of a high frequency approximation. Here we found that Coulomb repulsion mainly influences the magnitude of the current, while it renders shot noise, as characterized by the Fano factor, virtually unchanged. Our results indicate that for both effects under investigation, Coulomb repulsion does not play a major role.

Acknowledgements

We thank Peter Hänggi for interesting and helpful discussions. This work has been supported by DFG through SPP 1243 "Quantum transport on the molecular scale" and by the German Excellence Initiative via the "Nanosystems Initiative Munich (NIM)". FJK acknowledges support from the Elite Network of Bavaria via the International Doctorate Program "NanoBioTechnology".

References

1 Blick, R.H., Haug, R.J., Weis, J., Pfannkuche, D., von Klitzing, K., and Eberl, K. (1996) *Phys. Rev. B*, **53**, 7899.
2 van der Wiel, W.G., Fujisawa, T., Oosterkamp, T.H., and Kouwenhoven, L.P. (1999) *Physica B* (Amsterdam), **272**, 31.
3 van der Wiel, W.G., De Franceschi, S., Elzerman, J.M., Fujisawa, T., Tarucha, S., and Kouwenhoven, L.P. (2003) *Rev. Mod. Phys.*, **75**, 1.
4 Khrapai, V.S., Ludwig, S., Kotthaus, J.P. Tranitz, H.P., and Wegscheider, W. (2006) *Phys. Rev. Lett.*, **97**, 176803.
5 Cui, X.D. et al. (2001) *Science*, **294**, 571.
6 Reichert, J., Ochs, R., Beckmann, D., Weber, H.B., Mayor, M., and von Löhneysen, H. (2002) *Phys. Rev. Lett.*, **88**, 176804.
7 Nitzan, A. (2001) *Annu. Rev. Phys. Chem.*, **52**, 681.
8 Hänggi, P., Ratner, M., and Yaliraki, S. (2002) *Chem. Phys.*, **281**, 111.
9 Tien, P.K. and Gordon, J.P. (1963) *Phys. Rev.*, **129**, 647.
10 Iñarrea, J., Platero, G., and Tejedor, C. (1994) *Phys. Rev. B*, **50**, 4581.
11 Blick, R.H., Haug, R.J., van der Weide, D.W., von Klitzing, K., and Eberl, K. (1995) *Appl. Phys. Lett.*, **67**, 3924.
12 Brune, P., Bruder, C., and Schoeller, H. (1997) *Phys. Rev. B*, **56**, 4730.
13 Platero, G. and Aguado, R. (2004) *Phys. Rep.* **395**, 1.
14 Kohler, S., Lehmann, J., and Hänggi, P. (2005) *Phys. Rep.* **406**, 379.
15 Kohler, S., Lehmann, J., Strass, M., and Hänggi, P. (2004) *Adv. Solid State Phys.*, **44**, 151.
16 Kouwenhoven, L.P., Johnson, A.T., van der Vaart, N.C., Harmans, C.J.P.M., and Foxon, C.T. (1991) *Phys. Rev. Lett.*, **67**, 1626.
17 Pothier, H., Lafarge, P., Urbina, C., Esteve, D., and Devoret, M.H. (1992) *Europhys. Lett.*, **17**, 249.
18 Stafford, C.A. and Wingreen, N.S. (1996) *Phys. Rev. Lett.*, **76**, 1916.
19 Switkes, M., Marcus, C.M., Campman, K., and Gossard, A.C. (1999) *Science*, **283**, 1905.
20 Wagner, M. and Sols, F. (1999) *Phys. Rev. Lett.* **83**, 4377.
21 Hazelzet, B.L., Wegewijs, M.R., Stoof, T.H., and Nazarov, Yu.V. (2001) *Phys. Rev. B*, **63**, 165313.
22 Lehmann, J., Kohler, S., Hänggi, P., and Nitzan, A. (2002) *Phys. Rev. Lett.*, **88**, 228305.
23 Arrachea, L. (2004) *Phys. Rev. B*, **70**, 155407.
24 Arrachea, L. (2005) *Phys. Rev. B*, **72**, 125349.
25 Blaauboer, M. and Fricot, C.M.L. (2005) *Phys. Rev. B*, **71**, 041303(R).
26 Cota, E., Aguado, R., and Platero, G. (2005) *Phys. Rev. Lett.*, **94**, 107202.
27 Sanchez, R., Cota, E., Aguado, R., and Platero, G. (2006) *Phys. Rev. B*, **74**, 035326.
28 Scheid, M., Pfund, A., Bercioux, D., and Richter, K. (2007) *Phys. Rev. B*, **76**, 195303.
29 Sánchez, R., Kaiser, F.J., Kohler, S., Hänggi, P. and Platero, G. (2008) *Physica E*, **40**, 1276.
30 Cohen, G., Hod, O., and Rabani, E. (2007) *Phys. Rev. B*, **76**, 235120.
31 Strass, M., Hänggi, P., and Kohler, S. (2005) *Phys. Rev. Lett.*, **95**, 130601.
32 Sánchez, R., Kohler, S., and Platero, G. (2008) *New J. Phys.*, **10**, 115013.
33 Lehmann, J., Camalet, S., Kohler, S., and Hänggi, P. (2003) *Chem. Phys. Lett.*, **368**, 282.
34 Kleinekathöfer, U., Li, G., Welack, S., and Schreiber, M. (2006) *Europhys. Lett.*, **75**, 139.
35 Camalet, S., Lehmann, J., Kohler, S., and Hänggi, P. (2003) *Phys. Rev. Lett.*, **90**, 210602.
36 Kohler, S., Camalet, S., Strass, M., Lehmann, J., Ingold, G.-L., and Hänggi, P. (2004) *Chem. Phys.*, **296**, 243.
37 Li, G.-Q., Schreiber, M., and Kleinekathöfer, U. (2007) *EPL*, **79**, 27006.
38 Grabert, H. and Devoret, M.H. (eds) (1992) *Single Charge Tunneling*, Vol. 294 of *NATO ASI Series B*, Plenum, New York.
39 Gurvitz, S.A. and Prager, Ya.S. (1996) *Phys. Rev. B*, **53**, 15932.
40 Petrov, E.G. and Hänggi, P. (2001) *Phys. Rev. Lett.*, **86**, 2862.
41 Petrov, E.G., May, V., and Hänggi, P. (2002) *Chem. Phys.*, **281**, 211.

42 Lehmann, J., Ingold, G.-L., and Hänggi, P. (2002) *Chem. Phys.*, **281**, 199.
43 Bruder, C. and Schoeller, H. (1994) *Phys. Rev. Lett.*, **72**, 1076.
44 Stoof, T.H. and Nazarov, Yu.V. (1996) *Phys. Rev. B*, **53**, 1050.
45 Lehmann, J., Kohler, S., May, V., and Hänggi, P. (2004) *J. Chem. Phys.*, **121**, 2278.
46 Elattari, B. and Gurvitz, S.A. (2002) *Phys. Lett. A*, **292**, 289.
47 Bagrets, D.A. and Nazarov, Yu.V. (2003) *Phys. Rev. B*, **67**, 085316.
48 Kießlich, G., Wacker, A., and Schöll, E. (2003) *Phys. Rev. B*, **68**, 125320.
49 Novotný, T., Donarini, A., Flindt, C., and Jauho, A.-P. (2004) *Phys. Rev. Lett.*, **92**, 248302.
50 Flindt, C., Novotný, T., and Jauho, A.-P. (2004) *Phys. Rev. B*, **70**, 205334.
51 Belzig, W. (2005) *Phys. Rev. B*, **71**, 161301(R).
52 Koch, J., Raikh, M.E., and von Oppen, F. (2005) *Phys. Rev. Lett.*, **95**, 056801.
53 Aghassi, J., Thielmann, A., Hettler, M.H., and Schön, G. (2006) *Phys. Rev. B*, **73**, 195323.
54 Kaiser, F.J. and Kohler, S. (2007) *Ann. Phys. (Leipzig)*, **16**, 702.
55 Camalet, S., Kohler, S., and Hänggi, P. (2004) *Phys. Rev. B*, **70**, 155326.
56 Fano, U. (1947) *Phys. Rev.*, **72**, 26.
57 Blanter, Ya.M. and Büttiker, M. (2000) *Phys. Rep.*, **336**, 1.
58 Grabert, H. and Ingold, G.-L. (2002) *Europhys. Lett.*, **58**, 429.
59 Risken, H. (1989) *The Fokker–Planck Equation*, Vol. 18 of *Springer Series in Synergetics*, 2nd edn, Springer, Berlin.
60 Reulet, B., Senzier, J., and Prober, D.E. (2003) *Phys. Rev. Lett.*, **91**, 196601.
61 Ankerhold, J. and Grabert, H. (2005) *Phys. Rev. Lett.*, **95**, 186601.
62 Kaiser, F.J., Hänggi, P., and Kohler, S. (2006) *Eur. Phys. J. B*, **54**, 201.
63 Grifoni, M. and Hänggi, P. (1998) *Phys. Rep.*, **304**, 229.
64 Novotný, T. (2002) *Europhys. Lett.*, **59**, 648.
65 Oosterkamp, T.H., Fujisawa, T., W.G. van der Wiel, Ishibashi, K., Hijman, R.V., Tarucha, S., and Kouwenhoven, L.P. (1998) *Nature*, **395**, 873.
66 Brouwer, P.W. (1998) *Phys. Rev. B*, **58**, R10135.
67 Strass, M., Kohler, S., Lehmann, J., and Hänggi, P. (2005) *Mol. Cryst. Liq. Cryst.*, **426**, 59.
68 Grossmann, F., Dittrich, T., Jung, P., and Hänggi, P. (1991) *Phys. Rev. Lett.*, **67**, 516.
69 Großmann, F. and Hänggi, P. (1992) *Europhys. Lett.*, **18**, 571.
70 Creffield, C.E., and Platero, G. (2002) *Phys. Rev. B*, **65**, 113304.

12
Pattern Formation and Time Delayed Feedback Control at the Nanoscale
Eckehard Schöll

12.1
Introduction

Nonlinear and chaotic space-time patterns arise in a variety of nonlinear dissipative dynamic systems in physics, chemistry, and biology [1–5]. In semiconductors driven far from thermodynamic equilibrium by an applied voltage, such patterns occur as current density and electric field distributions, providing an abundance of examples for complex or chaotic dynamics and self-organized spatiotemporal patterns [6–8]. Of particular current interest are state-of-the-art semiconductor structures whose structural and electronic properties vary on a nanometer scale [9]. In these nanostructures, nonlinear charge transport mechanisms are given, for instance, by quantum mechanical tunneling through potential barriers or by thermionic emission of hot electrons which have enough kinetic energy to overcome the barrier. A further important feature connected with potential barriers and quantum wells in such semiconductor structures is the ubiquitous presence of space charges. This, according to Poisson's equation, induces a further feedback between the charge carrier distribution and the electric potential distribution governing the transport. This mutual nonlinear interdependence is particularly pronounced in the cases of semiconductor heterostructures (consisting of layers of different materials) and low-dimensional nanostructures where abrupt junctions between different materials on an atomic length scale cause conduction band discontinuities resulting in potential barriers and wells. The local charge accumulation in these potential wells, together with nonlinear transport processes across the barriers have been found to provide a number of nonlinearities and instabilities in current-voltage characteristics [8].

In particular, instabilities are very likely to occur in the case of negative differential conductance, that is, if the current I decreases with increasing voltage U, and vice versa. The actual electric response depends upon the attached circuit which generally contains, even in the absence of external load resistors, unavoidable resistive and reactive components such as lead resistances, lead inductances, package inductances, and package capacitances. These reactive components give rise to ad-

---

Nonlinear Dynamics of Nanosystems. Edited by Günter Radons, Benno Rumpf, and Heinz Georg Schuster
Copyright © 2010 WILEY-VCH Verlag GmbH & Co. KGaA, Weinheim
ISBN: 978-3-527-40791-0

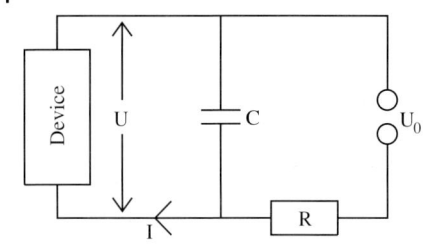

Figure 12.1 A semiconductor device operated in a circuit with load resistor R and capacitor C, and applied bias voltage U_0.

ditional degrees of freedom which are described by Kirchhoff's equations of the circuit. If, for instance, a circuit is considered which contains a capacitance C parallel to the semiconductor device, and a load resistance R_L and a bias voltage U_0 in series with the device (Figure 12.1), then Kirchhoff's laws lead to

$$U_0 = R_L I_0(t) + U(t), \qquad (12.1)$$

$$I_0(t) = I(t) + C \frac{dU}{dt}. \qquad (12.2)$$

Hence, the temporal behavior of the voltage is determined by the circuit equation

$$\frac{dU(t)}{dt} = \frac{1}{C}\left(\frac{U_0 - U}{R_L} - I\right). \qquad (12.3)$$

If a semiconductor element with negative differential conductance is operated in a reactive circuit, oscillatory instabilities may be induced by these reactive components, even if the relaxation time of the semiconductor is much smaller than that of the external circuit so that the semiconductor can be described by its stationary $I(U)$ characteristic and simply acts as a nonlinear resistor. Here, we will consider the more interesting case where the semiconductor itself introduces an internal unstable temporal degree of freedom, leading to self-sustained current or voltage oscillations. Such self-sustained oscillations under time-independent external bias will be discussed in the following. Examples for internal degrees of freedom are the charge carrier density, the electron temperature, or a junction capacitance within the device. Equation (12.3) is then supplemented by a dynamic equation for this internal variable. It should be noted that the same class of models is also applicable to describe neural dynamics in the framework of the Hodgkin–Huxley equations [10].

The global $I(U)$ characteristic must be distinguished from the local current density j versus electric field F. Two important cases of *negative differential conductivity* (NDC) are described by an N-shaped or an S-shaped $j(F)$ characteristic, and denoted by NNDC and SNDC, respectively. However, more complicated forms like Z-shaped, loop-shaped, or disconnected characteristics are also possible [8]. NNDC and SNDC are associated with voltage- or current-controlled instabilities, respectively. In the NNDC, case the current density is a single-valued function of the field, but the field is multivalued: the $F(j)$ relation has three branches in a certain range of j. The SNDC case is complementary in the sense that F and j are interchanged. In the case of NNDC, the NDC branch is often but not always, depending upon external circuit and boundary conditions, unstable against the formation of nonuniform field profiles along the charge transport direction (*electric field domains*). While in the SNDC case, *current filamentation* generally occurs, that is, the current density becomes nonuniform over the cross-section of the current flow and forms a conducting channel, a current filament. The elementary structures which make up these self-organized patterns are stationary or moving *fronts* representing the boundaries of the high-field domain or high-current filament. These primary self-organized spatial patterns may themselves become unstable in secondary bifurcations leading to periodically or chaotically breathing, rocking, moving, or spiking filaments or domains, or even solid-state turbulence and spatiotemporal chaos [6–8, 11–13].

The control of such complex spatiotemporal scenarios is of utmost importance. For instance, chaotic current or voltage oscillations should be avoided for a reliable operation of semiconductor nanostructure devices. Moreover, there is need to control the frequencies and the regularity of noisy oscillations, or stabilize steady states. The whole issue of controlling unstable or noisy states has evolved into a growing field of applied nonlinear science [14]. This field has various aspects comprising stabilization of unstable periodic orbits embedded in a deterministic chaotic attractor, which is generally referred to as *chaos control*, stabilization of unstable fixed points, or control of the coherence and timescales of stochastic motion. Various methods of control, going well beyond the classical control theory [15–17], have been developed since the ground-breaking work of Ott, Grebogi and Yorke [18]. One scheme where the control force is constructed from time-delayed signals [19] has turned out to be very robust and universal in application. It has been used in a large variety of systems in physics, chemistry, biology, and medicine, and in purely temporal dynamics as well as in spatially extended systems. Moreover, it has recently been shown to be applicable also to noise-induced oscillations and patterns [20, 21]. This is an interesting observation in the context of ongoing research on the constructive influence of noise in nonlinear systems [22–26].

An important aspect of time-delayed feedback control (*time-delay autosynchronization*, TDAS) is the noninvasiveness of this control method, that is, the control force vanishes when the target orbit is reached. An extension to multiple time-delays (*extended time-delay autosynchronization*, ETDAS) has been proposed by Socolar et al. [27], and analytical insight into those schemes has been gained by several theoretical studies, for example, [28–40] as well as by numerical bifurcation analysis, for example, [41, 42]. Such self-stabilizing feedback control schemes with different couplings of the control force have also been applied to spatiotemporal patterns resulting from various models of semiconductor oscillators, for example, impact ionization driven Hall instability [43], and semiconductor nanostructures described by an N-shaped [44–46], S-shaped [47–50], or Z-shaped [51] $j(F)$ characteristics.

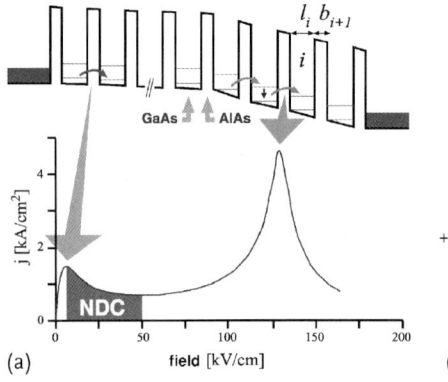

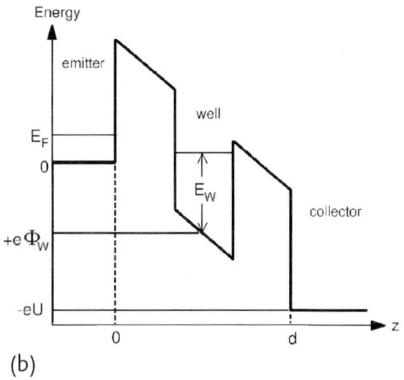

Figure 12.2 Schematic energy profiles of two nanostructures. (a) Superlattice exhibiting domain formation. The associated current density (j) versus field (F) characteristic shows negative differential conductivity (NDC). The low-field domain corresponds to sequential tunneling between equivalent levels of adjacent quantum wells (low-field peak of the $j(F)$ characteristic), while the high-field domain corresponds to resonant tunneling between different levels of adjacent wells (high-field peak). (b) Schematic potential profile of the double barrier resonant tunneling structure (DBRT). E_F and E_w denote the Fermi level in the emitter, and the energy level in the quantum well, respectively. U is the voltage applied across the structure.

Time-delayed feedback control has also been applied to purely noise-induced oscillations and patterns in a regime where the deterministic system rests in a steady state. In this way, both the coherence and the mean frequency of the oscillations has been controlled in various nonlinear systems [20, 21, 52–58], including semiconductor nanostructures of N-type [59, 60], S-type [61], and Z-type [62, 63].

In the following, we use two paradigmatic models of state-of-the-art semiconductor nanostructures (Figure 12.2) where time-delayed feedback control should be easy to implement experimentally by a feedback loop in the electronic circuit:

1. Electron transport in semiconductor superlattices exhibits strongly nonlinear spatiotemporal dynamics. Complex scenarios including chaotic motion of multiple fronts and domains have been found under time independent bias conditions [64], showing signs of universal front dynamics [65, 66]. Deterministic chaos of traveling field domains can be controlled by time-delayed feedback control [46], as well as noise-induced front patterns [59, 60, 67, 68].

2. Charge accumulation in the quantum-well of a double-barrier resonant-tunneling diode (DBRT) may result in lateral spatiotemporal patterns of the current density and chaos [69]. Unstable deterministic current density patterns, for example, periodic breathing or spiking modes [51], as well as noise-induced breathing patterns [62, 63, 70, 71]. can be stabilized by a delayed feedback loop.

12.2
Control of Chaotic Domain and Front Patterns in Superlattices

Semiconductor superlattices [72] have been demonstrated to give rise to self-sustained current oscillations ranging from several hundred MHz [73, 74] to 150 GHz at room temperature [75]. In any case, a superlattice constitutes a highly nonlinear system [8, 76–78], and instabilities are likely to occur. Indeed, chaotic scenarios have been found experimentally [79–81] and described theoretically in periodically driven [82] as well as in undriven systems [64]. For a reliable operation of a superlattice as an ultra-high frequency oscillator, such unpredictable and irregular conditions should be avoided, which might not be easy in practice.

Here, we focus on simulations of dynamic scenarios for superlattices under fixed time independent external voltage in the regime where self-sustained dipole waves are spontaneously generated at the emitter. The dipole waves are associated with traveling field domains and consist of electron accumulation and depletion fronts which generally travel at different velocities and may merge and annihilate. Depending on the applied voltage and the contact conductivity, this gives rise to various oscillation modes as well as different routes to chaotic behavior [64, 66].

We use a model of a superlattice based on sequential tunneling of electrons. In the framework of this model, electrons are assumed to be localized at one particular well and only weakly coupled to the neighboring wells. The tunneling rate to the next well is lower than the typical relaxation rate between the different energy levels within one well. The electrons within one well are then in quasi-equilibrium and transport through the barrier is incoherent. The resulting tunneling current density $J_{m \to m+1}(F_m, n_m, n_{m+1})$ from well m to well $m+1$ only depends on the electric field F_m between both wells and the electron densities n_m and n_{m+1} in the wells (in units of cm^{-2}). A detailed microscopic derivation of the model has been given elsewhere [76]. A typical result for the current density versus electric field characteristic is depicted in Figure 1.1a in the spatially homogeneous case, that is, $n_m = n_{m+1} = N_D$, with donor density N_D.

In the following, we will adopt the total number of electrons in each well as the dynamic variables of the system. The dynamic equations are then given by the continuity equation

$$e\frac{dn_m}{dt} = J_{m-1 \to m} - J_{m \to m+1} \quad \text{for} \quad m = 1, \ldots N , \tag{12.4}$$

where N is the number of wells in the superlattice, and $e < 0$ is the electron charge.

The electron densities and the electric fields are coupled by the following discrete version of Gauss's law

$$\epsilon_r \epsilon_0 (F_m - F_{m-1}) = e(n_m - N_D) \quad \text{for} \quad m = 1, \ldots N , \tag{12.5}$$

where ϵ_r and ϵ_0 are the relative and absolute permittivities, and F_0 and F_N are the fields at the emitter and collector barrier, respectively.

The applied voltage between emitter and collector gives rise to a global constraint

$$U = -\sum_{m=0}^{N} F_m d, \quad (12.6)$$

where d is the superlattice period.

The current densities at the contacts are chosen such that dipole waves are generated at the emitter. For this purpose, it is sufficient to choose Ohmic boundary conditions:

$$J_{0 \to 1} = \sigma F_0, \quad (12.7)$$

$$J_{N \to N+1} = \sigma F_N \frac{n_N}{N_D}, \quad (12.8)$$

where σ is the Ohmic contact conductivity, and the factor n_N/N_D is introduced in order to avoid negative electron densities at the collector. Here, we make the physical assumption that the current from the last well to the collector is proportional to the electron density in the last well. In principle, it is possible to use a more realistic exponential emitter characteristic [83] or calculate the boundary conditions using microscopic considerations, but the qualitative behavior is not changed.

In our computer simulations, we use a superlattice with $N = 100$ periods, $Al_{0.3}Ga_{0.7}As$ barriers of width $b = 5$ nm and GaAs quantum wells of width $w = 8$ nm, doping density $N_D = 1.0 \times 10^{11}$ cm^{-2} and scattering induced broadening $\Gamma = 8$ meV at $T = 20$ K. If the contact conductivity σ is chosen such that the intersection point with the homogeneous N-shaped current density versus field characteristic is at a sufficiently low current value, accumulation and depletion fronts are generated at the emitter. Those fronts form a traveling high-field domain, with a leading electron depletion front and trailing accumulation front. For fixed voltage U, (12.6) imposes constraints on the lengths of the high-field domains and thus on the front velocities. If N_a accumulations fronts and N_d depletion fronts are present, the respective front velocities v_a and v_d must obey $v_d/v_a = N_a/N_d$. Since the front velocities are monotonic functions of the current density [84], this also fixes the current. If the accumulation and depletion fronts have different velocities, they may collide in pairs and annihilate. With decreasing contact conductivity or increasing voltage, chaotic scenarios arise where the annihilation of fronts of opposite polarity occurs at irregular positions within the superlattice [64], leading to complex bifurcation diagrams.

The transition from periodic to chaotic oscillations is enlightened by considering the space-time plot for the evolution of the electron densities (Figure 12.3a). At $U = 1.15$ V, chaotic front patterns with irregular sequences of annihilation of front pairs occur.

We shall now introduce a time-delayed feedback loop (ETDAS) to control the chaotic front motion and stabilize a periodic oscillation mode which is inherent in the chaotic attractor [46, 85]. As a global output signal which is coupled back in the

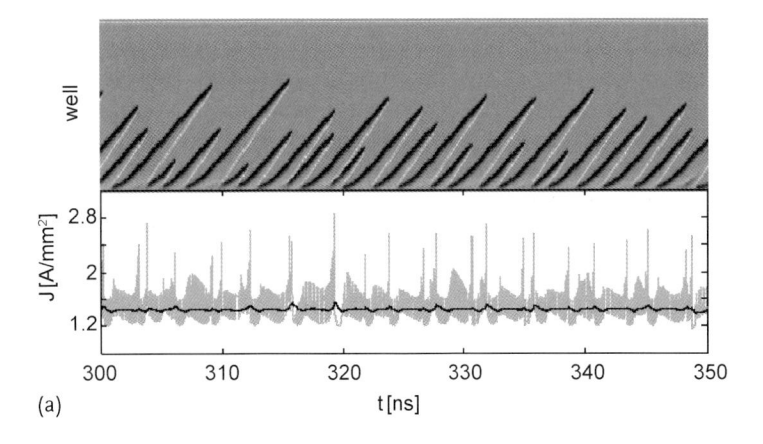

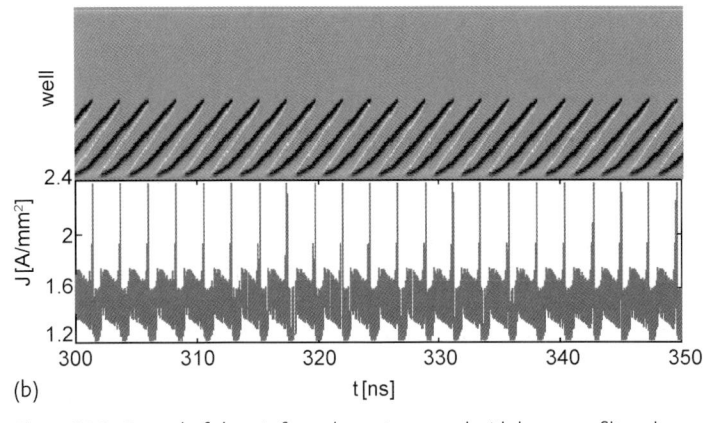

Figure 12.3 Control of chaotic front dynamics by extended time-delay autosynchronization. (a) Space-time plot of the uncontrolled charge density, and current density J versus time. (b) Same with global voltage control with low-pass filtered current density (denoted by the black curve in panel (a)). $U = 1.15\,\text{V}$, $\sigma = 0.5\,\Omega^{-1}\,\text{m}^{-1}$, $\tau = 2.29\,\text{ns}$, $K = 3 \times 10^{-6}\,\text{V mm}^2/\text{A}$, $R = 0.2$, $\alpha = 10^9\,\text{s}^{-1}$. Other parameters as in [46, 66].

feedback loop, it is natural to use the total current density

$$J = \frac{1}{N+1} \sum_{m=0}^{N} J_{m \to m+1}. \tag{12.9}$$

For the uncontrolled chaotic oscillations, J versus time (grey trace in Figure 12.3a) shows irregular spikes at those times when two fronts annihilate. Note that the grey current time trace is modulated by fast small-amplitude oscillations (due to well-to-well hopping of depletion and accumulation fronts in our discrete model) which are not resolved temporally in the plot. They can be averaged out by considering an exponentially weighted current density (black curve in Figure 12.3a), which corresponds to a low-pass filter:

$$\overline{J}(t) = \alpha \int_0^t J(t') e^{-\alpha(t-t')} dt', \quad (12.10)$$

with a cut-off frequency α.

The information contained in the low-frequency part of the current (Figure 12.3a, black curve) is then used as input in the extended multiple-time autosynchronization scheme. The voltage U across the superlattice is modulated by multiple differences of the filtered signal at time t and at delayed times $t - \tau$

$$U(t) = U_0 + U_c(t) \quad (12.11)$$

$$\begin{aligned} U_c(t) &= -K\left[\overline{J}(t) - \overline{J}(t-\tau)\right] + R\,U_c(t-\tau) \\ &= -K \sum_{\nu=0}^{\infty} R^{\nu} \left[\overline{J}(t-\nu\tau) - \overline{J}(t-(\nu+1)\tau)\right] \\ &= -K\left[\overline{J}(t) - (1-R) \sum_{k=1}^{\infty} R^{k-1} \overline{J}(t-k\tau)\right], \end{aligned} \quad (12.12)$$

where U_0 is a time-independent external bias, and U_c is the control voltage. K is the amplitude of the control force, τ is the delay time, and R is a memory parameter. A sketch of the entire control circuit is displayed in Figure 12.4a. Such a global control scheme is easy to experimentally implement. It is noninvasive in the sense that the control force vanishes when the target state of period τ has been reached. This target state is an unstable periodic orbit of the uncontrolled system. The period τ can be determined a priori by observing the resonance-like behavior of the mean control force versus τ. The result of the control is shown in Figure 12.3b. The front dynamics exhibit annihilation of front pairs at fixed positions in the superlattice, and stable periodic oscillations of the current are obtained.

In Figure 12.4b, the control domain is depicted in the parameter plane of R and K. A typical horn–like control domain similar to the ones known from other coupling schemes [48] is found. Control is achieved in a range of values of the control amplitude K, which is widened and shifted to larger K with increasing memory parameter R. Typically, the left-hand control boundary corresponds to a period-doubling bifurcation leading to chaos for smaller K, while the right-hand boundary is associated with a Hopf bifurcation. The shape of our control domain and its size resemble the results obtained analytically for diagonal control schemes where observables are measured and controlled locally. Thus, our control scheme is of similar control performance as local control.

In conclusion, time-delay autosynchronization represents a convenient and simple scheme for the self-stabilization of high-frequency current oscillations due to moving domains in superlattices under dc bias. This approach lacks the drawback of synchronization by an external ultrahigh-frequency forcing since it requires nothing but delaying of the global electrical system output by the specified time lag.

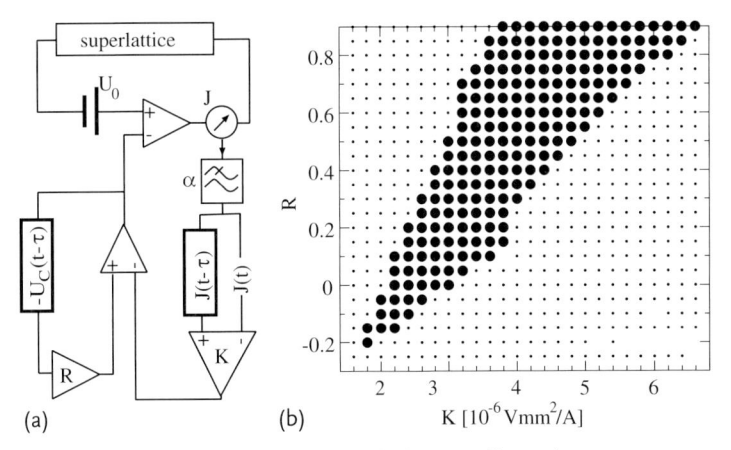

Figure 12.4 (a) Control circuit including the low-pass filter with cut-off frequency α and the time-delayed feedback loop (K) and its extension to multiple time delays (R). (b) Control domain for global voltage control. Full circles denote successful control, small dots denote no control. Parameters as in Figure 12.3 [46].

12.3
Control of Noise-Induced Oscillations in Superlattices

Theoretical and experimental research has recently shown that noise can have surprisingly constructive effects in many nonlinear systems. In particular, an optimal noise level may give rise to ordered behavior and even produce new dynamical states [26]. Well known examples are provided by stochastic resonance [86] in periodically driven systems, and by coherence resonance [22, 23] in autonomous systems. In spite of considerable progress on a fundamental level, useful applications of noise-induced phenomena in technologically relevant devices are still scarce. Here, we will demonstrate that noise can give rise to oscillating current and charge density patterns in semiconductor nanostructures even if the deterministic system exhibits only a steady state, and that these space-time patterns can be controlled by the time-delayed feedback scheme applied to purely deterministic chaotic front patterns in a superlattice mentioned in the previous section [59, 60, 67, 68].

We develop a stochastic model for the superlattice [67] approximating the random fluctuations of the current densities by additive Gaussian white noise $\xi_m(t)$ with

$$\langle \xi_m(t) \rangle = 0 , \quad \langle \xi_m(t) \xi_{m'}(t') \rangle = \delta(t-t') \delta_{mm'} \tag{12.13}$$

in the continuity equation (12.4)

$$e \frac{dn_m}{dt} = J_{m-1 \to m} + D\xi_m(t) - J_{m \to m+1} - D\xi_{m+1}(t) , \tag{12.14}$$

where D is the noise intensity. Since the inter-well coupling in our superlattice model is very weak and the tunneling times are much smaller than the characteristic time scale of the global current, these noise sources can be treated as uncorrelated both in time and space. Charge conservation is automatically guaranteed by adding a noise term ξ_m to each current density $J_{m-1\to m}$. The physical origin of the noise may be, for example, thermal noise, $1/f$ noise, or shot noise due to the randomly fluctuating tunneling times of discrete charges across the barriers. The latter is Poissonian and can be approximated [78, 87] by $D = (eJ_{m-1\to m}/A)^{1/2}$ which increases with decreasing current cross-section A. Therefore, this type of noise dominates for small devices. In the following, we summarize the global effect of noise by a constant D.

We choose the control parameters U and σ such that the deterministic system exhibits no oscillations but is very close to a bifurcation, yielding it very sensitive to noise. The transition from stationarity to oscillations in the system may occur either via a Hopf or a saddle-node bifurcation on a limit cycle, as depicted in the bifurcation diagram of Figure 12.5. The different nature of these two bifurcations is reflected in the effect noise has in each case.

In the vicinity of the lower bifurcation line in Figure 12.5, slightly below a Hopf bifurcation marked by the small rectangle in Figure 12.5 [59], the only stationary solution in the absence of noise is a stable fixed point which corresponds to

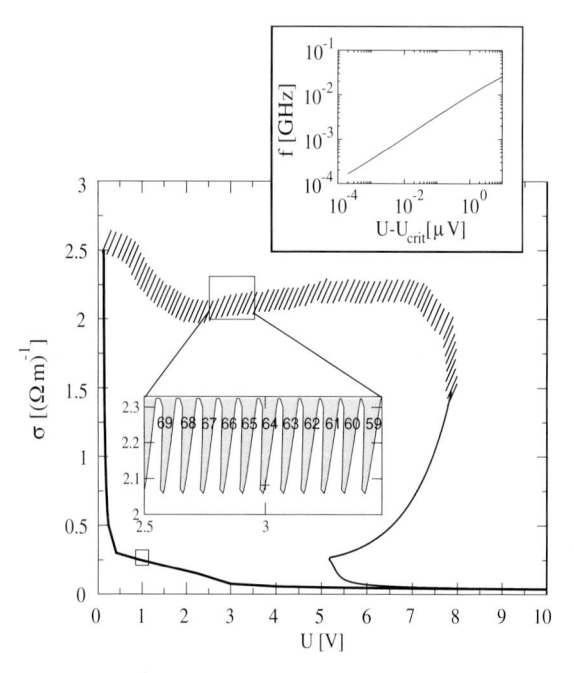

Figure 12.5 Bifurcation diagram in the (σ, U) plane. Thick and hatched lines mark the transition from stationary to moving fronts via a Hopf or a saddle-node bifurcation on a limit cycle, respectively. The inset shows a blow-up of a small part of the hatched line revealing its sawtooth-like structure. Dark and white correspond to stationary and moving fronts, respectively, where the numbers denote the positions of the stationary accumulation front in the superlattice. Upper inset shows the frequency f of the limit cycle which is born above the critical point (marked by a cross in the lower inset) as function of U [67].

a stationary *depletion front* near the emitter, associated with a stationary current density flowing through the device. With increasing noise intensity ($D > 0$), the current density exhibits small irregular oscillations around the steady state. In the spatiotemporal picture, no significant front motion is observed, but the depletion front as a whole starts to "wiggle" around its deterministically fixed position. In order to investigate the effect of time-delayed feedback on the coherence and time scales of the noise-induced oscillations, the scheme proposed by (12.11) has been applied with $R = 0$ and cut-off frequency $\alpha = 1\,\text{GHz}$ [59]. The application of control with delay time τ chosen equal to the basic period of the Hopf bifurcation certainly improves the coherence of the current signal, and the main peak in the Fourier power spectral density (shortly referred to as spectrum) becomes narrower. The position of the main spectral peak shifts in dependence of τ in the same piecewise linear way as in simple generic models of a Hopf bifurcation [20, 52].

A more interesting regime is the noise-induced transition from stationary to moving fronts (and associated high-field domains) [67]. This scenario corresponds to a global saddle-node bifurcation on a limit cycle (cross in the lower inset of Figure 12.5). Keeping σ fixed and increasing the voltage U, a limit cycle of approximately constant amplitude and increasing frequency is born. This happens through the collision of a stable fixed point and a saddle-point. Plotting the frequency of these oscillations versus the bifurcation parameter U, we obtain the characteristic *square-root scaling law* (upper inset of Figure 12.5) that governs a saddle-node bifurcation on a limit cycle. At the critical point U_{crit}, the frequency of the oscillations tends to zero. This corresponds to an infinite period of oscillation and therefore this bifurcation is also known as saddle-node infinite period bifurcation or SNIPER [88, 89].

We now prepare the system at the stable fixed point which corresponds to a stationary *accumulation front* (Figure 12.6a), and introduce noise. As the noise intensity is increased, the behavior of the system changes dramatically (Figure 12.6b): the accumulation front remains stationary only for a while, until a pair of a depletion and another accumulation front (i.e., a charge dipole with a high-field domain in between) is generated at the emitter. As is known from the deterministic system, this dipole injection critically depends upon the emitter current [66]. Here, it is triggered by noise at the emitter (the same scenarios occur if noise is applied locally only to the wells near the emitter). Because of the global voltage constraint, (12.6), the growing dipole field domain between the injected depletion and accumulation fronts requires the high-field domain between the stationary accumulation front and the collector to shrink, and hence, that accumulation front starts moving towards the collector. For a short time, there are two accumulation fronts and one depletion front in the sample, thereby forming a tripole [90], until the first accumulation front reaches the collector and disappears. When the depletion front reaches the collector, the remaining accumulation front must stop moving because of the global constraint. This happens at the position where the first accumulation front was initially localized. After some time, noise generates another dipole at the emitter and the same scenario is repeated.

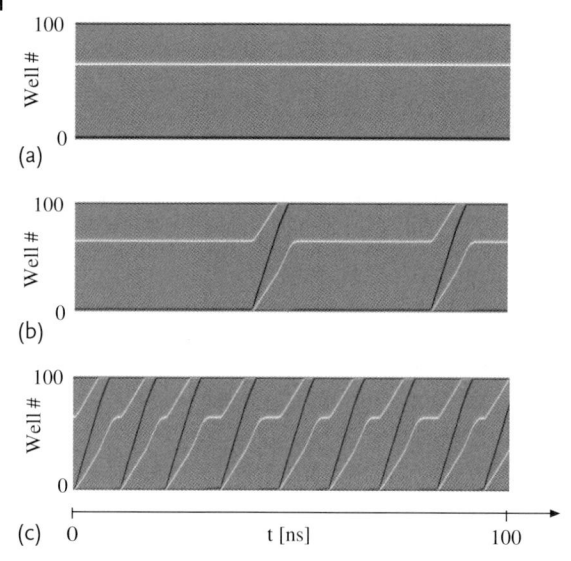

Figure 12.6 Noise-induced front motion: Space-time plots of the electron density for (a) $D = 0$ (no noise), (b) $D = 0.5\,\text{A s}^{1/2}/\text{m}^2$, (c) $D = 2.0\,\text{A s}^{1/2}/\text{m}^2$. Light and dark shading corresponds to electron accumulation and depletion fronts, respectively. The emitter is at the bottom. Parameters: $U = 2.99\,\text{V}$, $\sigma = 2.082\,101\,2488\,\Omega^{-1}\,\text{m}^{-1}$, $N_D = 10^{11}\,\text{cm}^{-2}$, $T = 20\,\text{K}$, $N = 100$ GaAs wells of width $w = 8\,\text{nm}$, and $\text{Al}_{0.3}\text{Ga}_{0.7}\text{As}$ barriers of width $b = 5\,\text{nm}$, other parameters as in [66, 67].

There are two distinct time scales in the system. One is related to the time the depletion front takes to STOP travel through the superlattice. The other timescale is associated with the time needed for a new depletion front to be generated at the emitter. These two time scales are also visible in the noise-induced current oscillations, see Figure 12.7a. The time series of the current density are in the form of a pulse train with two characteristic times: the activation time, which is the time needed to excite the system from this stable fixed point (time needed for a new depletion front to be generated at the emitter) and the excursion time which is the time needed to return from the excited state to the fixed point (time the depletion front needs to travel through the device). Low noise is associated with large activation times and small, almost constant, excursion times, while as the noise level increases activation times become smaller and at sufficiently large D vanish. At low D, the spike train looks irregular and the interval between excitations (mean interspike-interval $\langle T \rangle$) is relatively large and random in time. At moderate noise, the spiking is rather regular, suggesting that the mean interspike-interval does not substantially vary. Further increase of noise results in a highly irregular spike train with very frequent spikes.

To gain deeper insight into the effect noise has on the time scales and coherence of the system, we determine the interval between two consecutive excitations and calculate the mean interspike-interval $\langle T \rangle$. In Figure 12.7b (top), the decrease of $\langle T \rangle$ as a function of D is shown, demonstrating that the mean interspike-interval

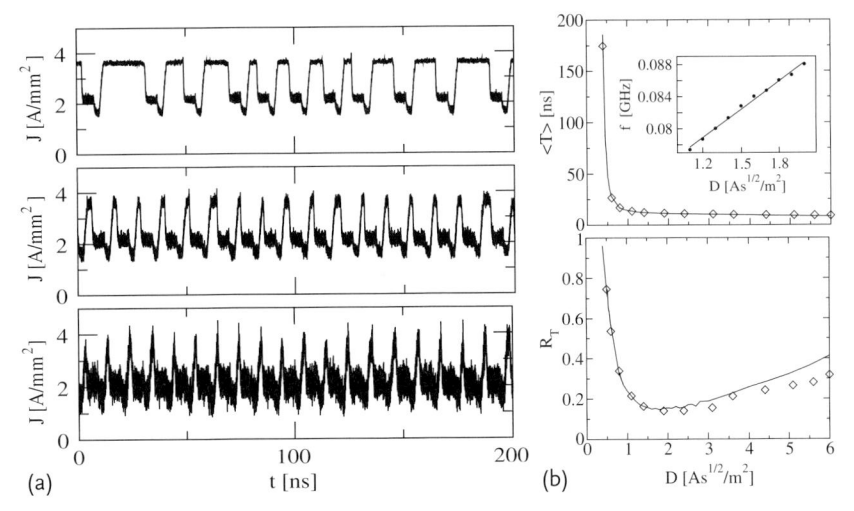

Figure 12.7 (a) Three noise realizations of the current density $J(t)$. From top to bottom, $D = 0.8$, $D = 2.0$ and $D = 5.0\,\text{A s}^{1/2}/\text{m}^2$. (b) Mean interspike-interval (top) and its normalized fluctuations R_T (bottom) vs. noise intensity. The inset shows the peak frequency vs. D. Parameters as in Figure 12.6 [67].

is strongly controlled by the noise intensity especially at lower values of the latter. This is very important in terms of experiments, where noise can induce oscillations by forcing stationary fronts to move. The corresponding spectral peak frequency f shows a linear scaling for small D. As a measure for coherence, we use the normalized fluctuations of pulse duration [23]

$$R_T = \frac{\left(\langle T^2 \rangle - \langle T \rangle^2\right)^{1/2}}{\langle T \rangle}. \tag{12.15}$$

This quantity, as seen in Figure 12.7b (bottom), is a non-monotonic function of D, exhibiting a minimum at moderate noise intensity. This is the well known phenomenon of *coherence resonance* and is strongly connected to excitability.

An alternative measure of coherence is the correlation time t_{cor} given by the formula [91]:

$$t_{\text{cor}} = \frac{1}{\psi(0)} \int_0^\infty |\psi(s)|\,\text{d}s, \tag{12.16}$$

where $\psi(s)$ is the autocorrelation function of the current density signal $J(t)$,

$$\psi(s) = \langle (J(t) - \langle J \rangle)(J(t-s) - \langle J \rangle) \rangle, \tag{12.17}$$

and $\psi(0)$ is its variance.

Next, we will focus on controlling this noise-induced domain motion by time-delayed feedback. To this purpose we add the feedback control voltage to the time-independent external voltage U_0 as in (12.11), but with $R = 0$:

$$U = U_0 - K(\overline{J}(t) - \overline{J}(t - \tau)),\qquad (12.18)$$

where $\overline{J}$ is the low-pass filtered total current density (12.10).

First, we switch off the noise in order to investigate the effect that time-delayed feedback has on its own. We observe a delay-induced limit cycle (traveling field domain) which we track in the $K - \tau$ parameter plane. The resulting phase diagram can be seen in Figure 12.8. For values of K and τ in the grey area, delay control generates oscillations. In the white area, no delay-induced limit cycle exists, but only a stationary domain. The bifurcation line separating the two regimes denotes a global, homoclinic bifurcation [8]. For a fixed value of τ, the period of these oscillations decreases with increasing control strength K, as shown in the inset of Figure 12.8. The logarithmic scaling of the period is characteristic of a homoclinic bifurcation. Thus, the frequency goes to zero at the bifurcation point.

Now, we turn on the noise and observe how control influences the noise-induced dynamics in the regime where delay induces a limit cycle (marked by the cross in the grey area in Figure 12.8). The corresponding time series without control exhibits irregular noise-induced oscillations (Figure 12.9a). With delayed feedback control, the time series looks much more regular reflecting the delay-induced limit cycle which is just smeared out a little by noise (Figure 12.9b). The coherence resonance effect is thus destroyed. This is visible both in t_{cor} and R_T. Figure 12.10a and b show these coherence measures for $K = 0$ (black circles) and for $K \neq 0$ (grey

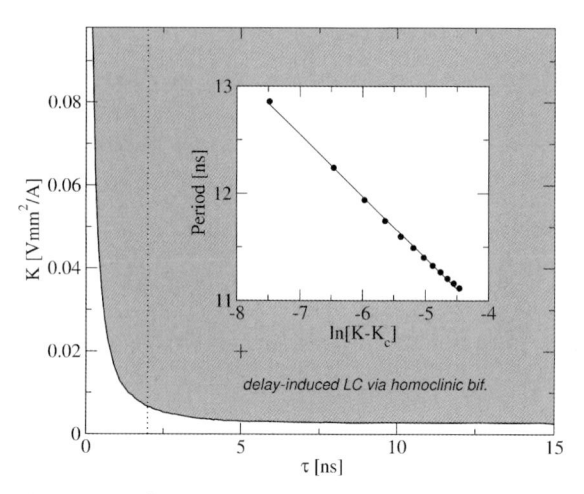

Figure 12.8 Bifurcation diagram of delay-induced limit cycles, corresponding to traveling domains, in the $K - \tau$ control parameter plane (marked by grey shading) for $D = 0$. Inset shows the scaling of the period of the delay-induced limit cycle above the homoclinic bifurcation in dependence on K for a fixed value of $\tau = 2$ ns denoted by the vertical dashed line. Full circles: simulation, line: linear fit. $K_c \approx 0.006437$ V mm^2/A. Parameters as in Figure 12.6 [60].

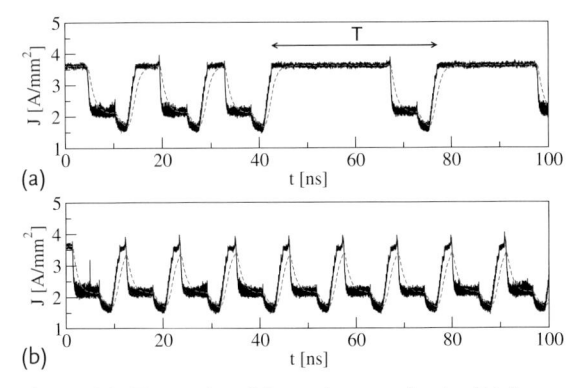

Figure 12.9 Time series of the total current density $J(t)$ for noise-induced front patterns ($D = 0.8\,\text{A}\,\text{s}^{1/2}/\text{m}^2$). (a) $K = 0$ and (b) $K = 0.02\,\text{V}\,\text{mm}^2/\text{A}$, $\tau = 5$ ns. T marks the pulse duration in (a). Dashed curves correspond to filtered signal $\overline{J}(t)$. Parameters as in Figure 12.6 [60].

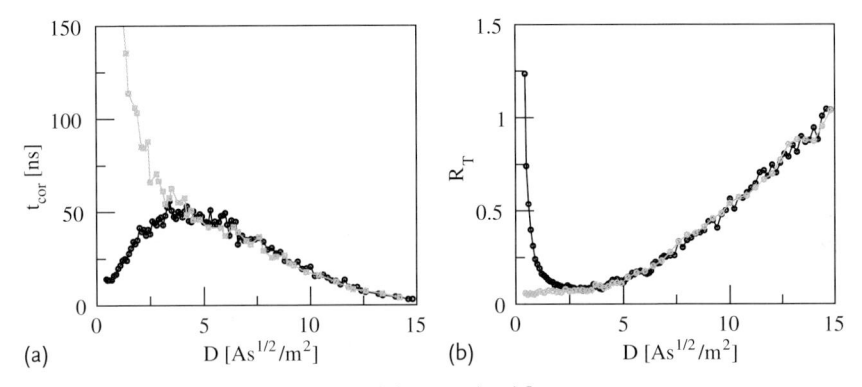

Figure 12.10 (a) Correlation time t_{cor} and (b) normalized fluctuations of pulse duration R_T vs. noise intensity D. Black circles correspond to free system ($K = 0$) while grey (cyan) squares are calculated for $K = 0.02\,\text{V}\,\text{mm}^2/\text{A}$, $\tau = 5$ ns marked by the cross in Figure 12.8. Parameters as in Figure 12.6 [60].

squares). For $K \neq 0$, both t_{cor} and R_T depend monotonically upon the noise intensity (grey squares). Maximum coherence is now achieved at low noise where the delay-induced domain motion is most regular, and the coherence monotonically decreases with increasing noise intensity. For small noise intensity, the correlation time and the fluctuation of the pulse duration is dramatically improved by the delayed feedback.

By fixing the control parameters such that they lie outside the delay-induced limit cycle regime, that is, in the white area in Figure 12.8, we observe almost no change in the presence of delay since the feedback amplitude K is too small (Figure 12.11a). In the right panel of Figure 12.11a, the correlation time is plotted versus the time delay. It exhibits a slight modulation with a period close to the

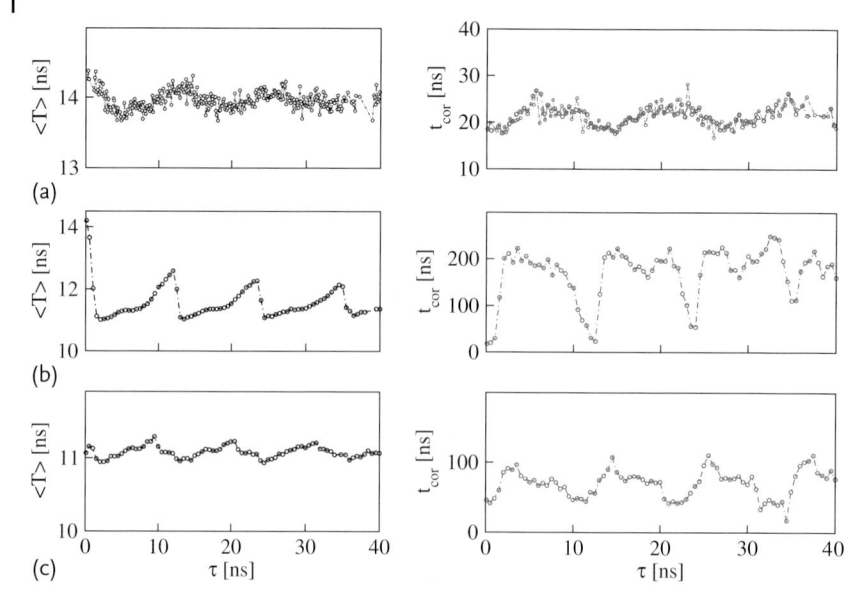

Figure 12.11 (Color online) Mean interspike interval $\langle T \rangle$ (left) and correlation time t_{cor} (right) in dependence on the time delay τ. (a) Control strength $K = 0.002\,\text{V}\,\text{mm}^2/\text{A}$ and noise intensity $D = 1.0\,\text{A}\,\text{s}^{1/2}/\text{m}^2$, (b) $K = 0.02\,\text{V}\,\text{mm}^2/\text{A}$ and $D = 1.0\,\text{A}\,\text{s}^{1/2}/\text{m}^2$ and (c) $K = 0.02\,\text{V}\,\text{mm}^2/\text{A}$ and $D = 2.5\,\text{A}\,\text{s}^{1/2}/\text{m}^2$. Averages over 30 time series realizations of length $T = 1600\,\text{ns}$ have been used for the calculation of t_{cor} and averages over 1000 periods for $\langle T \rangle$. Parameters as in Figure 12.6 [68].

period of the noise-induced oscillations [67], $\langle T \rangle = 14.5\,\text{ns}$, and reaches minimum values for $\tau = n\langle T \rangle, n \in \mathbb{N}$. Overall, however, it remains close to the control-free value, $t_{cor} = 19.76\,\text{ns}$. At $K = 0.02\,\text{V}\,\text{mm}^2/\text{A}$ inside the delay-induced limit cycle regime (grey), this modulation is much stronger and has a period close to the delay-induced period ($T = 11\,\text{ns}$). In addition, one can better distinguish between non-optimal and optimal values of τ at which the correlation time attains maximum values. This is shown in the right panel of Figure 12.11b. For a higher noise intensity (Figure 12.11c, right panel), the effect is similar but weaker.

Next, we are interested in how the time scales are affected by the delay. We express the time scales through the mean interspike interval $\langle T \rangle$ and look at its dependence upon the time delay τ for a fixed value of the noise intensity, $D = 1.0\,\text{A}\,\text{s}^{1/2}/\text{m}^2$, and control strength $K = 0.002\,\text{V}\,\text{mm}^2/\text{A}$, chosen outside of the delay-induced oscillations regime. As shown in the left panel of Figure 12.11a, $\langle T \rangle$ is slightly modulated due to the delay with a period close to the noise-induced mean period ($\langle T \rangle \approx 14.5\,\text{ns}$) [67]. In the left panel of Figure 12.11b, a value of K inside the delay-induced oscillations regime is used, $K = 0.02\,\text{V}\,\text{mm}^2/\text{A}$. For $\tau = 0$, the mean interspike interval is equal to the noise-induced period, $\langle T \rangle \approx 14.5\,\text{ns}$ [67]. As the time delay increases, and the delay-induced bifurcation line is crossed, $\langle T \rangle$ sharply drops to the value of 11 ns which corresponds to the period induced by the delay. By further increase of τ, $\langle T \rangle$ rises a little above 12 ns. Then, for $\tau = 11$ ns, the mean interspike interval decreases again and the same scenario is repeated with a modulation period very close to the delay-induced period.

There is some resemblance to the piecewise linear dependence of $\langle T \rangle$ upon τ reported in other excitable systems, for example, the FitzHugh–Nagumo model in [20, 52, 58] or the Oregonator model of the Belousov–Zhabotinsky reaction (under correlated noise and nonlinear delayed feedback) in [55] which, like our system, is also spatially extended. The difference to our present analysis is that in those models, the case of a delay-induced limit cycle was excluded. An explanation for the entrainment of the time scales by the delayed feedback in case of systems below a Hopf bifurcation [20, 21, 52, 62] was given on the basis of a linear stability analysis. It was shown that the basic period is proportional to the inverse of the imaginary part of the eigenvalue of the fixed point which itself depends linearly upon τ for large time delays. The effect of noise and delay in excitable systems was also studied analytically in [58, 92] based on waiting time distributions and renewal theory.

In conclusion, noise is able to force field domains to move through the superlattice when prepared in the regime of stationary domains below a global bifurcation (saddle-node bifurcation on a limit cycle). The signature of this global deterministic bifurcation is still visible in the nonlinear stochastic system, and makes the system excitable and thus sensitive to fluctuations. Maximum coherence is observed at an optimum noise level, that is, *coherence resonance* occurs. Time-delayed feedback induces traveling domains, the frequency of which increases with the control force for a fixed time delay, in a homoclinic bifurcation. Coherence resonance is no longer maintained due to the delay-imposed frequency which is now robust against noise. For small noise intensity (less than the maximum of coherence resonance), the regularity of the current oscillations is strongly enhanced by time-delayed feedback.

12.4
Control of Chaotic Spatiotemporal Oscillations in Resonant Tunneling Diodes

In this section, we consider a double-barrier resonant tunneling diode (DBRT), see Figure 1.1b, which exhibits a Z-shaped (bistable) current-voltage characteristic [8, 93]. The DBRT is a semiconductor nanostructure which consists of one GaAs quantum well sandwiched between two AlGaAs barriers along the z-direction. The quantum well defines a two-dimensional electron gas in the x-y plane. We include the lateral redistribution of electrons in the quantum well plane giving rise to filamentary current flow [94–96]. Complex chaotic scenarios including spatiotemporal breathing and spiking oscillations have been found in a simple deterministic reaction-diffusion model with one lateral dimension x [69] as well as with two lateral dimensions x, y [97]. We extend this model (in the one-dimensional case) to include control terms and obtain the following equations [51] where we use dimensionless variables throughout

$$\frac{\partial a}{\partial t} = f(a, u) + \frac{\partial}{\partial x}\left(D(a)\frac{\partial a}{\partial x}\right) - KF_a(x, t) \quad (12.19)$$

$$\frac{du}{dt} = \frac{1}{\varepsilon}(U_0 - u - rJ) - KF_u(t). \quad (12.20)$$

Here, $u(t)$ is the inhibitor and $a(x, t)$ is the activator variable. In the semiconductor context, $u(t)$ denotes the voltage drop across the device and $a(x, t)$ is the electron density in the quantum well.

The net tunneling rate of the electrons through the two energy barriers (Figure 12.2b) is given by the balance of the incoming and outgoing current densities, that is, from the emitter into the quantum well j_{in} and from the quantum well into the collector j_{out}, respectively. It is modeled by the nonlinear, nonmonotonic function [69]

$$f(a, u) = j_{in} - j_{out}$$

$$j_{in} = \left[\frac{1}{2} + \frac{1}{\pi}\arctan\left(\frac{2}{\gamma}\left(x_0 - \frac{u}{2} + \frac{d}{r_B}a\right)\right)\right]$$

$$\times \left[\ln\left(1 + \exp\left(\eta_e - x_0 + \frac{u}{2} - \frac{d}{r_B}a\right)\right) - a\right]$$

$$j_{out} = a, \quad (12.21)$$

where d is the effective thickness of the double-barrier nanostructure, $r_B = (4\pi\epsilon\epsilon_0\hbar^2)/(e^2 m)$ is the effective Bohr radius in the semiconductor material, ϵ and ϵ_0 are the relative and absolute permittivity of the material, x_0 and γ describe the

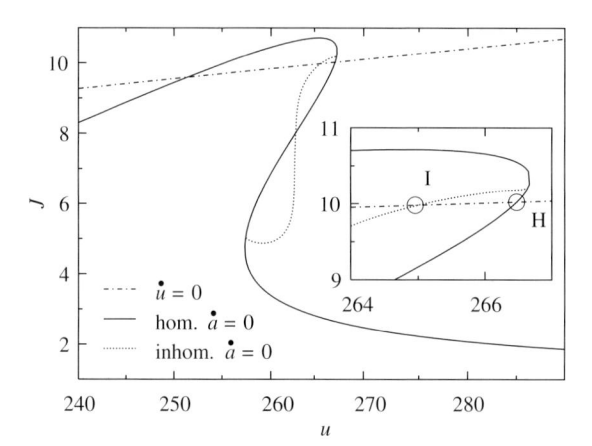

Figure 12.12 Current-voltage characteristic of the DBRT model. The null isoclines for the dynamical variables u (i.e., the load line, dash-dotted) and a in the case of a homogeneous $a(x)$ (solid) and in the case of inhomogeneous $a(x)$ (dotted) are shown. The inset shows an enlargement where I and H mark the inhomogeneous and the homogeneous fixed points of the system, respectively. $U_0 = -84.2895$, $r = -35$. Other parameters as in [51, 62, 70].

energy level and the broadening of the electron states in the quantum well, and η_e is the dimensionless Fermi level in the emitter, all in units of $k_B T$. Throughout this work, we use values of $\gamma = 6$, $d/r_B = 2$, $\eta_e = 28$ and $x_0 = 114$, corresponding to typical device parameters at 4 K [69].

The effective diffusion coefficient $D(a)$ resulting from the inhomogeneous lateral redistribution of carriers and from the change in the local potential due to the charge accumulated in the quantum well by Poisson's equation is given by [95]

$$D(a) = a \left(\frac{d}{r_B} + \frac{1}{1 - \exp(-a)} \right). \quad (12.22)$$

It describes the diffusion of the electrons within the quantum well perpendicular to the current flow. $J = \frac{1}{L} \int_0^L j \, dx$ gives the total current through the device, where $j(a, u) = \frac{1}{2}(j_{in} + j_{out}) = \frac{1}{2}(f(a, u) + 2a)$ is the local current density within the well.

Equation (12.20) represents Kirchhoff's law of the circuit (12.3) in which the device is operated. The external bias voltage U_0, the dimensionless load resistance $r \sim R_L$, and the time-scale ratio $\varepsilon = R_L C/\tau_a$ (where C is the capacitance of the circuit and τ_a is the tunneling time) act as control parameters. The one-dimensional spatial coordinate x corresponds to the direction transverse to the current flow. We consider a system of width $L = 30$ with Neumann boundary conditions $\partial_x a = 0$ at $x = 0, L$ corresponding to no charge transfer through the lateral boundaries.

Equations (12.19) and (12.20) contain control forces F_a and F_u for stabilizing time periodic patterns. The strength of the control terms is proportional to the control amplitude K, which gives one important parameter of each control scheme. Physically, the control forces may be realized by appropriate electronic control circuits. KF_u corresponds to an additional control voltage applied in series with the bias U_0, and KF_a may be implemented by a spatially extended lateral gate electrode which influences the two-dimensional electron gas in the quantum well locally or globally.

Without control, $K = 0$, one can calculate the null isoclines of the system. These are plotted in Figure 12.12 using the current-voltage projection of the originally infinite-dimensional phase space. There are three curves, the null isocline $\dot{u} = 0$ (i.e., the *load line*, dash-dotted) and two null isoclines $\dot{a} = 0$, one for a reduced system, including only spatially homogeneous states (solid), and one for the full system (dotted). We call the system *spatially homogeneous* if the space dependent variable $a(x, t)$ is uniformly distributed over the whole width of the device, that is, $a(x, t) = a(t)$ for all $x \in [0, L]$, otherwise it is called *spatially inhomogeneous*. Figure 12.12 shows the Z-shaped homogeneous current-voltage characteristic of the DBRT (solid curve) and the branch of inhomogeneous, filamentary states (dotted). The inset represents our special regime of interest for the following investigations. The spatially inhomogeneous fixed point marked 'I' in the inset of Figure 12.12 is determined by the intersection of the load line with the nullcline $\dot{a} = 0$ for inhomogeneous $a(x, t)$. With increasing ε, a supercritical Hopf bifurcation of the inhomogeneous steady state I occurs at $\varepsilon_{\text{Hopf}} \approx 6.469$ (cf. Figure 12.13). Below this value, I is stable. The neighboring intersection of the load line with the homogeneous null-

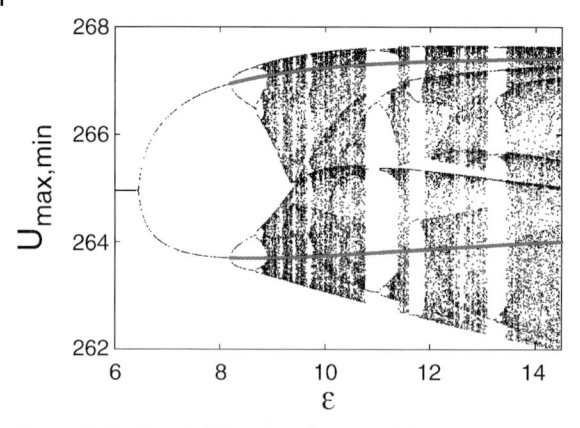

Figure 12.13 Chaotic bifurcation diagram of the resonant tunneling diode. The maxima and minima of the voltage oscillations are plotted versus the time-scale parameter ε ($r = -35$, $U_0 = -84.2895$, $K = 0$) [51].

cline (marked 'H') defines another spatially homogeneous fixed point. It is always unstable in a passive external circuit with load resistance $r > 0$ [94]. By choosing $r < 0$, which can be realized by an active circuit [98], the homogeneous fixed point can be stabilized with respect to completely homogeneous perturbations, though it is generally unstable against spatially inhomogeneous fluctuations, that is, it is a saddle-point. In the following, we assume $r < 0$.

The dynamics of the free system, that is, $K = 0$, develops temporally chaotic and spatially nonuniform states (spatiotemporal breathing or spiking) in appropriate parameter regimes [69] which can be corroborated by calculating the Lyapunov exponents [51]. A characteristic bifurcation diagram exhibiting a period-doubling route to chaos is shown in Figure 12.13. Figure 12.14 shows two examples of periodically (a) and chaotically (b) breathing current filaments. Note that the current density distribution is qualitatively similar to the electron density distribution in the quantum well. For any value of L, the system, due to the global coupling, only allows monotonic spatial profiles, that is, current filaments located at the boundary of the spatial domain [99]. In the semiconductor context, the time and length scales of our dimensionless variables are typically given by a few picoseconds (tunneling time) and 100 nm (diffusion length), respectively. Typical units of the electron density, the current density, and the voltage are 10^{10} cm^{-2}, 500 A/cm^2, and 0.35 mV, respectively.

We are now concerned with controlling unstable time periodic patterns $u_p(t) = u_p(t+\tau)$, $a_p(x,t) = a_p(x, t+\tau)$ which are embedded in a chaotic attractor. For that purpose, we apply control forces F_a and F_u which are derived from time-delayed differences of the voltage and the charge density. For example, we may choose $F_u = F_{vf}$ with the voltage feedback force

$$F_{vf}(t) = u(t) - u(t-\tau) + R F_{vf}(t-\tau) \tag{12.23}$$

(extended time-delay autosynchronization).

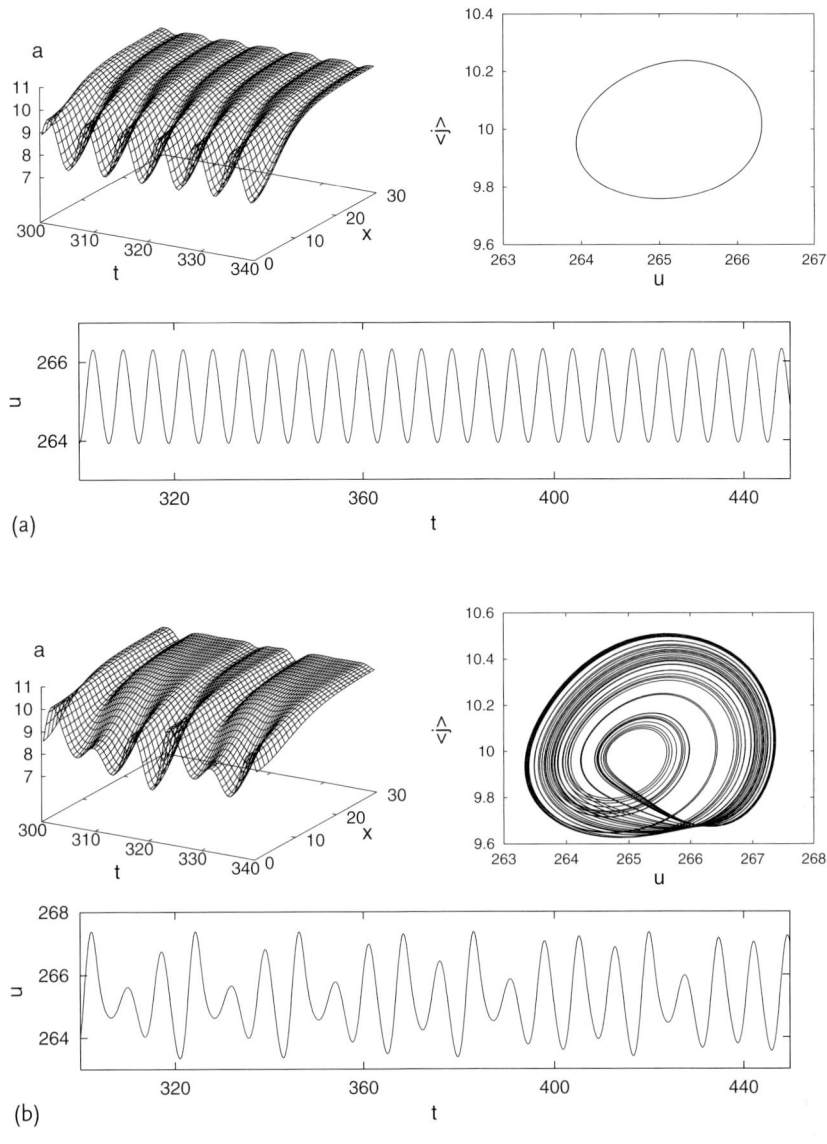

Figure 12.14 Spatiotemporal breathing patterns of the DBRT: electron density evolution, phase portrait, and voltage evolution for (a) $\varepsilon = 7.0$: periodic breathing, (b) $\varepsilon = 9.1$: chaotic breathing ($r = -35$, $U_0 = -84.2895$, $K = 0$). Time t and space x are measured in units of the tunneling time τ_a and the diffusion length l_a, respectively. Typical values at 4 K are $\tau_a = 3.3$ ps and $l_a = 100$ nm [51].

Here, we concentrate on the question of how the coupling of the control forces to the internal degrees of freedom influences the performance of the control. For our model, we consider different choices for the control force F_a. On the one hand, we may use a force which is based on the local charge density according to

$$F_{\text{loc}}(x, t) = a(x, t) - a(x, t - \tau) + R F_{\text{loc}}(x, t - \tau) . \qquad (12.24)$$

Whereas on the other hand, we propose a construction which is only based on its spatial average

$$F_{\text{glo}}(t) = \frac{1}{L} \int_0^L \left[a(x, t) - a(x, t - \tau) \right] dx + R F_{\text{glo}}(t - \tau) . \qquad (12.25)$$

We call the choice $F_a = F_{\text{loc}}$ a *local* control scheme in contrast to the *global* control scheme $F_a = F_{\text{glo}}$ which requires only the global average and does not depend explicitly on the spatial variable. The second option has considerable experimental advantages since the spatial average is related to the total charge in the quantum well and does not require a spatially resolved measurement.

In general, the analysis of the control performance of time-delayed feedback methods results in differential-difference equations which are hard to tackle. Stability of the orbit is governed by eigenmodes and the corresponding complex valued growth rates (Floquet exponents). A simple case exists (which we call *diagonal* control) where analytical results are available [29, 100], namely, for $F_a = F_{\text{loc}}$ and $F_u = F_{\text{vf}}$. It is a straightforward extension to a spatially extended system of an identity matrix for the control of discrete systems of ordinary differential equations (cf. [28]). Figure 12.15 shows successful control of the chaotic breathing oscillation of Figure 12.14b. After the control is switched on, the control force decays exponentially as $|F|_s \sim |\exp(\Lambda t)|$ to a new level which is about three orders of magnitude smaller than the uncontrolled level (Figure 12.15b). At the same time, the voltage signal becomes periodic (Figure 12.15a) and the chaotic attractor in the phase portrait collapses to a periodic orbit (Figure 12.15c).

In Figure 12.16, the regime of successful control in the (K, R) parameter plane and the real part of the Floquet spectrum $\Lambda(K)$ for $R = 0$ is depicted. The control domain has a typical triangular shape bounded by a flip instability (period-doubling, $Re(\Lambda) = 0$, $Im(\Lambda) = \pi/\tau$) to its left and by a Hopf (Neimark-Sacker) bifurcation to its right. Inclusion of the memory parameter R increases the range of K over which control is achieved. We observe that the numerical result fits very well with the analytical prediction.

To confirm the bifurcations at the boundaries, we consider the real part of the Floquet spectrum of the orbit subjected to control. Complex conjugate Floquet exponents show up as doubly degenerate pairs. The largest nontrivial exponent decreases with increasing K and collides at negative values with a branch coming from negative infinity. As a result, a complex conjugate pair develops and real parts increase again. The real part of the exponent finally crosses the zero axis giving rise to a Hopf (Neimark-Sacker) bifurcation. Our numerical simulations are in agreement with the analytical result.

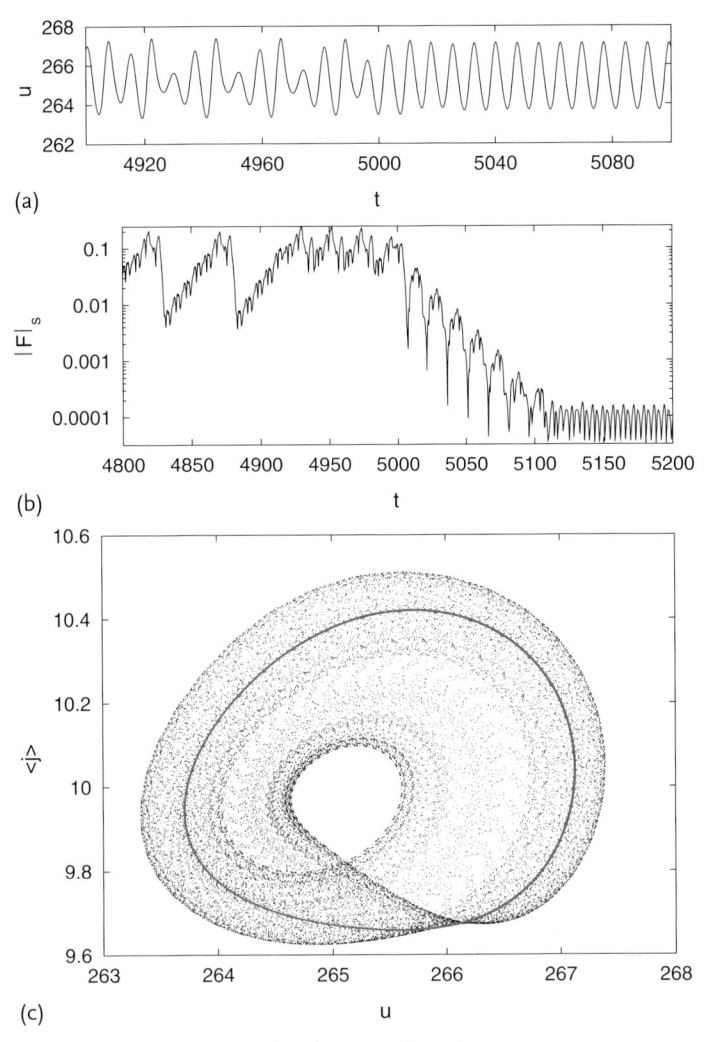

Figure 12.15 Diagonal control in the DBRT where the control force is switched on at $t = 5000$. (a) Voltage u vs. time. (b) Supremum of the control force vs. time. (b) Phase portrait (global current vs. voltage) showing the chaotic breathing attractor and the embedded stabilized periodic orbit (black cycle). Parameters: $r = -35$, $\varepsilon = 9.1$, $\tau = 7.389$, $K = 0.137$, $R = 0$ [51].

If we replace the local control force $F_a = F_{\text{loc}}$ by the global control $F_a = F_{\text{glo}}$, the corresponding control domain looks similar in shape as for diagonal control, although the size of the domain for the global scheme is drastically reduced [51]. The shift in the control boundaries is due to different branches of the Floquet spectrum crossing the ($Re\Lambda = 0$)-axis. It is now interesting to note that if we keep $F_a = F_{\text{glo}}$ as before, but remove the voltage feedback completely, the control domain is shifted

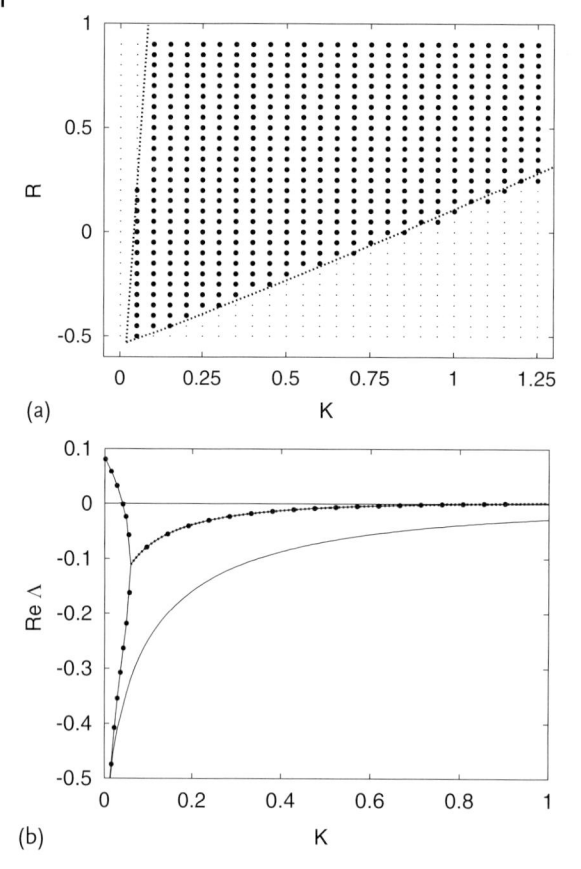

Figure 12.16 (a) Control domains in the (K, R) parameter plane for diagonal control of the unstable periodic orbit with period $\tau = 7.389$. Large dots: successful control in the numerical simulation, small grey dots: no control, dotted lines: analytical result for the boundary of the control domain according to [100]. (b) Leading real parts Λ of the Floquet spectrum for diagonal control in dependence on K ($R = 0$) [51].

to higher K values and at the same time is dramatically increased [51]. For local control without voltage feedback, the control regime is surprisingly even larger than for diagonal control. The reason for this behavior has been explained by a Floquet analysis [51].

From the practical point of view, the most relevant control scheme is the pure voltage control, namely, $F_u = F_{vf}$, $F_a = 0$, since the voltage variable may be conveniently measured and manipulated by an external electronic device. The corresponding control domain and Floquet exponents are shown in Figure 12.17. Here, the control regime is even somewhat smaller than in the case of global control with voltage feedback, but the shape of the control regime and the Floquet spectrum are

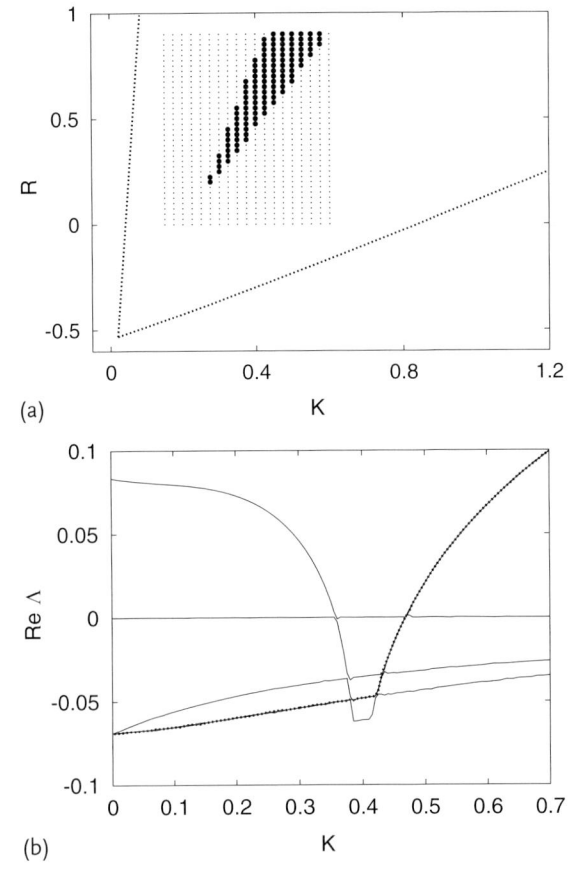

Figure 12.17 Same as Figure 12.16 for pure voltage control (b) ($R = 0.6$). The boundaries of the control domain for diagonal control from Figure 12.16 are shown in panel (a) as dotted lines for comparison [51].

qualitatively very similar. The Floquet spectrum reveals a feature which has already been noted in a general context: The influence of lower Floquet modes, which may cross over, can reduce the size of the control domain severely [101]. It should be noted that this voltage control scheme opens up the opportunity to conveniently study chaos control in a real world device.

Finally, we note that the period-one orbit can be stabilized by our control scheme throughout the whole bifurcation diagram, including chaotic bands and windows of higher periodicity, as marked by two solid lines in Figure 12.13 for diagonal control. Thus, our method represents a way of tracking desired unstable orbits. In our nanostructure model, this finds an interesting application since we obtain stable periodic self-sustained voltage oscillations in a whole range of operating conditions independent of parameter fluctuations.

12.5
Noise-Induced Spatiotemporal Patterns in Resonant Tunneling Diodes

In the previous section, we discussed the possibilities of controlling deterministic chaotic oscillations in the double barrier resonant tunneling diode (DBRT). Now, we will study the effects of noise in this system and investigate whether we can control noise-induced spatiotemporal oscillations by the same method of time-delayed feedback [62, 63, 70, 71].

We use the same model (12.19) and (12.20), but with two additional noise sources [70] and with voltage control [62]

$$\frac{\partial a(x,t)}{\partial t} = f(a,u) + \frac{\partial}{\partial x}\left(D(a)\frac{\partial a}{\partial x}\right) + D_a \xi(x,t)$$

$$\frac{\partial u(t)}{\partial t} = \frac{1}{\varepsilon}(U_0 - u - rJ) + D_u \eta(t) + F(t), \quad (12.26)$$

where $\xi(x,t)$ and $\eta(t)$ represent uncorrelated Gaussian white noise with noise intensities D_a and D_u, respectively:

$$\langle \xi(x,t) \rangle = \langle \eta(t) \rangle = 0 \quad (x \in [0, L]),$$
$$\langle \xi(x,t)\xi(x',t') \rangle = \delta(x-x')\delta(t-t'),$$
$$\langle \eta(t)\eta(t') \rangle = \delta(t-t'). \quad (12.27)$$

Here, we concentrate on the effects of *external* noise modeled by the additional noise voltage $D_u \eta(t)$ in the circuit equation. This term is easily accessible in a real circuit and the noise intensity D_u can be adjusted in a large parameter range using a noise generator in parallel with the supply bias, as realized experimentally, for example, in [102]. In typical dimensional units of $\varepsilon k_B T/e$ [69], $D_u = 1$ corresponds to a parallel noise voltage of 2 mV at temperature $T = 4$ K. Internal fluctuations of the local current density on the other hand, for example, shot noise [87], can not be tuned from the outside. Therefore, in the following, we keep this value fixed at a small noise amplitude of $D_a = 10^{-4}$, corresponding to a noise current density of the order of 50 mA/cm^2 which is within the range of Poissonian shot noise currents.

The control force $F(t)$ represents a control voltage which is constructed recursively from a time-delayed feedback loop with delay time τ, feedback strength $K \geq 0$, and memory parameter R, and can be written as

$$F(t) = K(u(t-\tau) - u(t)) + RF(t-\tau) \quad (12.28)$$

$$= K\sum_{n=0}^{\infty} R^n \left[u(t-(n+1)\tau) - u(t-n\tau)\right]. \quad (12.29)$$

We fix $\varepsilon = 6.2$ slightly below the Hopf bifurcation of the system ($D_a = D_u = 0$), which occurs at $\varepsilon_{\text{Hopf}} \approx 6.469$ (Figure 12.13). Although the deterministic systems rests in the stable spatially inhomogeneous steady state, noise can induce irregular spatiotemporal oscillations of the current density [70]. With increasing noise

12.5 Noise-Induced Spatiotemporal Patterns in Resonant Tunneling Diodes

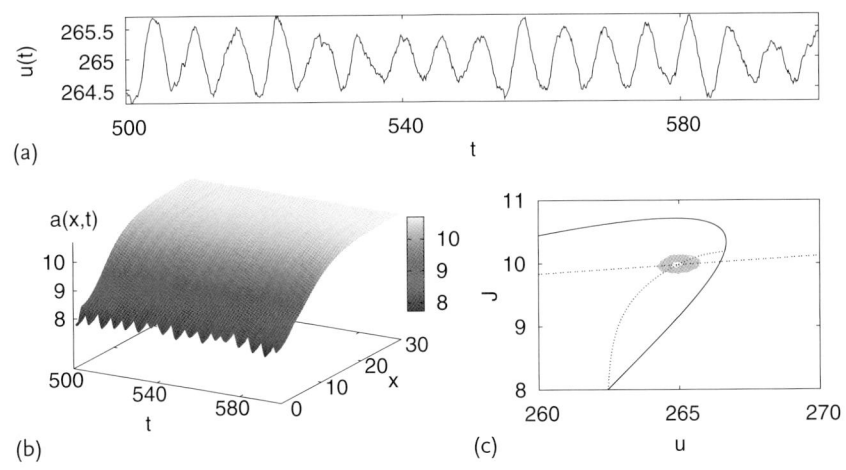

Figure 12.18 Stochastic spatiotemporal dynamics under multiple time-delayed feedback control. (a) Voltage time series $u(t)$ (in units of 0.35 mV); (b) charge carrier density $a(x,t)$ (in units of $10^{10}/\text{cm}^2$); (c) phase portrait of current J (in units of 500 A/cm^2) vs. voltage u. Space x and time t are scaled in units of 100 nm and 3.3 ps, respectively, corresponding to typical device parameters at 4 K [69]. Parameters are $U_0 = -84.2895$, $r = -35$, $\epsilon = 6.2$, $D_u = 0.1$, $D_a = 10^{-4}$, $K = 0.1$, $\tau = 6.3$, $R = 0.5$ [71].

intensity D_u, they become more and more spatially homogeneous, but at the same time, more temporally irregular. In the following, we shall study how these noise-induced oscillations are influenced by the control force.

Figure 12.18 shows simulations of the spatiotemporal dynamics under the influence of noise and delayed feedback. The voltage time series (a), the spatiotemporal charge density patterns (b), and the current-voltage projection of the infinite-dimensional phase space (c) are depicted. Noise induces small *spatially inhomogeneous* oscillations around the inhomogeneous steady state (*breathing current filaments*). In the J–u phase portrait (c), the spatially inhomogeneous steady state (fixed point) is determined by the intersection of the load line (null isocline $\dot{u} = 0$, dash-dotted) with the nullcline $\dot{a} = 0$ for inhomogeneous $a(x, t)$ (dotted). The neighboring intersection of the load line with the nullcline $\dot{a} = 0$ for homogeneous a (black solid curve) defines the second, spatially homogeneous fixed point which is a saddle. With increasing noise intensity (Figure 12.19), the oscillation amplitude becomes larger, the oscillations become more irregular, and finally, at even larger noise, the oscillations are more spatially homogeneous, that is, in the phase space they are more centered around the homogeneous fixed point (Figure 12.20).

We shall now investigate how the regularity and the time-scales of these noise-induced oscillations depend upon the feedback parameters K, R, τ.

To quantify the temporal regularity of the noise-induced oscillations, we evaluate the correlation time [91] calculated from the voltage signal,

$$t_{\text{cor}} \equiv \frac{1}{\sigma^2} \int_0^\infty |\Psi(s)|\, ds, \qquad (12.30)$$

where $\Psi(s) \equiv \langle (u(t) - \langle u \rangle)(u(t+s) - \langle u \rangle) \rangle_t$ is the autocorrelation function of the variable $u(t)$ and $\sigma^2 = \Psi(0)$ its variance.

To get a first impression as to whether or not this control force is able to change the temporal regularity of the noise-induced oscillations, we fix $D_u = 0.1$, $D_a = 10^{-4}$, as in Figure 12.18, and calculate the correlation time in dependence of the feedback strength K for two different delay times τ, and $R = 0$. From Figure 12.21, one can see that the qualitative result depends strongly upon the choice of the delay time. While for $\tau = 7$ the control loop strongly increases the correlation time with increasing K, it is also able to significantly decrease it for $\tau = 5$. The same can be seen from Figure 12.22. Here, the control with $K = 0.1$ and $\tau = 7$ strongly enhances the correlation time, compared with the uncontrolled case, over

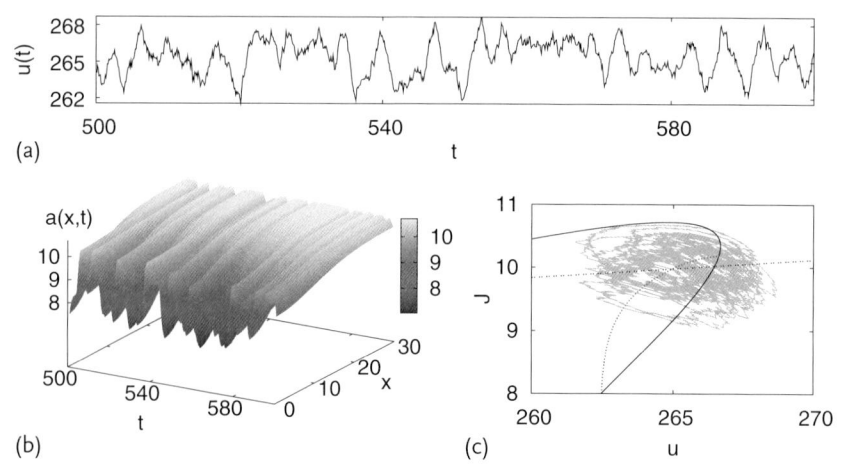

Figure 12.19 Same as Figure 12.18 for $D_u = 1.0$.

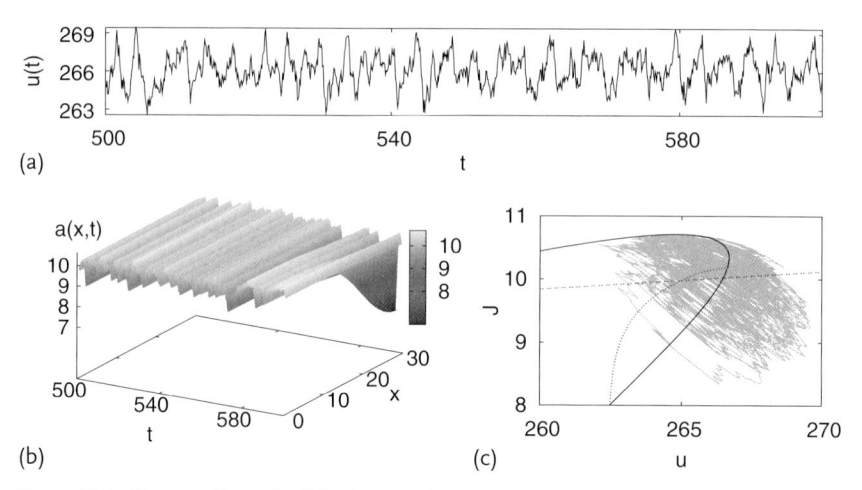

Figure 12.20 Same as Figure 12.18 for $D_u = 2.0$.

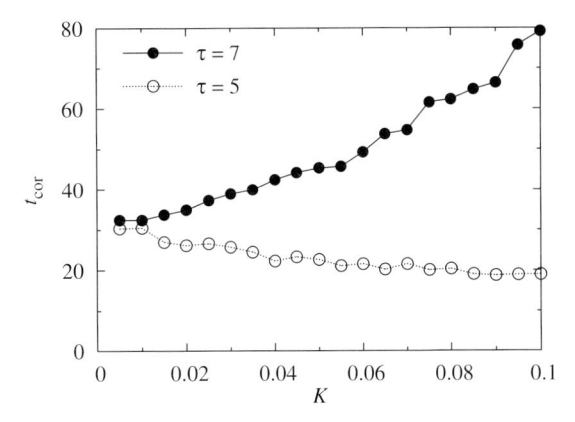

Figure 12.21 Correlation time vs. feedback strength K for $\tau = 5$ and $\tau = 7$. $D_u = 0.1$, $D_a = 10^{-4}$, $R = 0$. Averages from 100 time series of length $T = 10\,000$, other parameters as in Figure 12.18 [62].

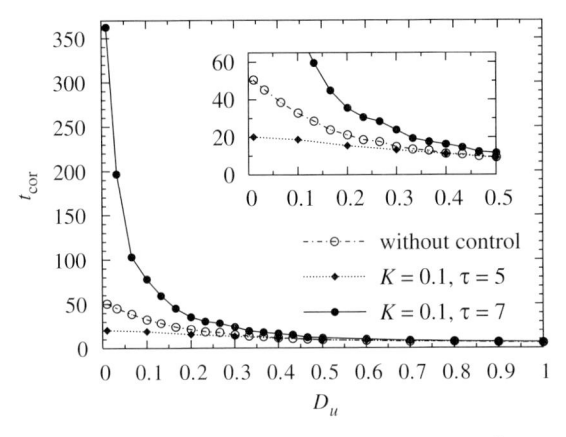

Figure 12.22 Correlation time vs. noise intensity D_u without control ($K = 0$) and with control and two different values of τ as indicated, $R = 0$. Averages from 100 time series of length $T = 10\,000$, parameters as in Figure 12.18. The inset shows a blow-up [62].

a relatively wide range of the noise intensity up to $D_u \approx 0.5$, whereas $\tau = 5$ decreases it within the same range. The difference in regularity for different values of τ and K also shows up in the spatiotemporal patterns and voltage time series (Figure 12.23), where (b) is clearly more regular than (a).

The role of the appropriate choice of the control delay τ becomes even clearer if we keep K fixed and calculate the correlation time in dependence of τ. The result is plotted in Figure 12.24a where one can clearly see the oscillatory character of the correlation time under variation of τ, which is characterized by the presence of "optimal" values of τ, corresponding to maximum regularity and "worst" values of

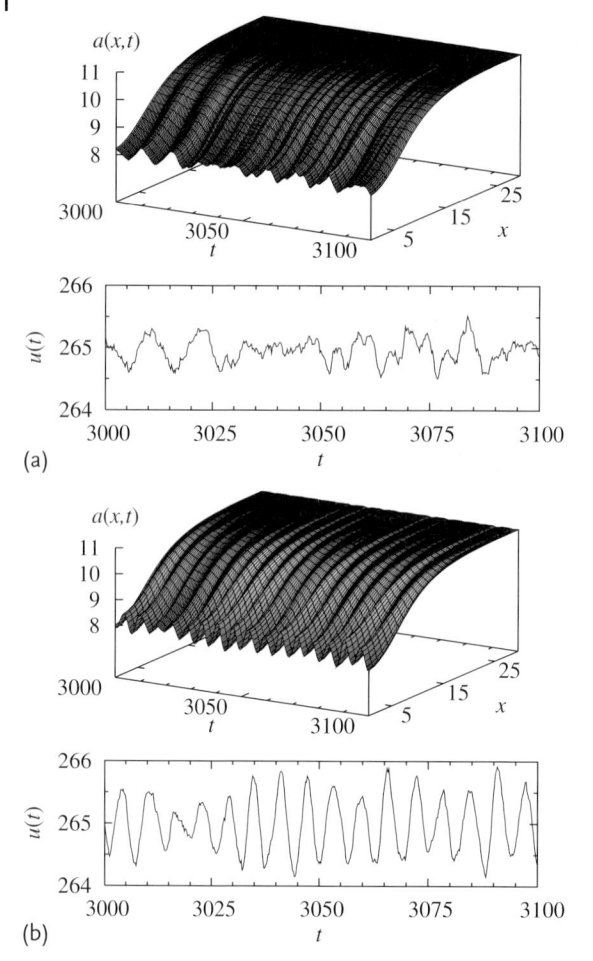

Figure 12.23 Spatiotemporal patterns $a(x, t)$ and voltage time series $u(t)$ for different values of the control strength K and delay time τ: (a) $\tau = 4.0$, $K = 0.4$, (b) $\tau = 13.4$, $K = 0.1$. $D_u = 0.1$, $D_a = 10^{-4}$, $R = 0$ and other parameters as in Figure 12.18 [62].

τ which are related to minimum regularity of the noise-induced dynamics. At the same time, it is shown that the control with $K = 0.1$ produces no effect upon the correlation time if the noise is too large (lower curve for $D_u = 1.0$).

The fact that noise-induced oscillations take place in the vicinity of the spatially inhomogeneous fixed point gives us a hint that some properties of these oscillations could relate to the stability of this fixed point. To gain some insight into how the control actually affects the systems dynamics around the spatially inhomogeneous fixed point, we linearize the system equations (12.26) for $D_u = D_a = 0$ and calculate the complex eigenvalues Λ_i at the fixed point. First of all, we calculate these eigenvalues from the spatially discretized system which we use for

12.5 Noise-Induced Spatiotemporal Patterns in Resonant Tunneling Diodes

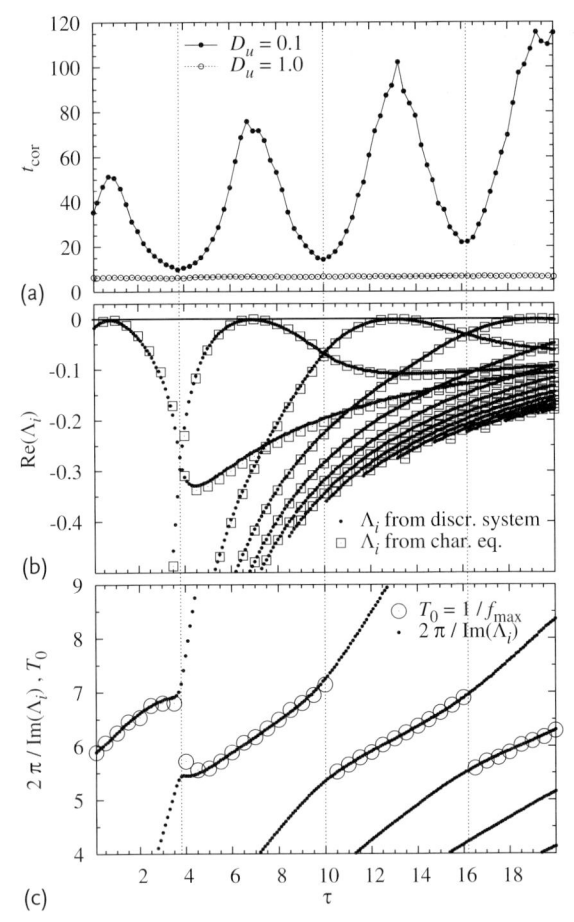

Figure 12.24 (a) Correlation time (12.30) for two different noise intensities in dependence of the feedback delay τ for $R = 0$. (b) Real parts of the eigenvalues Λ_i of the linearized deterministic system ($D_a = D_u = 0$) calculated at the spatially inhomogeneous fixed point for $K = 0.1$. The black dots are calculated from the spatially discretized system (set of ODEs) whereas the squares are calculated from (12.36) (see text). The vertical dotted lines mark values of τ at which the *leading* eigenvalue (i.e. the one with the largest real part) changes. (c) Eigenperiods $2\pi/\text{Im}(\Lambda_i)$ of the deterministic system and basic periods $T_0 := 1/f_{\max}$ of the noise-induced oscillations, where $f_{\max}$ denotes the frequency of the highest peak in the Fourier power spectral density of the noisy system with $D_u = 0.1$, $K = 0.1$ [62].

the numerical simulation. This discretized version is just a set of ordinary differential equations (ODEs) and the linearization and the eigenvalues can be easily computed.

In Figure 12.24b, one can see that the control with $K = 0.1$ does not change the stability of the inhomogenous fixed point since the real parts of all eigenvalues do not become positive within the given range of τ. Nevertheless, by increasing τ, the real parts of the eigenvalues intersect at particular values of τ (vertical dotted lines) and therefore the *leading* eigenvalue, that is, the *least stable* one or the one with the largest real part, changes at these values of τ. As one can see, these crossover points correspond to the minima of the correlation time in Figure 12.24a, whereas the local maxima of the real parts correspond to the maxima of the correlation time. This gives rise to a rather intuitive explanation for the behavior of the correlation time. The closer to zero the real part of an eigenvalue is, the weaker the attracting stability of the fixed point and the easier it is for the noise to excite the oscillating mode corresponding to this particular eigenvalue. On the other hand, at the intersection points of the real parts of the leading eigenvalue, these values have the largest distance from zero, meaning that the attracting stability of the fixed point is stronger. In addition, there are two different corresponding oscillating modes which are excited by the noise. Thus, the control cannot reach its optimal effect.

As a direct consequence, the main frequency which is activated by the noise switches exactly at these values of τ to the eigenfrequency of the corresponding leading eigenvalue. In Figure 12.24c, the eigenperiods are plotted as black dots in dependence of τ. The circles mark the positions of the highest peak in the Fourier power spectrum for the corresponding noisy system with $D_u = 0.1$. One can clearly see that these main periods switch from one branch to another exactly at the positions where the real parts of two different eigenvalues crossover.

In order to achieve a deeper understanding, we examine the stability properties of the inhomogeneous fixed point $(a_0(x), u_0)$ under the influence of the control force. We perform a linearization of the original continuous system (12.26) (with $D_u = D_a = 0$) around the inhomogeneous fixed point along the same lines as in [62, 99]. We use an exponential ansatz for the deviations from the fixed point $\delta a(x,t) \equiv a(x,t) - a_0(x) = e^{\Lambda t}\tilde{a}(x)$ and $\delta u(t) \equiv u(t) - u_0 = e^{\Lambda t}\tilde{u}$. The resulting coupled eigenvalue problem reads

$$\Lambda \tilde{a}(x) = \hat{H}\tilde{a}(x) + f_u(x)\tilde{u}, \tag{12.31}$$

$$\Lambda \tilde{u} = -\frac{r}{\varepsilon L}\int_0^L j_a(x)\tilde{a}(x)\,dx - \frac{1+rJ_u}{\varepsilon}\tilde{u}$$
$$- K\frac{1-e^{-\Lambda\tau}}{1-Re^{-\Lambda\tau}}\tilde{u}, \tag{12.32}$$

where we have introduced a self-adjoint linear operator $\hat{H}$. Its eigenfunctions Ψ_i and eigenvalues λ_i correspond to the voltage-clamped system, $\delta u = 0$. Furthermore,

$$f_u \equiv \left.\frac{\partial f}{\partial u}\right|_{a_0,u_0}, \quad j_a \equiv \left.\frac{\partial j}{\partial a}\right|_{a_0,u_0},$$

$$J_u = \frac{1}{L}\int_0^L \left.\frac{\partial j}{\partial u}\right|_{a_0,u_0} dx. \tag{12.33}$$

Due to the global constraint, (12.32) mixes the eigenmodes Ψ_i and both equations have to be solved simultaneously. An expansion of the eigenmodes $\tilde{a}$ of the full system in terms of the eigenmodes Ψ_i of the voltage-clamped system, keeping only the dominant eigenmode Ψ_0 with eigenvalue $\lambda_0 > 0$, leads to a characteristic equation for the eigenvalues Λ

$$\Lambda^2 + \left(\frac{1+rJ_u}{\varepsilon} - \lambda_0\right)\Lambda + (\Lambda-\lambda_0)K\frac{1-e^{-\Lambda\tau}}{1-Re^{-\Lambda\tau}} - \frac{\lambda_0}{\varepsilon}(1+r\sigma_d) = 0 , \quad (12.34)$$

where the static differential conductance at the inhomogeneous fixed point $\sigma_d \equiv \frac{dJ}{du}\big|_{a_0,u_0}$ has been introduced. In [62], a more detailed derivation of the characteristic equation is given for the special case $R = 0$. The extension to the case $R \neq 0$ is straightforward.

Without control, $K = 0$, (12.34) reduces to a characteristic polynomial of second order, which gives the well known conditions for stability of a filament [99]

$$A \equiv \frac{1+rJ_u}{\varepsilon} - \lambda_0 > 0 ,$$

$$C \equiv -\frac{\lambda_0}{\varepsilon}(1+r\sigma_d) > 0 . \quad (12.35)$$

A Hopf bifurcation occurs on the two-dimensional center manifold if $A = 0$.

With control, (12.34) can be expressed as

$$\Lambda^2 + A\Lambda + (\Lambda - B)K\frac{1-e^{-\Lambda\tau}}{1-Re^{-\Lambda\tau}} + C = 0 \quad (12.36)$$

with $B \equiv \lambda_0 > 0$. The parameters A, B, C can be calculated directly from (12.35) [62], yielding $A = 0.0447$, $B = 1.0281$ and $C = 1.1458$.

Using (12.36), we can calculate the domains of stability in the τ-K plane numerically for selected values of the memory parameter R. In order to find the curves containing the boundaries of stability of the inhomogeneous fixed point as a subset, we set $\Lambda = p + iq$ with $p = 0$ and separate (12.36) into real and imaginary parts:

$$\begin{aligned}
\left[BK - R\left(C - q^2\right)\right]\cos(q\tau) - q\left(AR + K\right)\sin(q\tau) &= BK + \left(q^2 - C\right)\\
q\left(AR + K\right)\cos(q\tau) + \left[BK + R\left(q^2 - C\right)\right]\sin(q\tau) &= q\left(A + K\right).
\end{aligned} \quad (12.37)$$

Using (12.37), the boundary of stability can be obtained from the set of parametric functions $K(q)$ and $\tau(q)$ using $q = \mathrm{Im}(\Lambda)$ as the curve parameter.

$$K(q) = \frac{\left(A^2 q^2 + (C - q^2)^2\right)(1+R)}{2BC - 2(A+B)q^2}$$

$$\tau(q) = \frac{1}{q}\left(\arcsin \Phi + 2\pi N\right)$$

$$\Phi = \frac{q(AB + C - q^2)(1-R)K}{(A^2 q^2 + (C - q^2)^2)R^2 + 2(-BC + (A+B)q^2)RK + (B^2 + q^2)K^2}$$

$$(12.38)$$

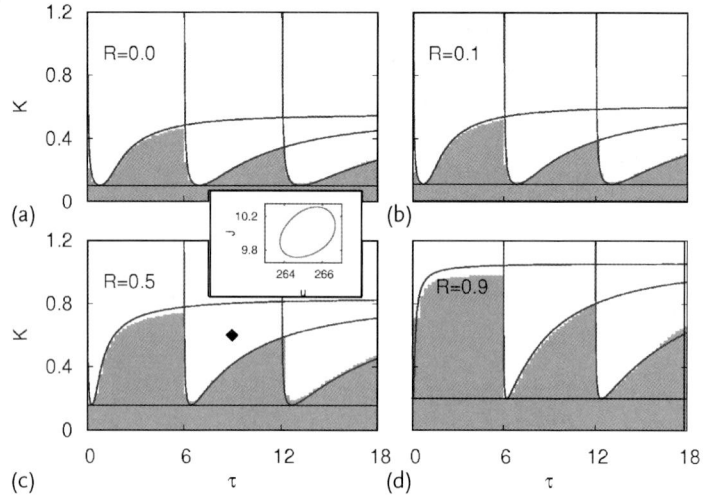

Figure 12.25 Stability domains of the inhomogeneous fixed point in the τ-K plane of the deterministic system (12.26) ($D_u = D_a = 0$), for selected values of the memory parameter R. Curved lines: Solutions of (12.36) with Re(Λ) = 0 calculated from (12.38). Gray region: regime of stability of the fixed point obtained from the numerical solution of (12.26). Black horizontal line: upper bound for K where the fixed point is stable for all values of τ, calculated from (12.39). The black diamond in panel (c) marks the parameter values for which a (J-u) phase portrait of the delay-induced limit cycle is shown in the inset [71].

Figure 12.25 shows these curves (12.38). The boundaries of stability where a delay-induced Hopf bifurcation of deterministic breathing oscillations occurs are a subset of these curves because the fixed point may already be unstable when a complex eigenvalue crosses the imaginary axis due to other unstable eigenvalue branches. The boundaries are in good agreement with the domain of stability obtained from dynamical simulations of the nonlinear system equations (12.26) ($D_u = D_a = 0$), shown as gray areas. The inset of panel (c) shows the delay-induced limit cycle in the J–u phase space for parameters outside the stability domain of the inhomogeneous fixed point. The stability domains increase significantly with increasing memory parameter R from (a) to (d). The modulation of their boundaries in dependence on τ results from the crossover of different eigenvalue branches, which is a typical feature of delay differential equations.

From (12.38), it is possible to calculate an upper bound K_c of K for which the stability properties of the uncontrolled deterministic system remain unchanged over the whole τ interval, meaning that no delay-induced Hopf bifurcation occurs.

$$K_c = \frac{A^2(A+B)(BC-G) + (AC+G)^2}{2(A+B)^2 G}(1+R)$$

$$\approx 0.1059(1+R)$$

$$G := \sqrt{A^2 C(B(A+B) + C)} \tag{12.39}$$

Figure 12.25 shows this upper bound plotted as a black horizontal line.

In Figure 12.26, the Fourier power spectral density $S_{uu}(f)$ obtained from the time series $u(t)$ is shown for different values of the delay time τ and the memory parameter R. The shape of the spectra $S_{uu}(f)$ alternates between broad and sharply peaked with varying τ. An excellent analytic approximation of the power spectral density can be obtained by a straightforward extension of the arguments in [53, 62] to multiple time-delayed feedback:

$$S_{uu}(\omega) = \frac{D'^2}{2\pi}\left[\left(-\omega^2 + BK\frac{(\cos(\omega\tau)-1)(R+1)}{1+R^2-2R\cos(\omega\tau)} - \frac{\omega K(1-R)\sin(\omega\tau)}{1+R^2-2R\cos(\omega\tau)} + C\right)^2 + \left(-A\omega + \omega K\frac{(\cos(\omega\tau)-1)(R+1)}{1+R^2-2R\cos(\omega\tau)} + BK\frac{(1-R)\sin(\omega\tau)}{1+R^2-2R\cos(\omega\tau)}\right)^2\right]^{-1},$$

(12.40)

which is shown as black curves in Figure 12.26.

At certain resonant values of τ (left column), the spectral peaks become extremely sharp. With increasing memory parameter R, the broad spectra prevail over larger intervals of τ, whereas the regime of sharply peaked spectra becomes smaller. Thus, multiple time feedback control exhibits more pronounced resonant features both in the frequency domain and in the delay time. Since a sharply peaked spectrum gives rise to long correlation times (which are in the linear regime proportional to the inverse spectral width), we expect the domains of strong correlation to shrink with increasing memory parameter and the domains of low correlation to increase. Extracting from the Fourier power spectral density, the autocorrelation function $\Psi(s) = \int_{-\infty}^{\infty} S_{uu}(f) e^{2\pi i f s} df$ and using (12.30), we obtain the correlation time t_{cor} in dependence of τ. This is shown in Figure 12.27 for different values of the memory parameter R. The feedback strength is kept at a constant value of $K = 0.1$, where the system is below the Hopf bifurcation for all values of τ and R.

For small memory parameter R, the correlation times alternate between high and low values with growing τ. For higher memory parameters R, the peaks in correlation time become narrower and sharpen up, and the domains of low correlation time increase. The stability of the inhomogeneous fixed point reveals a relation between properties of the controlled deterministic system and the noise-induced dynamics: maximum regularity of noise-induced oscillations is attained when the deterministic fixed point is least stable. This feature is maintained for all values of the memory parameter R. In the case of small R, the crossover of the real part of eigenvalue branches also determines the location of the minima in correlation time. In that case, two eigenmodes with the same stability (real part) but different frequencies are present in the system, resulting in rather irregular noise-induced dynamics. For large memory parameters, the broad domains of low correlation display many eigenmodes that are not well separated (stability-wise), causing irregular mixed dynamics.

In conclusion, the regularity of noise-induced oscillations measured by the correlation time exhibits sharp resonances as a function of the delay time τ, and can

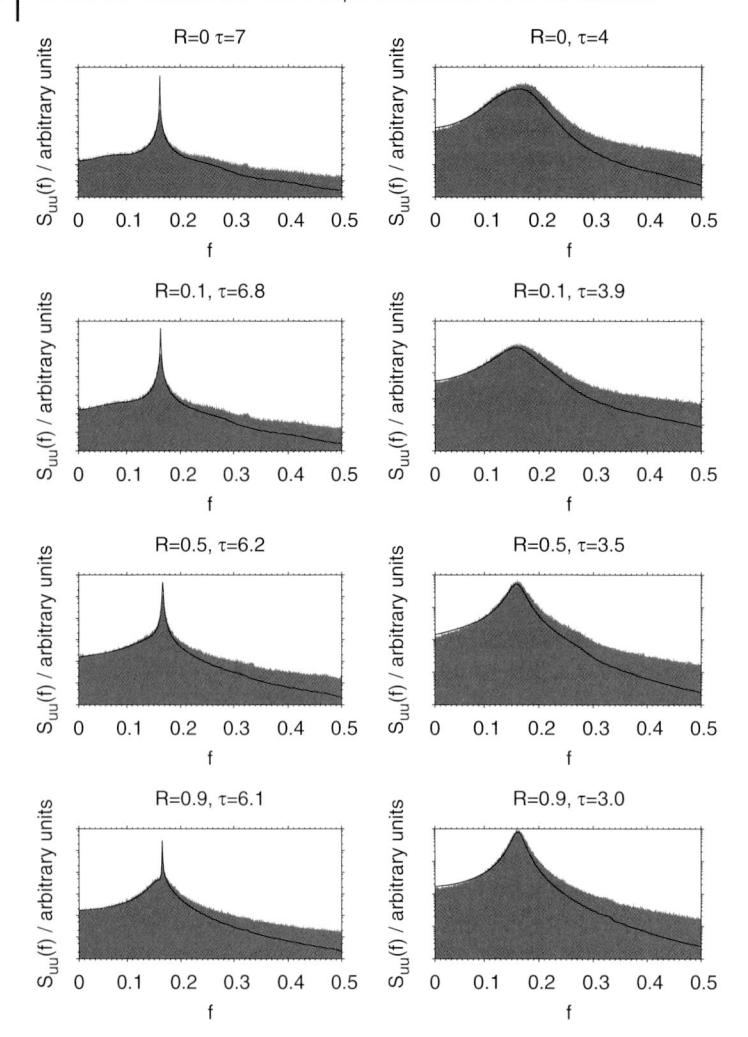

Figure 12.26 Power spectral density $S_{uu}(f)$ of the dynamical variable u in dependence of the frequency f for various delay times τ and memory parameters R ($K = 0.1$, $\epsilon = 6.2$, $D_u = 0.1$, $D_a = 10^{-4}$) [71].

be strongly increased by control with optimal choices of τ, whereas it decreases in a broad range of non-optimal values of τ. These resonant features are enhanced by multiple time-delayed feedback control, as compared to single time feedback. Similarly, the peaks in the power spectral density are sharper and exhibit stronger resonances in dependence on τ for multiple time feedback, whereas their position, namely, the main period of the oscillations, is less sensitive to variations in τ in wider intervals.

Furthermore, delay-induced bifurcations occur at some threshold values of the control amplitude K. For multiple time-delayed feedback, the regime of stability of

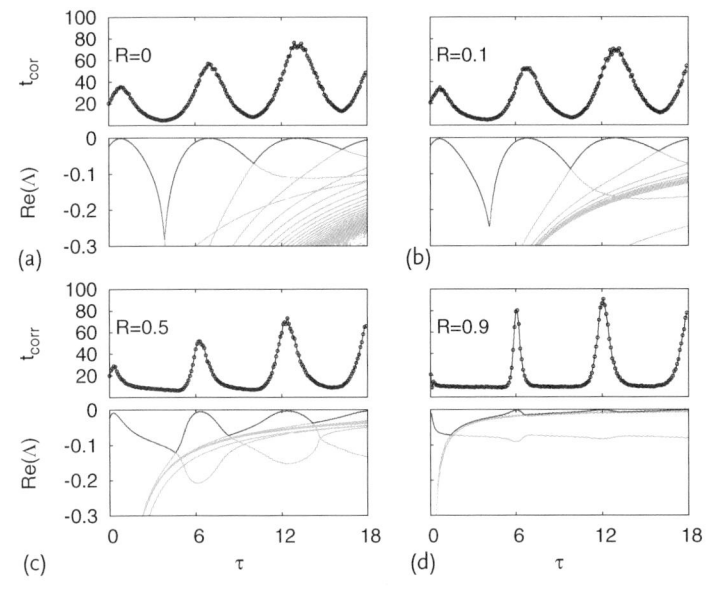

Figure 12.27 Correlation times t_{cor} (upper panels of (a)–(d)) and deterministic stability of the inhomogeneous fixed point, $Re(\Lambda)$ (lower panels of (a)–(d)), in dependence of the delay time τ for different values of the memory parameter R (a)–(d) and fixed $K = 0.1$. The dark-gray curves in the lower panels mark the leading eigenvalue which governs the overall stability of the fixed point. Parameters: $\epsilon = 6.2$, $D_u = 0.1$ and $D_a = 10^{-4}$ (in the panels showing t_{cor}) [71].

the stationary filamentary current pattern in the deterministic system is larger than for single time-delayed feedback.

12.6
Conclusion

We have investigated the complex spatiotemporal behavior of two semiconductor nanostructures, namely, the superlattice and the double barrier resonant tunneling diode (DBRT). The first exhibits nonlinear dynamics of interacting fronts, while the second demonstrates breathing and spiking of filamentary current density patterns characteristic of globally coupled reaction-diffusion systems. By applying time-delayed Pyragas-type feedback control to both deterministic and stochastic oscillations, we have been able to suppress deterministic chaos and control the regularity and the mean period of noise-induced dynamics.

As an example for the constructive influence of noise in nonlinear systems, we have shown that random fluctuations are able to induce quite coherent oscillations of the current density in a regime where the deterministic system exhibits a stable fixed point, thereby demonstrating the phenomenon of coherence resonance for systems close to, but below, a Hopf bifurcation (superlattice and DBRT) as well as close to, but below, a global saddle-node bifurcation on a limit cycle (superlattice). This extends the phenomena of noise-induced oscillations from purely time-dependent generic models, for example, [20], to space-time patterns. Moreover, we have shown that for the DBRT, the noise which is applied globally to a space-independent variable determines the type of the spatiotemporal pattern of these oscillations. While for small noise intensity the system demonstrates oscillations which are quite correlated in time, but spatially inhomogeneous, with increasing noise intensity the shape of the spatiotemporal pattern changes qualitatively until the system reaches a highly homogeneous state. Thus, the increase of spatial coherence is accompanied by the decrease of temporal correlation of the observed oscillations. In between these two situations, for intermediate noise strength one can observe complex spatiotemporal behavior resulting from the competition between homogeneous and inhomogeneous oscillations.

We have seen that delayed feedback can be an efficient method for manipulation of essential characteristics of chaotic or noise-induced spatiotemporal dynamics in a spatially discrete front system and in a continuous reaction-diffusion system. By variation of the time delay, one can stabilize particular unstable periodic orbits associated with space-time patterns, or deliberately change the timescale of oscillatory patterns and thus adjust and stabilize the frequency of the electronic device. Moreover, with a proper choice of feedback parameters, one can also effectively control the coherence of spatiotemporal dynamics and, for example, enhance or destroy it. Increase of coherence is possible up to a reasonably large intensity of noise. However, as the level of noise grows, the efficiency of the control upon the temporal coherence decreases.

The effects of the delayed feedback can be explained in terms of a Floquet mode analysis of the periodic orbits, or a linear stability analysis of the fixed point. For a better understanding of noise-induced patterns in the DBRT, we have derived the general form of the characteristic equation for the deterministic system close to, but below, a Hopf bifurcation. Both dependences, coherence and timescale versus τ, demonstrate an oscillatory character which can be explained by oscillations of the real and imaginary parts of the eigenvalues of the linearized system at the fixed point in the vicinity of which the noise-induced oscillations occur. The most coherent timescale corresponds to values of τ for which the real parts of the eigenvalues attain a maximum. In some sense, the noise excites the least stable eigenmode: the less stable an eigenmode is, the greater the coherence of the corresponding oscillations. Multiple time-delayed feedback control leads to more pronounced resonant features of noise-induced spatiotemporal current oscillations compared to single time feedback, rendering the system more sensitive to variations in the delay time. It also enlarges the regime in which the deterministic dynamical properties of the system are not changed by delay-induced bifurcations.

While these investigations have enlightened our basic understanding of nonlinear, spatially extended systems under the influence of time-delayed feedback and noise, they may also introduce relevant applications as nanoelectronic devices, such as oscillators and sensors.

Acknowledgements

This work was supported by DFG in the framework of Sfb 555. I am indebted to the stimulating collaboration and discussions with A. Amann, R. Aust, N. Baba, A. Balanov, O. Beck, V. Flunkert, J. Hizanidis, P. Hövel, N. Janson, W. Just, M. Kehrt, K. Lüdge, N. Majer, J. Pomplun, S. Popovich, P. Rodin, J. Schlesner, J. E. S. Socolar, G. Stegemann, J. Unkelbach, A. Wacker, M. Wolfrum, H. J. Wünsche, and S. Yanchuk.

References

1 Haken, H. (1983) *Synergetics, An Introduction*, Springer, Berlin, 3rd edn.
2 Haken, H. (1987) *Advanced Synergetics*, Springer, Berlin, 2nd edn.
3 Mikhailov, A.S. (1994) *Foundations of Synergetics Vol. I*, Springer, Berlin, 2nd edn.
4 Schimansky-Geier, L., Fiedler, B., Kurths, J., and Schöll, E. (eds) (2007) *Analysis and control of complex nonlinear processes in physics, chemistry and biology*, World Scientific, Singapore.
5 Schuster, H.G. (ed.) (2008) *Reviews of Nonlinear Dynamics and Complexity*, Wiley-VCH Verlag, Weinheim.
6 Schöll, E. (1987) *Nonequilibrium Phase Transitions in Semiconductors*, Springer, Berlin.
7 Shaw, M.P., Mitin, V.V., Schöll, E., and Grubin, H.L. (1992) *The Physics of Instabilities in Solid State Electron Devices*, Plenum Press, New York.
8 Schöll, E. (2001) *Nonlinear spatio-temporal dynamics and chaos in semiconductors*, Cambridge Univ. Press, Cambridge.
9 Bimberg, D. (2008) *Semiconductor Nanostructures*, Springer, Berlin.
10 Hodgkin, A.L. and Huxley, A.F. (1952) A quantitative description of membrane current and its application to conduction and excitation in nerve. *J. Physiol.*, **117**, 500.
11 Peinke, J., Parisi, J., Rössler, O.E., and Stoop, R. (1992) *Encounter with Chaos*, Springer, Berlin, Heidelberg.
12 Niedernostheide, F.J. (ed.) (1995) *Nonlinear Dynamics and Pattern Formation in Semiconductors and Devices*, Springer, Berlin.
13 Aoki, K. (2000) *Nonlinear dynamics and chaos in semiconductors*, Institute of Physics Publishing, Bristol.
14 Schöll, E. and Schuster, H.G. (eds) (2008) *Handbook of Chaos Control*, Wiley-VCH Verlag, Weinheim, 2nd edition.
15 Nijmeijer, H. and Schaft, A.V.D. (1996) *Nonlinear Dynamical Control Systems*, Springer, New York, 3rd edn.
16 Ogata, K. (1997) *Modern Control Engineering*, Prentice-Hall, New York.
17 Fradkov, A.L., Miroshnik, I.V., and Nikiforov, V.O. (1999) *Nonlinear and Adaptive Control of Complex Systems*, Kluwer, Dordrecht.
18 Ott, E., Grebogi, C., and Yorke, J.A. (1990) Controlling chaos. *Phys. Rev. Lett.*, **64**, 1196.
19 Pyragas, K. (1992) Continuous control of chaos by self-controlling feedback. *Phys. Lett. A*, **170**, 421.
20 Janson, N.B., Balanov, A.G., and Schöll, E. (2004) Delayed feedback as a means of control of noise-induced motion. *Phys. Rev. Lett.*, **93**, 010601.
21 Pomplun, J., Amann, A., and Schöll, E. (2005) Mean field approximation of time-delayed feedback control of noise-induced oscillations in the Van der Pol system. *Europhys. Lett.*, **71**, 366.
22 Hu, G., Ditzinger, T., Ning, C.Z., and Haken, H. (1993) Stochastic resonance without external periodic force. *Phys. Rev. Lett.*, **71**, 807.
23 Pikovsky, A. and Kurths, J. (1997) Coherence resonance in a noise-driven excitable system. *Phys. Rev. Lett.*, **78**, 775.

24 García-Ojalvo, J. and Sancho, J.M. (1999) *Noise in Spatially Extended Systems*, Springer, New York.

25 Lindner, B., García-Ojalvo, J., Neiman, A., and Schimansky-Geier, L. (2004) Effects of noise in excitable systems. *Phys. Rep.*, **392**, 321.

26 Sagués, F., Sancho, J.M., and García-Ojalvo, J. (2007) Spatiotemporal order out of noise. *Rev. Mod. Phys.*, **79**, 829.

27 Socolar, J.E.S., Sukow, D.W., and Gauthier, D.J. (1994) Stabilizing unstable periodic orbits in fast dynamical systems. *Phys. Rev. E*, **50**, 3245.

28 Bleich, M.E. and Socolar, J.E.S. (1996) Stability of periodic orbits controlled by time-delay feedback. *Phys. Lett. A*, **210**, 87.

29 Just, W., Bernard, T., Ostheimer, M., Reibold, E., and Benner, H. (1997) Mechanism of time-delayed feedback control. *Phys. Rev. Lett.*, **78**, 203.

30 Nakajima, H. (1997) On analytical properties of delayed feedback control of chaos. *Phys. Lett. A*, **232**, 207.

31 Pyragas, K. (2002) Analytical properties and optimization of time-delayed feedback control. *Phys. Rev. E*, **66**, 26207.

32 Just, W., Benner, H., and Schöll, E. (2003) Control of chaos by time-delayed feedback: a survey of theoretical and experimental aspects, in *Advances in Solid State Physics*, (ed. B. Kramer), Springer, Berlin, Vol. 43, pp. 589–603.

33 Hövel, P. and Socolar, J.E.S. (2003) Stability domains for time-delay feedback control with latency. *Phys. Rev. E*, **68**, 036206.

34 Hövel, P. and Schöll, E. (2005) Control of unstable steady states by time-delayed feedback methods. *Phys. Rev. E*, **72**, 046203.

35 Yanchuk, S., Wolfrum, M., Hövel, P., and Schöll, E. (2006) Control of unstable steady states by long delay feedback. *Phys. Rev. E*, **74**, 026201.

36 Dahms, T., Hövel, P., and Schöll, E. (2007) Control of unstable steady states by extended time-delayed feedback. *Phys. Rev. E*, **76**, 056201.

37 Amann, A., Schöll, E., and Just, W. (2007) Some basic remarks on eigenmode expansions of time-delay dynamics. *Physica A*, **373**, 191.

38 Fiedler, B., Flunkert, V., Georgi, M., Hövel, P., and Schöll, E. (2007) Refuting the odd number limitation of time-delayed feedback control. *Phys. Rev. Lett.*, **98**, 114101.

39 Just, W., Fiedler, B., Flunkert, V., Georgi, M., Hövel, P. and Schöll, E. (2007) Beyond odd number limitation: a bifurcation analysis of time-delayed feedback control. *Phys. Rev. E*, **76**, 026210.

40 Fiedler, B., Yanchuk, S., Flunkert, V., Hövel, P., Wünsche, H.J., and Schöll, E. (2008) Delay stabilization of rotating waves near fold bifurcation and application to all-optical control of a semiconductor laser. *Phys. Rev. E*, **77**, 066207.

41 Balanov, A.G., Janson, N.B., and Schöll, E. (2005) Delayed feedback control of chaos: Bifurcation analysis. *Phys. Rev. E*, **71**, 016222.

42 Hizanidis, J., Aust, R., and Schöll, E. (2008) Delay-induced multistability near a global bifurcation. *Int. J. Bifur. Chaos*, **18**, 1759.

43 Schöll, E. and Pyragas, K. (1993) Tunable semiconductor oscillator based on self-control of chaos in the dynamic Hall effect. *Europhys. Lett.*, **24**, 159.

44 Reznik, D. and Schöll, E. (1993) Oscillation modes, transient chaos and its control in a modulation-doped semiconductor double-heterostructure. *Z. Phys. B*, **91**, 309.

45 Cooper, D.P. and Schöll, E. (1995) Tunable real space transfer oscillator by delayed feedback control of chaos. *Z. Naturforsch.*, **50a**, 117.

46 Schlesner, J., Amann, A., Janson, N.B., Just, W., and Schöll, E. (2003) Self-stabilization of high frequency oscillations in semiconductor superlattices by time-delay autosynchronization. *Phys. Rev. E*, **68**, 066208.

47 Franceschini, G., Bose, S., and Schöll, E. (1999) Control of chaotic spatiotemporal spiking by time-delay autosynchronisation. *Phys. Rev. E*, **60**, 5426.

48 Beck, O., Amann, A., Schöll, E., Socolar, J.E.S., and Just, W. (2002) Comparison of time-delayed feedback schemes for spatio-temporal control of chaos in a reaction-diffusion system with global coupling. *Phys. Rev. E*, **66**, 016213.

49 Baba, N., Amann, A., Schöll, E., and Just, W. (2002) Giant improvement of time-delayed feedback control by spatiotemporal filtering. *Phys. Rev. Lett.*, **89**, 074101.

50 Just, W., Popovich, S., Amann, A., Baba, N., and Schöll, E. (2003) Improvement of time-delayed feedback control by periodic modulation: Analytical theory of Floquet mode control scheme. *Phys. Rev. E*, **67**, 026222.

51 Unkelbach, J., Amann, A., Just, W., and Schöll, E. (2003) Time-delay autosynchronization of the spatiotemporal dynamics in resonant tunneling diodes. *Phys. Rev. E*, **68**, 026204.

52 Balanov, A.G., Janson, N.B., and Schöll, E. (2004) Control of noise-induced oscillations by delayed feedback. *Physica D*, **199**, 1.

53 Schöll, E., Balanov, A.G., Janson, N.B., and Neiman, A. (2005) Controlling stochastic oscillations close to a Hopf bifurcation by time-delayed feedback. *Stoch. Dyn.*, **5**, 281.

54 Hauschildt, B., Janson, N.B., Balanov, A.G., and Schöll, E. (2006) Noise-induced cooperative dynamics and its control in coupled neuron models. *Phys. Rev. E*, **74**, 051906.

55 Balanov, A.G., Beato, V., Janson, N.B., Engel, H., and Schöll, E. (2006) Delayed feedback control of noise-induced patterns in excitable media. *Phys. Rev. E*, **74**, 016214.

56 Flunkert, V. and Schöll, E. (2007) Suppressing noise-induced intensity pulsations in semiconductor lasers by means of time-delayed feedback. *Phys. Rev. E*, **76**, 066202.

57 Pomplun, J., Balanov, A.G., and Schöll, E. (2007) Long-term correlations in stochastic systems with extended time-delayed feedback. *Phys. Rev. E*, **75**, 040101(R).

58 Prager, T., Lerch, H.P., Schimansky-Geier, L., and Schöll, E. (2007) Increase of coherence in excitable systems by delayed feedback. *J. Phys. A*, **40**, 11045.

59 Hizanidis, J., Balanov, A.G., Amann, A., and Schöll, E. (2006) Noise-induced oscillations and their control in semiconductor superlattices. *Int. J. Bifur. Chaos*, **16**, 1701.

60 Hizanidis, J. and Schöll, E. (2008) Control of noise-induced spatiotemporal patterns in superlattices. *Phys. Status. Solidi. (c)*, **5**, 207.

61 Kehrt, M., Hövel, P., Rodin, P., and Schöll, E. (2009) Stabilisation of complex spatiotemporal dynamics near a subcritical Hopf bifurcation. *Eur. Phys. J. B*, **68**, 557.

62 Stegemann, G., Balanov, A.G., and Schöll, E. (2006) Delayed feedback control of stochastic spatiotemporal dynamics in a resonant tunneling diode. *Phys. Rev. E*, **73**, 016203.

63 Schöll, E., Majer, N., and Stegemann, G. (2008) Extended time delayed feedback control of stochastic dynamics in a resonant tunneling diode. *Phys. Status. Solidi. (c)*, **5**, 194.

64 Amann, A., Schlesner, J., Wacker, A., and Schöll, E. (2002) Chaotic front dynamics in semiconductor superlattices. *Phys. Rev. B*, **65**, 193313.

65 Amann, A., Peters, K., Parlitz, U., Wacker, A., and Schöll, E. (2003) Hybrid model for chaotic front dynamics: From semiconductors to water tanks. *Phys. Rev. Lett.*, **91**, 066601.

66 Amann, A. and Schöll, E. (2005) Bifurcations in a system of interacting fronts. *J. Stat. Phys.*, **119**, 1069.

67 Hizanidis, J., Balanov, A.G., Amann, A., and Schöll, E. (2006) Noise-induced front motion: signature of a global bifurcation. *Phys. Rev. Lett.*, **96**, 244104.

68 Hizanidis, J. and Schöll, E. (2009) Control of coherence resonance in semiconductor superlattices. *Phys. Rev. E*, **78**, 066205.

69 Schöll, E., Amann, A., Rudolf, M., and Unkelbach, J. (2002) Transverse spatiotemporal instabilities in the double barrier resonant tunneling diode. *Physica B*, **314**, 113.

70 Stegemann, G., Balanov, A.G., and Schöll, E. (2005) Noise-induced pattern formation in a semiconductor nanostructure. *Phys. Rev. E*, **71**, 016221.

71 Majer, N. and Schöll, E. (2009) Resonant control of stochastic spatio-temporal dynamics in a tunnel diode by multiple time delayed feedback. *Phys. Rev. E*, **79**, 011109.

72 Esaki, L. and Tsu, R. (1970) Superlattice and negative differential conductivity in semiconductors. *IBM J. Res. Dev.*, **14**, 61.

73 Kastrup, J., Klann, R., Grahn, H.T., Ploog, K., Bonilla, L.L., Galán, J., Kindelan, M., Moscoso, M., and Merlin, R. (1995) Self-oscillations of domains in doped GaAs–AlAs superlattices. *Phys. Rev. B*, **52**, 13761.

74 Hofbeck, K., Grenzer, J., Schomburg, E., Ignatov, A.A., Renk, K.F., Pavel'ev, D.G., Koschurinov, Y., Melzer, B., Ivanov, S., Schaposchnikov, S., and Kop'ev, P.S. (1996) High-frequency self-sustained current oscillation in an Esaki–Tsu superlattice monitored via microwave emission. *Phys. Lett. A*, **218**, 349.

75 Schomburg, E., Scheuerer, R., Brandl, S., Renk, K.F., Pavel'ev, D.G., Koschurinov, Y., Ustinov, V.M., Zhukov, A.E., Kovsh, A.R., and Kop'ev, P.S. (1999) InGaAs/InAlAs superlattice oscillator at 147 GHz. *Electron. Lett.*, **35**, 1491.

76 Wacker, A. (2002) Semiconductor superlattices: A model system for nonlinear transport. *Phys. Rep.*, **357**, 1.

77 Bonilla, L.L. (2002) Theory of nonlinear charge transport, wave propagation, and self-oscillations in semiconductor superlattices. *J. Phys.: Condens. Matter*, **14**, R341.

78 Bonilla, L.L. and Grahn, H.T. (2005) Non-linear dynamics of semiconductor superlattices. *Rep. Prog. Phys.*, **68**, 577.

79 Zhang, Y., Kastrup, J., Klann, R., Ploog, K.H., and Grahn, H.T. (1996) Synchronization and chaos induced by resonant tunneling in GaAs/AlAs superlattices. *Phys. Rev. Lett.*, **77**, 3001.

80 Luo, K.J., Grahn, H.T., Ploog, K.H., and Bonilla, L.L. (1998) Explosive bifurcation to chaos in weakly coupled semiconductor superlattices. *Phys. Rev. Lett.*, **81**, 1290.

81 Bulashenko, O.M., Luo, K.J., Grahn, H.T., Ploog, K.H., and Bonilla, L.L. (1999) Multifractal dimension of chaotic attractors in a driven semiconductor superlattice. *Phys. Rev. B*, **60**, 5694.

82 Bulashenko, O.M. and Bonilla, L.L. (1995) Chaos in resonant-tunneling superlattices. *Phys. Rev. B*, **52**, 7849.

83 Amann, A., Schlesner, J., Wacker, A., and Schöll, E. (2003) Self-generated chaotic dynamics of field domains in superlattices, in *Proceedings of the 26th International Conference on the Physics of Semiconductors (ICPS-26), Edinburgh 2002* (eds J.H. Davies and A.R. Long).

84 Amann, A., Wacker, A., Bonilla, L.L., and Schöll, E. (2001) Dynamic scenarios of multi-stable switching in semiconductor superlattices. *Phys. Rev. E*, **63**, 066207.

85 Schlesner, J., Amann, A., Janson, N.B., Just, W., and Schöll, E. (2004) Self-stabilization of chaotic domain oscillations in superlattices by time-delayed feedback control. *Semicond. Sci. Technol.*, **19**, S34.

86 Gammaitoni, L., Hänggi, P., Jung, P., and Marchesoni, F. (1998) Stochastic resonance. *Rev. Mod. Phys.*, **70**, 223.

87 Blanter, Y.M. and Büttiker, M. (2000) Shot noise in mesoscopic conductors. *Phys. Rep.*, **336**, 1.

88 Andronov, A.A., Leontovich, E.A., Gordon, I.I., and Maier, A.G. (1971) *Theory of bifurcations of systems on a plane*, Vol. 2., Israel program for scientific translation, Jerusalem.

89 Guckenheimer, J. and Holmes, P. (1986) *Nonlinear Oscillations, Dynamical Systems, and Bifurcations of Vector Fields*, Springer, Berlin.

90 Amann, A., Wacker, A., and Schöll, E. (2002) Tripole current oscillations in superlattices. *Physica B*, **314**, 404.

91 Stratonovich, R.L. (1963) *Topics in the Theory of Random Noise*, Vol. 1, Gordon and Breach, New York.

92 Pototsky, A. and Janson, N.B. (2008) Excitable systems with noise and delay, with applications to control: Renewal theory approach. *Phys. Rev. E*, **77**, 031113.

93 Goldman, V.J., Tsui, D.C., and Cunningham, J.E. (1987) Observation of intrinsic bistability in resonant-tunneling structures. *Phys. Rev. Lett.*, **58**, 1256.

94 Meixner, M., Rodin, P., Schöll, E., and Wacker, A. (2000) Lateral current density fronts in globally coupled bistable semiconductors with S- or Z-shaped current voltage characteristic. *Eur. Phys. J. B*, **13**, 157.

95 Cheianov, V., Rodin, P., and Schöll, E. (2000) Transverse coupling in bistable resonant-tunneling structures. *Phys. Rev. B*, **62**, 9966.

96 Rodin, P. and Schöll, E. (2003) Lateral current density fronts in asymmetric

double-barrier resonant-tunneling structures. *J. Appl. Phys.*, **93**, 6347.
97 Stegemann, G. and Schöll, E. (2007) Two-dimensional spatiotemporal pattern formation in the double-barrier resonant tunneling diode. *New J. Phys.*, **9**, 55.
98 Martin, A.D., Lerch, M.L.F., Simmonds, P.E., and Eaves, L. (1994) Observation of intrinsic tristability in a resonant tunneling structure. *Appl. Phys. Lett.*, **64**, 1248.
99 Alekseev, A., Bose, S., Rodin, P., and Schöll, E. (1998) Stability of current filaments in a bistable semiconductor system with global coupling. *Phys. Rev. E*, **57**, 2640.
100 Just, W., Reibold, E., Benner, H., Kacperski, K., Fronczak, P., and Holyst, J. (1999) Limits of time-delayed feedback control. *Phys. Lett. A*, **254**, 158.
101 Just, W., Reibold, E., Kacperski, K., Fronczak, P., Holyst, J.A., and Benner, H. (2000) Influence of stable Floquet exponents on time-delayed feedback control. *Phys. Rev. E*, **61**, 5045.
102 Sherstnev, V.V., Krier, A., Balanov, A.G., Janson, N.B., Silchenko, A.N., and McClintock, P.V.E. (2003) The stochastic laser: mid-infrared lasing induced by noise. *Fluct. Noise Lett.*, **3**, 91.

Part V Optic-Electronic Coupling

13
Laser-Assisted Electron Transport in Nanoscale Devices

Ciprian Padurariu, Atef Fadl Amin, and Ulrich Kleinekathöfer

Electron transport in nanoscale devices such as molecular wires and quantum dots has become an intensively studied field of research in recent years [15]. On the theoretical side, the formalism of open quantum systems has been well developed [9, 17, 50, 58, 70], primarily considering quantum dissipation for cases of quantum systems coupled to a heat baths. In recent years, the focus has shifted partially to transport systems, as will be described below. To a large extent this is motivated by the experimental progress in manipulation of atoms and molecules with nanoscale precision [13, 26, 32, 48, 67]. Externally shaped or self-assembled structures of nanometer dimensions and high structural complexity are now being produced in a rather routine manner. For example, efforts to reliably produce and characterize devices where a single molecule acts as a current rectifier have increased considerably in the past decades [59]. Furthermore, developments in lithography have made it possible to create metallic islands consisting of a small number of atoms. These can be connected to metallic leads and allow the passing of electric current [6, 20]. Such small clusters of atoms, or quantum dots, exhibit distinctly quantum mechanical features, and due to their orbital-like distribution of quantum levels have been compared to artificial atoms. Other particularly interesting aspects of nanotechnology are the possibilities of increasing the efficiency of information storage and information processing using molecular electronic devices in which the motion and the spin of single electrons can be manipulated [72, 74].

In the present chapter, the focus is on theoretical descriptions of the transport through molecular junctions and quantum dots using time-dependent methods. The motivation stems from the idea of combining electron transport with coherent optical manipulation of the quantum dynamics of molecules and quantum dots [8, 41, 42, 53]. The increased precision and ease in the technique of shaping laser pulses allows increasingly accurate control of molecular quantum mechanical evolution by optical coupling. The laser field leads to a (possibly ultra-short) time-dependent perturbation of the system, thus inducing nonequilibrium dynamics that provide interesting opportunities. In the future, optically activated molecular devices could create current pulses of femtosecond temporal widths, speeds greatly exceeding those of conventional electronics. In addition to building new

---

Nonlinear Dynamics of Nanosystems. Edited by Günter Radons, Benno Rumpf, and Heinz Georg Schuster
Copyright © 2010 WILEY-VCH Verlag GmbH & Co. KGaA, Weinheim
ISBN: 978-3-527-40791-0

and promising opto-electronic devices, the combination of laser manipulation of quantum systems with the coupling to external leads may prove an important technique in spectroscopy. By this means, an optically induced current or a current induced fluorescent light signal could be transformed into information regarding the molecule included in the device, or information about the characteristics of the device itself [22, 23].

This contribution starts by describing the theoretical basics of open quantum systems before detailing the calculation of current and shot noise in transport systems. The quantum master equation (QME) developed is then applied to the so-called single resonant level model. In the subsequent section, numerical results for the transport properties are presented. Finally, a summary is provided in the final section.

13.1
Open Quantum Systems

There have been many theories developed capable of describing the many body interactions involved in electron transport through various junctions [10, 17, 58, 70]. Most of these theories stem from the field of solid state physics. Theories that describe the dynamics of molecules, including their interactions with the electromagnetic field of laser light, usually have their basis in molecular and chemical physics. This field is traditionally directed towards the dynamics of molecules and optical spectroscopy (see, e.g., [65]). The domain of interest here is situated at the junction of these two research fields. The study of molecules, or even of some discrete quantum levels, incorporated in electron transport devices borrows techniques from both solid state physics and molecular and chemical physics. The intention of this contribution is to show how the description of electron transport through molecules and quantum dots can profit from its connection to these fields.

The approach used to investigate a quantum mechanical transport system is to treat the degrees of freedom of the molecule or quantum dot that are actively involved in transport as a reduced quantum system. This reduced system is coupled through electric contacts to the leads. The leads, which are considered massive, thus act as electron reservoirs that may accept or donate electrons from and towards the relevant system (see Figure 13.1). At this point in the analysis, there are different ways in which to proceed. Nonequilibrium Green's function theory [33, 53] is a commonly used formalism. Recently, other accurate formalisms have been applied to transport situations. These include the real time path integral approach [55, 69], the multilayer multiconfiguration time-dependent Hartree method [68], and the numerical renormalization group approach [2]. These approaches are promising but computationally demanding, and further tests of their applicability ranges are required. In another class of approaches, QMEs are derived in a perturbative treatment of the coupling between the relevant system and environment. Many of the studies, including those in this contribution, are based on a second order formalism. In recent years, the perturbative approaches have begun to go beyond second

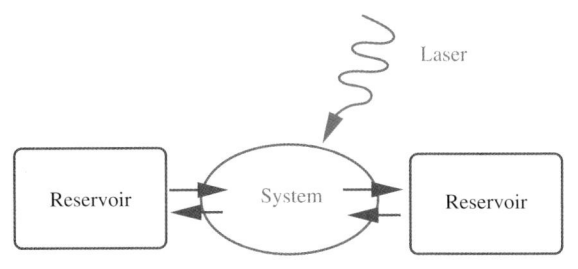

Figure 13.1 The concept of open quantum systems is illustrated schematically. The influence of the external field on the reduced system dynamics is suggested by the curly arrow.

order in the molecule-lead coupling. For example, cotunneling events only show up in perturbation expansions beyond the lowest order [66]. The hierarchical equations of motion method provides a systematic approach towards obtaining the higher order contributions. This formalism was first developed for the coupling to bosonic baths [64], but was recently extended to fermionic reservoirs as well [34, 35].

13.1.1
Quantum Master Equation Approach

In what follows, the goal is to derive an effective theoretical method for determining the time evolution of an open quantum system. Throughout this chapter, atomic units are employed, such that $\hbar = 1$. Therefore, $\hbar$ does not appear explicitly in the equations shown. In general, the full system density matrix obeys the Liouville–von Neumann equation of motion

$$\frac{\partial}{\partial t} \rho(t) = i\left[\mathcal{H}, \rho(t)\right]. \tag{13.1}$$

This equation can be derived directly from the Schrödinger equation, to which it is basically equivalent. It completely describes the full dynamics of the system. When treating an open quantum system, the complete Hamiltonian is decomposed as

$$\mathcal{H} = \mathcal{H}_S + \mathcal{H}_R + \mathcal{H}_{SR}. \tag{13.2}$$

The reservoir and the system-reservoir coupling part of the Hamiltonian are considered to be time independent. In contrast, the Hamiltonian governing the dynamics of the relevant system may be time dependent, providing the possibility to consider laser driving.

The reservoir Hamiltonian describes the large number of modes of the electronic lead states, which are assumed to remain in equilibrium. The electrons in the lead are well approximated to behave harmonically as an ideal free electron gas. For the purposes of this contribution, the spin degree of freedom of electrons plays no role in the dynamics and may thus be neglected. It can be included in a straightforward manner as, for example, in [39]. The reservoir Hamiltonian, in second quantization, reads

$$\mathcal{H}_R = \sum_k \omega_k c_k^\dagger c_k \, , \qquad (13.3)$$

where k labels all the states in the reservoir and ω_k are the corresponding energies.

The structure of the relevant system Hamiltonian $\mathcal{H}_S$ depends strongly on the structure of the molecule or quantum dot and defines the model to be described. It includes as a main component the energies of the system states

$$\mathcal{H}_S = \sum_\xi \omega_\xi c_\xi^\dagger c_\xi \, , \qquad (13.4)$$

where ξ labels the states of the system and ω_ξ, the corresponding energies. Off-diagonal elements coupling various system states may also be included as well as time-dependent components of the energies ω_ξ, introduced, for example, in order to model a time-dependent gate voltage.

Finally, the system-reservoir coupling $\mathcal{H}_{SR}$ is intended to describe electron tunneling between states of the reservoir and the system. Here we restrict ourselves to the standard bilinear coupling between the relevant system and the contacting leads, although this is not a necessary restriction [37, 52]. This yields

$$\mathcal{H}_{SR} = \sum_{k,\xi} V_{k\xi} c_k^\dagger \otimes c_\xi + \text{H.c.} \, , \qquad (13.5)$$

where $|V_{k\xi}|^2$ is the tunneling amplitude characterizing the probability for an electron to tunnel between states k of the reservoir and ξ of the system. The coupling Hamiltonian may also include other types of coupling terms, as will be described later.

The convention throughout this chapter will be that indices S, R, and SR will denote operators related to the relevant system, reservoir, and system-reservoir coupling, respectively. Apart from the Hamiltonians, other superoperators that depend on the Hamiltonian will exhibit the same indices. For example, consider a general superoperator $\mathcal{O}(\mathcal{H})$. The notation $\mathcal{O}_S$ will denote $\mathcal{O}(\mathcal{H}_S)$. Similarly, the indices will appear to differentiate the density matrices of the reduced and reservoir parts of the system, and also to specify the states over which a trace acts.

The Liouville–von Neumann equation is considerably difficult to solve because it involves the density matrix defined in the full Fock space of system states. The dimension of the Fock space is infinite, or at least very large since it contains all states of the reservoir.

The formal solution of the Liouville–von Neumann equation can be written in the form of a propagation operator $U(t, t')$,

$$\rho(t) = \vec{T} \exp\left(i \int_{t_0}^t d\tau \mathcal{H}(\tau)\right) \rho(t_0) \exp\left(-i \int_{t_0}^t d\tau \mathcal{H}(\tau)\right)$$
$$= U(t, t_0) \rho(t_0) \, , \qquad (13.6)$$

$$U(t, t') \bullet = \vec{T} \exp\left(i \int_{t'}^t d\tau \mathcal{H}(\tau)\right) \bullet \exp\left(-i \int_{t'}^t d\tau \mathcal{H}(\tau)\right) . \qquad (13.7)$$

$\vec{\mathcal{T}}$ denotes the time ordering operator in the positive time direction. Even though the solution seems simple and compact, there is no easy way to keep track of the complete system evolution induced by $U(t, t')$.

Furthermore, due to its infinite number of degrees of freedom, the reservoir cannot be driven out of the equilibrium by the interaction with the small relevant system [11]. It is thus inefficient to treat the reservoir states on the same level of theory as the relevant system states. Therefore, one projects the Liouville–von Neumann equation onto the states of the relevant part of the system. Formally, this can be achieved with the use of a projection operator. The choice of such an operator is not unique and several methods for its construction have been discussed in the literature. For example, the use of the Argyres–Kelley projection operator [4] has become standard, with

$$\mathcal{P}(\mathcal{O}) = \mathcal{O}_R \otimes \text{tr}_R \{\mathcal{O}\} \,, \tag{13.8}$$

and $\text{tr} \{\mathcal{O}_R\} = 1$ in order to insure that $\mathcal{P}^2 = \mathcal{P}$. Such a projector transforms any arbitrary operator into a form which is separable in the parts corresponding to the relevant system and the reservoir. Its action on the full density matrix is particularly illustrating for its use, with the result

$$\mathcal{P}(\rho(t)) = \rho_R \otimes \text{tr}_R \{\rho(t)\} = \rho_R \otimes \rho_S(t) \,. \tag{13.9}$$

This projection operator can be used to transform (13.1) into an equation of motion for the relevant part of the density matrix. Such an equation has the form

$$\frac{\partial}{\partial t} \rho_S(t) = \text{i} \left[\mathcal{H}_S(t), \rho_S(t) \right] + \mathcal{D}(t) \,, \tag{13.10}$$

where $\mathcal{D}(t)$ is a term describing the dissipation due to system-reservoir coupling. The interpretation of (13.10) is simple. The reduced density matrix evolves according to the Liouville–von Neumann equation generated by the corresponding system Hamiltonian, together with an additional term generated by the presence of the reservoir. The latter term should vanish if the system and the reservoir are decoupled, that is, if $\mathcal{H}_{SR} = 0$.

13.1.2
Time-Local and Time-Nonlocal Master Equations

Using the projection operator given in (13.8), two distinct forms of the dissipation term have been developed in the literature. One is the so-called time-nonlocal (TNL) formalism based on the Nakajima–Zwanzig identity [56, 75] and the time-local (TL) approach is based on the Tokuyama–Mori identity [62]. Both approaches lead to exact dissipation terms $\mathcal{D}(t)$, providing QMEs which are analytically equivalent to the Liouville–von Neumann equation. This is expected, since introducing the projection does nothing else but reformulate the initial problem, focusing on the relevant degrees of freedom.

Except for the simplest cases, the equations arising after applying the projection operator are very difficult to implement and solve. In order to proceed, one usually invokes the assumption of weak system-reservoir coupling. The explicit derivation of the dissipation term and the concrete method of applying the theory of small perturbations is presented elsewhere [38, 50, 58]. The results are reviewed below.

In second order perturbation theory, the TNL dissipation term takes the form

$$\mathcal{D}_{TNL}(t) = -\operatorname{tr}_R \left\{ \mathcal{L}_{SR}(t) \int_{t_0}^{t} dt' \, U_S(t,t') \otimes U_R \left(\mathcal{L}_{SR}(t') \rho_S(t') \otimes \rho_R \right) \right\} . \tag{13.11}$$

It can be written in a more compact form by introducing the dissipation kernel

$$\mathcal{K}(t,t') \bullet = \operatorname{tr}_R \left\{ \mathcal{L}_{SR}(t) \left(U_S(t,t') \otimes U_R \left(\mathcal{L}_{SR}(t') (\bullet \otimes \rho_R) \right) \right) \right\} . \tag{13.12}$$

The dissipation kernel is an operator that acts only on the Fock space of the relevant system states. This is insured by the trace over the reservoir degrees of freedom. In terms of this kernel, the dissipation takes the form

$$\mathcal{D}_{TNL}(t) = -\int_{t_0}^{t} dt' \, \mathcal{K}(t,t') \rho_S(t') . \tag{13.13}$$

It is now clear why this equation is nonlocal in time. The dissipation term at time t depends on the state of the system at all previous times t'. This is contrasted by the TL dissipation term

$$\mathcal{D}_{TL}(t) = -\operatorname{tr}_R \left\{ \mathcal{L}_{SR}(t) \int_{t_0}^{t} dt' \, U_S(t,t') \otimes U_R \left(\mathcal{L}_{SR}(t') \rho_S(t) \otimes \rho_R \right) \right\}$$

$$= -\int_{t_0}^{t} dt' \, \mathcal{K}(t,t') \rho_S(t) . \tag{13.14}$$

Here, the dissipation depends on the system reduced density matrix only at time t, that is, at the same time as the dissipation term $\mathcal{D}_{TL}(t)$. Note that the TL dissipation term can be derived from the TNL counterpart by making the substitution [18]

$$\rho_S(t') \to U_S^\dagger(t,t') \rho_S(t) . \tag{13.15}$$

Comparisons between the TL and TNL approach have been performed [9, 37, 61], but a clear proof of the superiority of either method has yet to arise. The initial conditions for a specific problem may decide which method is more suitable. For sufficiently weak coupling, the results obtained using both methods of course agree, since they are equivalent to second order in the system environment coupling. The TNL and TL approaches differ by different resummations of higher order terms. In the derivation of the dissipation terms, it was assumed that the reduced density matrix corresponding to the reservoir ρ_R and the relevant system ρ_S are initially uncorrelated. If this assumption is not satisfied, the initial correlations give rise to an extra dissipation term which complicates the calculations [12]. This fact may render a TNL approach artificial, as past correlations do not exist at the start of the system-reservoir interaction. Thus, throughout this chapter, the study will focus on the TL equation of motion.

Together with the TL dissipation term already described, the equation of motion for the reduced density matrix given in general form by (13.10) becomes

$$\frac{\partial}{\partial t} \rho_S = i\mathcal{L}_S \rho_S - \int_{t_0}^{t} dt' \mathcal{K}(t, t') \rho_S(t). \tag{13.16}$$

At a glance, it can be noticed that the QME, (13.16), is an integro-differential equation, since the intermediary time t' is still present in the dissipation kernel $\mathcal{K}(t, t')$. In order to solve such an equation numerically, it is more convenient to transform it to a system of ordinary differential equations (ODEs). The system of ODEs will involve the reduced density matrix, together with several auxiliary operators that describe the evolution of the dissipation kernel. In what follows, the kernel will be reformulated to a suitable shape, and with the use of a numerical decomposition it will be split into the auxiliary operators desired for obtaining a closed system of ODEs.

First consider the shape of $\mathcal{K}(t, t')$ given by (13.12). The goal is to split the entire operator into a direct product of a system and a reservoir part. The propagator $U(t, t')$ is already split into its corresponding system and reservoir parts, so the focus can be directed to the coupling Hamiltonian involved in $\mathcal{L}_{SR}$. The structure of the coupling Hamiltonian is

$$\mathcal{H}_{SR} = \sum_{x} K_x \otimes \phi_x, \tag{13.17}$$

with K_x an operator defined on the states of the relevant system and ϕ_x an operator defined on the states of the reservoir. For the explicit coupling Hamiltonian given by (13.5), such a decomposition is easily realized using as the system part $K_1 = c_\xi$ and $K_2 = c_\xi^\dagger$, and as the bath part $\phi_1 = \sum_k V_{k,\xi} c_\xi^\dagger$ and $\phi_2 = \sum_k V_{k,\xi} c_\xi$. The index ξ, as used in (13.5), labels the system states. Thus, the coupling Hamiltonian is readily split into a direct sum of system and reservoir operators. The index x incorporates three indices, $x \equiv \{\xi, i, \alpha\}$, where ξ counts system states, $i \in \{1, 2\}$ for the system operator, $i = 1$, or its Hermitian conjugate, $i = 2$, and α labels the corresponding leads, when more than one lead is connected to the dot.

It is now possible to write the dissipation kernel as a direct product of system and reservoir operators, yielding

$$\mathcal{K}(t, t') \bullet = \sum_{xx'} \left[K_x, U_S(t, t') [K_{x'}, \bullet] \right] \otimes \text{tr}_R \left\{ [\phi_x, U_R [\phi_{x'}, \rho_R]] \right\}. \tag{13.18}$$

The trace in this equation transforms the reservoir operator into a scalar function of time and thus, the direct product is transformed into a simple multiplication by a scalar function. The resulting function of time is called the reservoir correlation function. It can be interpreted as the correlation strength between transitions involving the various levels of the system and the quantum levels of the reservoir.

The correlation function is specific to a reservoir and also specific to the quantum levels of the system involved in the electron tunneling

$$C_{xx'}(t) = \text{tr}_R \{U_R(\phi_x) \phi_{x'} \rho_R\} . \tag{13.19}$$

In terms of the correlation function, the dissipation kernel takes the form

$$\mathcal{K}(t,t') \bullet = \sum_{xx'} \left[K_x, C_{xx'}(t-t') U_S(t,t') K_{x'} \bullet \right]$$
$$+ \left[K_{x'}, C_{x'x}(t-t') U_S(t,t') K_x \bullet \right]^\dagger . \tag{13.20}$$

The second term in the sum is obtained from the first term by interchanging x and x' and taking the Hermitian conjugate.

In the next step, it is also easy to write the dissipation term $\mathcal{D}_{\text{TL}}(t)$ by plugging (13.20) into (13.14). In order to simplify the resulting form of the dissipation term, it is convenient to define auxiliary operators $\Lambda_{xx'}(t)$ as

$$\Lambda_{xx'}(t) = \int_{t_0}^{t} dt' \, C_{xx'}(t-t') U_S(t,t') K_{x'} . \tag{13.21}$$

Using these auxiliary operators the dissipation term becomes

$$\mathcal{D}_{\text{TL}}(t) = -\sum_{xx'} \left[K_x, \Lambda_{xx'}(t) \rho_S(t) - \rho_S(t) \Lambda^\dagger_{x'x}(t) \right] . \tag{13.22}$$

This is the form of the dissipation term that transforms the most general form of the TL QME, (13.10), into an ODE that can be implemented and easily solved numerically using the wide array of efficient ODE solvers. All that is needed is the time evolution of the auxiliary operators, so that the system of ODEs becomes closed.

In order to obtain the equation of motion for the auxiliary operators, it is convenient to take the time derivative of (13.21),

$$\frac{\partial}{\partial t} \Lambda_{xx'}(t) = C_{xx'}(0) K_{x'}$$
$$+ \int_{t_0}^{t} dt' \, \frac{\partial}{\partial t} C_{xx'}(t-t') U_S(t,t') K_{x'}$$
$$- i \int_{t_0}^{t} dt' \, C_{xx'}(t-t') \mathcal{L}_S \, U_S(t,t') K_{x'} . \tag{13.23}$$

The equation of motion above takes a particularly simple form if the correlation function can be numerically decomposed into a sum of exponentials [37, 52],

$$C_{xx'}(t) = \sum_k a^k_{xx'} e^{\gamma^k_{xx'} t} . \tag{13.24}$$

The coefficients $a^k_{xx'}$ and $\gamma^k_{xx'}$ are time-independent complex numbers. The decomposition has the role of reshaping the time dependence of the correlation function to a purely exponential form, which allows further analytic treatment of (13.23).

Equation 13.24 should not be interpreted as a limitation of the theory. Indeed, it does not introduce any significant additional approximation. Any useful correlation function can be approximated arbitrarily well by the series implied in (13.24). For functions that are smooth, everywhere bounded, and admit a Fourier series, the exponential $\gamma_{xx'}^k$ will take only imaginary values, recovering the corresponding series. For smooth functions with exponential divergences, the diverging parts can be arbitrarily well approximated by the real part of the coefficients $\gamma_{xx'}^k$. From a practical point of view, it is imperative to reduce the number of terms in the decomposition. This can be achieved by using numerical optimization algorithms, in order to achieve the smallest approximation with the best truncation of the series (see the example in [52] for an illustration).

The form of the correlation function given by (13.24) inspires a decomposition of the lambda operators

$$\Lambda_{xx'}^k(t) = \int_{t_0}^{t} dt' \, a_{xx'}^k \exp\left\{\gamma_{xx'}^k (t-t')\right\} U_S(t,t') K_{x'}, \tag{13.25}$$

such that $\Lambda_{xx'}(t) = \sum_k \Lambda_{xx'}^k(t)$. The equation of motion for each individual $\Lambda_{xx'}^k(t)$ can be easily read from (13.23) as

$$\frac{\partial}{\partial t} \Lambda_{xx'}^k(t) = a_{xx'}^k K_{x'} + \gamma_{xx'}^k \Lambda_{xx'}^k(t) - i\mathcal{L}_S \Lambda_{xx'}^k(t). \tag{13.26}$$

The set of ODEs (13.26) completely describes the time evolution of the auxiliary operators, and together with the equation of motion for the reduced density matrix

$$\frac{\partial}{\partial t} \rho_S = i\mathcal{L}_S \rho_S - \sum_{xx'} \left[K_x, \Lambda_{xx'}(t)\rho_S(t) - \rho_S(t)\Lambda_{x'x}^\dagger(t) \right], \tag{13.27}$$

a complete set of ODEs governing the time evolution of the system is obtained.

13.1.3
Full Counting Statistics

Charge transfer through molecules and quantum dots occurs via electron tunneling between the leads and the states of the relevant system. The discrete nature of the charge flow and the distinct quantum mechanical features encountered render the electron transport a stochastic process. In order to completely describe charge transfer, one must count the number of electrons that have tunneled to and from any of the contacting leads. This turns out to be a difficult task that has attracted a lot of theoretical [7, 57], and recently also experimental, interest [29–31]. Its solution involves keeping track of the probabilities $P_n(t)$ that n electrons have tunneled through a specific junction at a given time. Solving the statistics of the charge transfer process and its cumulants is the task of this section.

It will be shown that the formalism employed to solve the dynamics of the reduced density matrix can be extended to provide a powerful tool for handling electron transfer statistics. Here, the time dependence of the particle current and the current noise will be calculated explicitly. The equation of motion for the reduced density matrix derived above can be projected onto those states involving a precise number of tunneled electrons, thus providing equations of motion for the *n*-resolved reduced density matrices. Further, this chapter presents an efficient method for side stepping the numerical complications arising from keeping track of the density matrices describing all the possible numbers of tunneled electrons up to infinity. The transfer of electrons or other relevant quantum particles such as Cooper pairs, phonons, and excitons is described mainly by the two quantities current and shot noise. These quantities are also very important measurables from an experimental point of view. In recent years the full counting statistics has also attracted a lot of attention.

The operator representing the transferred charge for one of the contacts is

$$\mathcal{Q}(t) = e(\mathcal{N}(t) - \mathcal{N}(t_0)) , \qquad (13.28)$$

where $\mathcal{N}$ is the number operator in the respective lead and e is the elementary charge. Using atomic units with $e = 1$, the elementary charge may be omitted throughout the derivation. Note that the sign in front of the charge is defined to be positive if electrons have tunneled into the lead, and negative if electrons from the lead have tunneled into the relevant system.

The current is defined as the rate of the particle transfer, that is, the time derivative of the number of transferred particles. The current operator, describing the current flow through one of the contacts is given by

$$\mathcal{I}(t) = \frac{\partial}{\partial t}\mathcal{Q}(t) = \frac{\partial}{\partial t}\mathcal{N}(t) . \qquad (13.29)$$

Another important point here is that the current and charge operators are defined on the full Fock space of the system plus reservoir states.

Finally, the shot noise describes the current fluctuations and may be understood as a charge diffusion coefficient

$$S(t) = \langle \mathcal{I}^2(t) \rangle - \langle \mathcal{I}(t) \rangle^2 = \frac{\partial}{\partial t}\left(\langle \mathcal{Q}^2(t) \rangle - \langle \mathcal{Q}(t) \rangle^2\right) , \qquad (13.30)$$

where $\langle \bullet \rangle = \mathrm{tr}\{\bullet \rho\}$ and the trace and ρ refer to all the states of the full system. Note that the shot noise is not an operator but a time-dependent scalar function. Similar measurable scalar quantities, the transferred charge and the current, can be obtained averaging the corresponding operators

$$Q(t) = \langle \mathcal{Q}(t) \rangle , \qquad (13.31)$$

$$I(t) = \langle \mathcal{I}(t) \rangle . \qquad (13.32)$$

Due to the discrete nature of the charge carriers, it is more convenient to express the quantities charge, current, and noise in terms of the probability that *n* electrons tunneled through the junction $P_n(t)$. The transferred charge, in atomic units, is given by

$$Q(t) = \sum_{n=-\infty}^{\infty} n\, P_n(t) . \qquad (13.33)$$

The fact that n assumes positive as well as negative values reflects once more that the electron counting is done in one specific direction. For example, the transfer of particles is considered positive when particles are transferred from the reservoir to the relevant system. $P_{-1}(t)$ then designates the probability that in total one electron has tunneled in the reverse direction, that is, from the system to the reservoir.

Equation 13.33 expresses the charge transferred through a specific junction as a probabilistic quantity, emphasizing the stochastic nature of the tunneling process. Similar expressions can be obtained for the current and the shot noise by inserting (13.33) into (13.32) and (13.30), respectively

$$I(t) = \sum_{n=-\infty}^{\infty} n\, \frac{\partial}{\partial t} P_n(t), \qquad (13.34)$$

$$S(t) = \frac{\partial}{\partial t} \left(\sum_{n=-\infty}^{\infty} n^2\, P_n(t) - \left(\sum_{n=-\infty}^{\infty} n\, P_n(t) \right)^2 \right). \qquad (13.35)$$

Relations (13.34) and (13.35) reveal the information that can be extracted from the probabilities $P_n(t)$. These probabilities can be interpreted as the population summed over all the states in the Fock subspace involving the tunneling of precisely n electrons at a given time t. If we denote by $\mathcal{F}_n$ the corresponding subspace and by $|\psi_n\rangle$ the wavefunctions of the states in $\mathcal{F}_n$, then $P_n(t)$ is given by

$$P_n(t) = \sum_{\{|\psi_n\rangle\} \in \mathcal{F}_n} \langle \psi_n(t) | \psi_n(t) \rangle . \qquad (13.36)$$

Here, the number of tunneled electrons n can be interpreted as an additional quantum number. Thus, it is possible to define a density matrix $\rho_n(t)$ corresponding to the states in $\mathcal{F}_n$. The use of n-resolved density matrices for solving the full counting statistics of electron transfer processes has been proposed in [28]. Several elegant implementations for solving the statistical cumulants have been implemented in the cases of quantum interference phenomena [27, 54] and the quantum shuttle [21, 60]. While these methods treat steady state properties of transport, a generalized method has been developed that solves the time dependence of the statistical cumulants [36]. The derivations presented in this chapter follow along the lines of the latter method.

In order to transform between the full system density matrix and the n-resolved density matrix $\rho_n(t)$, it is necessary to define an additional projection operator $\mathcal{P}_n$, such that $\mathcal{P}_n \rho(t) = \rho_n(t)$. The properties of the $\mathcal{P}_n$ projector stem from its definition and its intended use. It should provide a flexible, formally correct method of focusing on the dynamics of the states in any subspace $\mathcal{F}_n$ of the Fock space. Thus,

$\mathcal{P}_n$ should commute with any operator that does not directly influence the number of tunneled electrons, such as the Hamiltonian of the system or reservoir, the number operators in any states, and others. The nontrivial use of $\mathcal{P}_n$ results from the action of the system-reservoir coupling Hamiltonian $\mathcal{H}_{SR}$, which introduces the process of tunneling at the theoretical level. The quantum number n counts electrons passing through only one of the contacts of the quantum dot. Even if several leads are in contact with the quantum device, the operators describing tunneling through different contacts do not mix. This allows us to consider each contact individually and calculate the statistics of each corresponding transport process individually.

In order to understand the properties of the projector $\mathcal{P}_n$, it is useful to study an illustrating example. Consider, for simplicity, a quantum dot with several quantum states, in contact with one electron reservoir. The coupling Hamiltonian may be the one described by (13.5). We may arbitrarily choose to count the electrons jumping from the reservoir to the states of the quantum dot, that is, n increases for tunneling into the dot. The operator inducing tunneling in the positive direction can be abbreviated $\mathcal{V}_+$ and its Hermitian conjugate $\mathcal{V}_-$

$$\mathcal{V}_- = \sum_{k,\xi} V_{k\xi} c_k^\dagger \otimes c_\xi , \qquad (13.37)$$

$$\mathcal{V}_+ = \sum_{k,\xi} V_{k\xi} c_k \otimes c_\xi^\dagger , \qquad (13.38)$$

and

$$\mathcal{H}_{SR} = \mathcal{V}_+ + \mathcal{V}_- . \qquad (13.39)$$

The two operators $\mathcal{V}_+$ and $\mathcal{V}_-$ act as ladder operators for the number of tunneled electrons, and

$$\mathcal{V}_+ |\psi_n\rangle \propto |\psi_{n+1}\rangle , \qquad (13.40)$$

$$\mathcal{V}_- |\psi_n\rangle \propto |\psi_{n+1}\rangle , \qquad (13.41)$$

with $\forall |\psi_n\rangle \in \mathcal{F}_n$ and $|\psi_{n\pm 1}\rangle \in \mathcal{F}_{n\pm 1}$.

$\mathcal{V}_+$ and $\mathcal{V}_-$ make the transition between neighboring Fock subspaces, and thus do not commute with the projection operators $\mathcal{P}_n$. The commutation relations are analogous to those between the creation/annihilation operators and the number operator

$$\mathcal{V}_+ \mathcal{P}_n = \mathcal{P}_{n+1} \mathcal{V}_+ , \qquad (13.42)$$

$$\mathcal{V}_- \mathcal{P}_n = \mathcal{P}_{n-1} \mathcal{V}_- . \qquad (13.43)$$

In what follows, the goal will be to derive an equation of motion for the n-resolved density matrices by using $\mathcal{P}_n$ to project (13.27) onto the subspace $\mathcal{F}_n$. Knowing the

n-resolved density matrix operators, the probabilities $P_n(t)$ are given by the trace of $\rho_n(t)$, as suggested by (13.36),

$$P_n(t) = \text{tr}\{\rho_n(t)\}. \tag{13.44}$$

Because the trace does not involve off-diagonal elements, the following relation can be used

$$\text{tr}\{\rho_n(t)\} = \text{tr}\{\mathcal{P}_n(\rho_S(t) \otimes \rho_R)\}. \tag{13.45}$$

The operator $\mathcal{P}_n$ has no effect on the reservoir part of the density matrix ρ_R, as it was assumed that ρ_R has no dynamics and remains in equilibrium, that is, ρ_R is invariant with the number of tunneled electrons. Taking the trace over the reservoir degrees of freedom in (13.45) first, the equation is for $P_n(t)$

$$P_n(t) = \text{tr}_S\{\rho_{S,n}(t)\}, \tag{13.46}$$

where $\rho_{S,n}(t) = \mathcal{P}_n \rho_S(t)$, that is, it is sufficient to add the quantum number n to the reduced system density matrix.

For simplicity, we will assume here that the tunneling can occur in only one state of the quantum dot, eliminating the need to use ξ in (13.42) and (13.43) to keep track of the states. The index α, which labels the contacts, will contribute significantly only for the contact where we count the tunneled electrons. Auxiliary operators originating from all other contacts commute with the projection operator $\mathcal{P}_n$. In order to keep the illustration simple, it will be assumed that only one electron reservoir is in contact with the quantum dot, thus eliminating the need to include the α index.

Particularly important is the index $i \in \{1, 2\}$, which specifies if $\Lambda_{xx'}(t)$ involves electron tunneling into or out of the relevant system states. Explicitly showing the summation over this index, the projection operator applied to the dissipation term yields

$$\mathcal{P}_n \mathcal{D}_{\text{TL}}(t) = -\mathcal{P}_n \left(\left[K_1, \Lambda_{12}(t) \rho_S(t) - \rho_S(t) \Lambda_{21}^\dagger(t) \right] \right) \\ - \mathcal{P}_n \left(\left[K_2, \Lambda_{21}(t) \rho_S(t) - \rho_S(t) \Lambda_{12}^\dagger(t) \right] \right), \tag{13.47}$$

where $K_1 = c^\dagger$ denotes the creation operator in the system state involved in the tunneling event. Furthermore, one has the relation $K_2 = K_1^\dagger = c$.

The commutation relations between the projector and the various operators are

$$\begin{aligned}
\mathcal{P}_n K_1 &= K_1 \mathcal{P}_{n-1}, \\
\mathcal{P}_n K_2 &= K_2 \mathcal{P}_{n+1}, \\
\mathcal{P}_n \Lambda_{12} &= \Lambda_{12} \mathcal{P}_{n+1}, \\
\mathcal{P}_n \Lambda_{21} &= \Lambda_{21} \mathcal{P}_{n-1}, \\
\mathcal{P}_n \Lambda_{12}^\dagger &= \Lambda_{12}^\dagger \mathcal{P}_{n-1}, \\
\mathcal{P}_n \Lambda_{21}^\dagger &= \Lambda_{21}^\dagger \mathcal{P}_{n+1}.
\end{aligned} \tag{13.48}$$

Using these relations, (13.47) becomes

$$\begin{aligned}\mathcal{P}_n \mathcal{D}_{\text{TL}}(t) = &-\rho_{S,n}\left(\Lambda_{21}^\dagger(t)K_1 + \Lambda_{12}^\dagger(t)K_2\right)\\&-(K_1\Lambda_{12}(t) + K_2\Lambda_{21}(t))\rho_{S,n}\\&+ K_1\rho_{S,n-1}\Lambda_{21}^\dagger(t) + \Lambda_{12}(t)\rho_{S,n+1}K_1\\&+ K_2\rho_{S,n+1}\Lambda_{12}^\dagger(t) + \Lambda_{21}(t)\rho_{S,n-1}K_2\,.\end{aligned} \quad (13.49)$$

As can be seen in (13.49), the dissipation part contributes several terms that relate to $\rho_{S,n}$, which can be interpreted as energy due to electrons tunneling in and out of the reduced system. The rest of the terms describe effective electron tunneling into the system $\rho_{S,n+1}$ and out of the system $\rho_{S,n-1}$. The latter terms are important for the statistics of the electron transport process. In fact, it will turn out that these terms are directly related to the current of particles through the respective junction. In anticipation, the two corresponding operators can be denoted by

$$\mathcal{I}_+ \bullet = K_1 \bullet \Lambda_{21}^\dagger(t) + \Lambda_{21}(t) \bullet K_2\,, \quad (13.50)$$

$$\mathcal{I}_- \bullet = K_2 \bullet \Lambda_{12}^\dagger(t) + \Lambda_{12}(t) \bullet K_1\,. \quad (13.51)$$

The interpretation of $\mathcal{I}_+$ and $\mathcal{I}_-$, as will be seen later, is that they are the current operators for particles going into and out of the system, respectively.

Using the additional abbreviation introduced above, the equation of motion of the n-resolved reduced density matrix is given by

$$\begin{aligned}\frac{\partial}{\partial t}\rho_{S,n} = &\,i\mathcal{L}_S\rho_{S,n} - \rho_{S,n}\left(\Lambda_{21}^\dagger(t)K_1 + \Lambda_{12}^\dagger(t)K_2\right)\\&-(K_1\Lambda_{12}(t) + K_2\Lambda_{21}(t))\rho_{S,n}\\&+ \mathcal{I}_+(\rho_{S,n-1}) + \mathcal{I}_-(\rho_{S,n+1})\,.\end{aligned} \quad (13.52)$$

This equation provides a complete method of solving the dynamics of the tunneling process. A similar equation has been reported in [49], starting from a slightly modified QME.

In terms of the n-resolved density matrices, the first and second statistical moments of the particle transfer can be written as

$$\sum_{n=-\infty}^{\infty} n\, P_n(t) = \sum_{n=-\infty}^{\infty} n\, \text{tr}\{\rho_{S,n}\}\,, \quad (13.53)$$

$$\sum_{n=-\infty}^{\infty} n^2\, P_n(t) = \sum_{n=-\infty}^{\infty} n^2\, \text{tr}\{\rho_{S,n}\}\,. \quad (13.54)$$

While useful for the in depth understanding of the interpretation to be given to the various operators involved in the equation of motion, (13.52) has limited applicability in the implementation and solving of the transport dynamics. The obvious practical problem that arises is the necessity to solve all n equations of motion, that is, to keep track of an infinite number of density matrices. Therefore, in order to obtain the correct expression of the moments, additional techniques need to be developed. This will be the goal of the next two sections.

The time-dependent electron current, as given by (13.34), is the time derivative of the first statistical moment. In order to extract the statistical moments from the equation of motion of the density matrix, it is useful to define the moment generating function [36, 49]

$$\phi(\chi, t) = \mathrm{tr}\{\exp(i\chi\mathcal{N})\,\rho(t)\}\,, \tag{13.55}$$

where $\mathcal{N}$ is the number operator counting the tunneling electrons and the trace is over all the system plus reservoir degrees of freedom.

The variable χ, the so-called counting variable, is the dual of the number of transferred particles, in the sense that χ and n form a conjugate pair similar to the relation between coordinate and momentum, or time and frequency [5]. As can be guessed from (13.55), χ and n are related via the Fourier transform.

The definition of the moment generating function $\phi(\chi, t)$ justifies its name. Indeed, taking its Taylor series with respect to χ around $\chi = 0$, one obtains the moments as the coefficients of the polynomial

$$\phi(\chi, t) = \sum_{k=0}^{\infty} \frac{(i\chi)^k}{k!} \left(\sum_{n=-\infty}^{\infty} n^k P_n(t) \right). \tag{13.56}$$

It turns out to be more useful to find the time evolution of the moment generating function instead of the n-resolved density matrices. In fact, it is convenient to define the moment generating operator in the Fock space of the reduced system as

$$F(\chi, t) = \mathrm{tr}_R\{\exp(i\chi\mathcal{N})\,\rho_S(t) \otimes \rho_R\}\,, \tag{13.57}$$

whose trace obviously recovers the moment generating function. The equation of motion of $F(\chi, t)$ and its Taylor series components will provide a general solution for the calculation of statistical moments of electron transport.

The time derivative of the moment generating operator is given by

$$\frac{\partial}{\partial t} F(\chi, t) = \mathrm{tr}_R \left\{ \exp(i\chi\mathcal{N}) \frac{\partial}{\partial t} (\rho_S(t) \otimes \rho_R) \right\}. \tag{13.58}$$

The equation of motion (13.27) in the form developed for (13.52) is inserted into (13.58), and the straightforward operator algebra involved in the commutation of $\mathcal{N}$ with the operators $\mathcal{I}_+$ and $\mathcal{I}_-$ is performed, yielding [36]

$$\frac{\partial}{\partial t} F(\chi, t) = \{\mathcal{R}_S(t) + (\exp(i\chi) - 1)\mathcal{I}_+ + (\exp(-i\chi) - 1)\mathcal{I}_-\} F(\chi, t)\,, \tag{13.59}$$

where $\mathcal{R}_S(t) = \mathcal{L}_S(t) - \int_{t_0}^{t} dt' \mathcal{K}(t, t')$ is used for the brevity of notation. The operator $\mathcal{R}_S(t)$ is chosen such that $\partial/(\partial t)\rho_S(t) = \mathcal{R}_S(t)\rho_S(t)$. Equation 13.59 provides an efficient means of calculating the statistical moments. In the following, it will be shown explicitly how to calculate the first two moments needed for the expression of the current and the shot noise. The generalization to the calculation of higher moments is straightforward.

As suggested by (13.56), the information in the moment generating operator can be extracted by taking the Taylor series of (13.59) with respect to the counting variable χ. The Taylor series of the moment generating operator is given by

$$F(\chi, t) = \sum_{k=0}^{\infty} \frac{(i\chi)^k}{k!} F_k(t), \qquad (13.60)$$

where $F_0(t)$ is nothing else but the reduced system density matrix and, by comparison with (13.56), $\mathrm{tr}\{F_k(t)\} = \sum_{n=-\infty}^{\infty} n^k P_n(t)$.

To zeroth order in χ one obtains

$$\frac{\partial}{\partial t} F_0(t) = \mathcal{R}_S(t) F_0(t). \qquad (13.61)$$

This was expected because $F_0(t) = \rho_S(t)$. The zeroth order recovers the QME for the density matrix.

The first order term in χ of the Taylor expansion of (13.59) provides the first statistical moment, whose time derivative provides the current operator, that is,

$$\frac{\partial}{\partial t} F_1(t) = \mathcal{R}_S(t) F_1(t) + (\mathcal{I}_+(t) - \mathcal{I}_-(t)) F_0(t). \qquad (13.62)$$

Finally, the current is given by

$$I(t) = \mathrm{tr}\left\{\frac{\partial}{\partial t} F_1(t)\right\} = \mathrm{tr}\{(\mathcal{I}_+(t) - \mathcal{I}_-(t)) F_0(t)\}, \qquad (13.63)$$

where $\mathrm{tr}\{\mathcal{R}_S(t)\mathcal{O}\} = 0$ was used for any operator $\mathcal{O}$, that is, the QME describes a dynamical behavior that preserves the trace.

The current operator, as defined by (13.29), is given by

$$\mathcal{I}(t) = \mathcal{I}_+(t) - \mathcal{I}_-(t), \qquad (13.64)$$

where again it was used that $F_0(t) = \rho_S(t)$.

The shot noise involves the use of the second statistical moment of the transport process. This moment is contained in the second order term of the Taylor expansion of $F(\chi, t)$. The equation of motion corresponds to the second order expansion of (13.59),

$$\frac{\partial}{\partial t} F_2(t) = \mathcal{R}_S(t) F_2(t) + 2(\mathcal{I}_+(t) - \mathcal{I}_-(t)) F_1(t) + (\mathcal{I}_+(t) + \mathcal{I}_-(t)) F_0(t). \qquad (13.65)$$

Comparing this QME with (13.35), the shot noise, in terms of the Taylor components of $F(\chi, t)$, is

$$S(t) = \mathrm{tr}\left\{\frac{\partial}{\partial t} F_2(t)\right\} - 2\,\mathrm{tr}\{F_1(t)\}\,\mathrm{tr}\left\{\frac{\partial}{\partial t} F_1(t)\right\}. \qquad (13.66)$$

Replacing $\mathrm{tr}\left\{\frac{\partial}{\partial t} F_1(t)\right\} = \mathrm{I}(t)$ and inserting (13.65) into the above equation, the noise becomes

$$S(t) = \mathrm{tr}\left\{2(\mathcal{I}_+(t) - \mathcal{I}_-(t) - \mathrm{I}(t))F_1(t) + (\mathcal{I}_+(t) + \mathcal{I}_-(t))F_0(t)\right\}. \tag{13.67}$$

Note that the noise depends on the full dynamics of $F_1(t)$ as opposed to the current, which can be directly extracted from the time dependence of the reduced system density matrix. In order to solve the dynamics of the noise, or charge diffusion as suggested by the definition (13.30), it is necessary to solve the equation of motion for $F_1(t)$ in addition to the equation of motion for the reduced density matrix. However, this is feasible and the method described provides an elegant and efficient method to solve the statistics of an arbitrary transport process. The method proves particularly superior to other methods developed in the case of relevant systems with many quantum states.

Finally, it will prove more convenient (see [36]) to define a quantity $F_{1\perp}(t) = F_1(t) - \rho_S(t)\,\mathrm{tr}\left\{F_1(t)\right\}$ that has the property of being perpendicular to the reduced density matrix operator, and additionally that $\mathrm{tr}\left\{F_{1\perp}(t)\right\} = 0$. The equation of motion for $F_{1\perp}(t)$ can be easily derived from (13.62), yielding

$$\frac{\partial}{\partial t} F_{1\perp}(t) = \mathcal{R}_S(t) F_{1\perp}(t) + (\mathcal{I}_+(t) - \mathcal{I}_-(t) - \mathrm{I}(t))F_0(t). \tag{13.68}$$

In terms of $F_{1\perp}(t)$, the shot noise takes the form

$$S(t) = \mathrm{tr}\left\{2(\mathcal{I}_+(t) - \mathcal{I}_-(t))F_{1\perp}(t) + (\mathcal{I}_+(t) + \mathcal{I}_-(t))\rho_S(t)\right\}. \tag{13.69}$$

The above equation can be directly implemented and used for the practical computation of the shot noise.

In addition, the methods described give access to the calculation of the so-called Fano factor, a dimensionless quantity measuring the relative noise strength. It is defined as

$$F = \lim_{t \to \infty} \frac{S(t)}{|\mathrm{I}(t)|}. \tag{13.70}$$

The Fano factor provides information related to the nature of the transport mechanism [5]. It is equal to one for a Poissonian stochastic process, that is, a process where the tunneling of one electron is completely independent of any other tunneling processes. A process is called super-Poissonian if $F > 1$, meaning that the electrons are bunching, so that when one tunneling process occurs it is more likely that a second tunneling process follows soon after. The process is called sub-Poissonian if $F < 1$, meaning that the electrons are anti-bunching, that is, the tunneling of one electron reduces the probability of another tunneling process.

13.2
Model System Describing Molecular Wires and Quantum Dots

A simple model describing a molecular wire consists of a series of N localized sites, representing empty orbitals that can accommodate electrons. In real molecu-

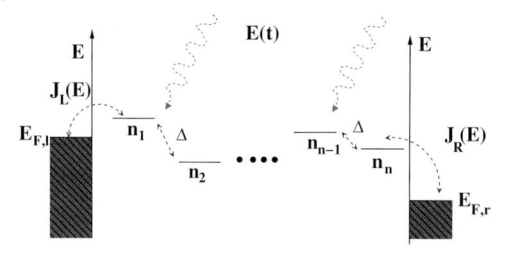

Figure 13.2 Scheme of the tight-binding model for a molecular wire. The sites of the wire, each with energy E_n, are coupled by a constant coupling Δ. The chemical potentials of the leads are denoted by $E_{F,l}$ and $E_{F,r}$, respectively. The energies of the sites can be manipulated by a time-dependent field $E(t)$.

lar wires, such orbitals may overlap so that there is a finite probability that electrons can tunnel between the sites. For simplicity, in the following we will assume that the energy associated with the tunneling processes between all sites is a fixed quantity Δ.

The molecular wire can be connected to two metallic leads that act as reservoirs of electrons. Here, we consider that the leads couple weakly to the wire, so that only the closest site is involved in direct tunneling into and out of the leads. The convention we adopt is that site number one is coupled to the left lead and site number N is coupled to the right lead. Also by convention, we choose the left lead as the source of electrons, having a high chemical potential, while the right lead will act as a drain, with a lower chemical potential (see Figure 13.2).

An additional important assumption made in the following derivations is that the on site electron–electron interaction is strong enough such that not more than one electron can occupy the site at any time. This justifies the assumption that the spin degree of freedom of electrons can be disregarded.

In the following, it is described in detail how to employ the methods developed in the previous theoretical sections. The particular case of a single site wire is solved analytically for illustration in Section 13.3, and numerical results obtained using single and double site wires are presented in Section 13.4.

As described above, the full Hamiltonian can be split into a system part, a reservoir part, and a coupling part. The system Hamiltonian is given by the Hamiltonian of the N-sites molecular wire in the orbital tight-binding description assumed above [42, 58],

$$\mathcal{H}_S = \sum_{n,n'=1}^{N} (\mathcal{H}_S)_{nn'} c_n^\dagger c_{n'}, \tag{13.71}$$

$$(\mathcal{H}_S)_{nn'} = E_n \delta_{n,n'} - \Delta (\delta_{n+1,n'} + \delta_{n,n'+1}). \tag{13.72}$$

The reservoir part is described by a Hamiltonian identical to the one given by (13.3), $\mathcal{H}_R = \sum_{k,\alpha} \omega_{k\alpha} c_{k\alpha}^\dagger c_{k\alpha}$, where $\mathbf{k}$ labels the continuum of states in the leads and $\alpha \in \{L, R\}$, designates the left (L) or right lead (R), respectively. The occupation of the reservoir states at equilibrium is described by the Fermi–Dirac distribution function. Thus, the expectation value of the number operator corresponding to a reservoir state is

$$\text{tr}_R \left\{ c_{k\alpha}^\dagger c_{k'\alpha'} \rho_R \right\} = n_F (\omega_{k\alpha} - \mu_\alpha) \delta_{k,k'} \delta_{\alpha,\alpha'}, \tag{13.73}$$

where the Fermi function is given by

$$n_F(\omega) = \frac{1}{\exp(\omega) - 1}, \tag{13.74}$$

and μ_α is the chemical potential of lead α. $n_F(\omega_{k\alpha} - \mu_\alpha)$ represents the probability that the reservoir state with energy $\omega_{k\alpha}$ is filled with an electron, while $n_F(\mu_\alpha - \omega_{k\alpha})$ represents the probability that the same reservoir state is empty.

The coupling Hamiltonian is given by

$$\mathcal{H}_{SR} = \sum_{k \in L, k' \in R} \left(V_{kL} c_1^\dagger c_{kL} + V_{k'R} c_N^\dagger c_{k'R} + \text{H.c.} \right). \tag{13.75}$$

It is of the same type as the coupling in (13.5), describing the tunneling processes between the left lead states and the first site, and the tunneling processes between the right lead states and the Nth site of the wire.

In order to apply the theory developed in the previous section, and thus solve the system dynamics within second order perturbation theory in the tunneling amplitudes $V_{k\alpha}$, there are a few steps that need to be followed. The immediate goal is to find the correlation functions, defined by (13.19). The coupling Hamiltonian can be split in the spirit of (13.17), by defining

$$K_1 = c_1^\dagger \quad \text{and} \quad \phi_1 = \sum_{k \in L} V_{kL} c_{kL}, \tag{13.76}$$

$$K_2 = c_1 \quad \text{and} \quad \phi_2 = \sum_{k \in L} \bar{V}_{kL} c_{kL}^\dagger, \tag{13.77}$$

$$K_3 = c_N^\dagger \quad \text{and} \quad \phi_1 = \sum_{k \in R} V_{kR} c_{kR}, \tag{13.78}$$

$$K_4 = c_N \quad \text{and} \quad \phi_2 = \sum_{k \in R} \bar{V}_{kR} c_{kR}^\dagger, \tag{13.79}$$

where the overbar denotes complex conjugation. Focusing on the correlation functions in the left lead, we have

$$C_{12}(t) = \text{tr}_R \left\{ U_R(\phi_1) \phi_2 \rho_R \right\}$$

$$= \text{tr}_R \left\{ \sum_{k,k' \in L} V_{kL} \bar{V}_{k'L} U_R(c_{kL}) c_{k'L}^\dagger \rho_R \right\} \quad \text{and} \tag{13.80}$$

$$C_{21}(t) = \text{tr}_R \left\{ U_R(\phi_2) \phi_1 \rho_R \right\}$$

$$= \text{tr}_R \left\{ \sum_{k,k' \in L} V_{kL} \bar{V}_{k'L} U_R\left(c_{kL}^\dagger\right) c_{k'L} \rho_R \right\}. \tag{13.81}$$

The diagonal correlation functions $C_{11}(t)$ and $C_{22}(t)$ vanish since they involve the square of the creation and annihilation operators, respectively, which are fermionic operators.

The action of the propagation operator on the creation or annihilation operators is equivalent to their interaction picture form [10], namely

$$U_R\left(c_{k\alpha}^\dagger\right) = \exp(i\omega_{k\alpha}t)\, c_{k\alpha}^\dagger, \tag{13.82}$$

$$U_R(c_{k\alpha}) = \exp(-i\omega_{k\alpha}t)\, c_{k\alpha}. \tag{13.83}$$

Using the fermionic commutation relations of (13.73), the correlation functions become

$$C_{12}(t) = \sum_{k\in L} |V_{kL}|\, n_F(\mu_L - \omega_{kL}) \exp(-i\omega_{kL}t), \tag{13.84}$$

$$C_{21}(t) = \sum_{k\in L} |V_{kL}|\, n_F(\omega_{kL} - \mu_L) \exp(i\omega_{kL}t). \tag{13.85}$$

The final information needed to completely define the problem concerns the reservoir density of states and the tunneling amplitudes $V_{k\alpha}$, $\alpha \in \{L, R\}$. This information is encoded in the spectral density [70]

$$J_\alpha(\omega) = \pi \sum_{k,\alpha} |V_{k\alpha}|^2 \delta(\omega - \omega_k). \tag{13.86}$$

In terms of the left lead spectral function, the correlation functions take the form

$$C_{12}(t) = \int_{-\infty}^{\infty} \frac{d\omega}{\pi}\, J_L(\omega)\, n_F(\mu_L - \omega)\, e^{-i\omega t}, \tag{13.87}$$

$$C_{21}(t) = \int_{-\infty}^{\infty} \frac{d\omega}{\pi}\, J_L(\omega)\, n_F(\omega - \mu_L)\, e^{i\omega t}. \tag{13.88}$$

The integral form of the correlation function is a more suggestive indication that the states of the leads form a continuum. In the negative region of the spectrum, the spectral density should be defined such that it vanishes, $J(\omega) = 0$, $\omega \in (-\infty, 0]$. This is not a concern, since the spectral density has a physical meaning only in the positive region of the spectrum.

In complete analogy, the right lead correlation functions can be recovered as

$$C_{34}(t) = \int_{-\infty}^{\infty} \frac{d\omega}{\pi}\, J_R(\omega)\, n_F(\mu_R - \omega)\, e^{-i\omega t}, \tag{13.89}$$

$$C_{43}(t) = \int_{-\infty}^{\infty} \frac{d\omega}{\pi}\, J_R(\omega)\, n_F(\omega - \mu_R)\, e^{i\omega t}. \tag{13.90}$$

The correlation functions are very important quantities for the dynamical problem. Their physical interpretation can be easily understood. For example, $C_{12}(t)$ is directly proportional to the probability that one electron tunnels from the first site of the wire to the left lead, while $C_{21}(t)$ provides the probability of tunneling in the reverse direction, from the left lead states to the first site.

In order to proceed in applying the method developed in Section 13.1.1, the correlation functions should be decomposed in the spirit of (13.24). For this model and, in general, for electron tunneling correlations resulting from the bilinear coupling Hamiltonian in (13.5), it turns out that the decomposition of the correlation functions is equivalent to the decomposition of the spectral density into a sum of Lorentzian functions

$$J_\alpha(\omega) = \sum_{k=1}^{N} p_{k,\alpha} \frac{\Gamma_{k,\alpha}^2}{(\omega - \Omega_{k,\alpha})^2 + \Gamma_{k,\alpha}^2}, \tag{13.91}$$

where $\Omega_{k,\alpha}$ is the position of the peak of the k-th Lorentzian, $\Gamma_{k,\alpha}$ controls the width, and $p_{k,\alpha}$ is the energy scale of the system-reservoir coupling, determining the height of the Lorentzian function.

The quantity $p_{k,\alpha}$ is a measure of lead-quantum dot coupling strength. It is now possible to quantify the concept of weak coupling. Through numerical investigations [37], it was found that at room temperature the parameter regime where second order perturbation theory in the coupling strength can be considered to converge occurs when the order of magnitude of the coupling is $\sum_{k=1}^{N} p_{k,\alpha} \propto 10^{-2}$eV.

Inserting the decomposed form of the spectral density function into (13.87), (13.88), (13.89), and (13.90), and applying the theorem of residues by closing the integral along the real axis with a semicircle in the complex plane, the correlation functions become

$$C_{\alpha-}(t) = \sum_{k=1}^{N} p_{k,\alpha} \Gamma_{k,\alpha} n_F \left(\mu_\alpha - \Omega_{k,\alpha}^-\right) e^{-i\Omega_{k,\alpha}^- t}$$

$$- \frac{2i}{\beta} \sum_{k=0}^{\infty} J_\alpha(\bar{\nu}_{k,\alpha}) e^{-i\bar{\nu}_{k,\alpha} t}, \tag{13.92}$$

$$C_{\alpha+}(t) = \sum_{k=1}^{N} p_{k,\alpha} \Gamma_{k,\alpha} n_F \left(\Omega_{k,\alpha}^+ - \mu_\alpha\right) e^{i\Omega_{k,\alpha}^+ t}$$

$$- \frac{2i}{\beta} \sum_{k=0}^{\infty} J_\alpha(\nu_{k,\alpha}) e^{i\nu_{k,\alpha} t}, \tag{13.93}$$

with the convenient notation $C_{L-}(t) = C_{12}(t)$, $C_{L+}(t) = C_{21}(t)$ and $C_{R-}(t) = C_{34}(t)$, $C_{R+}(t) = C_{43}(t)$. The abbreviations $\beta = 1/(k_B T)$, $\Omega^{\pm}_{k,\alpha} = \Omega_{k,\alpha} \pm i\Gamma_{k,\alpha}$, and $\nu_{k,\alpha} = i(2\pi k + \pi)/\beta + \mu_\alpha$ have been used.

The above equations give the correlation functions in the desired form, described by (13.24). The terms involving $\nu_{k,\alpha}$ are formally infinitely many, resembling the well known Matsubara terms. The sum may be truncated at a finite value depending on the system temperature [1]. In the following it will be assumed for simplicity that the spectral density of the two leads has a Lorentzian shape, that is, $N = 1$ in

(13.92) and (13.93). Thus, J_L and J_R are given by

$$J_\alpha(\omega) = p_\alpha \frac{\Gamma_\alpha^2}{(\omega - \Omega_\alpha)^2 + \Gamma_\alpha^2} \,. \tag{13.94}$$

Directed electron transport is normally induced by a voltage drop between the left and the right lead. The bias voltage can be modeled by a difference in the chemical potential of the two leads. It will be assumed by convention and without loss of generality, that the left lead has the higher chemical potential, that is, $\mu_L - \mu_R = e V_b$, where V_b is the bias voltage.

If the energy $e V_b$ is much larger than the thermal energy $k_B T$, it can be assumed that the chemical potentials of the left and right lead may be taken to infinity, that is, $\mu_L = \infty$ and $\mu_R = -\infty$. The result is that the states of the left lead are all occupied by electrons, while the states of the right lead are all empty. Electron transport can thus occur only from the left lead to the quantum dot, and from the quantum dot to the right lead. Tunneling from the quantum dot and the left lead, or from the right lead and the dot level is forbidden. For the rest of this chapter we will restrict ourselves to this special case, but numerical solutions are easily done for finite bias as well [71].

This is mathematically expressed through the vanishing of the corresponding tunneling strengths, that is, vanishing of the corresponding correlation functions. Thus, since $n_F(\omega - \mu_L) = 1$ and $n_F(\omega - \mu_R) = 0$, in comparison to (13.87), (13.88), (13.89), and (13.90), the correlations $C_{12} = 0$ and $C_{43} = 0$ vanish. Only two correlation functions remain, one modeling electron hopping from the left lead to the first site C_{21}, and from the Nth site to the right lead C_{34}. They take the form

$$C_{21}(t) = \int_{-\infty}^{\infty} \frac{d\omega}{\pi} J_L(\omega) e^{i\omega t} \,, \tag{13.95}$$

$$C_{34}(t) = \int_{-\infty}^{\infty} \frac{d\omega}{\pi} J_R(\omega) e^{-i\omega t} \,. \tag{13.96}$$

It can be observed that in this simple model, the correlation functions are the Fourier transforms of the lead spectral densities.

Using the Lorentzian form of the spectral functions and applying the residue theorem to the integrals in (13.95) and (13.96), the correlation functions become

$$C_{21}(t) = p_L \Gamma_L \exp\left(i(\Omega_L + i\Gamma_L) t\right) \,, \tag{13.97}$$

$$C_{34}(t) = p_R \Gamma_R \exp\left(-i(\Omega_R - i\Gamma_R) t\right) \,. \tag{13.98}$$

The correlation functions have been brought to the desired form, where the time dependence is placed in the exponent. The coefficients in (13.24) can be trivially recovered as

$$a_L = p_L \Gamma_L \,, \quad \gamma_L = i\Omega_L - \Gamma_L \,, \tag{13.99}$$

$$a_R = p_R \Gamma_R \,, \quad \gamma_R = -i\Omega_R - \Gamma_R \,. \tag{13.100}$$

Using (13.26), the equation of motion for the non-vanishing auxiliary operators $\Lambda_{21}(t)$ and $\Lambda_{34}(t)$ can be easily derived, yielding

$$\frac{\partial}{\partial t} \Lambda_{21}(t) = p_L \Gamma_L K_1 + (i\Omega_L - \Gamma_L) \Lambda_{21}(t) - i\mathcal{L}_S(\Lambda_{21}(t)), \qquad (13.101)$$

$$\frac{\partial}{\partial t} \Lambda_{34}(t) = p_R \Gamma_R K_4 + (-i\Omega_R - \Gamma_R) \Lambda_{34}(t) - i\mathcal{L}_S(\Lambda_{34}(t)). \qquad (13.102)$$

The initial condition imposed on the system that the system and reservoir states are uncorrelated at time $t = 0$ implies that all auxiliary operators initially vanish, that is, $\Lambda_{21}(0) = \Lambda_{34}(0) = 0$. The above equations, together with the initial condition, completely define the time evolution of the auxiliary operators. Using (13.27), the evolution of the reduced density matrix is straightforward to solve. Further analytic treatment requires explicit knowledge of the number of sites of the wire.

13.3
The Single Resonant Level Model

The QME approach developed in the previous sections is employed here for a simple test system. The single resonant level model has attracted a lot of interest from the theoretical point of view [3, 14, 19, 63, 73]. The model describes a one level quantum dot in contact with two leads that play the role of the electron reservoirs (see Figure 13.3). In the Coulomb blockade regime, this model can be considered as a one site molecular wire. Even though this model has been extensively studied in recent years [33], it remains an interesting study case for the comparison of different approximate methods. The fact that there exists an exact analytic solution [33], obtained by the use of nonequilibrium Green's functions techniques, provides a good reference for testing the approximate methods developed here. These are second order in the coupling Hamiltonian, and thus are accurate only for weak reservoir-system coupling.

The relevant system consists of only one quantum level. Thus, the corresponding Fock subspace is spanned by two state vectors, denoted as $|0\rangle$ for the state when the system is unoccupied, and $|1\rangle$ when the system is occupied by one electron. It may be convenient to represent the two states of the dot by the vectors $|0\rangle \equiv (1\,0)^T$ and $|1\rangle \equiv (0\,1)^T$. In this basis, the dot annihilation operator is given in matrix

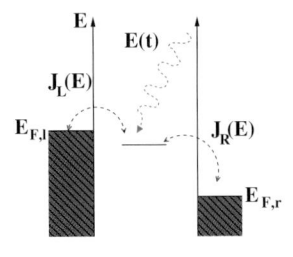

Figure 13.3 The single resonant level model. The two chemical potentials of the leads are denoted by $E_{f,L}$ and $E_{f,R}$, respectively. The site energy, denoted by $E_D = \omega_D$ ($\hbar = 1$), is situated in the bias window, that is, $E_{f,L} > E_D > E_{f,R}$. The energy of the site can be manipulated by a time-dependent electric field $E(t)$.

representation by

$$d = \begin{pmatrix} 0 & 1 \\ 0 & 0 \end{pmatrix}, \qquad (13.103)$$

obeying $d \ket{1} = \ket{0}$. The creation operator is recovered by taking the Hermitian conjugate of d and satisfies $d^\dagger \ket{0} = \ket{1}$. Using this matrix representation of the reduced system, the Hamiltonian takes the form

$$\mathcal{H}_S = \begin{pmatrix} 0 & 0 \\ 0 & \omega_D \end{pmatrix}. \qquad (13.104)$$

Equations 13.101 and (13.102), derived for a general molecular wire, become for this example

$$\frac{\partial}{\partial t} \Lambda_{21}(t) = p_L \Gamma_L \begin{pmatrix} 0 & 0 \\ 1 & 0 \end{pmatrix} + (i\Omega_L - \Gamma_L) \Lambda_{21}(t)$$

$$- i \begin{pmatrix} 0 & 0 \\ 0 & \omega_D \end{pmatrix} \Lambda_{21}(t) + i \Lambda_{21}(t) \begin{pmatrix} 0 & 0 \\ 0 & \omega_D \end{pmatrix}, \qquad (13.105)$$

$$\frac{\partial}{\partial t} \Lambda_{34}(t) = p_R \Gamma_R \begin{pmatrix} 0 & 1 \\ 0 & 0 \end{pmatrix} + (-i\Omega_R - \Gamma_R) \Lambda_{34}(t)$$

$$- i \begin{pmatrix} 0 & 0 \\ 0 & \omega_D \end{pmatrix} \Lambda_{34}(t) + i \Lambda_{34}(t) \begin{pmatrix} 0 & 0 \\ 0 & \omega_D \end{pmatrix}. \qquad (13.106)$$

It can be observed that all the terms in the left hand side (LHS) of (13.105) and (13.106) are proportional to the auxiliary operator, except for the first. This observation, together with the initial conditions, lead to the conclusion that the auxiliary operators have only one non-vanishing entry, that is,

$$\Lambda_{21}(t) = \begin{pmatrix} 0 & 0 \\ \Lambda_L(t) & 0 \end{pmatrix}, \qquad (13.107)$$

$$\Lambda_{34}(t) = \begin{pmatrix} 0 & \Lambda_R(t) \\ 0 & 0 \end{pmatrix}, \qquad (13.108)$$

where $\Lambda_L(t)$ and $\Lambda_R(t)$ are scalar functions. They obey simple ODEs that can be derived from (13.105) and (13.106), producing

$$\frac{\partial}{\partial t} \Lambda_L(t) = p_L \Gamma_L + (i\Omega_L - \Gamma_L - i\omega_D) \Lambda_L(t), \qquad (13.109)$$

$$\frac{\partial}{\partial t} \Lambda_R(t) = p_R \Gamma_R + (-i\Omega_L - \Gamma_R + i\omega_D) \Lambda_R(t). \qquad (13.110)$$

Using the abbreviations $\Omega^{\pm}_{\alpha} = \Omega_{\alpha} \pm i\Gamma_{\alpha}$, the solutions to the above equations are

$$\Lambda_{\text{L}}(t) = p_{\text{L}} \Gamma_{\text{L}} \frac{\exp\left(i\left(\Omega^{+}_{\text{L}} - \omega_{\text{D}}\right) t\right) - 1}{i\left(\Omega^{+}_{\text{L}} - \omega_{\text{D}}\right)}, \tag{13.111}$$

$$\Lambda_{\text{R}}(t) = p_{\text{R}} \Gamma_{\text{R}} \frac{\exp\left(i\left(-\Omega^{-}_{\text{R}} + \omega_{\text{D}}\right) t\right) - 1}{i\left(-\Omega^{-}_{\text{R}} + \omega_{\text{D}}\right)}. \tag{13.112}$$

The expressions above provide the time evolution of the auxiliary operators. It may be noted that in the TL formalism, the equation of motion of the auxiliary operators is independent of the reduced density matrix. In the TNL formalism, the reduced density matrix enters the expression of the auxiliary operators, and thus also becomes involved in their equation of motion. This makes finding analytic solutions to the problem a more difficult task.

Having the time evolution of the auxiliary operators, it is easy to use (13.27), and thus provide the equation of motion of the reduced system density matrix. By simply plugging in and solving the algebra, it turns out that the 2×2 density matrix has only two non-vanishing entries. These are the diagonal entries $(\rho_{\text{S}})_{11} \equiv \rho_0$ and $(\rho_{\text{S}})_{22} \equiv \rho_1$, which represent the probabilities that the quantum dot level is empty (ρ_0), or filled with one electron (ρ_1). The two probabilities must obey the normalization condition $\rho_0 + \rho_1 = 1$ and, therefore, only the equation for one of the probabilities is needed.

The ODE governing the dynamics of the probability that there is an electron on the dot is given by

$$\frac{\partial}{\partial t} \rho_1(t) = 2\,\text{Re}\left[\Lambda_{\text{L}}(t)\right] - 2\,\text{Re}\left[\Lambda_{\text{L}}(t) + \Lambda_{\text{R}}(t)\right] \rho_1(t). \tag{13.113}$$

The equation of motion for $\rho_1(t)$ is of the type $\partial/(\partial t)\, \rho_1(t) = q(t) - p(t)\rho_1(t)$, where $q(t) = 2\,\text{Re}\left[\Lambda_{\text{L}}(t)\right]$ and $p(t) = 2\,\text{Re}\left[\Lambda_{\text{L}}(t) + \Lambda_{\text{R}}(t)\right]$ are known functions. Such an equation can be solved analytically using an integrating factor. The general solution is reads

$$\rho_1(t) = \frac{\int dt\, q(t) \exp\left(\int^{t} d\tau\, p(\tau)\right) + \mathcal{C}}{\exp\left(\int^{t} d\tau\, p(\tau)\right)}, \tag{13.114}$$

where $\mathcal{C}$ is an integration constant that is fixed by the initial value $\rho_1(0)$. The integrals involved in (13.114) can be solved analytically, but the results are too lengthy to be presented. Even though the analytic expression is available, it is easier to describe $\rho_1(t)$ using numerical methods. Finally, knowing the probability that the quantum dot level is filled, the probability that it is empty is simply given by $\rho_0(t) = 1 - \rho_1(t)$.

The current as a function of time can be easily derived using techniques developed in Section 13.1.3. The expression of the current operators $\mathcal{I}_{\pm}$ can be derived directly from (13.50) and (13.51). For the left lead

$$\mathcal{I}_+ \bullet = K_1 \bullet \Lambda_{21}^\dagger(t) + \Lambda_{21}(t) \bullet K_2 , \tag{13.115}$$

$$\mathcal{I}_- \bullet = 0 , \tag{13.116}$$

and for the right lead

$$\mathcal{I}_+ \bullet = 0 , \tag{13.117}$$

$$\mathcal{I}_- \bullet = K_4 \bullet \Lambda_{34}^\dagger(t) + \Lambda_{34}(t) \bullet K_3 . \tag{13.118}$$

The current is given by (13.63) and therefore reads

$$\begin{aligned} I_L(t) &= \mathrm{tr}\left\{ K_1 \, \rho_S \, \Lambda_{21,L}^\dagger(t) + \Lambda_{21,L}(t) \, \rho_S \, K_2 \right\} \\ &= 2\, \mathrm{Re}\left[\Lambda_L(t) \right] \rho_0(t) , \end{aligned} \tag{13.119}$$

$$\begin{aligned} I_R(t) &= \mathrm{tr}\left\{ -K_2 \, \rho_S \, \Lambda_{12,R}^\dagger(t) - \Lambda_{12,R}(t) \, \rho_S \, K_1 \right\} \\ &= -2\, \mathrm{Re}\left[\Lambda_R(t) \right] \rho_1(t) . \end{aligned} \tag{13.120}$$

The interpretation of the result is clear. The current through the left lead is positive, indicating that electrons tunnel into the dot, while the right current is negative, indicating that electrons tunnel into the lead. The left current going into the dot is proportional to the probability that the dot is empty, since this is the state in which an electron can tunnel, while the reverse is true for the right current.

The total current is the difference between the left and the right current, divided by two to avoid double counting. Considering the direction from left to right to be positive, the total current is given by

$$I_{\text{total}}(t) = \mathrm{Re}\left[\Lambda_L(t) \rho_0(t) + \Lambda_R(t) \rho_1(t) \right] . \tag{13.121}$$

This current is positive, indicating that electrons indeed tunnel in the direction from the left to the right lead.

The shot noise is another important quantity that describes the transport process which can be calculated using the methods developed in Section 13.1.3. In order to apply (13.69) and calculate the noise, it is necessary to first solve the equation of motion for $F_{1\perp}(t)$, which is given by (13.68).

Studying the structure of the operators in (13.68) and knowing that initially $F_{1\perp}(0) = 0$, it can be observed that the only non-vanishing entries of the 2×2 matrix are the diagonal ones. This is analogous to the case of ρ_S. Since we have defined $F_{1\perp}(t)$ such that, by construction, $\mathrm{tr}\{F_{1\perp}(t)\} = 0$, it will be sufficient to know only $(F_{1\perp}(t))_{11} \equiv f_1$. The other non-zero entry will be $(F_{1\perp}(t))_{22} \equiv f_2 = -f_1$.

It is straightforward to find the equation of motion for f_1 from (13.68). It is only necessary to plug in the corresponding matrices in a similar fashion as for the equation of motion of ρ_S, and then to carry out the algebra. The resulting equation reads

$$\frac{\partial}{\partial t} f_{1L}(t) = -2\, \mathrm{Re}\left[\Lambda_L(t) + \Lambda_R(t) \right] f_{1L}(t) - 2\, \mathrm{Re}\left[\Lambda_L(t) \right] (\rho_0(t))^2 , \tag{13.122}$$

for the noise in the left lead, and

$$\frac{\partial}{\partial t} f_{1R}(t) = -2\,\text{Re}\left[\Lambda_L(t) + \Lambda_R(t)\right] f_{1R}(t) - 2\,\text{Re}\left[\Lambda_R(t)\right] (\rho_1(t))^2 \,, \quad (13.123)$$

for the noise in the right lead.

The solutions to the above equations can be directly used to calculate the shot noise by inserting the appropriate form of the operators in (13.69). The resulting form of the noise in the left and right leads is

$$S_L(t) = 4\,\text{Re}\left[\Lambda_L(t)\right] f_{1L}(t) + 2\,\text{Re}\left[\Lambda_L(t)\right] \rho_0(t) \,, \quad (13.124)$$

$$S_R(t) = 4\,\text{Re}\left[\Lambda_R(t)\right] f_{1R}(t) + 2\,\text{Re}\left[\Lambda_R(t)\right] \rho_1(t) \,. \quad (13.125)$$

The total noise may be defined as half the sum of the left and right noise. Thus

$$S_{\text{total}}(t) = 2\,\text{Re}\left[\Lambda_L(t) f_{1L}(t) + \Lambda_R(t) f_{1R}(t)\right] + I_{\text{total}}(t) \,. \quad (13.126)$$

The Fano factor, defined in (13.70), is given by

$$F = \lim_{t \to \infty} \left(1 + 2\,\frac{\text{Re}\left[\Lambda_L(t) f_{1L}(t) + \Lambda_R(t) f_{1R}(t)\right]}{\text{Re}\left[\Lambda_L(t) \rho_0(t) + \Lambda_R(t) \rho_1(t)\right]}\right) \,. \quad (13.127)$$

The question whether the dynamics of the transport is sub- or super-Poissonian can be answered by observing that ρ_0 and ρ_1 are positive real numbers, while f_{1L} and f_{1R} are negative reals, as can be observed from (13.122) and (13.123). The real parts of the auxiliary operators are also positive, and thus the second term of the Fano factor is negative overall. This means that $F < 1$ and that the dynamics is sub-Poissonian.

The fact that the dynamics is sub-Poissonian, that is, that the electrons are anti-bunching, is physically reasonable. The mechanism of anti-bunching originates from the fact that the quantum level of the dot can accommodate at most one electron at any time. Thus, after one tunneling event occurs from the left lead to the dot, a second tunneling in the same direction is hindered by the presence of the previous electron. It is the finite time that the electron needs to tunnel to the right lead, and thus leave the quantum dot empty, that induces the sub-Poissonian character of the dynamics.

The main value for this simple model of the calculation presented in the section above stems from the possibility to resolve in time the transient dynamics. This resolution is between the initial nonequilibrium state of the system, that is, the uncorrelated state, and the long time equilibrium dynamics. The calculation of transients and the possibility to see current and noise build up in time are two of the arguments for the strength of the theoretical method described here.

In order to illustrate the transient behavior of the system, that is, the population, current, and noise, it is sufficient to discuss the fundamental quantity which determines the others – the real part of the auxiliary operators given by (13.111) and

(13.112). Their real part is given by

$$\mathrm{Re}\left[\Lambda_\mathrm{L}(t)\right] = \frac{p_\mathrm{L}\,\Gamma_\mathrm{L}}{\Gamma_\mathrm{L}^2 + \Delta_\mathrm{L}^2}\left[e^{-\Gamma_\mathrm{L} t}\left(\Delta_\mathrm{L}\sin(\Delta_\mathrm{L} t) - \Gamma_\mathrm{L}\cos(\Delta_\mathrm{L} t)\right) + \Gamma_\mathrm{L}\right],$$

(13.128)

$$\mathrm{Re}\left[\Lambda_\mathrm{R}(t)\right] = \frac{p_\mathrm{R}\,\Gamma_\mathrm{R}}{\Gamma_\mathrm{R}^2 + \Delta_\mathrm{R}^2}\left[e^{-\Gamma_\mathrm{R} t}\left(\Delta_\mathrm{R}\sin(\Delta_\mathrm{R} t) - \Gamma_\mathrm{R}\cos(\Delta_\mathrm{R} t)\right) + \Gamma_\mathrm{L}\right],$$

(13.129)

with the abbreviation $\Delta_\mathrm{L,R} = \omega_\mathrm{D} - \Omega_\mathrm{L,R}$.

It can be seen that the time dependence is dominated by the exponential factor $e^{-\Gamma_\mathrm{L,R} t}$. The auxiliary operators grow from zero, at $t = 0$, to their limit as $t \to \infty$, and are $\mathrm{Re}\left[\Lambda_\mathrm{L,R}(t \to \infty)\right] = J_\mathrm{L,R}(\omega_\mathrm{D})$. The time of transit between these two values defines the transient, and the time scale is clearly $\tau_\mathrm{transient} = \Gamma_\mathrm{L,R}^{-1}$. The transient time scale may be given by the inverse of the coupling strength $p_\mathrm{L,R}^{-1}$ if the spectral density function is too broad. The latter time scale is a measure of the charging time and thus limits the speed at which the system can reach equilibrium.

An important observation is that the model discussed involves shaped spectral densities. If the wide band limit is invoked, that is, spectral density is considered flat with $J_a(\omega) = p_a$, the time corresponding to the transient behavior becomes infinitesimal. This occurs because the constant spectral density is equivalent to a Lorentzian density with $\Gamma_a \to \infty$.

In this limit, all quantities describing transport are constant in time, and this model has been solved using many techniques [7, 16] that are exact, that is, that do not need the assumption of weak tunneling. Thus, it is possible in the wide band limit to test the results yielded by the technique presented here.

Indeed, by straightforward computation, the populations are given by

$$\rho_0 = \frac{p_\mathrm{R}}{p_\mathrm{L} + p_\mathrm{R}},$$

(13.130)

$$\rho_1 = \frac{p_\mathrm{L}}{p_\mathrm{L} + p_\mathrm{R}}.$$

(13.131)

The total current becomes

$$I_\mathrm{total} = 2\frac{p_\mathrm{L} p_\mathrm{R}}{p_\mathrm{L} + p_\mathrm{R}}.$$

(13.132)

Finally, the noise and the Fano factor yield

$$S_\mathrm{total} = \left(1 - 2\frac{p_\mathrm{L} p_\mathrm{R}}{(p_\mathrm{L} + p_\mathrm{R})^2}\right) I_\mathrm{total},$$

(13.133)

$$F = 1 - 2\frac{p_\mathrm{L} p_\mathrm{R}}{(p_\mathrm{L} + p_\mathrm{R})^2}.$$

(13.134)

All these results are in perfect agreement with results well known from the literature [7, 16]. It is interesting to observe that the existence of the transient be-

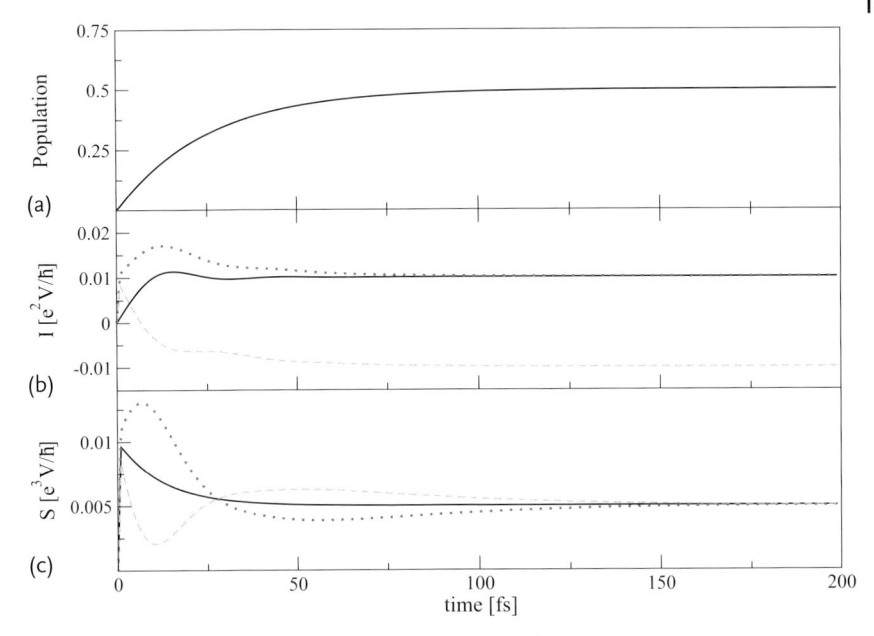

Figure 13.4 Time evolution of the population (a), current (b), and shot noise (c) for the single resonant level model. In the plots for the current and for the noise, the net quantities (black solid lines) are plotted together with the corresponding quantities measured on the left (dotted lines) and on the right hand side (dashed lines).

havior depends on the spectral density, that is, on the structure of the leads. Such a statement has subtle physical implications and may cast a shadow of doubt on the physical validity of the wide band limit that has recently become a widely used assumption.

As an example, Figure 13.4 depicts the time evolution of the population, current, and noise, including the transient behavior and the equilibrium values reached for longer times. The spectral densities of the left and right lead have been assumed to take a reasonably broad Lorentzian shape, as described by (13.94). The Lorentzian is described by the parameters $p_L = p_R = 0.01$ eV, $\Gamma_L = \Gamma_R = 5$ eV, and $\Omega_L = \Omega_R = \omega_D = 50$ eV. Note that the actual value of $\Omega_{L,R}$ or of ω_D has no impact on the dynamics. It is only their difference that is important. Here, the peak of the Lorentzian was aligned with the energy of the sites. It can be observed in Figure 13.4 that the equilibrium values of the population, current, and noise agree with the values calculated in the wide band limit, (13.132) and (13.133). The Fano factor is found to be $F = 0.5$ for sufficiently large times. The equilibrium is reached in a timescale on the order of $t \approx 100$ fs, that is, on the order of $p_{L,R}^{-1}$. For the relatively broad spectral function chosen, it was expected that the equilibrium time is proportional to the time needed to charge the dot.

13.4
Influence of Laser Pulses

In the previous section, the time-independent single resonant level model was studied, while here this is extended to two site models and the influence of laser pulses. The time evolution of a two site molecular wire is plotted in Figure 13.5. The quantities can be easily implemented and evaluated numerically by starting from (13.101) and (13.102), and plugging $\{|0,0\rangle, |1,0\rangle, |0,1\rangle, |1,1\rangle\}$ in the Hamiltonian of the system, written in the occupation number representation basis, yielding

$$\mathcal{H}_S = \begin{pmatrix} 0 & 0 & 0 & 0 \\ 0 & E_1 & -\Delta & 0 \\ 0 & -\Delta & E_2 & 0 \\ 0 & 0 & 0 & E_1 + E_2 \end{pmatrix}. \tag{13.135}$$

After solving the dynamics of the auxiliary operators, the relevant quantities can be calculated in a straightforward manner following the steps taken for the single site wire in the previous sections [71].

As for the case of the single resonant level model, the difference between the Fermi energies of the left and right lead is assumed to be infinite, and the lead

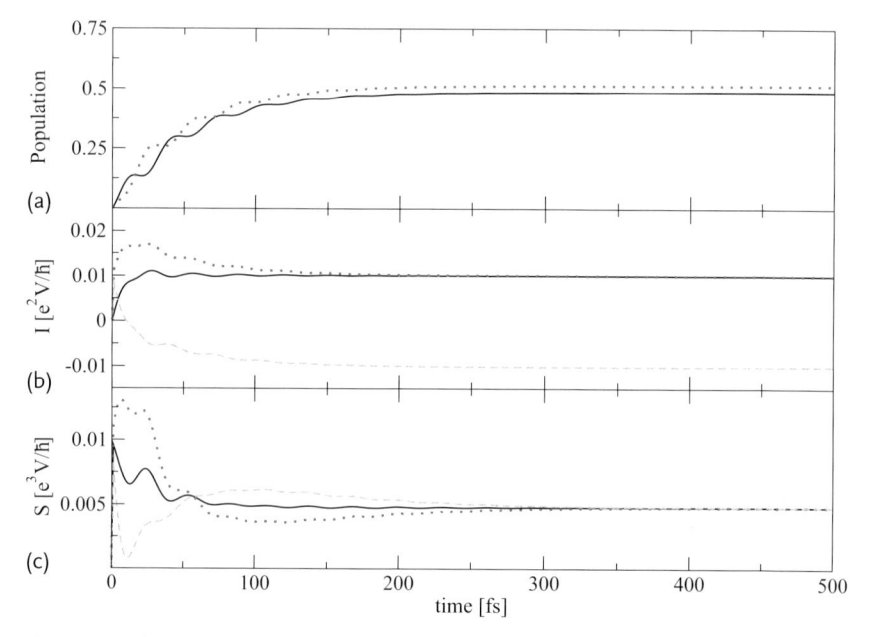

Figure 13.5 The same time evolution as shown in Figure 13.4, but for a system with two sites. In the plot of the populations, the solid black line represents the evolution of the population of site 1 and the dotted line represents the evolution of the population of site 2.

spectral densities were chosen to be identical to the single site case. For the results shown in Figure 13.5, the energy levels of the system were aligned with the peak of the Lorentzian spectral functions $\Omega_L = \Omega_R = E_1 = E_2 = 50$ eV. The intersite hopping parameter was chosen to be $\Delta = 0.1$ eV, and the wire-lead coupling to be $p_{L,R} = 0.1\Delta$.

One can observe that the population transient depicted in Figure 13.5 shows small oscillations which do not appear in the case of the single site wire. This can be attributed to interference of waves going from site 1 to site 2 and vice versa. The equilibrium population is close to one half in each site, showing a small difference between site 1 and site 2 due to their imperfect coupling. The total population in the wire is equal to one. At equilibrium, the current and the noise are the same as in the single site wire, owing to the tight-binding assumption. The corresponding equilibrium values are limited by the weakest coupling. Thus, if the intersite coupling becomes weaker than the wire-lead coupling, there will be a significant difference between single site and double sites wires.

The main advantage of being able to resolve the complete dynamics of the system, including the transient behavior, is that one can study perturbations that drive the system away from equilibrium. An important example of such a perturbation is the electric component of a strong laser field. It is straightforward to include external field driving in the system, as long as the laser is focused on the molecular wire and does not influence the reservoir or the system-reservoir coupling. For example, the following field-induced term can be added to the system Hamiltonian [42]

$$\mathcal{H}_{\text{field}} = \mu\, E(t), \quad (13.136)$$

where the wire electrical dipole operator μ acts on the states of the reduced system and $E(t)$ is an arbitrary time-dependent electrical field strength. In the system described, the field can be used to manipulate the energies of the sites, shifting them up or down in a controlled manner. The corresponding dipole operator is given by

$$\mu = c^\dagger c, \quad (13.137)$$

for the single site wire and

$$\mu = \frac{1}{2}\left(c_1^\dagger c_1 - c_2^\dagger c_2\right), \quad (13.138)$$

for the double sites case. The minus sign in expression (13.138) describes an asymmetric coupling, that is, while the first site gains energy from the field, the second site loses energy. A similar energy modulation can be achieved in quantum dots by applying a time-dependent gate voltage to each site. Possible effects of field-induced resonant excitations, for example ionization, have been neglected. In the theoretical framework designed, the laser field is introduced as an additional term to the system Hamiltonian, $H'_S = H_S + H_{\text{field}}$, and within the present TL formalism it is treated non-perturbatively [44]. The equations of motion and all the techniques presented for the case without laser field are carried out identically.

It has been shown that a field-induced modulation of the site energies in a molecular wire can coherently enhance the interference. For example, for a spe-

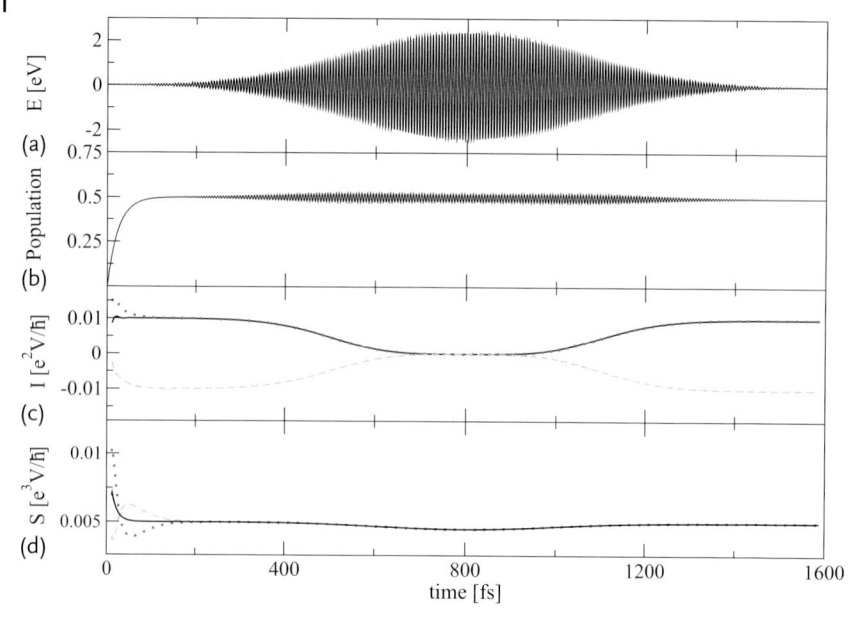

Figure 13.6 Single site system as in Figure 13.4, but with laser pulse as given in (a) and a bias voltage of 0.4 eV at temperature $T = 4$ K. The curves representing currents and noise have been obtained by averaging the corresponding quantities over four pulse periods in order to suppress the strong oscillations.

cific choice of the pulse amplitude and frequency, the electron transport through the wire can be suppressed, an effect known as coherent destruction of tunneling (CDT). It was first investigated by Grossmann et al. [24, 25]. They found that tunneling can be quenched in a periodically driven quantum system. In the context of molecular wires, this phenomenon can be explained using Floquet theory for the case of a periodic laser field [43], and it occurs for certain amplitudes of the laser field at fixed frequencies [42, 71]. As shown earlier, CDT also works in combination with short laser pulses to switch the current [40, 45]. The CDT effect is demonstrated in Figure 13.6 for a single site wire. As discussed in [45], for a single site wire CDT is only effective for bias voltages that are small compared to the pulse amplitude, since otherwise photon-assisted states wash out this effect. The population and noise shown in Figure 13.6 for a site which is energetically in the middle of the conduction window is rather independent of the bias voltage. In contrast, for a small bias voltage as in Figure 13.6 the current is suppressed by CDT, while in the infinite bias limit there is only a very small effect of the laser pulse on the current. The parameters characterizing the leads in Figure 13.6 are identical to those used to produce Figure 13.4. The laser pulse used had a Gaussian shape given by the expression

$$E(t) = E(t_0) \exp\left\{-\frac{(t-t_0)^2}{2\sigma}\right\} \cos(\omega(t-t_0)) \,. \tag{13.139}$$

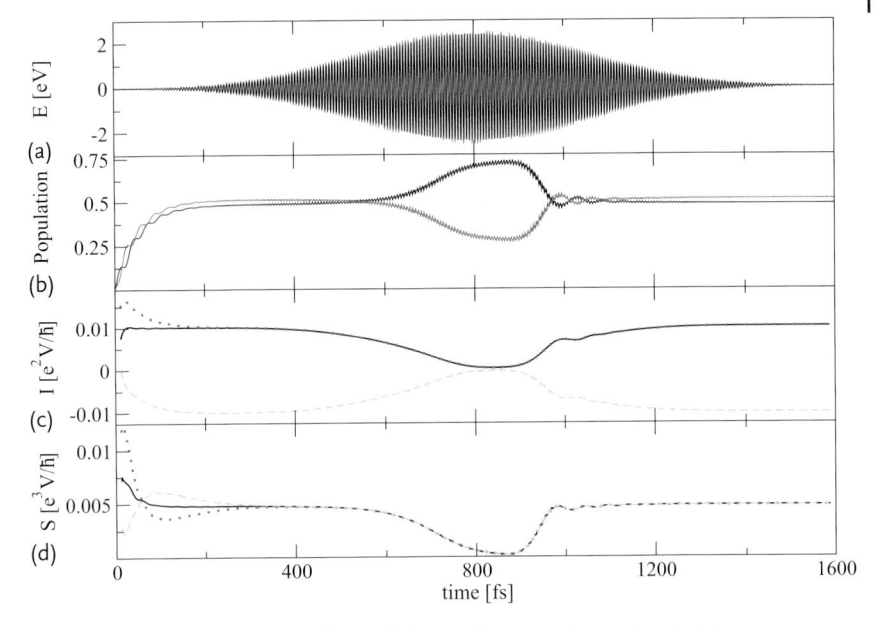

Figure 13.7 Two site system as in Figure 13.5 but with laser pulse as given in (a).

The parameters characterizing the pulse were as follows. The pulse is centered at $t_0 = 800$ fs, its width is given by $\sigma = 250$ fs^2, its intensity is $E(t_0) = 2.405$ eV, and the carrier frequency by $\omega = 1$ eV. The ratio of laser amplitude $E(t_0)$ and laser frequency ω has to be a root of the zeroth order Bessel function $J_0(E(t_0)/\omega)$ for CDT to be most effective. The pulse is turned on at some time after the system reaches equilibrium. It must be noted that the current and the noise oscillate strongly with the same period as the high frequency driving field.

From Figure 13.6 one can observe that the pulse action suppresses the current for a short period of time. Once the pulse strength decreases, at times larger than t_0 the system relaxes back to equilibrium. It is interesting to note that the noise does not deviate much from its equilibrium value, even when the current is strongly suppressed. This is a clear indication that the suppression is due to interference and not due to a blockade of the electron tunneling process.

For comparison, the phenomenon of CDT is presented for a wire with two sites in Figure 13.7. The parameters characterizing the wire are identical to those used in Figure 13.5, while the parameters characterizing the pulse are the same as in Figure 13.6. Here also, the current and noise are averaged over four pulse periods. The first striking difference between the CDT behavior in the two models can be observed by comparing the transients of the populations. For the double site wire, the fact that the intersite coupling is finite creates an asymmetry in the population distribution of the two sites. The opposite shifts induced in the site energies fuel this asymmetry, hindering the relaxation process between the sites. As a result, a strong localization of charge is observed in the first site. In this case, the external field suppresses the tunneling of the charges between the two sites of the wire, but

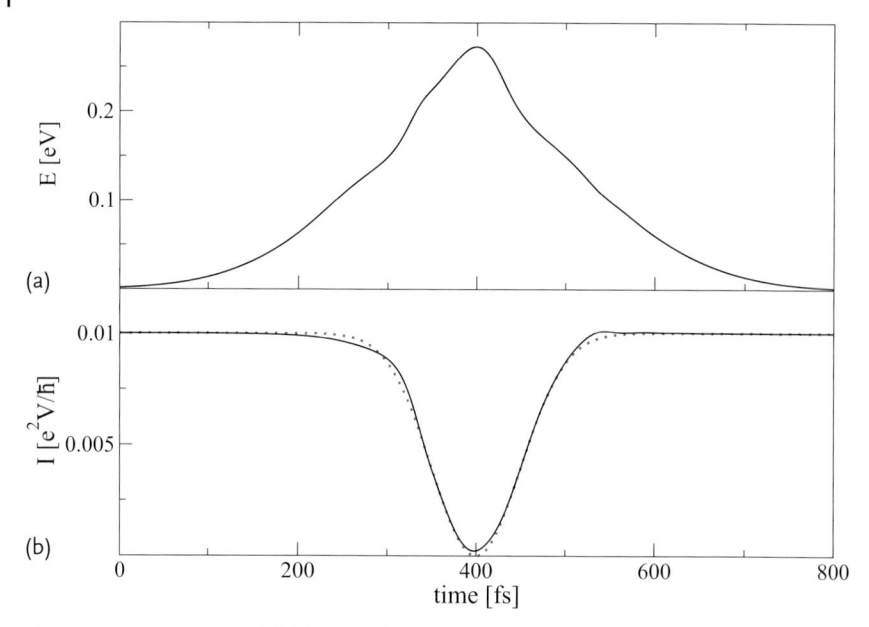

Figure 13.8 Optimal control field current for a molecular wire model with two sites obtained by optimal control theory. The electric field in (a) induces the current shape in (b), displayed as a solid black line. The target current is depicted as dotted line.

not between wire and leads. Due to this effect, charge accumulates on the first site. The noise in the case of the wire with two sites supports this conclusion. Because the current between the two sites is blocked completely, the noise also goes to zero at the maximum of the external field. This is different from the case of one site in which the current between the leads and the site is effectively zero due to the same amount of in- and outflowing electrons [46]. This nonzero partial currents result in a non-vanishing noise value.

For the current suppression in the CDT case described above, a predefined pulse was employed and then the corresponding current was calculated. The inverse situation is also possible. One first defines a desired electron flow pattern and then determines the corresponding laser field, which achieves this goal by applying optimal control theory. Using this technique optimal laser pulses can be achieved, which suppress the current [46, 47] or reverse the spin current [1]. Figure 13.8 depicts an example of an optimal control calculation for a target charge current showing current suppression for a short period of time. The electric field acts on the two site wire system, identical to the one in Figures 13.5 and 13.7. In this example no high frequency carrier field is present, but such a field can be added as well [47]. Accordingly, the electric pulse found through the optimal control algorithm does not oscillate. This example of optical control of the current through a molecular wire illustrates the interesting new possibilities offered by opto-electronic devices.

13.5
Summary and Outlook

This chapter presents a theoretical study focusing on the process of electron transport through small quantum systems in contact with metallic leads, with the goal of describing nanoscale electronic devices. The main result presented is the TL QME for the reduced density matrix and the elegant techniques necessary for constructing its solution. The properties of the electron transport are extracted from the system evolution using the theory of full counting statistics. Everything is done consistently, using the assumption of weak system-reservoir coupling. Extension towards stronger system-environment coupling can be done through the hierarchical method [34, 35] including, for example, cotunneling events.

The advantages of the method described over other popular transport methods are the easy possibility of calculating transient behaviors, which provides access to the full time dependence of the current and noise. This allows one to treat time-dependent external fields, providing the possibility of optical manipulation of electron transport. Since the external field is treated non-perturbatively in the present theory, full freedom in the choice of pulse shape and intensity is provided.

The outlook of this work is oriented towards describing models increasingly resembling the quantum features of realistic molecules, such as an improved description of laser-matter interaction. An interesting and popular subject at the moment is the interplay between electron tunneling and vibrational degrees of freedom of the molecule [51]. The electron-phonon coupling and the discrete phonon spectrum of the molecule constitute a complex problem that continues to elude a complete theoretical formulation.

References

1 Amin, A.F., Li, G.-Q., Phillips, A.H., and Kleinekathöfer, U. (2009) Coherent control of the spin current through a quantum dot. *Eur. Phys. J. B*, **68**, 103–109.

2 Anders, F.B. (2008) Steady-state currents through nanodevices: A scattering-states numerical renormalization-group approach to open quantum systems. *Phys. Rev. Lett.*, **101**, 066804.

3 Anderson, P.W. (1961) Localized magnetic states in metals. *Phys. Rev.*, **124**, 41.

4 Argyres, P.N. and Kelley, P.L. (1964) Theory of spin resonance and relaxation. *Phys. Rev.*, **134**, A98.

5 Bagrets, D.A. and Nazarov, Y.V. (2003) Full counting statistics of charge transfer in coulomb blockade systems. *Phys. Rev. B*, **67**, 085316.

6 Björk, M.T., Fuhrer, A., Hansen, A.E., Larsson, M.W., Fröberg, L.E., and Samuelson, L. (2005) Tunable effective g factor in InAs nanowire quantum dots. *Phys. Rev. B*, **72**, 201307.

7 Blanter, Y.M. and Büttiker, M. (2000) Shot noise in mesoscopic conductors. *Phys. Rep.*, **336**, 1.

8 Brandes, T. (2005) Coherent and collective quantum optical effects in mesoscopic systems. *Phys. Rep.*, **408**, 315–474.

9 Breuer, H.P. and Petruccione, F. (2002) *The Theory of Open Quantum Systems*. Oxford Univ. Press, Oxford.

10 Bruus, H. and Flensberg, K. (2004) *Many-Body Quantum Theory in Condensed Matter Physics*. Oxford Univ. Press, Oxford.

11 Caldeira, A.O. and Leggett, A.J. (1981) Influence of dissipation on quantum tun-

neling in macroscopic systems. *Phys. Rev. Lett.*, **46**, 211.

12 Čápek, V. and Kleinekathöfer, U. (2002) On homogeneous generalized master equations. *J. Phys. A: Math. Gen.*, **35**, 5521–5524.

13 Chen, F., Hihath, J., Huang, Z., Li, X., and Tao, N.J. (2007) Measurement of single-molecule conductance. *Ann. Rev. Phys. Chem.*, **58**, 535–564.

14 Cini, M. (1980) Time-dependent approach to electron transport through junctions: General theory and simple applications. *Phys. Rev. B*, **22**, 5887.

15 Cuniberti, G., Fagas, G., and Richter, K. (eds) (2005) Introducing Molecular Electronics, in *Lecture Notes in Physics*. Springer, Berlin.

16 De Jong, M.J.M. and Beenakker, C.W.J. (1995) Semiclassical theory of shot-noise suppression. *Phys. Rev. B*, **51**, 16867.

17 Dittrich, T., Hänggi, P., Ingold, G.-L., Kramer, B., Schön, G., and Zwerger, W. (1998) *Quantum Transport and Dissipation*. Wiley-VCH Verlag, Weinheim.

18 Egorova, D., Thoss, M., Domcke, W., and Wang, H. (2003) Modeling of ultrafast electron-transfer processes: Validity of multi-level Redfield theory. *J. Chem. Phys.*, **119**, 2761.

19 Fano, U. (1961) Effects of configuration interaction on intensities and phase shifts. *Phys. Rev.*, **124**, 1866.

20 Fasth, C., Fuhrer, A., Björk, M.T., and Samuelson, L. (2005) Tunable double quantum dots in InAs nanowires defined by local gate electrodes. *Nano Lett.*, **5**, 1487.

21 Flindt, C., Novotný, T., and Jauho, A.-P. (2004) Current noise in a vibrating quantum dot array. *Phys. Rev. B*, **70**, 205334.

22 Galperin, M. and Nitzan, A. (2006) Optical properties of current carrying molecular wires. *J. Chem. Phys.*, **124**, 234709.

23 Galperin, M., Ratner, M.A., and Nitzan, A. (2009) Raman scattering from nonequilibrium molecular conduction junctions. *Nano Lett.*, **9**, 758–762.

24 Grossmann, F., Dittrich, T., Jung, P., and Hänggi, P. (1991) Coherent destruction of tunneling. *Phys. Rev. Lett.*, **67**, 516–519.

25 Grossmann, F. and Hänggi, P. (1992) Localization in a driven two-level dynamics. *Europhys. Lett.*, **18**, 571–576.

26 Guhr, D.C., Rettinger, D., Boneberg, J., Erbe, A., Leiderer, P., and Scheer, E. (2007) Influence of laser light on electronic transport through atomic-size contacts. *Phys. Rev. Lett.*, **99**, 086801.

27 Gurvitz, S.A. (2005) Quantum interference in resonant tunneling and single spin measurements. *IEEE Trans. Nanotechnol.*, **4**, 45.

28 Gurvitz, S.A. and Prager, Y.S. (1996) Microscopic derivation of rate equations for quantum transport. *Phys. Rev. B*, **53**, 15932.

29 Gustavsson, S., Leturcq, R., Ihn, T., Ensslin, K., Reinwald, M., and Wegscheider, W. (2007) Measurements of high-order noise correlations in a quantum dot with a finite bandwidth detector. *Phys. Rev. B*, **75**, 75314.

30 Gustavsson, S., Leturcq, R., Simovic, B., Schleser, R., Ihn, T., Studerus, P., Ensslin, K., Driscoll, D.C., and Gossard, A.C. (2006) Counting statistics of single-electron transport in a quantum dot. *Phys. Rev. Lett.*, **96**, 076605.

31 Gustavsson, S., Leturcq, R., Simovic, B., Schleser, R., Studerus, P., Ihn, T., Ensslin, K., Driscoll, D.C., and Gossard, A.C. (2006) Counting statistics and super-poissonian noise in a quantum dot: Time resolved measurements of electron transport. *Phys. Rev. B*, **74**, 195305.

32 Hanson, R., Kouwenhoven, L.P., Petta, J., Tarucha, S., and Vandersypen, L. (2007) Spins in few-electron quantum dots. *Rev. Mod. Phys.*, **79**, 1217.

33 Haug, H. and Jauho, A.-P. (2007) *Quantum Kinetics in Transport and Optics of Semiconductors*, 2nd edn. Springer, Berlin.

34 Jin, J., Zheng, X., and Yan, Y. (2008) Exact dynamics of dissipative electronic systems and quantum transport: Hierarchical equations of motion approach. *J. Chem. Phys.*, **128**, 234703.

35 Jin, J.S., Welack, S., Luo, J.Y., Li, X.Q., Cui, P., Xu, R.X., and Yan, Y.J. (2007) Dynamics of quantum dissipation systems interacting with fermion and boson grand canonical bath ensembles: Hierachical

equations of motion approach. *J. Chem. Phys.*, **126**, 134113.

36 Kaiser, F.J., and Kohler, S. (2007) Shot noise in non-adiabatically driven nanoscale conductors. *Ann. Phys. (Leipzig)*, **16**, 702.

37 Kleinekathöfer, U. (2004) Non-Markovian theories based on the decomposition of the spectral density. *J. Chem. Phys.*, **121**, 2505.

38 Kleinekathöfer, U. (2009) Time-local quantum master equations and their applications to dissipative dynamics and molecular wires, in *Energy Transfer Dynamics in Biomaterial Systems* (eds Burghard, I., May, V. Micha, D.A., and Bittner, E.R.) *Ser. Chem. Phys.*, Springer, New York.

39 Kleinekathöfer, U., Li, G.-Q., Welack, S., and Schreiber, M. (2006) Coherent destruction of the current through molecular wires using short laser pulses. *Phys. Status. Solidi. (b)*, **243**, 3775–3781.

40 Kleinekathöfer, U., Li, G.-Q., Welack, S., and Schreiber, M. (2006) Switching the current through model molecular wires with gaussian laser pulses. *Europhys. Lett.*, **75**, 139–145.

41 Kohler, S. and Hänggi, P. (2007) Ultrafast stop and go. *Nat. Nanotech.*, **2**, 675–676.

42 Kohler, S., Lehmann, J., and Hänggi, P. (2005) Driven quantum transport on the nanoscale. *Phys. Rep.*, **406**, 379.

43 Lehmann, J., Camalet, S., Kohler, S., and Hänggi, P. (2003) Laser controlled molecular switches and transistors. *Chem. Phys. Lett.*, **368**, 282.

44 Li, G.-Q., Kleinekathöfer, U., and Schreiber, M. (2008) Treatment of laser-field effects on a molecular wire and its coupling to the leads. *J. Lumin.*, **128**, 1078–1080.

45 Li, G.-Q., Schreiber, M., and Kleinekathöfer, U. (2007) Coherent laser control of the current through molecular junctions. *EPL*, **79**, 27006-1–6.

46 Li, G.-Q., Schreiber, M., and Kleinekathöfer, U. (2008) Suppressing the current through molecular wires: Comparison of two mechanisms. *New J. Phys.*, **10**, 085005.

47 Li, G.-Q., Welack, S., Schreiber, M., and Kleinekathöfer, U. (2008) Tailoring current flow patterns through molecular wires using shaped optical pulses. *Phys. Rev. B*, **77**, 075321-1–5.

48 Lortscher, E., Weber, H.B., and Riel, H. (2007) Statistical approach to investigating transport through single molecules. *Phys. Rev. Lett.*, **98**, 176807.

49 Luo, J.Y., Li, X.-Q., and Yan, Y. (2007) Calculation of the current noise spectrum in mesoscopic transport: A quantum master equation approach. *Phys. Rev. B*, **76**, 085325.

50 May, V. and Kühn, O. (2000) *Charge and Energy Transfer in Molecular Systems*. Wiley-VCH Verlag GmbH, Berlin.

51 May, V. and Kühn, O. (2008) Photoinduced removal of the franck-condon blockade in single-electron inelastic charge transmission. *Nano Lett.*, **8**, 1095–1099.

52 Meier, C. and Tannor, D.J. (1999) Non-Markovian evolution of the density operator in the presence of strong laser fields. *J. Chem. Phys.*, **111**, 3365.

53 Meir, Y. and Wingreen, N.S. (1992) Landauer formular for the current through an interacting electron region. *Phys. Rev. Lett.*, **68**, 2512–2515.

54 Mozyrsky, D., Fedichkin, L., Gurvitz, S.A., and Berman, G.P. (2002) Interference effects in resonant magnetotransport. *Phys. Rev. B*, **66**, 161313.

55 Mühlbacher, L. and Rabani, E. (2008) Real-time path integral approach to nonequilibrium many-body quantum systems. *Phys. Rev. Lett.*, **100**, 176403.

56 Nakajima, S. (1958) On quantum theory of transport phenomena steady diffusion. *Prog. Theor. Phys.*, **20**, 948.

57 Nazarov, Yu.V. (2003) *Quantum Noise in Mesoscopic Physics*. Kluwer, Dordrecht.

58 Nitzan, A. (2006) *Chemical Dynamics in Condensed Phases Relaxation, Transfer and Reactions in Condensed Molecular Systems*. Oxford Univ. Press, Oxford.

59 Nitzan, A. and Ratner, M.A. (2003) Electron transport in molecular wire junctions. *Science*, **300**, 1384.

60 Novotný, T., Donarini, A., Flindt, C., and Jauho, A.-P. (2004) Shot noise of a quantum shuttle. *Phys. Rev. Lett.*, **92**, 248302.

61 Schröder, M., Schreiber, M., and Kleinekathöfer, U. (2007) Reduced dy-

namics of coupled harmonic and anharmonic oscillators using higher-order perturbation theory. *J. Chem. Phys.*, **126**, 114102–1–10.

62 Shibata, F., Takahashi, Y., and Hashitsume, N. (1977) A generalized stochastic Liouville equation. Non-Markovian versus memoryless master equations. *J. Stat. Phys.*, **17**, 171.

63 Stefanucci, G. and Almbladh, C.-O. (2004) Time-dependent partition-free approach in resonant tunneling systems. *Phys. Rev. B*, **69**, 195318.

64 Tanimura, Y. (2006) Stochastic Liouville, Langevin, Fokker–Planck and master equation approaches to quantum dissipative systems. *J. Phys. Soc. Jpn.*, **75**, 082001.

65 Tannor, D.J. (2007) *Introduction to Quantum Mechanics: A Time-Dependent Perspective*. Univ. Science Press, Sausalito.

66 Thielmann, A., Hettler, M.H., König, J., and Schön, G. (2005) Cotunneling current and shot noise in quantum dots. *Phys. Rev. Lett.*, **95**, 146806.

67 Van der Wiel, W.G., De Franceschi, S., Elzerman, J.M., Fujisawa, T., Tarucha, S., and Kouwenhoven, L.P. (2003) Electron transport through double quantum dots. *Rev. Mod. Phys.*, **75**, 1–22.

68 Wang, H. and Thoss, M. (2003) Multi-layer formulation of the multi-configuration time-dependent Hartree theory. *J. Chem. Phys.*, **119**, 1289.

69 Weiss, S., Eckel, J., Thorwart, M., and Egger, R. (2008) Iterative real-time path integral approach to nonequilibrium quantum transport. *Phys. Rev. B*, **77**, 195316.

70 Weiss, U. (1999) *Quantum Dissipative Systems*, 2nd edn. World Scientific, Singapore.

71 Welack, S., Schreiber, M., and Kleinekathöfer, U. (2006) The influence of ultra-fast laser pulses on electron transfer in molecular wires studied by a non-Markovian density matrix approach. *J. Chem. Phys.*, **124**, 044712–1–9.

72 Wolf, S.A., Awschalom, D.D., Buhrman, R.A., Daughton, J.M., von Molnar, S., Roukes, M.L., Chtchelkanova, A.Y., and Treger, D.M. (2001) Spintronics: A spin-based electronics vision for the future. *Science*, **294**, 1488–1495.

73 Zedler, P., Schaller, G., Kießlich, G., Emary, C., and Brandes, T. (2009) Weak coupling approximations in non-Markovian transport. *Phys. Rev. B*, **80**, 045309.

74 Zutic, I., Fabian J., and Das Sarma, S. (2004) Spintronics: Fundamentals and applications. *Rev. Mod. Phys.*, **76**, 323–410.

75 Zwanzig, R. (1960) Ensemble method in the theory of irreversibility. *J. Chem. Phys.*, **33**, 1338.

14
Two-Photon Photoemission of Plasmonic Nanostructures with High Temporal and Lateral Resolution

Michael Bauer, Daniela Bayer, Carsten Wiemann, and Martin Aeschlimann

14.1
Introduction

In modern science, the optical properties of nanostructures has become a topic attracting much interest within fundamental physics as well as in technical applications. Silver and gold nanoparticles with typical sizes of 5 to 150 nm can exhibit particularly strong optical extinction in the visible spectral range due to resonantly driven electron plasma oscillations, termed as localized surface plasmons (LSP). The resonance energy of the LSP depends critically on the size and the shape as well as on the material of the particle and the embedding environment [1–3]. This enables the spectral tuning of the resonance, a property which is of considerable interest in the context of future electronic and optical device applications. Due to the rapid advances in the fabrication of small particles [4] and nanowires [5], their optical properties are now used in a wide range of applications, including biosensors [6, 7], near-field microscopy [8] and new optical devices [9–11]. Furthermore, since plasmons are associated with large electromagnetic fields near the particle surface, they play an important role within nonlinear processes including surface-enhanced Raman scattering (SERS) [12], second and high harmonic generation [13–15] and multiphoton photoemission [16–20]. The limiting factor for applications is the energy loss of the collective electron oscillation due to the damping of the LSP, which is manifested in the plasmon decay time τ_{pl} [21].

The fundamental microscopic mechanisms of collective electron excitations in small particles as well as their decay are still far from being completely understood. As a pioneer in this field, Gustav Mie developed a first theory based on Maxwell's equations to explain the optical properties of spherical nanoparticles. Mie's theory easily describes red shifts and the lifetime broadening of the dipole plasmon resonance as the particle size is increased. It also explains the appearance of resonance contributions of higher multipolar order [1]. However, this theory is strictly valid only for single particles with a spherical geometry. Therefore, during the last decades many theoretical studies have focused on the properties of LSP in nanostructures of different shapes in order to gain insight, for example, into their optical

Nonlinear Dynamics of Nanosystems. Edited by Günter Radons, Benno Rumpf, and Heinz Georg Schuster
Copyright © 2010 WILEY-VCH Verlag GmbH & Co. KGaA, Weinheim
ISBN: 978-3-527-40791-0

response, the field distribution of the resonant modes as well as relevant decay channels and the coupling between neighboring particles [22–24].

A simple oscillator model describing the interaction of a light field and a nanoparticle can be discussed as follows [25]. The light field couples occupied and unoccupied single electron states which are separated by the photon energy $h\nu$. The induced polarizations of these different, coherently coupled transitions superpose to a macroscopic polarization which represents the collective response of the electronic system. This polarization field adds to the incident light field and causes a modification of the particle-internal field (Figure 14.1). The relation between the internal field and light field is described by the frequency dependent field enhancement factor $f(\omega)$ [16]. Figure 14.2 displays the phase shift $\varphi(\omega)$ of the induced polarization with respect to the light field and the amplitude of the field enhancement factor $f(\omega)$. For frequencies below the evident resonance peak, the internal field is small because the π-shifted polarization field destructively adds to the light field. While passing the resonance frequency, the polarization response undergoes a phase shift from $-\pi$ to 0. The extraordinary field enhancement at λ_{pl} (corresponding to the LSP resonance) is determined by the resonant response of the electron collective to the light field, adding up to an extremely large polarization field. Finally, in the short wavelength regime, the amplitude of the polarization field decreases since the electron collective is too inert to follow the oscillating light field.

Damping of the plasmon excitation is basically governed by two different decay channels (see Figure 14.3). First, the plasmon energy can be returned coherently to the external radiation field (radiation damping) as the oscillating polarization field must emit electromagnetic radiation. According to the optical far-field theory, this decay channel corresponds to the scattering of the incident excitation light field. The signals exploited by pure optical far-field detection techniques such as

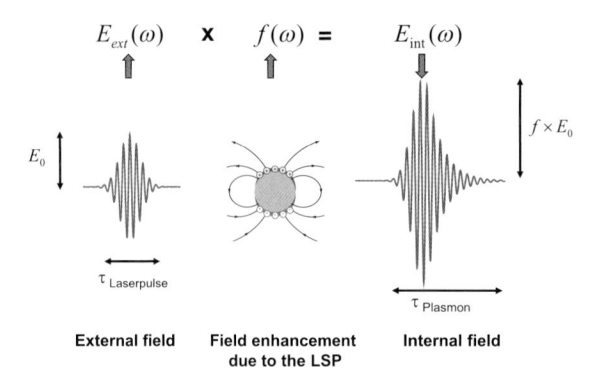

Figure 14.1 Schematic illustration of the interaction between an external light field pulse $E_{ext}(\omega)$ and a localized surface plasmonic excitation (LSP) resulting in a modified internal field $E_{int}(\omega)$. The amplitude $E_{int}(\omega)$ with respect to $E_{ext}(\omega)$ is determined by the field enhancement factor $f(\omega)$, the temporal characteristic of $E_{int}(\omega)$ is governed by the plasmon lifetime $\tau_{Plasmon}$.

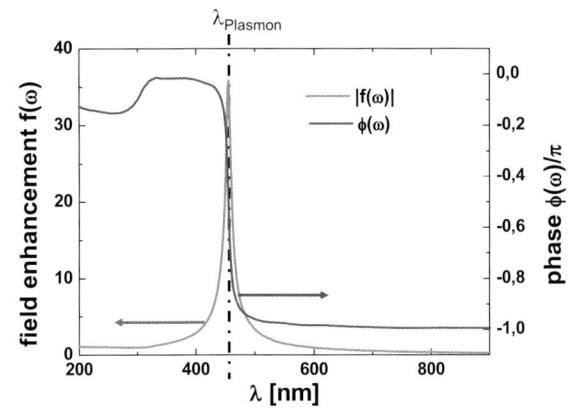

Figure 14.2 Field enhancement $f(\lambda)$ (dark grey line) and phase shift $\phi(\lambda)$ (light grey line) of $E_{int}(\omega)$ with respect to $E_{ext}(\omega)$ as function of the wavelength of the excitation field in the vicinity of the plasmon resonance $\lambda_{Plasmon}$ calculated in the dipole approximation for a small silver nanoparticle (diameters: $d_1 = d_2 = 5$ nm, $d_3 = 1$ nm, excitation along one of the two long axis).

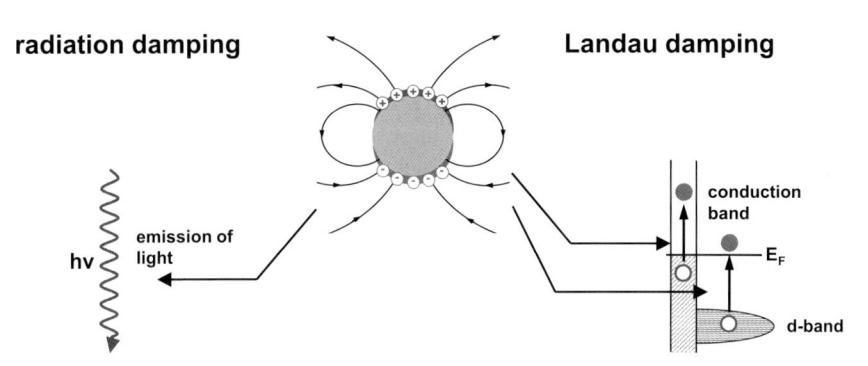

Figure 14.3 Schematic illustration of the relevant damping mechanisms of a localized surface plasmon adapted from reference [27]. For large particle diameters ($d > 10$ nm) the damping of the LSP is goverened by the coupling to the light field (radiation damping) and by the coupling to electron–hole pair excitations in the particle (Landau damping).

second harmonic generation [13, 14] and extinction spectroscopy [26] are due to the coupling to this radiation damping channel.

Furthermore, the decay of a plasmon is possible by the creation of electron–hole pairs and a subsequent transfer of energy to the internal degrees of freedom inside the particles (internal damping). This process results in a complete loss of coherence with respect to the exciting light field. In the far field, this damping channel is recognized as absorption. The involvement of single electron excitations in this process suggests that electron emission techniques such as photoemission may also be useful as probes for plasmonic properties. In this paper, we demonstrate that, in particular, two-photon photoemission (2PPE) is highly sensitive to plasmon excitations in metallic nanoparticles. A striking example is the study of particle-shape characteristics of the plasmon damping in elliptical nanoparticles as probed by means of the time-resolved 2PPE. Furthermore, we show that 2PPE in combination with the photoemission electron microscopy technique (PEEM) allows mapping of local near-field variations associated with plasmonic excitations with sub-diffraction (< 40 nm) resolution. Experimental examples illustrating the potential of this technique serve as the real-time observation of the field retardation in large nanoparticles and the plasmon-governed coupling of neighboring nanoparticles.

14.2
Experimental

Figure 14.4 shows the basic scheme of the time-resolved two-photon photoemission process for probing the decay of electronic excitations at a metal surface [28]. A first ultrashort laser pulse (pump) in interaction with the electronic subsystem at a given time t_0 populates an intermediate excited electron state E_I below the vacuum level. A second laser pulse (probe) incident at the time t_1 couples this excited state population to a detection state above the vacuum level, where it is addressed by an electron-sensitive detector such as an electron energy analyzer or a photoemission electron microscope. A successive and controlled increment of the temporal delay between both pulses enables one to record the time-evolution of the depopulation of the intermediate state. For instance, for an electron gas of a metal, the

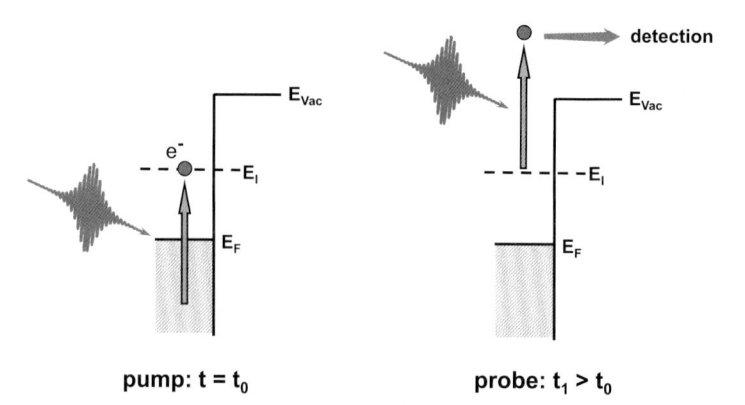

Figure 14.4 Scheme of a 2PPE process in a metal as used for time-resolved spectroscopy with femtosecond resolution; a first photon pulse (pump pulse) induces an electronic excitation in the metal which is probed by a second, temporally delayed photon via emission into the vacuum. The successive increase of the temporal delay $t_1 - t_0$ allows probing the depopulation dynamics of the electronic excitation.

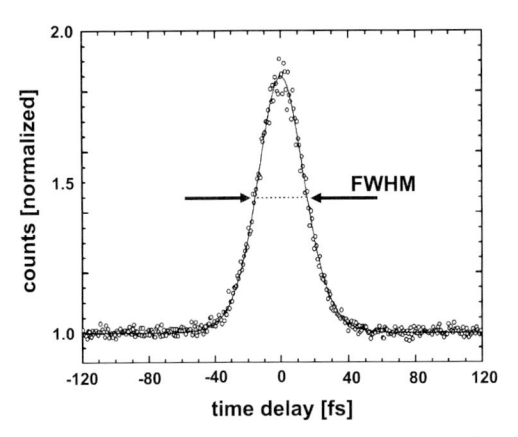

Figure 14.5 2PPE cross-correlation trace as recorded within a TR-2PPE experiment; the 2PPE yield increases as pump and probe pulse overlap in time due to the non-linear character of the 2PPE process. A finite lifetime of the involved electronic excitation delays the response of the system and gives rise to an increase of the cross-correlation FWHM.

depopulation is governed by inelastic electron–electron scattering processes and is characterized by the inelastic lifetime τ_{ee}.

A typical experimental time-resolved 2PPE trace as a function of the time delay between a pump pulse and cross-polarized probe pulses is shown in Figure 14.5. The shape of this cross-correlation trace is a convolution of the two laser pulses and the exponential decay of the probed intermediate state E_1 determined by the inelastic lifetime τ_{ee}. A deconvolution of τ_{ee} can be performed by a fit of simulated correlation traces to this data set. For bulk electron excitations, these simulations are performed within a rate equation model which corresponds to the solution of the Liouville–von Neumann equations of a three-level system within the density matrix formalism in the limit of rapid dephasing [29–31]. For qualitative statements on τ_{ee} and for comparing studies, it is often sufficient to analyze the broadening of the full width at half maximum (ΔFWHM) of the cross-correlation trace, which increases as the lifetime of the intermediate state τ_{ee} increases.

Thus far, only single electron states have been considered for the description of the 2PPE process. In the following, we will discuss to what extent the 2PPE process is also sensitive to the collective electron excitations in nanoparticles.

2PPE is a second order process and, therefore, the measured electron current j_{2PPE} is proportional to the fourth power of the electric field ($j_{2PPE}(r) \propto |E_{int}^4|$) acting on the electrons. In the case of a plasmon-resonant excitation of a nano-sized particle, this (particle internal) field is determined by the local field enhancement $f(\omega)$ as governed by the properties of the LSP. It is this relation which makes the two-photon photoemission a versatile tool in the investigation of plasmonic excitations. Later, we will see that besides a high-field enhancement, an efficient transfer of energy from the LSP resonance to the single electron excitation spectrum is a necessary condition to generate a sufficiently high photoemission signal.

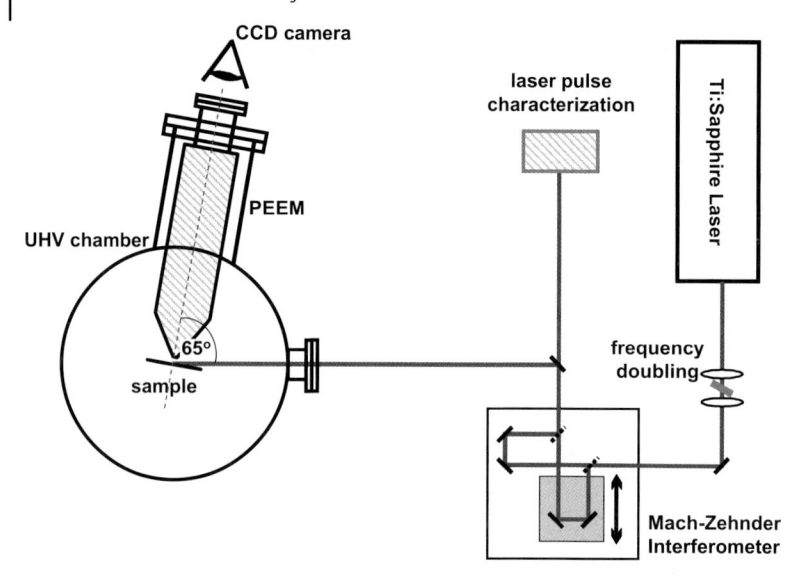

Figure 14.6 Principle scheme of the setup for the time-resolved 2PPE experiment: a femtosecond laser system delivers the required ultrafast light pulses. A nonlinear crystal is used for frequency doubling of the light so that LSP-resonances in silver nanoparticles can be excited resonantly. The temporal delay between pump and probe pulse is adjusted by an optical delay stage in the Mach–Zehnder interferometer. The sample is mounted in an UHV-chamber which is equipped with an electron energy analyzer or a photoemission electron microscope.

The involvement of the LSP in the 2PPE process also affects the shape of the cross-correlation trace in time-resolved experiments. Next to τ_{ee}, the inelastic lifetime of single electron excitation, also the LSP-lifetime τ_{LSP}, now contributes to the broadening ΔFWHM. A reasonable quantitative deconvolution of both quantities from the cross-correlation trace is a rather complex task as has been shown, for instance, in [25]. However, as we will see in the following, the elaborate application of the different experimental degrees of freedom will also provide interesting insights into the LSP dynamics and the single electron dynamics even though only changes in the FWHM of the cross-correlation are considered.

The setup of the time-resolved 2PPE experiments used for our studies is shown in Figure 14.6. Pump and probe laser pulses are delivered from the frequency doubled output (photon energy of 3.1 eV) of a femtosecond Ti:Sapphire laser system (repetition frequency 80 MHz, temporal pulse width 20 fs). A Mach–Zehnder interferometer allows one to adjust the difference in optical pathway between pump and probe pulse with an accuracy of more than 100 nm corresponding to a timing accuracy of < 0.3 fs. The collinear pulse pair is then focused onto the sample surface and excites the electrons in a 2PPE process which are subsequently detected by a suitable detection unit.

Two different types of electron detectors have been used for our studies. Spectroscopic measurements have been performed with a cylindrical sector electron energy analyzer (Focus CSA 200) with an energy resolution of 80 meV. For the plasmonic systems under consideration, it allows one to investigate the energy dependence of the electron–hole pair excitation spectrum subsequent to the LSP decay within lateral integrating measurements.

The second electron detector is a photoemission electron microscope (PEEM, Focus IS-PEEM), which enables one to focus on details of an *individual* nanoparticle [32, 33]. The (electrostatic) PEEM maps the lateral distribution of the electrons photoemitted from the sample at a lateral resolution in the sub-30 nm regime. Note that the experimental configuration of the PEEM system restricts the incident angle of the laser light with respect to the surface normal to a fixed angle of 65° (see Figure 14.6). This detail will be of relevance in the interpretation of time-resolved PEEM data discussed later in Section 14.3.2. For these PEEM experiments, a mercury vapor UV source (high energy cut-off 4.9 eV) next to the laser light is also available. It allows imaging of the surface by linear photoemission directly above the work function threshold of silver, at about 4.5 eV photon energy.

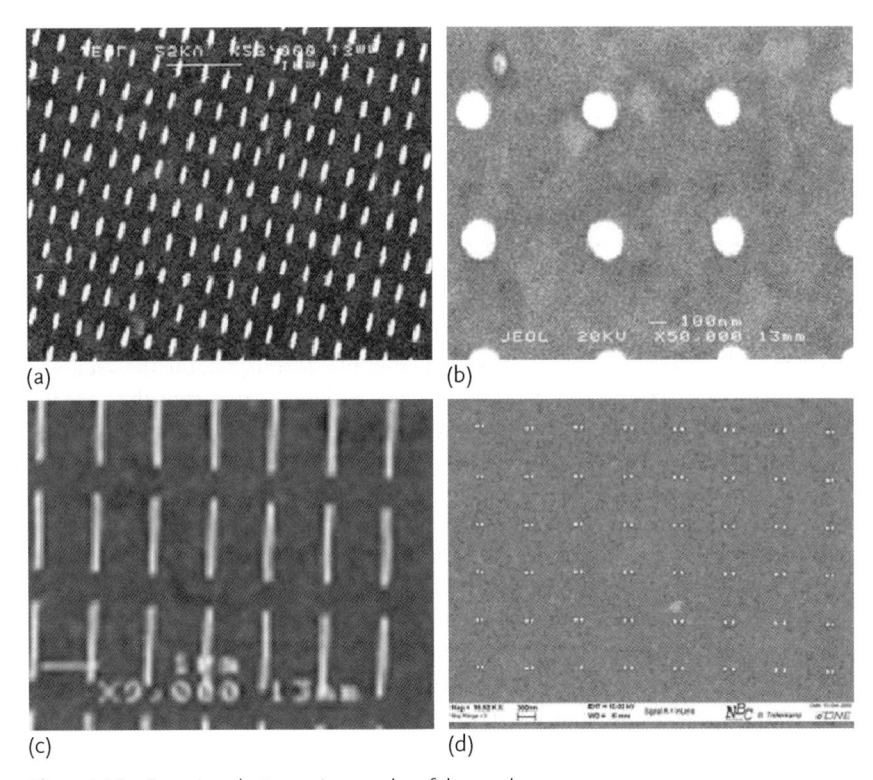

Figure 14.7 Scanning electron micrographs of the used nanoparticle arrays: (a) elliptical silver nanoparticles (long axis: 140 nm, short axis: 60 nm, height: 50 nm); (b) silver nano-dots (diameter: 200 nm, height: 50 nm); (c) silver nanowires (length: 1.6 μm, width: 60 nm, height: 50 nm); (d) silver nano-dot pairs (diameter: 50 nm, height: 40 nm, interparticle spacing: 130 nm, grating constant: 740 nm).

The nanostructured samples have been prepared by electron-beam lithography in a lift-off process. This technique enables a controlled and flexible design of metallic nanoparticles with respect to their shape and size. It allows for a tuning of the characteristic LSP resonance frequencies to the wavelength regime accessible by our laser system. Figure 14.7 shows SEM images of the different silver nanostructures deposited on ITO covered glass substrates as they are used in this study. The dimensions of the elliptical-shaped silver nanoparticles in Figure 14.7a are 140 nm (long axis), 60 nm (short axis) and 50 nm (height). They comprise versatile samples for the investigation of variations in the LSP decay in respect to resonant or off-resonant excitation. The silver nano-dot array (Figure 14.7b diameter: 200 nm, height: 50 nm) and the silver nanowire array (Figure 14.7c length: 1.6 μm, width: 60 nm, height: 50 nm) are used to illustrate the potential of the time-resolved PEEM technique to map retardation effects associated with a plasmon excitation at a nanometer resolution. Studies of the plasmon-induced coupling between neighboring nanoparticles are possible with nano-dot pairs of varying center-to-center spacing. Figure 14.7d shows an example of 50 nm dimers (height: 40 nm) at an interparticle spacing of 130 nm (grating constant: 740 nm).

14.3
Results and Discussion

14.3.1
Localized Surface Plasmons Probed by TR-2PPE

Figure 14.8 presents measured (black line) and calculated extinction spectra of the array of elliptically shaped silver nanoparticles, as shown in Figure 14.7a. The experiments were performed at normal light incidence using unpolarized light. The calculations are based on a numerical model described in [34]. Three different resonances at 431 nm, 450 nm and 795 nm are predicted corresponding to plasmon excitations along the z-axis (perpendicular to the surface plane), the in-plane short axis, and the in-plane long axis, respectively (see Figure 14.8 for details). The experimental configuration (perpendicular light incidence) only allows a coupling of the light field to the two in-plane resonances. The resonance energies of these two modes are almost perfectly reproduced by the calculations, whereas the broadening of the resonances is somewhat underestimated. This indicates the presence of damping mechanisms in the nanoparticles which are not taken into account in the simulation, for example, the interaction between particle and substrate and an enhanced internal damping due to a finite defect density in the particle itself [17].

The 400 nm laser-light used for the TR-2PPE experiment is close to resonance to the in-plane short axis mode of the particle. In contrast, a coupling to the long-axis mode is only possible under off-resonant conditions. Therefore, the polarization state of the laser light (p- or s-polarized) enables one to experimentally prepare resonant and off-resonant excitation conditions. Resonant conditions are obviously achieved with the polarization vector (electric field vector) oriented along the in-

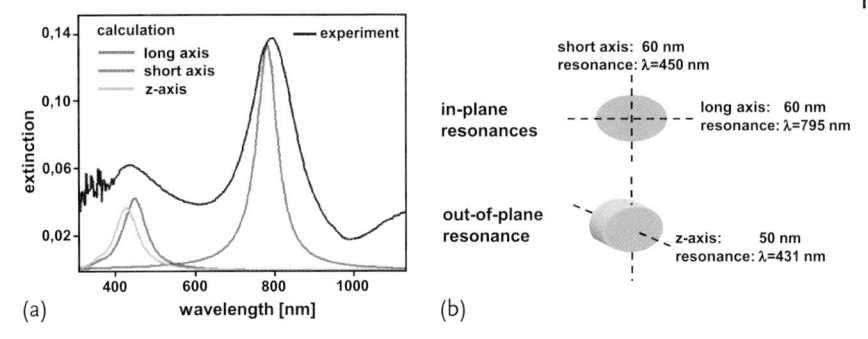

Figure 14.8 Measured and calculated extinction spectra of the investigated array of elliptical nanoparticles (see Figure 14.7a); resonant and off-resonant excitation conditions in the 2PPE experiments are realized by coupling of the 400 nm light pulses to the in-plane resonance along the long axis (off-resonant excitation) and the in-plane resonance along the short axis (resonant excitation), respectively.

plane short axis, whereas off-resonant conditions are achieved with a polarization vector oriented along the long axis.

The high sensitivity of the two-photon photoemission yield to a resonant excitation of the LSP becomes evident from the experimental date shown in Figure 14.9b. The displayed data points correspond to the 2PPE yield at varying polarization angles of the incident laser light with respect to the long or off-resonant axis of the nanoparticle. We observe a clear oscillation in the yield, where the yield maxima and minima coincide with the orientations of the electric field vector along the short and long axis, respectively. The same periodicity is observed in the time-resolved 2PPE data. The triangles in Figure 14.9a display the corresponding variation of the FWHM as obtained from $sech^2$ fits to the experimental cross-correlation traces. As discussed in Section 14.2, differences in the FWHM of the correlation trace can be assigned to variations in the femtosecond dynamics associated with the LSP excitation. For resonant (short-axis) excitation, the cross-correlation FWHM is maximum, indicating a long LSP lifetime. For smaller off-resonant excitation, FWHM points to a more efficient LSP decay.

TR-2PPE experiments at varying intermediate state energies $E-E_F$ allow one to illustrate the different mechanisms governing the inelastic lifetime τ_{ee} of single electron excitations and the plasmon lifetime τ_{LSP} associated with the decay of a collective electron mode. For metals, τ_{ee} exhibits a characteristic energy dependence, as has been shown in the past in several theoretical and experimental works [25, 28, 35–37]. As an example, Figure 14.10a shows TR-2PPE data of the lifetime of single electron excitations in a polycrystalline silver sample. τ_{ee} increases monotonously as the intermediate state energy decreases. In contrast, the plasmon lifetime τ_{LSP} is expected to exhibit no dependence on the probed intermediate state energy ($E-E_F$). Figure 14.10b shows 2PPE cross-correlation FWHM data for varying intermediate state energies ($E-E_F$) measured for the elliptical nanoparticles under resonant and off-resonant conditions. Both curves exhibit an energy depen-

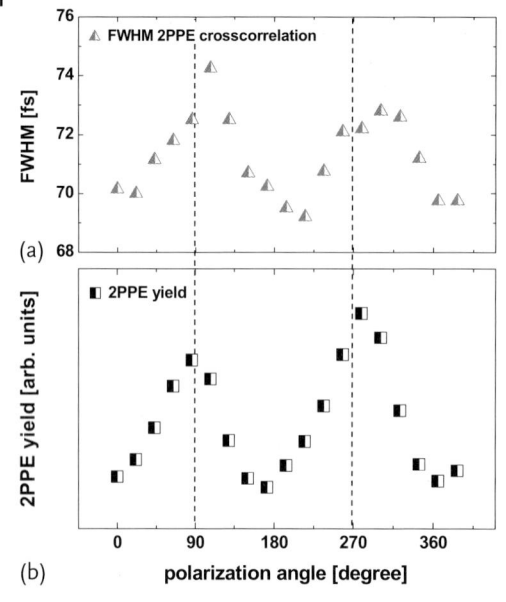

Figure 14.9 2PPE data of the elliptical nanoparticles – polarization scans: (a) cross-correlation FWHM of time-resolved 2PPE data as function of polarization angle; the periodic modulation in the 2PPE signal is evident and indicates strong differences in the decay of the LSP-induced particle field for resonant and off-resonant conditions. (b) 2PPE yield as function of polarization angle; the periodic modulations (which are in phase with the changes in the FWHM) arise from changes in the field enhancement as the excitation is tuned from resonant to off-resonant conditions.

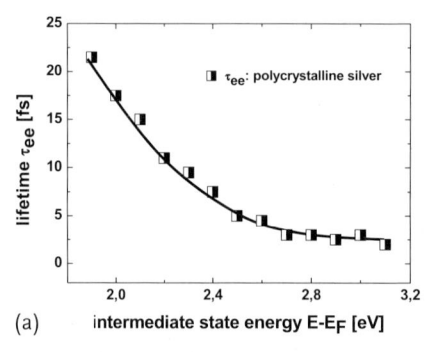

Figure 14.10 (a) Inelastic lifetime τ_{ee} of electron excitations in polycrystalline bulk silver as function of intermediate state energy $E_I = E - E_F$ (see Figure 14.4) as measured within a time-resolved 2PPE study [35]. (b) cross-correlation FWHM data from TR-2PPE measurements of the elliptical nanoparticle array as function of $E_I = E - E_F$ for LSP resonant and LSP off-resonant excitation conditions; the offset between the two traces arises from the finite lifetime $\tau_{Plasmon}$ of the LSP.

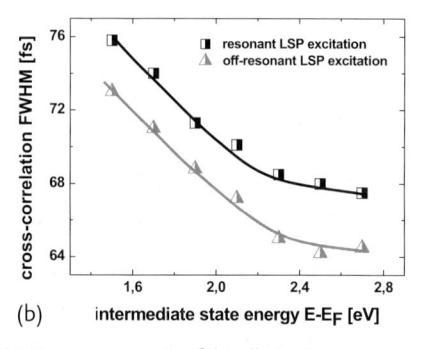

dence characteristic for the single electron decay τ_{ee} in silver. At the same time, the resonant and off-resonant curves keep a constant FWHM displacement along the abscissa (time axis) caused by the energy-independent broadening of the internal electric field pulse under resonant conditions. The offset is about 3 fs and is of the same order as the plasmon decay time determined from line width analysis of the optical extinction spectrum ($1/\Gamma \approx 2$ fs, see Figure 14.8).

14.3.2
Single Particle Plasmon Spectroscopy by Means of Time-Resolved Photoemission Microscopy

In order to study local variations in electron dynamics on nanometer scales, a technique capable of a high lateral resolution is required, such as photoemission electron microscopy. In combining the high temporal resolution of the time-resolved 2PPE technique and the high lateral resolution of the PEEM, we succeeded in mapping local variations in the LSP dynamics even within a single nanoparticle. Figure 14.11 shows PEEM images of a 2D array of silver nano-dots (diameter: 200 nm, height: 50 nm, grating constant: 650 nm) recorded with a mercury vapor lamp in 1PPE ($h\nu = 4.9$ eV, Figure 14.11a) and the second harmonic of the laser in 2PPE ($h\nu = 3.1$ eV, (Figure 14.11b), respectively.

The homogeneous response of the nanoparticle array to the UV excitation, as visible in Figure 14.11a, is clear evidence for the accurate lithography process. In contrast, the 2P-PEEM image (Figure 14.11b) exhibits a distinct brightness variation among the particles pointing in the first instance to considerable variations in the LSP excitation conditions. However, a detailed analysis of the data and the comparison of images taken at different excitation wavelengths show that these inhomogeneities are caused by the internal defect structure of the different particles rather then differences in the collective electron response [17].

Figure 14.11c shows the result from a time-resolved PEEM scan of the identical area of the sample. In this depiction, the grey-scale coded FWHM-value of the

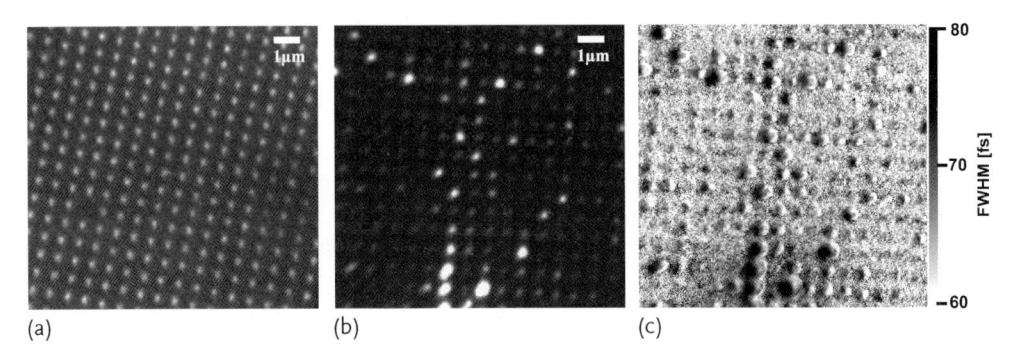

Figure 14.11 PEEM images of the silver nano-dot array from Figure 14.7b (diameter: 200 nm, height: 50 nm, grating constant: 650 nm) (a) Conventional PEEM image recorded in threshold one-photon photoemission using the mercury vapor lamp (4.9 eV). (b) 2P-PEEM image of the same sample recorded with a photon energy of 3.1 eV. (c) Lifetime map deduced from a pixel-wise analysis of the cross-correlation traces of a TR-PEEM scan. The map displays the corresponding gray-scale coded FWHM values of the traces.

cross-correlation trace of every image pixel is plotted as a measurement for the local femtosecond dynamics. This lifetime map allows for the identification of local variations in the ultrafast response between neighboring particles as well as particle internal variations in the ultrafast response in an intuitive way. The central cross-section of the 2PPE intensity distribution and the FWHM distribution of an individual particle is displayed in Figure 14.12a. Surprisingly, the FWHM-trace and 2PPE intensity profile do not match one another. Instead, both traces are laterally shifted with respect to each other. Furthermore, it seems as if the decay of the LSP varies considerably across the particle. This view is quantitatively confirmed by the data plotted in Figure 14.12b. Here, three cross-correlation traces recorded along the central particle cross-section are shown, including corresponding cross-correlation simulations based on a rate equation model that describes a depopulation process. These fits yield a maximum variation of the decay characteristics as large as 10 fs. However, such an interpretation is against the intuition that the LSP, as a collective excitation of the electron gas, is a global and characteristic property of the entire nanoparticle. Upon locally probing the LSP excitation with the external laser field, we instead create a situation of laterally varying interference conditions between the external and the internal (plasmonic) field, quite similar to the explanation given in [39] for the observation of stationary emission maxima and minima along self-assembled nanowires. In this study, silver wires of varying length are imaged in two-photon photoemission PEEM using a setup very similar to the experiment described herein. The authors assign the occurrence as well as the periodicity of the emission pattern to the constructive and destructive interference between a propagating surface plasmon wave in the wire and the external light field which is incident at an oblique angle of 74° with respect to the surface normal. Since the phase velocity of the surface plasmon is reduced in comparison to the phase velocity of the light field propagating in vacuum, the relative phase between the two fields changes along the wire, forming a stationary beating pattern of

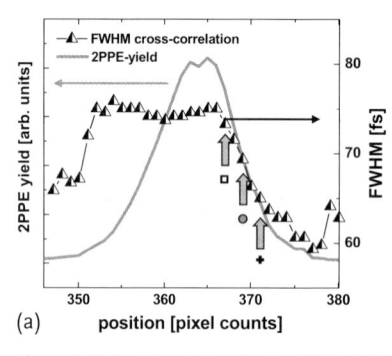

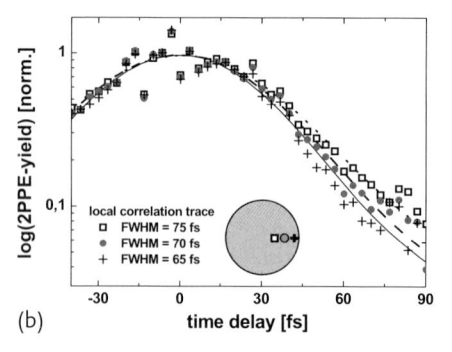

Figure 14.12 (a) variation in the 2PPE yield and the cross-correlation FWHM across a single nanoparticle; note the relative shift between yield and FWHM (b) 2PPE cross-correlation traces as measured at three different points at the right edge of the particle; the respective postitions are also indicated by the arrows in (a). The solid, dashed and dotted lines are sech^2-fits to the experimental data and allow one to determine the FWHM value of the respective traces in a reliable manner.

alternating constructive and destructive interference. In our case, the experimental situation is similar to the extent that the detected photoemission is due to a local field which results from the superposition of the external light field and the light induced polarization field within the nanoparticle. However, the dimension of the particles is, in our case, much smaller than the periodicity P of the induced beating pattern ($P \approx 2.5\,\mu$m). Therefore, the mapped variations correspond to a small fraction of a π-shift between the two phases and are consequently much less developed. The overall shape of time-resolved correlation traces is also influenced by the varying phase between light and plasmon field and changes systematically across the particle, which in turn, results in a systematic variation of the best-fit sech^2 FWHM parameter. The grey-scale coded display of the FWHM in Figure 14.11c shows that it is indeed a systematic effect since each of the imaged nanoparticles shows a distinct red-blue contrast from left to right.

Further evidence for this interpretation is obtained within time-resolved PEEM measurement at extremely high, so-called interferometric temporal resolution. In these experiments, the temporal delay between the pump and the probe pulse is adjusted at accuracy much better than the oscillation period ($T \approx 1.5$ fs) of the electric field vector of the laser light. This enables mapping of the variations in the two-photon photoemission arising from the constructive and destructive interference of the pulses as a function of the temporal delay. By these means, the relative phase between the (total) external field and the induced plasmonic excitation is also changed, giving rise to modifications in the beating pattern. Figure 14.13a shows experimental results from reference [40] obtained from a single nanoparticle within a interferometric time-resolved PEEM measurement. Five PEEM images are shown which have been recorded at incremental temporal delay steps of 130 fs. The modifications in the interference pattern are clearly visible, as the phase relation between the external light field and internal plasmonic field are changed. A quantitative comparison of the 2PPE yields from two different areas of the particle for a temporal delay covering five oscillation periods of the light field is shown in Figure 14.12b. From this data, we find a net phase shift of $\Delta\varphi \approx \pi/30$ in the 2PPE response from the two different areas.

As mentioned above [39], the interference patterns are more pronounced for extended nanostructures, such as nanowires. Figure 14.14 shows 1P-PEEM and a 2P-PEEM images and a corresponding lifetime map of an array of nanowires introduced in Figure 14.7c.

The 1P-PEEM images show clear internal intensity variations, indicating a structural inhomogeneity of the nanowires. Also, the 2PPE image shows distinct brightness variations along the wires. Note that for the 2PPE measurements, the nanowires have been aligned perpendicular to the direction of incidence of the excitation laser pulse. Strikingly, the endings of the wires almost exclusively exhibit a local maximum in the 2PPE yield. Additionally, most of the wires show three further emission maxima. These local variations are also reproduced in the corresponding lifetime map (Figure 14.14c). The FWHM of the cross-correlation traces varies along the wire on length-scales of about 100 nm, roughly the value also determining the width of the wires. We conclude that the individual nanowires have

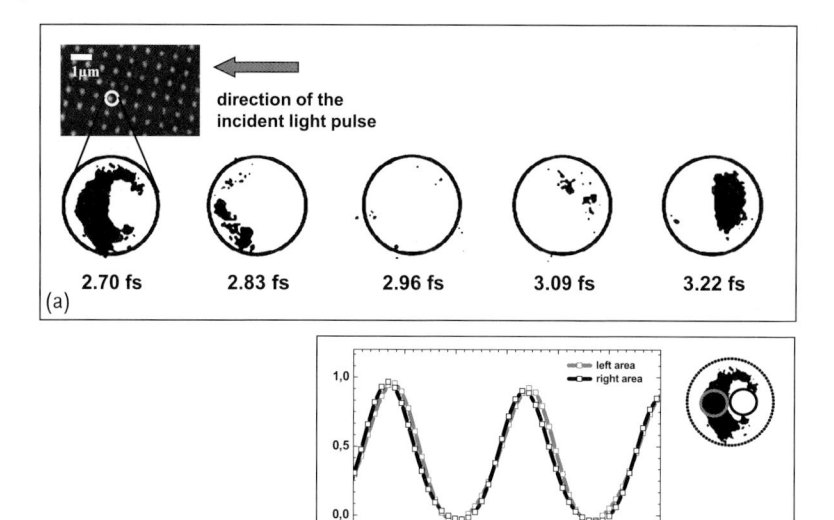

Figure 14.13 Results from interferometric TR-PEEM measurements of a single nanoparticle; (a) on timescales much shorter than the oscillation period of the electric field of the incident light field, the local 2PPE yield distribution clearly changes within the nanoparticle; these changes are induced by the interference between pump and probe pulse and the particle internal field due to the excitation of the LSP mode; (b) quantitative representation of the data in (a) over several oscillation periods. The phase-shift in the response between the two selected areas is about $\pi/30$.

Figure 14.14 PEEM images of the silver nanowires from Figure 14.7c (length: 1.6 µm). (a) 1P-PEEM image taken with the mercury vapor lamp; (b) 2P-PEEM image recorded at 400 nm with the femtosecond laser; (c) corresponding TR-PEEM lifetime map.

a structural inhomogeneity. The wires obviously decompose into conglomerates of small silver particles during the preparation process. The brightness variations originate from LSP excitations in the individual conglomerates as well as from local defects along the wire. This result is in good agreement with the findings by Kubo *et al.* [41] and by Cherula *et al.* [42], and is corroborated by scanning near-field microscopy (SNOM) data by Ditlbacher *et al.* [43], who showed that propagating plasmon modes leading to standing wave patterns are only supported in single crystalline nanowires. Note, however, that the FWHM variations perpendicular to the long wire axis are very similar to those observed for the nano-dot array in Figure 14.11 and can be interpreted in terms of the model introduced before.

In a final example, we would like to show how a static 2P-PEEM measurement can be used to monitor the dipolar coupling of LSP modes excited in two neighboring particles. The interaction between both particles is governed by the relative phase of the LSP excitations and, hence, by the interparticle distance and angle of light incidence. Here, we map to what extent the dipolar field of one particle of a particle dimer modulates the local 2PPE signal from the other particle. Figure 14.15 shows 1P-PEEM (Figure 14.15a) and 2P-PEEM (Figure 14.15b) images of an array of silver particle dimers (diameter: 50 nm, height: 40 nm, grating constant: 740 nm) with a center-to-center distance of 130 nm. The 2PPE data were collected under resonance conditions with respect to the perpendicular single particle LSP resonance at a photon energy of 3.1 eV (400 nm, p-polarized light incident parallel to the dimer axis).

In the 1P-PEEM image, the individual particles are clearly resolved. For the 2P-PEEM image, we observe strong local variations of the photoemission signal similar to the findings of the array of individual nano-dots. A randomly distributed contribution to these inhomogeneities arises once again from particle to particle

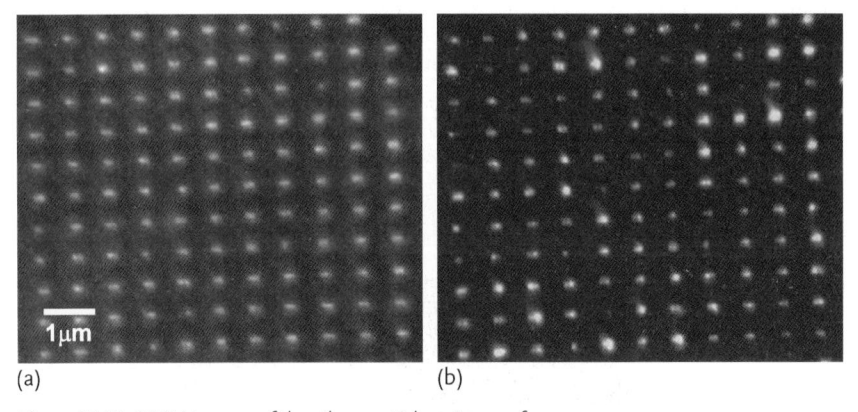

Figure 14.15 PEEM images of the silver particle pair array from Figure 14.7d (diameter: 50 nm, height: 40 nm, grating constant: 740 nm) with a center-to-center distance of 130 nm. (a) 1P-PEEM image taken at a photon energy of 4.9 eV; (b) 2P-PEEM image taken at a photon energy of 3.1 eV (p-polarized light, light is incident from the left).

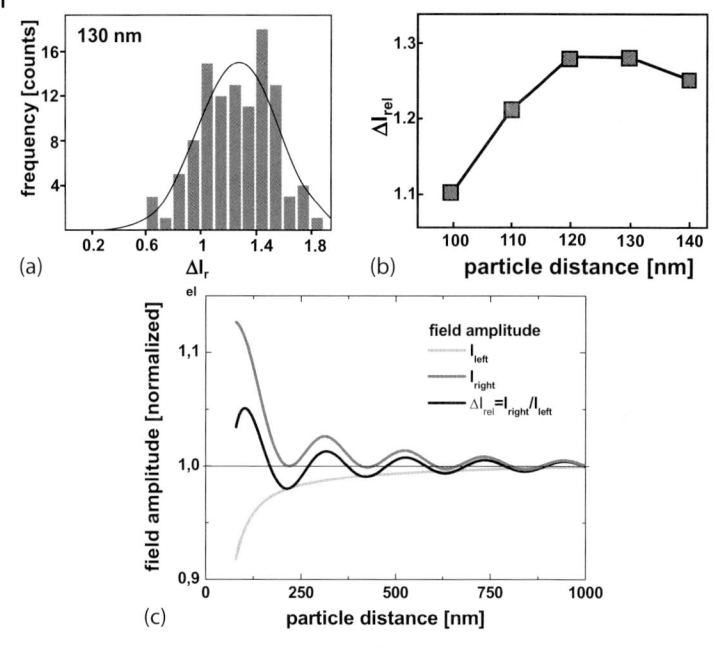

Figure 14.16 (a) Frequency distribution of the ratio in the 2PPE yield between left and right particle of particle pairs with a distance of 130 nm (left); (b) Mean value of the 2PPE yield ratio as function of the distance; (c) calculated field amplitude of left and right particles in a particle pair (grey lines) as modified by the dipolar interaction and the resulting ratio in the field amplitude (black line).

variations in the local defect density (see also Figure 14.11b). In contrast, a dipolar particle–particle coupling should give rise to a well determined difference in the 2PPE yield between two neighboring particles. To be able to discriminate between random and systematic intensity variations, we performed a statistical analysis of the 2PPE intensity ratio between the left and right particles of the dimers. The average brightness values in a 7×7 pixel region of interest centered on the left and right particle within each pair is extracted from the image. In a further step, we calculate the relative count rate difference between left and right particles $\Delta I_{\text{rel}} = \bar{I}_{\text{r}}/\bar{I}_{\text{l}}$. Figure 14.16a shows the distribution of the frequency of occurrence of ΔI_{rel} as determined from a sample area covering 100 individual dimers.

A positive value of ΔI_{rel} corresponds to the situation where the right particle is brighter than the left particle. The histogram gives evidence that on average, the right particle shows a 28% yield enhancement in comparison to the left particle. Note that the left particle is located towards the direction of light incidence. Calculations based on a dipole model can qualitatively reproduce the trend of asymmetry [44]. Further measurements have been conducted for particle dimers at varying intra-pair distances between 100 and 140 nm. The ΔI_{rel} values are shown as functions of particle distance in Figure 14.16b. For all distances, we observe an asymmetry in the photoemission yield with preferential photoemission from the right particle ($\Delta I_{rel} > 0$). Overall, we find a monotonous variation in ΔI_{rel} with a local maximum at about 120 nm. The results of the corresponding model calculation are shown in Figure 14.16c. For excitation at a fixed wavelength (in the present example $\lambda = 400$ nm), a periodic distance modulation of ΔI_{rel} is predicted at a periodicity of the order of the wavelength of the excitation field. It arises from the relative phase change of the particle fields at a given position as the particle distance increases and the consequent modulation of the interference conditions. The experimental data shown in Figure 14.16b correspond to the first oscillation maximum of this modulation. At sufficiently large distances, the differences in the particle yields will disappear due to the finite range of the dipolar plasmon field. Note that once again, the dipolar coupling between the particles has been identified on the basis of a statistical analysis of the particle array. This approach delivers very reproducible and clear results even though the photoemission signal from a single dimer is considerably blurred by the sample inhomogeneities. The 2PPE yield analysis gives direct evidence for the dipole induced coupling between neighboring particles.

14.4
Conclusion

Time-resolved 2PPE is a well established method of investigating the relaxation dynamics of optically excited electrons. In contrast to pure optical methods, the 2PPE directly addresses the electronic system and is therefore well suited to investigate the complex interplay between collective and single electron excitations on a microscopic level. By using well designed, elliptically shaped nanoparticles, switching between resonant and off-resonant excitation conditions is achieved by the rotation of the polarization vector of the incident light pulse. The presented time-resolved 2PPE data allow insight into the ultrafast decay dynamics of collective electron oscillation. The results presented in this context confirm the model developed by Pfeiffer et al. [25], which treats the plasmon resonance as a modification of the internal electric field with respect to amplitude, phase and temporal structure.

The combination of TR-2PPE and a photoemission electron microscope (PEEM) permits a spatial resolution well below the optical diffraction limit at a time resolution in the femtosecond regime. The direct imaging capability of the PEEM instrument allows access to the spatiotemporal dynamics of the plasmon-resonance-enhanced electric fields in the vicinity of metallic nanostructures. In comparison to other microscopy techniques, such as scanning near-field optical microscopy, the PEEM lacks the need to scan the sample surface, enabling an efficient parallel data acquisition. The presented data underlines the capabilities of TR-PEEM in visualizing the ultrafast dynamics of energy flow through nanoscopic devices. Illustrative results on static and dynamical properties of near-fields in the vicinity of single particles, nanowires and particle pairs have been presented and discussed. In the case of extended silver nanoparticles, the direct observation of the phase propagation of a plasmon mode was demonstrated. In polycrystalline silver nanowires, the

PEEM method has been used to clearly identify structural distortions. Finally, we succeeded in imaging characteristic LSP field modulations that are due to the dipolar coupling between neighboring particles, and have center-to-center distances as small as 100 nm. In conclusion, we would like to stress that employing TR-PEEM as a tool for characterizing metal nanostructures shows potential for becoming a key technique in the field of ultrafast nano-optics.

Acknowledgments

The authors would like to thank O. Oksana for assisting in finishing this article. Furthermore, we would like to thank the Nano-Bio Center at the University of Kaiserslautern for their support in preparing the silver nanoparticle samples. This work was supported by the Deutsche Forschungsgemeinschaft through SPP 1093 and the DFG Graduiertenkolleg 792.

References

1 Mie, G. (1908) *Ann. Phys.*, **25**, 377.
2 Kreibig, U. and Vollmer, M. (1995) *Optical Properties of Metal Clusters*, Vol. 25, Springer, Berlin.
3 Bohren, C.F. and Huffmann, D.R. (1983) *Absorption and Scattering of Light by Small Particles*, John Wiley & Sons Ltd, New York.
4 Hohenau, A., Ditlbacher, H., Lamprecht, B., Krenn, J.R., Leitner, A., and Aussenegg, F.R. (2006) *Microelectron. Eng.*, **83**, 1464.
5 Graff, A., Wagner, D., Ditlbacher, H., and Kreibig, U. (2005) *Eur. Phys. J. D*, **34**, 263.
6 Elghanian, R., Storhoff, J.J., Mucic, R.C., Letsinger, R.L., and Mirkin, C.A. (1997) *Science*, **277**, 1078.
7 Lyon, L.A., Musick, M.D., and Natan, M.J. (1998) *Anal. Chem.*, **70**, 5177.
8 Gan, Y. (2007) *Rev. Sci. Instrum.*, **78**, 081101.
9 Weeber, J.C., Dereux, A., Girad, C., Krenn, J.R., and Goudonnet, J.-P. (1999) *Phys. Rev. B*, **60**, 9061.
10 Quitten, M., Leitner, A., Krenn, J.R., and Aussenegg, F.R. (1998) *Opt. Lett.*, **23**, 1331.
11 Salerno, M., Krenn, J.R., Lamprecht, B., Schider, G., Ditlbacher, H., Félidj, H., Leitner, A., and Aussenegg, F.R. (2002) *Opto-Electron. Rev.*, **10**(3), 217.
12 Moskovits, M. (1985) *Rev. Mod. Phys.*, **57**, 783.
13 Simon, H.J. and Chen, Z. (1989) *Phys. Rev. B*, **39**, 3077.
14 Bouhelier, A., Beversluis, M., Hartschuh, A., and Novotny, L. (2003) *Phys. Rev. Lett.*, **90**, 013903.
15 Kim, S., Jin, J., Kim, Y.-J., Park, I.-Y., Kim, Y., and Kim, S.-W. (2008) *Nature*, **453**, 757.
16 Scharte, M., Porath, R., Ohms, T., Aeschlimann, M., Krenn, J.R., Dittelbacher, H., Aussenegg, F.R., and Liebsch, A. (2001) *Appl. Phys. B*, **73**, 305.
17 Wiemann, C., Bayer, D., Rohmer, M., Aeschlimann, M., and Bauer, M. (2007) *Surf. Sci.*, **601**, 4714.
18 Lange, J., Bayer, D., Rohmer, M., Wiemann, C., Gaier, O., Aeschlimann, M., and Bauer, M. (2006) *Proc. SPIE*, **6195**, 61950Z.
19 Rohmer, M., Ghaleh, F., Aeschlimann, M., Bauer, M., and Hövel, H. (2007) *Eur. Phys. J. D*, **45**, 491.
20 Cinchetti, M., Gloskowskii, A., Nepjiko, S.A., and Schönhense, G. (2005) *Appl. Phys. Lett.*, **95**, 047601.
21 Bosbach, J., Hendrich, C., Stietz, E., Vartanyan, T., and Träger, F. (2002) *Phys. Rev. Lett.*, **89**, 257404.
22 Kottmann, J.P. and Martin, O.J.F. (2001) *Opt. Lett.*, **26**, 1096.

23 Kottmann, J.P. and Martin, O.J.F. (2001) *Opt. Express*, **8**, 655.
24 Zhao, L.L., Kelly, K.L., and Schatz, G.C. (2003) *J. Phys. Chem B*, **107**, 7343.
25 Merschdorf, M., Kennerknecht, C., and Pfeiffer, W. (2004) *Phys. Rev. B*, **70**, 193401.
26 Lamprecht, B., Schider, G., Lechner, R.T., Ditlbacher, H., Krenn, J.R., Leitner, A., and Aussenegg, F.R. (2000) *Phys. Rev. Lett.*, **84**, 4721.
27 Sönnichsen, C., Franzl, T., Wilk, T., von Plessen, G., and Feldmann, J. (2002) *Phys. Rev. Lett.*, **88**, 077402.
28 Schmuttenmaer, C.A., Aeschlimann, M., Elsayed-Ali, H.E., Miller, R.J.D., Mantell, D.A., Cao, J., and Gao, Y. (1994) *Phys. Rev. B*, **50**, 8957.
29 Loudon, R. (1973) *The Quantum Theory of Light*, Oxford Univ. Press, Oxford.
30 Hertel, T., Knoesel, E., Wolf, M., and Ertl, G. (1996) *Phys. Rev. Lett.*, **76**, 535.
31 Zhukov, V.P., Andreyev, O., Hoffmann, D., Bauer, M., Aeschlimann, M., Chulkov, E.V., and Echenique, P.M. (2004) *Phys. Rev. B*, **70**, 233106.
32 Swiech, W., Fecher, G.H., Zieten, C., Schmidt, O., Schönhense, G., Grzelakowski, K., Schneider, C.M., Frömter, R., Oepen, H.P., and Kirschner, J. (1997) *J. Electron. Spectroscop. Relat. Phenom.*, **84**, 171.
33 Schmidt, O., Bauer, M., Wiemann, C., Porath, R., Scharte, M., Andreyev, O., Schönhense, G., and Aeschlimann, M. (2002) *Appl. Phys. B*, **74**, 223.
34 Kuwata, H., Tamaru, H., Esumi, K., and Miyano, K. (2003) *Appl. Phys. Lett.*, **83**, 4625.
35 Aeschlimann, M., Bauer, M., Pawlik, S., Knorren, R., Bouzerar, G., and Bennemann, K.H. (2000) *Appl. Phys. A*, **71**, 485.
36 Zhukov, V.P., Aryasetiawan, F., Chulkov, E.V., de Gurtubay, I.G., and Echenique, P.M. (2001) *Phys. Rev. B*, **64**, 195122.
37 Chulkov, E.V., Borisov, A.G., Gauyacq, J.P., Sánchez-Portal, D., Silkin, V.M., Zhukov, V.P., and Echenique, P.M. (2006) *Chem. Rev.*, **106**, 4160.
38 Cinchetti, M., Valdaitsev, D.A., Gloskovskii, A., Oelsner, A., Nepijko, S.A., and Schönhense, G. (2004) *J. Electron. Spectrosc. Relat. Phenom.*, **137–140**, 249.
39 Meyer zu Heringdorf, F.-J., Chelaru, L.I., Möllenbeck, S., Thien, D., and Horn-von Hoegen, M. (2007) *Surf. Sci.*, **601**, 4700.
40 Bauer, M., Wiemann, C., Lange, J., Bayer, D., Rohmer, M., and Aeschlimann, M. (2007) *Appl. Phys. A*, **88**, 473.
41 Kubo, A., Onda, K., Petek, H., Sun, Z.J., Jung, Y.S., and Kim, H.K. (2005) *Nano Lett.*, **5**, 1123.
42 Cherula, L.I., Horn-von Hoegen, M., Thien, D., and Meyer zu Heringdorf, F.-J. (2006) *Phys. Rev. B*, **73**, 115416.
43 Ditlbacher, H., Hohenau, A., Wagner, D., Kreibig, U., Rogers, M., Hofer, F., Aussenegg, F.R., and Krenn, J.R. (2005) *Phys. Rev. Lett.*, **95**, 257403.
44 Wiemann, C. (2006) Dissertation, University Kaiserslautern.

15
Dynamics and Nonlinear Light Propagation in Complex Photonic Lattices

Bernd Terhalle, Patrick Rose, Dennis Göries, Jörg Imbrock, and Cornelia Denz

15.1
Introduction

Periodic structures in one dimension have been known for more than 100 years as Bragg filters. In the past twenty years, they have also been realized in two or three dimensions and are currently known as photonic crystal structures. Being periodic structures, they support wave dynamics equivalent to the transport dynamics of electrons in semiconductors [1]. Classic optical effects as Bragg reflection, interference, and diffraction effects dominate the light propagation in these structures. However, their linear dynamics is already fundamentally different as compared to their counterparts in homogeneous media.

When considering light propagation in periodic structures, we are interested in the propagation dynamics. The change in the diffraction features is crucial when considering them as long as we are using continuous waves. Here, the diffraction relation (k_z vs. k_x, k_y) plays the role of the dispersion relation (ω vs. k) that is well known in the temporal domain. In periodic wave systems, the linear modes in a waveguide structure are extended Bloch modes, with a transmission spectrum consisting of allowed bands separated by forbidden gaps.

This means that there exists a range of propagation constants β which are not allowed. Since each localized wavepacket consists of an ensemble of such modes, the band geometry determines the group dynamics. As an example, a grating structure can modify the spreading of a narrow beam in much the same sense as it can affect the dispersive behavior of a temporal pulse.

In a nonlinear lattice, the propagation dynamics will be further modified by the interplay between periodicity and nonlinearity. An intriguing example is the Kerr effect, in which the refractive index is changed as a function of the light intensity and directly depends on the spatial distribution of the light. In the case of a narrow beam, the light modifies the refractive index locally, thereby inducing a defect in the periodic structure of the lattice. Such a defect naturally has localized modes, whose propagation constants lie inside a gap. When these nonlinear modes induce the defect and populate it self-consistently, the wavepacket becomes

Nonlinear Dynamics of Nanosystems. Edited by Günter Radons, Benno Rumpf, and Heinz Georg Schuster
Copyright © 2010 WILEY-VCH Verlag GmbH & Co. KGaA, Weinheim
ISBN: 978-3-527-40791-0

self-localized and its diffractive broadening is eliminated. Showing a spatial profile that remains constant and stable during propagation, the beam is considered to be a "discrete soliton" or a "gap soliton" [2]. By now, discrete and gap spatial solitons have been studied in many different systems such as photorefractive materials, conjugated polymers [3], Bose–Einstein condensates [4], and nonlinear waveguide arrays [2].

In nonlinear optics, the induction of periodic refractive index structures in photorefractive materials [5] has been utilized to demonstrate a large variety of nonlinear localization effects. Apart from supporting lattice solitons, a nonlinearity can also couple different Bloch modes in a lattice, giving rise to such fundamental phenomena as modulation instability and spontaneous pattern formation known from homogeneous media.

Moreover, nonlinear mode coupling can occur not only within a band, but also between different bands. Novel waveguiding features arise from these effects, and the potential of spatial band engineering is increasingly being explored. Examples of such features that appear in photorefractive lattices are soliton trains [6], Zener tunneling, and Bloch oscillations [7] as well as vortex solitons [8–10].

The advantage of the optical induction technique is given by the electro-optic properties of photorefractive crystals such as strontium barium niobate (SBN), which allow highly nonlinear, reconfigurable refractive index patterns to be achieved at very low power levels. While in the past only comparatively simple geometries such as diamond, square [11–13], or hexagonal [14] lattices were studied, special attention is currently paid to more complex photonic structures such as modulated waveguide arrays [15], lattice interfaces [16], or double-periodic one-dimensional photonic lattices [17]. In addition, the propagation of more complex waves such as optical vortices or vortex solitons [18] have also received attention. In general, complex or multiperiodic structures as well as complex wavefronts interacting with these lattices are of immense interest since they offer many exciting possibilities to engineer the diffraction properties of light.

In this chapter, we review some of our results on realizing complex two-dimensional photonic lattices and their linear and nonlinear interactions with complex wavefronts. The latter carry phase dislocations that lead to novel stabilization mechanisms of complex waves.

15.2
Wave Propagation in Periodic Photonic Structures

Linear wave propagation in photonic lattices can be described using the Floquet–Bloch theory [19], with the corresponding dispersion/diffraction relation $k_z = k_z(k_x, k_y)$ characterized by bands of allowed propagation constants separated by forbidden gaps. The localization of a wave by compensation of diffraction is obtained by defect states, which can either be imposed by construction, as in solid- and hollowcore photonic crystal fibers [20], or through the self-focusing effect of an optical nonlinearity [2].

15.2.1
Linear Propagation

As a simple example of a periodic structure, we consider a two-dimensional square photonic lattice (Figure 15.1a). A typical band structure (in k-space) for such a lattice is depicted in Figure 15.1c.

It can be created exploiting the 2π periodicity and using the standard procedure of representing dynamics only within the irreducible part of the first Brillouin zone, which is determined by the high symmetry points Γ, X, and M (Figure 15.1b).

Since each mode of the system is an extended Bloch wave with its own propagation constant given by the eigenvalue and direction given by the normal gradient to the transmission band, different modes acquire their own individual phase as they propagate. Any wave, or wavepacket, entering the lattice is decomposed in these Bloch modes. Therefore, the accumulation of different relative phases during propagation will affect the waveform of the propagating wave significantly, resulting in an output that may considerably differ from the input waveform.

A typical example of linear wave propagation in photonic lattices is diffraction of a localized beam. Because the second derivative of the diffraction relation gives the relative spread or convergence of adjacent waves, the diffractive properties of a wavepacket are governed by the band curvature. In regions of convex curvature, the beam acquires a convex wavefront during propagation resulting in normal discrete diffraction, with wave behavior analogous to that in homogeneous media. By contrast, a group of modes in concave regions of band curvature will evolve anomalously, acquiring a concave wavefront during propagation. Therefore, because the propagation characteristics are strongly affected by the incident angle of a probe beam, changing the angle controls its diffraction properties. This feature is known as diffraction management [21] in analogy to dispersion management of optical pulses with gratings in fibers.

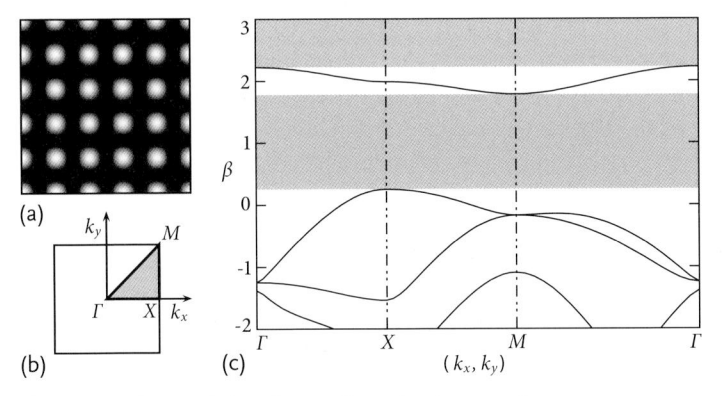

Figure 15.1 Characteristics of a two-dimensional square photonic lattice. (a) Intensity distribution of the lattice wave; (b) irreducible Brillouin zone; (c) band structure.

The effect of the lattice on the wave propagation depends on the transverse size of the input beam relative to the lattice spacing, and on its internal structure. For example, a broad Gaussian beam launched on-axis into the lattice excites modes from different bands and stays mostly Gaussian as it propagates. In contrast, coupling a narrow beam into the fundamental guided mode of a single period of the lattice excites Bloch modes primarily from the first band. In this case, the beam undergoes "discrete diffraction" characterized by intense side lobes with little or no light in the central starting lattice location [22]. This effect is a result of coupling between neighboring sites of the lattice combined with interference effects.

15.2.2
Nonlinear Propagation

One would expect nonlinear effects to considerably change light propagation in a lattice. This is due to the fact that in the nonlinear regime, Bloch modes of the underlying linear lattice can undergo modulational instabilities, while focusing effects can counteract the diffractive tendencies of a narrow beam [2, 23]. Therefore, localized wave propagation or spatial solitons can appear in periodic lattices.

The influence of the nonlinearity again depends on the band curvature of the Floquet–Bloch mode. In regions of convex curvature and thus normal diffraction, a focusing nonlinearity can compensate the convex curvature of the wavefront, while regions of concave curvature and thus anomalous diffraction require a defocussing nonlinearity [24, 25] to create localization. In turn, nonlinear dynamics can be manipulated by engineering the band structure, in much the same way as the management of linear diffraction [26].

A spatial soliton can only propagate in a stable manner if the propagation constant of the beam deviates from the linear bands of the transmission spectrum in such a way that the propagation constant is located in a gap. In this case only, a localized state rather than an extended Bloch mode is created.

The so-called lattice soliton lies in the semi-infinite gap above the first band (Figure 15.1c). A different type of soliton can be created at the edge of the Brillouin zone (high symmetry points X and M in Figure 15.1c), where its propagation constant lies between the first and second band, while the Bragg condition gives this soliton a staggered phase profile. This soliton is called a spatial "gap soliton".

For modes originating from the first band, the curvature dictates that a defocussing nonlinearity is necessary for the formation of bright gap solitons, and the propagation constant decreases into the gap with increasing nonlinearity. For modes originating from the second band that have a different mode structure, the propagation constant increases into the gap.

Dynamics in lattices can be even richer because nonlinear effects can couple defect states between bands or gaps [27, 28]. Moreover, indirect gaps may occur in which the minimum of one band (upper branch) occurs at a different k-vector than the maximum of the adjacent band (lower branch). The situation becomes even more complex when the transverse dimensions involved exhibit an anisotropy, resulting in different propagation behavior in different directions. This may result in effects of localization in one transverse direction due to Bragg reflection (semi-infinite band gap), and due to gap soliton formation (first gap) in the other direction. As a consequence, soliton mobility can be implemented in anisotropic lattices [29].

15.3
Optically-Induced Photonic Lattices in Photorefractive Media

A simple and well known procedure to generate the desired intensity distribution for optical induction of two-dimensional photonic lattices utilizes the interference of several plane waves inside the biased photorefractive crystal [5, 11]. The periodicity of the induced patterns can be controlled by the interference angle, whereas the modulation depth depends on the externally applied electric field.

More flexibility in changing the lattice parameters can be achieved by using a programmable spatial light modulator to create the diffraction free propagating transversely periodic lattice wave [12]. For example, we will demonstrate below that this configuration can be used to study photonic lattices of triangular shape which would otherwise require the use of six plane waves, and are consequently inconvenient for induction by interference.

15.3.1
Mathematical Description of Photorefractive Photonic Lattices

Assuming a temporal steady state and neglecting photovoltaic effects, one can describe the optical induction of photonic lattices in a photorefractive medium as well as beam propagation in these structures by the set of equations

$$2i\frac{\partial A}{\partial z} + \nabla_\perp^2 A - \Gamma E_{sc}(I) A = 0, \qquad (15.1)$$

$$E_{sc}(I) = -\frac{\partial \varphi}{\partial x}, \qquad (15.2)$$

$$\nabla^2 \varphi + \nabla \varphi \nabla \ln(1 + I) = E_{ext} \frac{\partial}{\partial x} \ln(1 + I). \qquad (15.3)$$

It is $\nabla_\perp^2 = \partial^2/\partial x^2 + \partial^2/\partial y^2$. $\Gamma = k^2 w_0^2 n_0^2 r_{eff}$ is proportional to the effective element of the linear electro-optic tensor r_{eff}. φ denotes the scalar potential of the electric screening field E_{sc}, and the externally applied electric field E_{ext} is directed along the c-axis of the crystal. $I = |A_{latt}|^2 + |A_{probe}|^2$ gives the total light intensity as a sum of the lattice wave and the probe beam intensity.

Equation 15.1 has been made dimensionless by introducing the length scales w_0 and $z_0 = k w_0^2$. k and n_0 denote the wave vector and the unperturbed refractive index, respectively. Throughout this chapter, we use $r_{eff} = 280\,\text{pm/V}$, $n_0 = 2.35$, and $w_0 = 10\,\mu\text{m}$.

One of the most important features of the photorefractive nonlinearity is that it is saturable, that is, $E_{sc}(I)$ does not grow infinitely as $I \to \infty$. As a consequence, the catastrophic self-focusing known from Kerr media cannot occur in photorefractive media [30]. Due to the externally applied electric field, (15.3) is also inherently anisotropic as can also be seen from the single partial x derivative on the right hand side. Furthermore, it is nonlocal in the sense that the change of the refractive index in each point depends on the intensity distribution in the whole transverse plane.

Although (15.3) can easily be solved in the transversely one-dimensional limit, there is no analytical solution for the two-dimensional case and one has to integrate numerically.

However, sometimes it is desirable to have an explicit, local expression for the photorefractive nonlinearity. Therefore, the solution of the one-dimensional case is often used in two dimensions as well, giving

$$E_{sc} = -\nabla\varphi = \frac{E_{ext} I}{1 + I}. \tag{15.4}$$

Obviously, this approximation is able to reproduce the saturability, but it cannot describe the intrinsic photorefractive anisotropy. As a result, (15.4) is often referred to as the isotropic approximation. Although this approximation has shown to be sufficient in special cases, the anisotropic effects cannot be neglected in general. Therefore, we will use (15.3) throughout this chapter and refer to it as the full anisotropic model. In fact, we will discuss below several examples clearly showing the importance of anisotropy for the lattice induction as well as for nonlinear light propagation inside the periodic refractive index structures.

15.3.2
Experimental Configuration for Photorefractive Lattice Creation

Figure 15.2 shows a schematic of a typical setup used in our experiments. A beam derived from a frequency-doubled Nd:YAG laser at a wavelength of $\lambda = 532$ nm is split into two beams using a beam splitter. The transmitted beam illuminates a programmable spatial light modulator (SLM1) to imprint a phase modulation in such a way to create the desired non-diffracting lattice forming wave [12] necessary for the induction of periodic structures. The modulated beam is then imaged at the input face of a $Sr_{0.60}Ba_{0.40}Nb_2O_6$ (SBN:Ce) crystal by a high numerical aperture telescope.

A half wave plate in front of the telescope is used to insure ordinary polarization of the lattice wave, thus enabling its effectively linear propagation. The crystal is additionally biased by an externally applied electric field and uniformly illuminated with a white-light source in order to control the dark irradiance. The second spatial light modulator (SLM2) combined with proper Fourier filtering is employed to achieve the desired amplitude and phase structure of an incident Gaussian probe beam. The polarization of the probe beam is extraordinary, so it propagates through the crystal in the nonlinear regime [5]. The strength of the nonlinearity is controlled by varying an applied external dc electric field. In order to visualize the phase structure of the probe beam, a third beam is derived from the laser. It is passed through a half wave plate to insure its extraordinary polarization and is subsequently sent

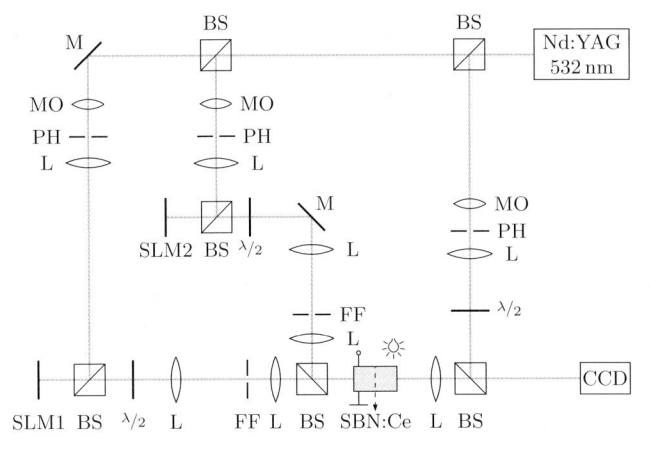

Figure 15.2 Experimental setup. BS: beam splitter, CCD: camera, FF: Fourier filtering, L: lens, M: mirror, MO: microscope objective, PH: pinhole, SLM: spatial light modulator.

directly to the CCD camera to record a phase interferogram with the probe beam. To observe the induced refractive index structure, the lattice can be illuminated with a broad plane wave, which is guided by the regions of high refractive index. As a consequence, the modulated intensity distribution at the output of the crystal qualitatively maps the induced refractive index change. When including a Fourier transform lens, the spectrum of the lattice can be analyzed too.

15.4
Complex Optically-Induced Lattices in Two Transverse Dimensions

As already discussed in Section 15.3.1, the induced refractive index change depends strongly on the anisotropy of the photorefractive response [12, 13]. In particular, the spatial orientation of the lattice wave with respect to the c-axis of the crystal significantly affects the structure of the resulting refractive index pattern. Therefore, the utilized lattice wave and the effectively induced refractive index change can have completely different symmetries.

Figure 15.3 illustrates these characteristics for the case of lattice waves with fourfold symmetry, which is one of the simplest examples of highly symmetric structures. Here, two lattice orientations can be distinguished. These are a square pattern with one high-symmetry axis orientated parallel to the c-axis (Figure 15.3, bottom), and a 45° tilted, so-called diamond pattern (Figure 15.3, top). The corresponding lattice fields can be described as $A_{\text{latt}}(X, Y) = A_0 \sin(X) \sin(Y)$ with $(X, Y) = (x, y)$ for the square pattern, and $(X, Y) = ((x + y)/\sqrt{2}, (x - y)/\sqrt{2})$ for the diamond case, respectively.

It has been shown that the square lattice results in an effectively one-dimensional refractive index structure consisting of vertical lines due to the photorefractive anisotropy, whereas the induced refractive index change for the diamond pattern

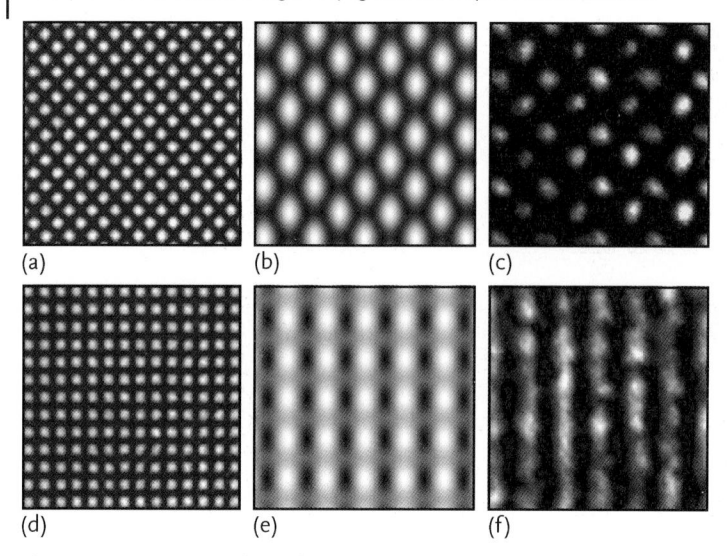

Figure 15.3 Structure analysis of the diamond (top) and square (bottom) lattice. (a), (d) Experimental lattice wave. (b), (e) Numerical simulation of the light-induced refractive index change. (c), (f) Guided wave.

contains well separated spots, thus forming a fully two-dimensional structure [12]. Therefore, in almost all experiments on optically induced lattices, the lattice wave used has been restricted to the diamond configuration (Figure 15.3, top) and the effects of anisotropy with respect to the electro-optic properties of the photorefractive crystal have been neglected. Recently, in the same spirit, the properties of hexagonal lattices and their potential to support discrete and gap solitons have also been investigated [14]. Again, in these experiments the orientation of the lattice wave has been chosen to minimize the effect of anisotropy.

To overcome these previous limitations of usable lattice configurations, the concept of optically-induced lattices can be extended to the more complex anisotropic lattices. In the following, we will introduce the highly anisotropic so-called triangular lattices and prove that they support discrete and gap solitons, even those of higher order.

15.4.1
Triangular Lattices

Akin to a lattice wave with fourfold symmetry, one can distinguish two orientations of the lattice wave of the triangular lattice and, therefore, two different configurations of the triangular lattice exist. These configurations are denoted as perpendicular and parallel, respectively. The corresponding lattice waves (Figure 15.4a and 15.4d) are given by

$$A_{\text{latt}}(X, Y) = A_0 \sin(2Y/\sqrt{3}) \sin(Y/\sqrt{3} + X) \sin(Y/\sqrt{3} - X), \qquad (15.5)$$

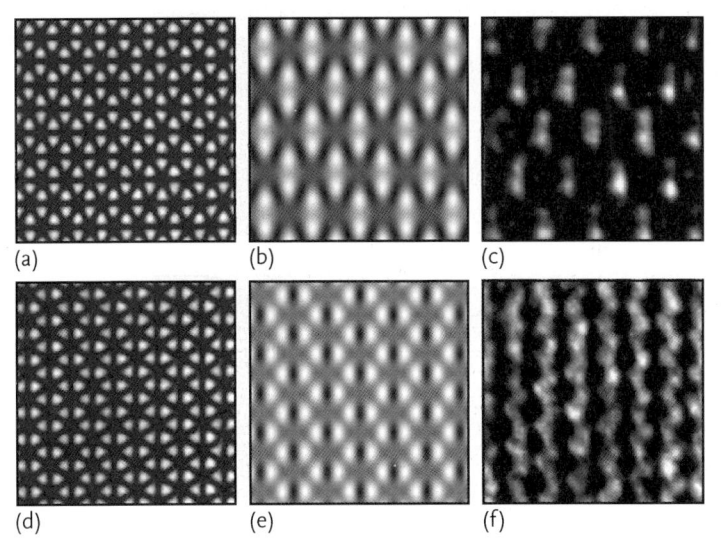

Figure 15.4 Structure analysis of the parallel (top) and perpendicular (bottom) triangular lattice. (a), (d) Experimental lattice wave. (b), (e) Numerical simulation of the light-induced refractive index change. (c), (f) Guided wave.

with $(X, Y) = (x, y)$ for the parallel lattice and $(X, Y) = (y, x)$ for the perpendicular orientation.

Similar to the diamond and square patterns, these lattices also show a strong anisotropy in the symmetry of the induced refractive index changes. Corresponding numerical simulations (Figure 15.4b and 15.4e) have been performed in the full anisotropic model described in Section 15.3.1.

For the parallel orientation (Figure 15.4, top), every two vertically neighboring out of phase lobes of the field distribution induce a focusing dipole island, and these islands essentially form a diamond pattern with angles of 60° (Figure 15.4b). In the same way, the triangular pattern with perpendicular orientation (Figure 15.4, bottom) induces a refractive index change comparable to the square pattern, with regions of high refractive index forming vertical lines (Figure 15.4e). The experimentally observed guided waves (Figure 15.4c and 15.4f) confirm these numerical results.

In the next step, we study the propagation of a Gaussian probe beam inside the two triangular lattice structures. The results for the parallel pattern are shown in Figure 15.5. It is clearly visible that the diffraction of the probe beam in the parallel lattice at low power shows a behavior similar to the diamond lattice forming a fully two-dimensional diffraction pattern (Figure 15.5a). Increasing the power of the probe beam, we observe the evolution from the described diffraction pattern to the strongly localized discrete solitons (Figure 15.5e).

Corresponding results for the perpendicular orientation are shown in Figure 15.6. Again, the low intensity diffraction pattern reveals the symmetry of the underlying refractive index structure, showing an effectively one-dimensional

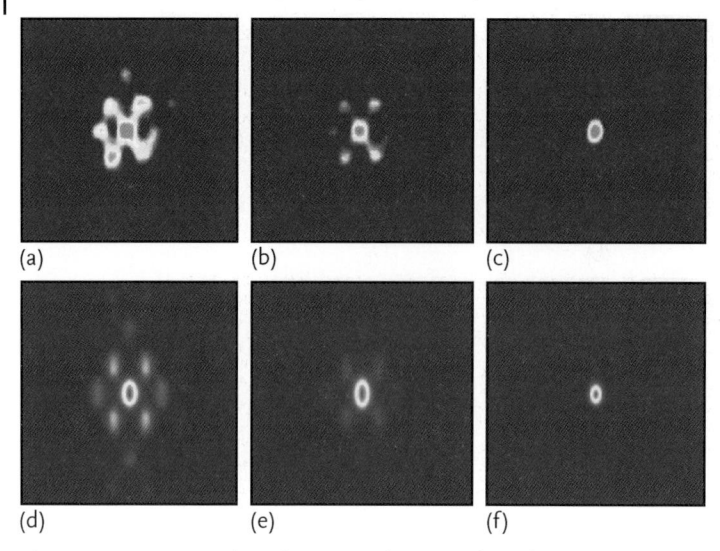

Figure 15.5 Experimental results (top) and numerical simulations (bottom) for the formation of fundamental discrete solitons in the parallel triangular lattice. (a), (d) Diffraction of the probe beam at low power. (b), (e) Localized state at moderate power. (c), (f) Discrete soliton.

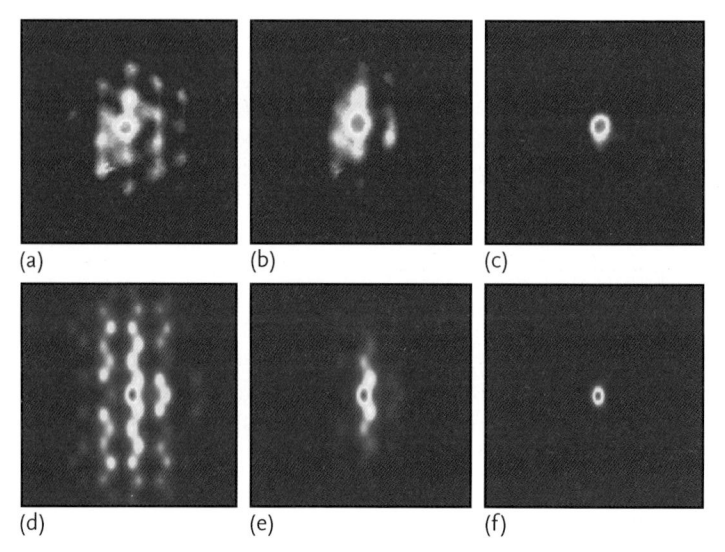

Figure 15.6 Experimental results (top) and numerical simulations (bottom) for the formation of fundamental discrete solitons in the perpendicular triangular lattice. (a), (d) Diffraction of the probe beam at low power. (b), (e) Localized state at moderate power. (c), (f) Discrete soliton.

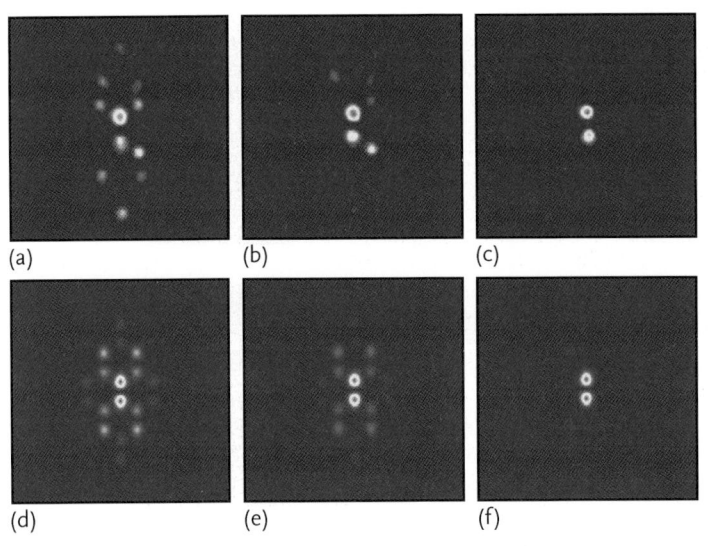

Figure 15.7 Experimental results (a–c) and numerical simulations (d–f) for the formation of dipole-mode gap solitons in the parallel triangular lattice. (a), (d) Diffraction of the probe beam at low power. (b), (e) Localized state at moderate power. (c), (f) Dipole-mode gap soliton.

diffraction pattern (Figure 15.6a). Also, with increased power the beam finally evolves into a discrete soliton as shown in Figure 15.6c and 15.6f.

In the following, we will expand the analysis to the propagation of dipole modes in the parallel triangular structure and show that it is possible to obtain a stable dipole structure, or a molecule of light, in this highly anisotropic lattice type.

In addition to the previously discussed fundamental discrete solitons in triangular photonic lattices, our numerical simulations reveal that the lattice in parallel orientation with its dipole-like islands of high refractive index gives rise to the formation of dipole-mode gap solitons [31].

To compare these numerical simulations to the experiment, we generate a dipole-like input beam and observe the output at the back face of the crystal for different probe beam powers. The experimental as well as the numerical results are summarized in Figure 15.7. At low probe beam powers the diffraction pattern consists of a central dipole surrounded by four side lobes, each forming a dipole itself (Figure 15.7a and 15.7b). With increased power the side lobes vanish and a stable dipole-mode gap soliton evolves (Figure 15.7c and 15.7f).

15.4.2
Multiperiodic Lattices

In addition to the previously discussed lattice geometries, more complex photonic structures such as modulated waveguide arrays [15], lattice interfaces [16], or double periodic one-dimensional photonic lattices [17] offer many exciting possibilities to engineer the diffraction properties of light. These structures open additional mini-gaps in the transmission spectrum and thereby facilitate the existence of new soliton families in nonlinear media [32].

In this spirit, we have implemented a novel approach for all optical induction of multiperiodic superlattices by superposition of several single periodic lattices [33]. Unfortunately, the first order approach to use the spatial light modulator for direct modulation of the lattice wave with a corresponding pattern is not successful for the induction of multiperiodic lattices. The reason is that lattice waves of different periodicities acquire different phase shifts during propagation, and their coherent superposition therefore leads to an intensity modulation in the direction of propagation due to interference. Consequently, a method of incoherent superposition is required.

A simple overlay of multiple mutually incoherent interference patterns is feasible for this purpose, but lacks the flexibility benefits offered by the usage of a spatial light modulator. One solution to this challenge will be demonstrated below. It is closely related to the multiplexing technology known from holographic data storage, where several different approaches like wavelength, angular, and phase code multiplexing [34–36] are utilized in this context. These methods allow a superposition of different refractive index patterns inside the volume of a photorefractive crystal. Therefore, they can serve as a basis for the induction of multiperiodic photonic superlattices.

In contrast to the commonly used sequential recording scheme, we decided to use the method of incremental multiplexing [37]. Thus, we are able to induce the superimposed lattices with different modulation depths by simply adjusting their relative illumination times. In fact, this enhances the flexibility of the presented induction process even more.

We first implement our method for the induction of one-dimensional stripe patterns (Figure 15.8). The induced refractive index structures are subsequently analyzed in Fourier space using the well established Brillouin zone spectroscopy technique [38]. This analysis is performed by imaging the partially spatially incoherent output of a rotating diffuser onto the front face of the crystal, and monitoring the output spectrum using a Fourier transform lens and a CCD camera.

If only one lattice period is used during the induction process, the Brillouin zone pictures show two dark lines marking the borders of the first Brillouin zone of the corresponding lattice. This is demonstrated in Figure 15.8d and 15.8e for two different lattice periods. The corresponding real space images of the lattice wave are shown in Figure 15.8a and 15.8b. Figure 15.8c and 15.8f depict the induction of a one-dimensional superlattice as a superposition of the two former lattices. The arrows in Figure 15.8c indicate the alternating sequence of the two single periodic lattice waves. Both waves are sent into the crystal in an alternating scheme for about 2 s, respectively. In this case, Figure 15.8f clearly shows four dark lines corresponding to the Brillouin zone structure of the double periodic one-dimensional superlattice induced by the superposition of the two single periodic structures via incremental recording.

15.4 Complex Optically-Induced Lattices in Two Transverse Dimensions | **439**

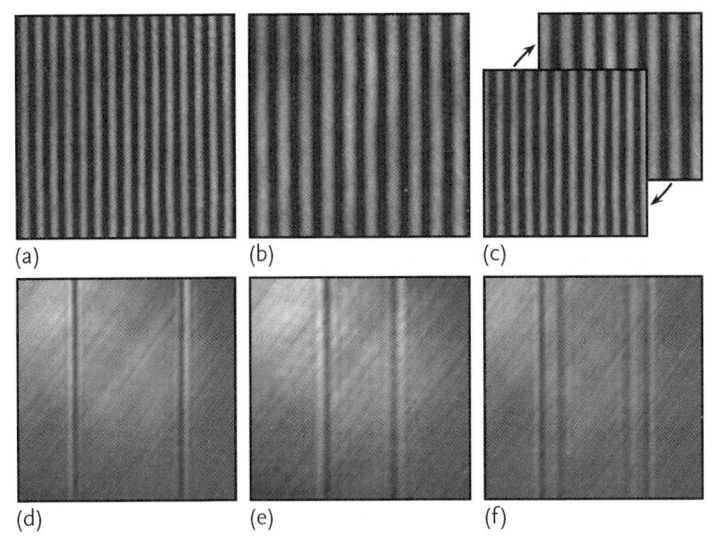

Figure 15.8 Lattice wave (a–c) and Brillouin zone spectroscopy (d–f) of a one-dimensional multiperiodic lattice. (a), (d) Stripe pattern with lattice period of 15 μm. (b), (e) Stripe pattern with lattice period of 24 μm. (c), (f) Incremental multiplexing of stripe patterns with lattice periods of 15 μm and 24 μm.

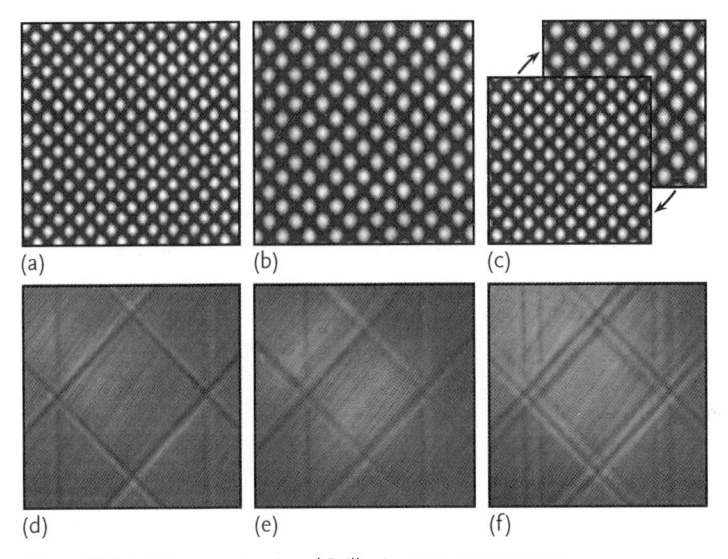

Figure 15.9 Lattice wave (a–c) and Brillouin zone spectroscopy (d–f) of a two-dimensional multiperiodic lattice. (a), (d) Diamond pattern with lattice period of 17 μm. (b), (e) Diamond pattern with lattice period of 28 μm. (c), (f) Incremental multiplexing of diamond patterns with lattice periods of 17 μm and 28 μm.

In addition to the optical induction of one-dimensional superlattices, the method can easily be extended to achieve multiperiodic structures in two transverse dimensions as well [33]. Figure 15.9 shows the corresponding results for the superposition of diamond lattices with different lattice constants. As before, Figure 15.9d and 15.9e show the first Brillouin zone of the single periodic lattices, and the corresponding Brillouin zone picture of the multiplexed superstructure is depicted in Figure 15.9f. Again, the superposition of the two single periodic lattices is clearly visible by the two dark squares indicating the first Brillouin zones of the two single periodic structures.

The only fundamental restriction on the successively multiplexed structures is their diffraction-free propagation through the medium. Therefore, the method of holographic multiplexing may also be extended to induce more sophisticated refractive index structures, for example asymmetric lattices that are a superposition of many single periodic lattices of different symmetries. Due to its simplicity and high flexibility, the presented method can serve as a novel tool for the investigation of several fascinating effects of nonlinear wave propagation in multiperiodic photonic lattices.

15.5
Vortex Clusters

Some of the most spectacular observations of nonlinear dynamics of coherent light waves in periodic potentials relate to the properties of vortices and vortex flows in optical lattices [8, 9, 39, 40]. Dramatic changes of light diffraction or tunneling of matter waves in media with periodically modulated parameters indicate novel directions for manipulating waves with complex phase structures. Self-trapped phase singularities [18] in the form of isolated discrete vortices have been predicted theoretically [41–44] and generated experimentally in square photonic lattices [8, 9, 45].

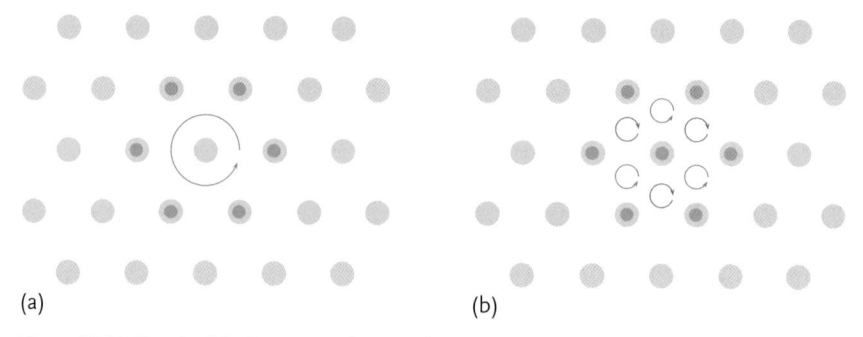

Figure 15.10 Sketch of the investigated vortex clusters in hexagonal photonic lattices. (a) Ring-shaped cluster containing six lobes and one phase singularity (topological charge $m = 1, 2$). (b) Multivortex cluster containing seven lobes and six phase singularities (total topological charge $m = 0$).

Here, we focus on the propagation of vortex clusters, that is, clusters of weakly coupled fundamental solitons with a superimposed vortex phase, and investigate the propagation dynamics in hexagonal photonic lattices. In particular, we study the stability properties of ring-shaped 6-lobe clusters (Figure 15.10a) as well as the more complex 7-lobe multivortex clusters shown in Figure 15.10b.

15.5.1
Necessary Stability Criterion

The stability of a vortex cluster is determined by the intensity flows between the individual lobes. In order to obtain a stable intensity distribution as well as phase profile, all the intensity flows within the cluster must be balanced. Provided that the lattice is deep enough, two adjacent lobes can be treated approximately as bell-shaped intensity distributions having homogeneous phases ϕ_1 and ϕ_2, and the coupling can be assumed to be only via evanescent fields. To derive a general criterion for the stability of vortex clusters, we choose, without loss of generality, a one-dimensional notation with the lobes located at $x = \pm \delta x$. Thus, the total optical amplitude can be written as

$$A(x) = A_1 e^{i\phi_1} e^{-\zeta(x+\delta x)} + A_2 e^{i\phi_2} e^{\zeta(x-\delta x)} \quad \text{for } |x| \ll \delta x. \tag{15.6}$$

The real constants A_1 and A_2 are proportional to the maximum amplitudes of the lobes. The intensity flow between the lobes

$$J_x = 2 \operatorname{Im}(A \partial_x A^*) = 4 \zeta A_1 A_2 e^{-2\delta x} \sin(\phi_1 - \phi_2) \tag{15.7}$$

is proportional to the sine of their phase differences. The same result is obtained in two transverse dimensions after integrating along the y axis.

In general, the intensity flows within a cluster are balanced if the intensities of the lobes do not change during propagation. Therefore, the sum of all intensity flows must vanish for each lobe, yielding [46]

$$\sum_{i=1}^{N} c_{ij} \sin(\phi_i - \phi_j) \stackrel{!}{=} 0 \quad \forall\, 1 \leq j \leq N. \tag{15.8}$$

All constants have been collected in the coupling coefficients c_{ij}, where i and j denote the respective lobes.

15.5.2
Compensation of Anisotropy in Hexagonal Photonic Lattices

Studying the propagation of vortex clusters in optically induced hexagonal photonic lattices, it is crucial to consider the photorefractive anisotropy. We consider the lattice field in the general form of three interfering plane waves

$$A_{\text{latt}} = \exp(2 i k_x x/3) + \exp(-i k_x x/3 + i k_y y) + \exp(-i k_x x/3 - i k_y y), \tag{15.9}$$

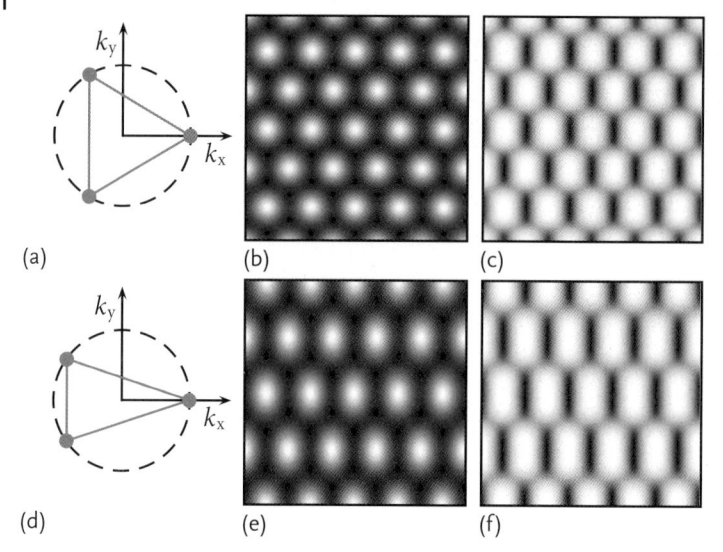

Figure 15.11 Structure analysis of symmetric (a–c) and stretched hexagonal lattice (d–f). (a), (d) Sketch of the Fourier image. (b), (e) Numerically calculated lattice wave intensity. (c), (f) Calculated refractive index change.

and start with an intensity profile of exact hexagonal symmetry when the spatial frequencies k_x and k_y obey the simple relation $k_x = \sqrt{3}k_y$. In Fourier space, the three interfering beams are represented by spots, forming an equilateral triangle as indicated in Figure 15.11a.

Due to the anisotropic nature of the nonlinear response of the crystal, the induced refractive index structure (Figure 15.11c) does not preserve the symmetry of the lattice wave (Figure 15.11b). In particular, the modulation of the refractive index is much stronger along the c-axis than along the diagonals, making the resulting optical coupling between refractive index maxima very asymmetric. As a result, the flow condition (15.8) cannot be fulfilled.

To counteract this effect of the anisotropy and to enable stable vortex clusters, the lattice can be stretched along the vertical direction such that the optical coupling between lattice maxima is closer to that of the original hexagonal symmetry of the lattice [10]. This is demonstrated in Figure 15.11d–f.

15.5.3
Ring-Shaped Vortex Clusters

Perhaps the most counterintuitive result to emerge from the consideration of hexagonal lattices is that in the simplest six site configuration (Figure 15.10) double charge vortices may become stable, while single charge vortices are always unstable [47, 48]. This is in agreement with the stability properties of vortex solitons in modulated Bessel lattices [49]. This is particularly surprising as higher

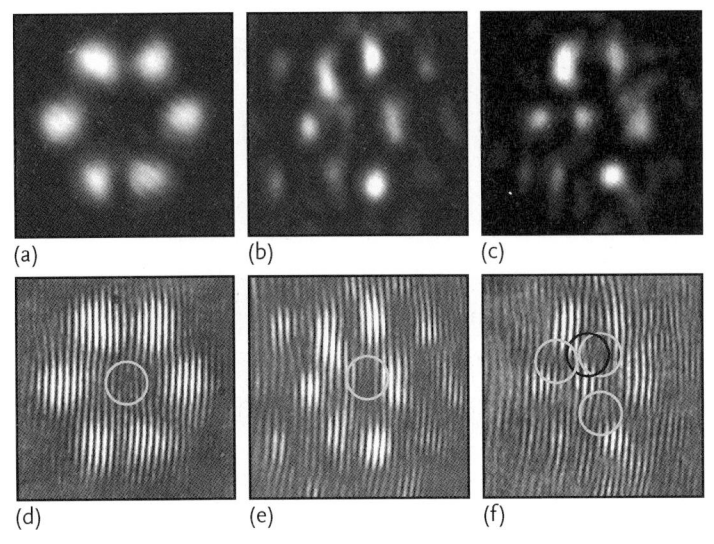

Figure 15.12 Experimentally obtained intensity (a–c) and phase profile (d–f) for the single charge vortex. Circles indicate positions of vortices with topological charge $m = +1$ (light) or $m = -1$ (dark). (a), (b) Intensity and phase distribution of an input single-charge vortex beam. (c), (d) Beam profile and phase at the output for low input power. (e), (f) Output for high input power.

charge vortices are typically unstable in homogeneous nonlinear systems [18]. Here, we demonstrate experimentally the stability of a double charge vortex in contrast to the corresponding single charge vortex state which is unstable under the same conditions. As described in Section 15.3.2, the lattice is formed with the help of one spatial light modulator. We use the second phase modulator to impose either a 2π or 4π phase winding on an input modulated six site beam for the generation of single and double charge vortices, respectively. The characteristics of the beams are otherwise identical, and thus any differences in the dynamics are due solely to the different input phases. We selectively vary the beam intensity to effectively move from the linear to the nonlinear regime.

The single charge vortex input is shown in Figure 15.12a. Its intensity distribution has a form of a necklace with six intensity peaks whose positions correspond to the lattice sites or index maxima. At low input power, the beam undergoes discrete diffraction and a complete loss of the initial six site input state (Figure 15.12b). At high power, the initial six site intensity profile changes significantly after propagation (Figure 15.12c), showing strong intensity modulations and even filling in the central lattice site. Furthermore, in the phase profile multiple vortices are seen to appear, further indicating a breakdown of the single charge state (circles in Figure 15.12f).

In the case of the double charge vortex (Figure 15.13), we again observe a discrete diffraction with low input power (see Figure 15.13b), however the result changes

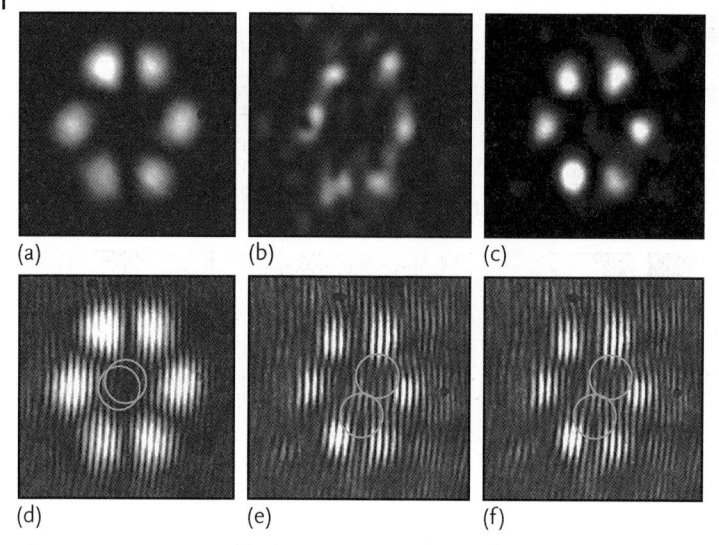

Figure 15.13 Experimentally obtained intensity (a–c) and phase profile (d–f) for the double-charge vortex. Circles indicate positions of vortices with topological charge $m = +1$ (light) or $m = -1$ (dark). (a), (b) Intensity and phase distribution of the double-charge input. (c), (d) Beam profile and phase at the output for low input power. (e), (f) Discrete double-charge vortex soliton.

dramatically when the power is increased (see Figure 15.13c). We now observe that the six site input structure is preserved in the nonlinear propagation. Interestingly, while the overall phase winding is still 4π, it can be clearly seen that the initial double charge singularity has split into two single charge vortices. This splitting of the higher order singularity has been shown to be due to an inherent topological instability in the higher phase winding. This topological breakdown in the linear or low power part of the field further indicates that the stability of the 4π phase winding across the six sites is due to the interplay of the nonlinearity and the local phase of the high power sites suppressing the development of a dynamical instability. However, this stability is critically dependent on the symmetry of the lattice, with a decrease in the lattice stretching and, thus, a corresponding decrease in the symmetry of the underlying modulated refractive index, leading to a dynamical instability in the double charge state as well. The phase interferogram in Figure 15.13f also indicates an additional pair of single charge vortices of the opposite charge inside the vortex structure (not marked by circles). However, this additional pair does not affect the stability of the 4π phase winding, and it can be fully attributed to inevitable experimental noise in this region of low light intensity. To corroborate our experimental results, we have performed numerical simulations using the full anisotropic model as described in Section 15.3.1.

First we consider the case of a six site initial state with a single charge vortex phase of the form shown in Figure 15.14a, with either low or high power propagat-

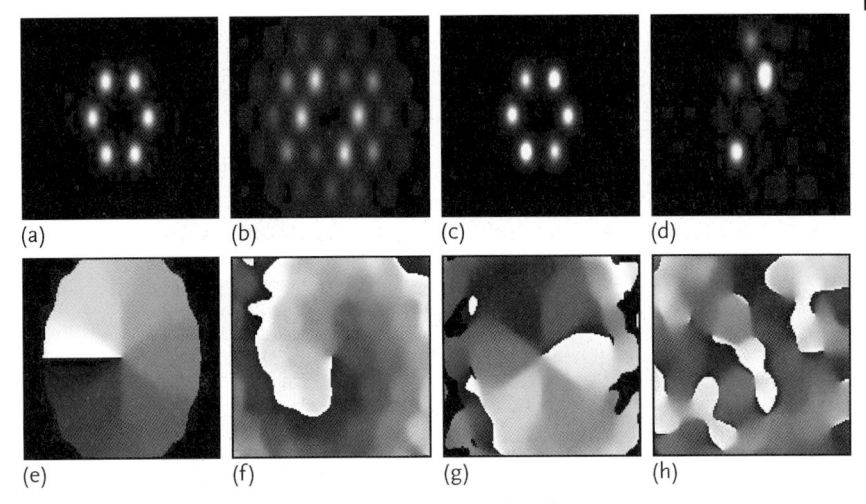

Figure 15.14 Numerical simulation of the intensity (a–d) and phase profile (e–h) for a single charge vortex. (a), (e) Input. (b), (f) Beam profile at $z = 20$ mm for low input power. (c), (g) Beam profile at $z = 20$ mm for high input power. (d), (h) High power output at $z = 280$ mm.

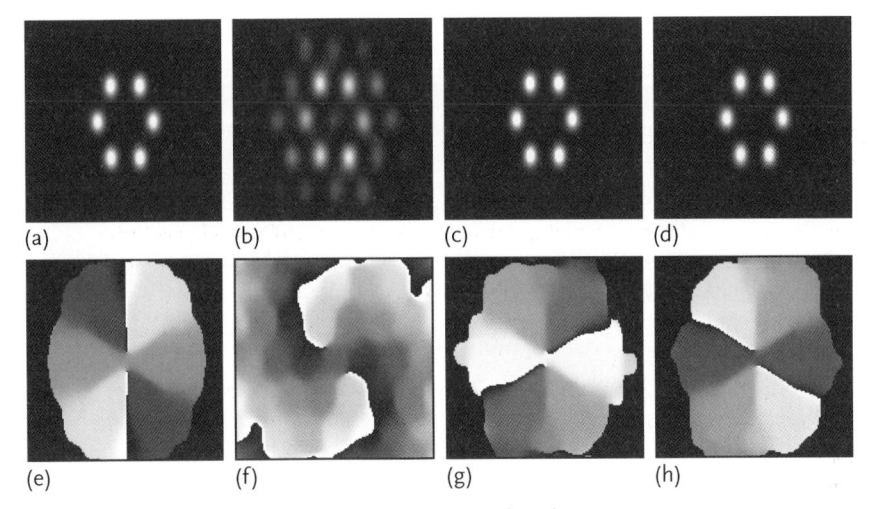

Figure 15.15 Numerical simulation of the intensity (a–d) and phase profile (e–h) for a double charge vortex. (a), (e) Input. (b), (f) Beam profile at $z = 20$ mm for low input power. (c), (g) Double charge discrete vortex soliton at $z = 20$ mm. (d), (h) Double-charge discrete vortex soliton at $z = 280$ mm.

ing a distance of 20 mm in the lattice. For the low input power case of Figure 15.14b we see that, as in the experiment, the vortex beam undergoes diffraction. If a high input power is instead considered as in Figure 15.14c, the vortex maintains much of its form. Some intensity fluctuations are evident, and more importantly the vortex phase has deteriorated, showing breakdown of the initial single charge vortex circulation. It must be noted that the breakup is clearly less than that observed in the experiment, and this discrepancy is attributed to the higher anisotropy of the experimental lattice leading to a larger instability growth rate. In our numerical simulations, the strong instability becomes evident for longer propagation distances as shown in Figure 15.14d for $z = 280$ mm. In Figure 15.15, we consider the same input beam intensities but change the phase to that of a double charge vortex. The low power output (Figure 15.15b and 15.15f) appears similar to the single charge case, exhibiting diffraction and break up of the vortex. In contrast, the high power output in Figure 15.15c and 15.15d appears unchanged in the intensity profile, with a well pronounced corresponding double charge vortex phase (Figure 15.15g and 15.15h). Similar to the experimental results, the separation of the double charge phase singularity into two single charge singularities is observed. However, the phase circulation around a contour tracing the six high intensity sites is well defined and equals 4π.

15.5.4
Multivortex Clusters

Multivortex coherent states appear naturally in systems with repulsive interparticle interactions where they can be confined by external potentials. For attractive interaction, multivortex structures are known to be unstable, and they have only been observed as infinite periodic waves [13]. However, it was predicted theoretically that photonic lattices with threefold symmetry can support stable multivortex spatially localized states [50], in sharp contrast to earlier studied square lattices [46]. Here, we present the experimental observation of topologically stable spatially localized multivortex solitons generated in optically induced hexagonal photonic lattices [10].

To generate a multivortex probe beam, we focus three extraordinarily polarized beams onto the front face of the crystal in such a way that they have the same symmetry as the induced lattice. In the real space depicted in Figure 15.16a, this arrangement results in an input probe beam having the form of seven distinctive spots, forming a hexagonal pattern with the same periods as the lattice and containing six vortices.

At low input powers, the diffraction of the probe beam leads to a broad output distribution as shown in Figure 15.16b. However, at high powers the structure becomes localized and the output intensity distribution features seven well pronounced spots closely resembling the input. To show the topological structure of multivortex solitons, we record the phase interferograms of the reference beam and the probe beam at low and high intensities, respectively. It is clearly visible that at low power and thus in the linear regime, the initial phase profile becomes strongly distorted. While the six initial vortices can still be found in the output field, their positions are changed. In contrast, for high input power of the probe beam (Figure 15.16c and 15.16f) in the nonlinear regime, the beam intensity not only becomes self-trapped but the phase profile (Figure 15.16f) also retains exactly the same hexagonal vortex pattern of the input beam.

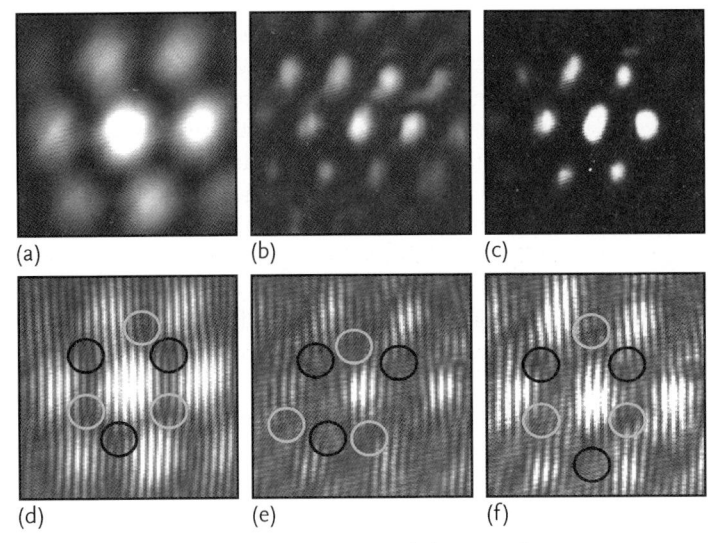

Figure 15.16 Intensity distributions (a–c) and phase interferograms (d–f) of a multivortex soliton in a hexagonal lattice. Positions of the vortices are indicated by dark circles for the topological charge $m = -1$ and light ones for $m = +1$. (a), (d) Probe beam input. (b), (e) Output intensity at low power. (c), (f) Multivortex soliton.

15.6
Summary and Outlook

In this chapter we have investigated the nonlinear propagation dynamics of complex light fields in periodic photonic structures, and demonstrated that the optical induction in photorefractive materials provides a powerful tool for the realization of complex photonic structures. These structures can offer a variety of possibilities for investigating linear and nonlinear dynamics of wave propagation in periodic potentials. By transferring the multiplexing techniques known from holographic data storage to the field of optically induced photonic structures, this variety increases even further. As examples of such complex wave dynamics, we have investigated discrete and dipole mode gap solitons in triangular lattices as well as vortex clusters carrying one or more phase singularities in hexagonal structures.

Two-dimensional reconfigurable photorefractive lattices have already become quite well understood, and show fascinating effects of complex nonlinear physics as we demonstrated above. However, three dimensional (3d) reconfigurable structures that are able to show advanced nonlinear features as, for example, slow and stopped light have not been realized until now. Theoretically, it has been shown that by the interference of multiple beams as well as by the multiple exposure of two beam interference, all fourteen three-dimensional Bravais lattices could be generated [51, 52].

The generation of well defined reconfigurable three-dimensional nonlinear photonic lattices in photorefractive crystals therefore seems to be a promising technique. As the photonic lattices in higher spatial dimensions drastically influence the intersite coupling and wave scattering in anisotropic media, the optical induction approach is well suited to the fabrication of large area reconfigurable three-dimensional structures with flexible parameters in such media.

In view of actual nonlinear photonic device integration in the technological quest for the realization of all optically active devices, we envisage the investigation of many exciting nonlinear propagation, trapping, switching, and steering beam dynamics in these highly reconfigurable photonic lattices.

Acknowledgements

Many parts of the work presented in this chapter have been inspired or realized together with excellent collaboration partners. In particular, Anton Desyatnikov and Yuri S. Kivshar, Nonlinear Physics Centre, Research School of Physics and Engineering, Australian National University, Canberra, Australia and Tobias Richter and Friedemann Kaiser, Institut für Angewandte Physik, Technische Universität Darmstadt, Germany should be named in this context. The activities with our collaboration partners have mainly been supported by the German Academic Research Association and a binational dissertation grant from the German Academic Exchange Service (DAAD).

References

1 Yablonovich, E. (1987) Inhibited spontaneous emission in solid-state and electronics. *Phys. Rev. Lett.*, **58**, 2059.

2 Christodoulides, D.N. and Joseph, R.I. (1988) Discrete self-focusing in nonlinear arrays of coupled waveguides. *Opt. Lett.*, **13**, 794–796.

3 Su, W.P., Schieffer, J.R., and Heeger, A.J. (1979) Solitons in polyacetylene. *Phys. Rev. Lett.*, **42**, 1698.

4 Trombettoni, A. and Smerzi, A. (2001) Discrete solitons and breathers with dilute Bose–Einstein condensates. *Phys. Rev. Lett.*, **86**, 2353.

5 Efremidis, N.K., Sears, S., Christodoulides, D.N., Fleischer, J.W., and Segev, M. (2002) Discrete solitons in photorefractive optically induced photonic lattices. *Phys. Rev. E*, **66**, 046602.

6 Chen, Z., Martin, H., Eugenieva, E.D., Xu, J., and Bezryadina, A. (2004) Anisotropic enhancement of discrete diffraction and formation of two-dimensional discrete-soliton trains. *Phys. Rev. Lett.*, **92**, 143902.

7 Trompeter, H., Krolikowski, W., Neshev, D.N., Desyatnikov, A.S., Sukhorukov, A.A., Kivshar, Yu.S., Pertsch, T., Peschel, U., and Lederer, F. (2006) Bloch oscillations and Zener tunneling in two-dimensional photonic lattices. *Phys. Rev. Lett.*, **96**, 053903.

8 Neshev, D.N., Alexander, T.J., Ostrovskaya, E.A., Kivshar, Yu.S., Martin, H., Makasyuk, I., and Chen, Z. (2004) Observation of Discrete Vortex Solitons in Optically Induced Photonic Lattices. *Phys. Rev. Lett.*, **92**, 123903.

9 Fleischer, J.W., Bartal, G., Cohen, O., Manela, O., Segev, M., Hudock, J., and Christodoulides, D.N. (2004) Observation

of Vortex-Ring "Discrete" Solitons in 2D Photonic Lattices. *Phys. Rev. Lett.*, **92**, 123904.

10 Terhalle, B., Richter, T., Desyatnikov, A.S., Neshev, D.N., Krolikowski, W., Kaiser, F., Denz, C., and Kivshar, Y.S. (2008) Observation of multivortex solitons in photonic lattices. *Phys. Rev. Lett.*, **101**, 013903.

11 Fleischer, J.W., Segev, M., Efremidis, N.K., and Christodoulides, D.N. (2003) Observation of two-dimensional discrete solitons in optically induced nonlinear photonic lattices. *Nature*, **422**, 147.

12 Desyatnikov, A.S., Neshev, D.N., Kivshar, Y.S., Sagemerten, N., Träger, D., Jägers, J., Denz, C., and Kartashov, Y.V. (2005) Nonlinear photonic lattices in anisotropic nonlocal self-focusing media. *Opt. Lett.*, **30**, 869–871

13 Desyatnikov, A.S., Sagemerten, N., Fischer, R., Terhalle, B., Träger, D., Neshev, D.N., Dreischuh, A., Denz, C., Krolikowski, W., and Kivshar, Yu.S. (2006) Two-dimensional nonlinear self-trapped photonic lattices. *Opt. Express*, **14**, 2851–2863.

14 Rosberg, C.R., Neshev, D.N., Sukhorukov, A.A., Krolikowski, W., and Kivshar, Yu.S. (2007) Observation of nonlinear self-trapping in triangular photonic lattices. *Opt. Lett.*, **32**, 397–399.

15 Rosberg, C.R., Garanovich, I.L., Sukhorukov, A.A., Neshev, D.N., Krolikowski, W., Kivshar, Yu.S. (2006) Demonstration of all-optical beam steering in modulated photonic lattices. *Opt. Lett.*, **31**, 1498–1500.

16 Suntsov, S., Makris, K.G., Christodoulides, D.N., Stegemann, G.I., Hache, A., Morandotti, R., Yang, H., Salamo, G., and Sorel, M. (2006) Observation of discrete surface solitons. *Phys. Rev. Lett.*, **96**, 063901.

17 Smirnov, E., Rüter, C.E., Kip, D., Shandarova, K., and Shandarov, V. (2007) Light propagation in double periodic nonlinear photonic lattices in lithium niobate. *Appl. Phys. B*, **88**, 359.

18 Desyatnikov, A.S., Kivshar, Yu.S., and Torner, L. (2005) *Prog. Optics*, **47**, 291–391 (ed. Wolf, E.), Elsevier, Amsterdam.

19 Russell, P.St.J. (1986) Optics of Floquet–Bloch waves in dielectric gratings. *Appl. Phys. B*, **39**, 231.

20 Russell, P.St.J. (2003) Photonic crystal fibers. *Science*, **299**, 358.

21 Eisenberg, H.S., Silberberg, Y., Morandotti, R., and Aitchison, J.S. (2000) Diffraction Management. *Phys. Rev. Lett.*, **85**, 1863.

22 Somekh, S., Garmire, E., Yariv, A., Garvin, H.L., and Hunsperger, R.G. (1973) Channel optical waveguide directional couplers. *Appl. Phys. Lett.*, **22**, 46.

23 Eisenberg, H.S., Silberberg, Y., Morandotti, R., Boyd, A.R., and Aitchison, J.S. (1998) Discrete spatial optical solitons in waveguide arrays. *Phys. Rev. Lett.*, **81**, 3383.

24 Kivshar, Yu.S. (1993) Self-localization in arrays of defocussing waveguides. *Opt. Lett.*, **18**, 1147–1149.

25 Morandotti, R., Eisenberg, H.S., Silberberg, Y., Sorel, M., and Aitchison, J.S. (2001) Self-focusing and defocussing in waveguide arrays. *Phys. Rev. Lett.*, **86**, 3296.

26 Ablowitz, M.J. and Musslimani, Z.H. (2001) Discrete diffraction managed spatial solitons. *Phys. Rev. Lett.*, **87**, 254102.

27 Cohen, O., Schwartz, T., Fleischer, J.W., Segev, M., and Christodoulides, D.N. (2003) Multiband vector lattice solitons. *Phys. Rev. Lett.*, **91**, 113901.

28 Sukhorukov, A.A. and Kivshar, Yu.S. (2003) Multigap discrete vector solitons. *Phys. Rev. Lett.*, **91**, 113902.

29 Fischer, R., Träger, D., Neshev, D.N., Sukhorukov, A.A., Krolikowski, W., Denz, C. and Kivshar, Yu.S. (2006) Reduced-symmetry two-dimensional solitons in photonic lattices. *Phys. Rev. Lett.*, **96**, 023905.

30 Juul Rasmussen, J. and Rypdal, K. (1986) Blow-up in nonlinear Schroedinger equations – I. A general review. *Phys. Scr.*, **33**, 481.

31 Rose, P., Richter, T., Terhalle, B., Imbrock, J., Kaiser, F., and Denz, C. (2007) Discrete and dipole-mode gap solitons in higher-order nonlinear photonic lattices. *Appl. Phys. B*, **89**, 521.

32 Louis, P.J.Y., Ostrovskaya, E.A., and Kivshar, Yu.S. (2005) Dispersion control for matter waves and gap solitons in optical superlattices. *Phys. Rev. A*, **71**, 02361.

33 Rose, P., Terhalle, B., Imbrock, J., and Denz, C. (2008) Optically-induced photonic superlattices by holographic multiplexing. *J. Phys. D: Appl. Phys.*, **41**, 224004.

34 Rakuljic, G.A., Leyva, V., and Yariv, A. (1992) Optical data storage using orthogonal wavelength multiplexed volume holograms. *Opt. Lett.*, **17**, 1471–1473.

35 Mok, F.H. (1993) Angle-multiplexed storage of 5000 holograms in lithium niobate. *Opt. Lett.*, **18**, 915–917.

36 Denz, C., Pauliat, G., Roosen, G and Tschudi, T. (1991) Volume hologram multiplexing using a deterministic phase encoding method. *Opt. Commun.*, **85**, 171.

37 Taketomi, Y., Ford, J.E., Sasaki, H., Ma, J., Fainman, Y., and Lee, S.H. (1991) Incremental recording for photorefractive hologram multiplexing. *Opt. Lett.*, **16**, 1774–1776.

38 Bartal, G., Manela, O., Cohen, O., Fleischer, J.W., and Segev, M. (2005) Brillouin zone spectroscopy of nonlinear photonic lattices. *Phys. Rev. Lett.*, **94**, 163902.

39 Tung, S., Schweikhard, V., and Cornell, E.A. (2006) Observation of vortex pinning in Bose–Einstein condensates. *Phys. Rev. Lett.*, **97**, 240402.

40 Schweikhard, V., Tung, S., and Cornell, E.A. (2007) Vortex proliferation in the Berezinskii–Kosterlitz–Thouless regime on a two-dimensional lattice of Bose–Einstein condensates. *Phys. Rev. Lett.*, **99**, 030401.

41 Kevrekidis, P.G., Malomed, B.A., and Gaididei, Yu.B. (2002) Solitons in triangular and honeycomb dynamical lattices with the cubic nonlinearity. *Phys. Rev. E*, **66**, 016609.

42 Yang, J. and Musslimani, Z.H. (2003) Fundamental and vortex solitons in a two-dimensional optical lattice. *Opt. Lett.*, **28**, 2094–2096.

43 Baizakov, B.B., Malomed, B.A., and Salerno, M. (2003) Multidimensional solitons in periodic potentials. *Europhys. Lett.*, **63**, 642–648.

44 Yang, J. (2004) Stability of vortex solitons in a photorefractive optical lattice. *New J. Phys.*, **6**, 47.

45 Bartal, G., Manela, O., Cohen, O., Fleischer, J.W., and Segev, M. (2005) Observation of second-band vortex solitons in 2D photonic lattices. *Phys. Rev. Lett.*, **95**, 053904.

46 Alexander, T.J., Sukhorukov, A.A., and Kivshar, Yu.S. (2004) Asymmetric vortex solitons in nonlinear periodic lattices. *Phys. Rev. Lett.*, **93**, 063901.

47 Law, K.J.H., Kevrekidis, P.G., Alexander, T.J., Krolikowski, W., and Kivhsar, Yu.S. (2008) Stable higher-charge discrete vortices in hexagonal optical lattices. *Phys. Rev. A*, **79**, 025801.

48 Terhalle, B., Richter, T., Law, K.J., Göries, D., Rose, P., Alexander, T.J., Kevrekidis, P.G., Desyatnikov, A.S., Krolikowski, W., Kaiser, F., Denz, C., and Kivshar, Y.S. (2009) Observation of double-charge discrete vortex solitons in hexagonal photonic lattices. *Phys. Rev. A*, **79**, 043821.

49 Kartashov, Y.V., Ferrando, A., Egorov, A.A., and Torner, L. (2005) Soliton topology versus discrete symmetry in optical lattices. *Phys. Rev. Lett.*, **95**, 123902.

50 Alexander, T.J., Desyatnikov, A.S., and Kivshar, Yu.S. (2007) Multivortex solitons in triangular photonic lattices. *Opt. Lett.*, **32**, 1293–1295.

51 Cai, L.Z., Yang, X.L., Wang, Y.R. (2002) All fourteen Bravais lattices can be formed by interference of four noncoplanar beams. *Opt. Lett.*, **27**, 900–902.

52 Dwivedi, A., Xavier, J., Joseph, J., Singh, K. (2008) Formation of all fourteen Bravais lattices of three dimensional photonic crystal structures by a dual beam multiple-exposure holographic technique. *Appl. Opt.*, **47**, 1973–1980.

Index

a

accumulation front   330, 334, 335
action   9, 41, 170, 171, 181, 214, 279, 280, 283, 290–292, 294, 297, 299–302, 373, 380, 388, 401
adenosine diphosphate (ADP)   33–40, 42–44
adenosine triphosphate (ATP)   3, 32–44, 64, 68–70
ADP, *see* adenosine diphosphate
AFM, *see* atomic force microscopy
amplitude equation   221, 250, 258–261, 263
arrays of coupled Duffing resonators   250, 251, 253, 255, 257
atomic force microscopy (AFM)   102, 133, 267–278, 280–283
ATP, *see* adenosine triphosphate

b

ballistic systems   287, 303
basin of attraction   211
BCL amplitude equation   260, 261
bifurcation   51, 56, 145, 146, 151, 197, 209, 217, 230, 232, 234, 239, 246, 247, 258, 262, 327, 330, 332, 334, 335, 338, 340, 341, 343, 344, 346, 349, 350, 357–359, 361, 362
– amplifier 217
– points 232, 234, 338
– subcritical 242, 247, 258, 261, 262
– supercritical 242, 247, 258, 261, 262, 343
bistability region   234
breathing   25, 158, 327, 328, 341, 344–347, 351, 358, 361
– current filament 344, 351
buckling instability   226
butterfly plot   213, 214

c

capillary waves   121–128, 130–138, 140
Casimir effect   166–168, 193, 198

Casimir energy   168–171, 173, 174, 177, 181–184, 186, 187, 190, 194
Casimir force   168–170, 174–178, 184, 185, 192–198
CDT, *see* coherent destruction of tunneling
CFT, *see* crooks fluctuation theorem
chaotic front patterns   330, 333
chaotic oscillations   268, 270, 275, 330, 331, 350
charge detectors   221
chemical clocks   4, 53, 54, 56, 57, 59, 60, 66, 69, 70
coherence measures   338
coherence resonance   333, 337, 338, 341, 361
coherent current suppression   320, 322
coherent destruction of tunneling (CDT)   320, 400–402
coherent transport   303
collective response   222, 255, 408
conductance   287–291, 293, 303, 325, 326, 357
control circuit   332, 333
control domain   332, 333, 346–349
correlation time   8, 54, 99, 337, 339, 340, 351–356, 359
Coulomb blockade   307, 308, 391
counter propagating waves   259
coupled arrays   222
coupled Duffing resonators   221, 250, 251, 253, 255, 257
coupled resonators   222, 250, 251, 258, 260
critical drive amplitude   232, 249
Crooks fluctuation theorem (CFT)   31, 32, 68, 86–91, 96, 105
current filament   327

d

damped Duffing resonator   227–229, 231, 233–237, 239, 241, 243, 245, 247, 249

Nonlinear Dynamics of Nanosystems. Edited by Günter Radons, Benno Rumpf, and Heinz Georg Schuster
Copyright © 2010 WILEY-VCH Verlag GmbH & Co. KGaA, Weinheim
ISBN: 978-3-527-40791-0

DBRT, *see* double-barrier resonant tunneling diode
deformed surfaces   130, 169, 176
dewetting   132–140
diagonal control   332, 346–349
Dirichlet boundary condition   170
dissipation theorem   16, 76, 84–86, 103
double walled carbon nanotube (DWCNT)   21–23, 26, 28–30, 68
double-barrier resonant tunneling diode (DBRT)   328, 341–343, 345, 347, 350, 361, 362
doubly-clamped beam   205–207, 228, 237, 238
driven systems   31, 243, 259, 312, 314, 320, 333
Duffing equation   203, 226–228, 237, 238
Duffing parameter   225, 227, 230, 236, 237, 258
Duffing resonator   204, 205, 208, 209, 221, 223, 227–229, 231, 233–237, 239, 241–243, 245–247, 249
DWCNT, *see* double walled carbon nanotube
dynamic mode AFM   267, 268, 274–277, 283

e

eigenmodes   225–227, 252, 346, 356, 357, 359, 362
eigenvalues   7, 23, 168, 293, 315, 341, 354–359, 361, 362, 429
elastic restoring force   224, 251
electric current damping   251
electromigration   143–145, 147–152, 154, 155, 157, 158, 162
electromotive actuation   238
electron pumping   319, 322
electron transport   303, 312, 321, 322, 328, 369, 370, 372, 374, 376–378, 380, 382–384, 386, 388, 390, 392, 394, 396, 398–400, 402, 403
electronic circuit   69, 328
electrostatic force   223, 224
enhanced response   239, 241
enhancement of the response   239
EOM, *see* equations of motion
equations of motion (EOM)   76, 77, 79, 82, 85, 86, 90, 92, 101, 153, 155, 156, 162, 208, 210, 230, 250–257, 259, 261–263, 314, 315, 317, 371, 378, 382, 399
ESFT, *see* Evans-Searles transient fluctuation theorem
ETDAS, *see* extended time-delay autosynchronization

Euler instability   226
Euler-Bernoulli equation   225
Evans-Searles transient fluctuation theorem (ESFT)   79–82, 84, 85, 88, 95–99, 101, 105
excitability   337
extended time-delay autosynchronization (ETDAS)   327, 330, 331, 344
external electrostatic force   223
external potentials   9, 223, 237, 446

f

Fano factor   311, 313, 314, 319, 321, 322, 385, 395–397
Faraday waves   222, 259, 260
FEM, *see* field electron microscopy
field domain   327, 328, 330, 335, 338
field electron microscopy (FEM)   56, 57
field ion microscopy (FIM)   57–59
filament   32, 327, 357
FIM, *see* field ion microscopy
Floquet exponents   346, 348
Floquet theory   312, 322, 400
fluctuation theorem (FT)   13–16, 20, 31, 32, 44, 49, 51–54, 67–70, 75, 78, 79, 81, 86, 94–97, 99, 101, 103–106, 144, 145, 150
frequency pulling   258
frequency tuning   216
fronts   59, 152, 153, 159, 327–337, 339, 361, 362, 378, 432, 438, 446
FT, *see* fluctuation theorem

g

Gallavotti-Cohen fluctuation theorem (GCFT)   99–101
Gaussian white noise   27, 29, 333, 350
GCFT, *see* Gallavotti-Cohen fluctuation theorem
global bifurcation   341
global control   332, 346–348

h

Hamiltonian system   5, 68, 281, 282
Hodgkin-Huxley equations   326
homoclinic bifurcation   338, 341
Hopf bifurcation   56, 151, 332, 334, 335, 341, 343, 350, 357–359, 361, 362
hysteresis   204, 205, 207, 215, 216, 233, 242, 246, 247, 256, 258, 262, 270

i

IFT, *see* integrated fluctuation theorem
increased damping   242
instability threshold   161, 244, 246

instability tongue 237, 238, 242, 246, 247, 249
integrated fluctuation theorem (IFT) 94
interspike interval 340, 341

j

Jarzynski equality (JE) 86–89, 91, 92, 95, 96, 100–103, 411
JE, see Jarzynski equality
Josephson junctions 113, 116, 117, 211

k

Kirchhoff's equation 326
Kirchhoff's laws 326, 343

l

large arrays 250, 257–259, 261
lateral forces 176–180, 194
Lifshitz formula 188, 189
linear instability 145, 241, 242, 245
linear stability 145, 152, 153, 158, 232, 341, 362
Liouville equation 5, 66, 78, 84, 92
local control 332, 346–348
localized surface plasmons (LSP) 407–409, 411–418, 420, 421, 424
– resonance frequencies 414
Lorentzian response 231
low-pass filter 331–333, 338
LSP, see localized surface plasmons

m

magnitude of the response 231–233, 239, 240
marginally stable 243
maximum entropy production (MEP) approach 103, 104
mean interspike interval 340, 341
MEMS, see microelectromechanical systems
mesoscopic physics of phonons 221
microelectromechanical systems (MEMS) 203, 221–225, 233, 236, 237, 259
micromechanical resonators 221, 222, 224, 226, 228, 230, 232, 234, 236, 238, 240, 242, 244, 246, 248, 250, 252, 254, 256, 258, 260, 262
minimum work principle (MWP) 100
molecular dynamics simulations 26, 27, 29, 68, 121
multiperiodic lattices 437–439multiple-time delay autosynchronization, see extended time-delay autosynchronization

multiwalled carbon nanotubes (MWCNT) 9, 21, 29, 30
MWCNT, see multiwalled carbon nanotubes
MWP, see minimum work principle

n

nanoelectromechanical systems (NEMS) 165, 203–210, 212, 214, 216, 218, 221–225, 233, 236, 237, 251, 259
nanofluidics 121
nanomechanical Duffing resonator 227
nanomechanical resonator arrays 222
nanomechanical resonators 203, 221, 222, 232, 234
nanoscopic films 132, 133, 135, 137, 139
nanostructures 128, 162, 287, 288, 304, 325, 327, 328, 333, 341, 342, 349, 361, 407, 408, 410, 412, 414, 416, 418–420, 422–424
NDC, see negative differential conductivity
negative differential conductance 325, 326
negative differential conductivity (NDC) 326–328
NEMS, see nanoelectromechanical systems
Neumann boundary condition 170
noise reduction 222
noise squeezing 237, 239
noise-induced oscillations 328, 333, 335, 337–340, 351, 352, 354, 355, 359, 362
nondiagonal contribution 292–294, 297–299
nonequilibrium nanosystems 1–22, 24, 26, 28, 30, 32, 34, 36, 38, 40, 42, 44, 46, 48, 50, 52, 54, 56, 58, 60, 62, 64–71
nonequilibrium partition identity (NPI) 96, 101, 103
nonequilibrium steady state 2, 11, 13, 14, 21, 34, 86, 97
nonequilibrium work relations 76, 86, 87, 89, 91, 93–95, 97, 99, 103
nonlinear
– charge transport 325
– damping 223, 227, 230–233, 244–247, 249, 251, 259–261
– damping term 223
– external potential 223
– saturation 243, 245
nonlinearities due to geometry 224
NPI, see nonequilibrium partition identity

o

off-diagonal contributions 289, 303

p

pairwise summation (PWS) approximation 172–174, 177–180

parametric amplification 203, 237, 238
parametric drive 238, 239, 241, 243, 244, 247, 251, 257, 260
parametric driving 248, 258
parametric excitation 236–239, 241, 243, 245, 247, 249–255, 257, 258, 260
– of arrays 250, 251, 253, 255, 257
parametric instability 241, 242
parametrically-driven Duffing resonator 242, 249
pattern formation 152, 325, 326, 328, 330, 332, 334, 336, 338, 340, 342, 344, 346, 348, 350, 352, 354, 356, 358, 360, 362, 428
2PPE, see two-photon photoemission
PEEM, see photoemission electron microscope
period-doubling 197, 332, 344
perturbed hamiltonians 279, 281, 282
PFA, see proximity force approximation
photoemission electron microscope (PEEM) 410, 412–414, 417—421, 423, 424
photonic crystals 427, 428
photorefractive nonlinearity 431, 432
piezoelectric NEMS structures 237
plasmon decay 407, 417
power spectral density 335, 355, 359, 360
proximity force approximation (PFA) 168, 176, 178–180, 182–184, 186, 187
pump frequency 234, 237–241, 243, 245, 249, 259, 261
PWS, see pairwise summation approximation

q
QME, see quantum master equation
quality factors 203, 222, 228, 240, 241, 249, 267, 270, 272
quantum dots 290, 307–309, 318–322, 369, 370, 372, 377, 380, 381, 385, 387, 389–391, 393, 395, 399
quantum master equation (QME) 370, 371, 375, 376, 382, 384, 391, 403
quantum transport 112, 287, 288, 303, 308, 322

r
random matrix theory (RMT) 289, 303
– predictions 303
Rashba spin-orbit interaction (RSOI) 112, 119
reaction-diffusion model 341
reactive coupling 251
resonance peaks 207, 232, 233, 408
resonant level models 370, 391, 393, 395, 397, 398
resonant tunneling diode 341, 344, 350, 361

resonator displacements 256
response 15, 16, 43, 45, 49, 67, 69, 70, 76, 79, 84–86, 103, 131, 167, 189, 215, 221, 222, 228, 230–236, 239–241, 243–250, 252, 255–258, 260, 262, 275, 287, 288, 307, 308, 310, 312, 314, 316, 318–320, 322, 325, 408, 411, 417–420, 433, 442
– function 231, 232
– intensity 245–247, 249, 255–257
– of an array 252
responsivity 233
RMT, see random matrix theory
rocked ratchets 114–117
RSOI, see Rashba spin-orbit interaction

s
saddle-node bifurcation 230, 232, 234, 246, 247, 334, 335, 341, 362
saddle-node infinite period (SNIPER) bifurcation 335
saturation of the response 244, 260
SBN, see strontium barium niobate
scaled Duffing equation 227
scaled equations 235, 238, 259
scaled response functions 230
scaling 101, 155, 156, 160, 183, 184, 235, 238, 244, 247–249, 251, 260, 335, 337, 338
scanning near-field optical microscopy (SNOM) 421
scanning probe microscopy (SPM) 133, 135, 267
scanning tunneling microscopy (STM) 143, 267, 281
second law inequality (SLI) 95, 96, 104, 106
secular perturbation theory 221, 228, 229, 238, 243, 250, 252
secular terms 229, 230, 236, 238, 245, 248, 249, 253
semiclassical 288–292, 298, 299, 303, 304
– approximation 290
separatrix 209–211, 214, 234
SERS, see surface-enhanced Raman scattering
SHO, see simple harmonic oscillator
shot noise 303, 307, 308, 319, 322, 334, 350, 370, 378, 379, 384, 385, 394, 395, 397
signal amplification 234
signal enhancement 222
simple harmonic oscillator (SHO) 204, 229, 231
single mode oscillations 250, 261
single particle plasmon spectroscopy 417
SLI, see second law inequality
SNIPER, see saddle-node infinite period bifurcation

SNOM, *see* scanning near-field microscopy
solvability condition   229, 243, 259, 260
spatial solitons   428, 430
spatiotemporal chaos   161, 327
spin ratchets   112, 119
spin-orbit interaction   112, 119, 303
SPM, *see* scanning probe microscopy
squeezing of noise   234
squeezing of the noisy displacement   240
stability boundaries   262
steady state   2, 11, 13, 14, 20, 21, 34, 56, 86, 97–99, 101, 104, 222, 254, 261, 276–278, 328, 333, 335, 343, 350, 351, 379, 431
– fluctuation theorem 97
– nonequilibrium 2, 11, 13, 14, 21, 34, 86, 97
– solution 229, 230, 245, 248, 249, 254
step bunches   153–155, 157, 158, 160
stiffening nonlinearity   204, 205, 226, 232
STM, *see* scanning tunneling microscopy
stochastic hydrodynamics   122, 123, 125–127
stochastic process   7, 9, 11, 13, 16, 28, 37, 41, 44, 46, 47, 50, 51, 53, 66, 67, 377, 385
strontium barium niobate (SBN)   428, 432, 433
subcritical bifurcation   242, 247, 258, 261, 262
supercritical bifurcation   242, 247, 258, 261, 262, 343
superlattice   328–330, 332–334, 336, 341, 361, 362, 438
surface steps   143, 144, 146, 148, 150, 152, 154, 156, 158, 160, 162
surface-enhanced Raman scattering (SERS)   407
SWCNT, *see* single walled carbon nanotube
symmetry-breaking terms   227

t

TDAS, *see* time-delay autosynchronization
thermal noise   113, 121, 122, 124, 132, 133, 135–140, 193, 212, 334
thermal switching   234
thermodynamic limit   75, 89
threshold tongue   246

time dependent fields   86, 96
time scale   3, 6–8, 23–25, 37, 135, 138, 155, 161, 162, 195, 198, 243, 334, 396
time-delay autosynchronization (TDAS)   327, 331, 332, 344
time-delayed feedback   268, 271, 273–276, 327, 328, 330, 333, 335, 338, 341, 346, 350, 351, 359–362
– control 268, 271, 273–276, 327, 328, 351, 360, 362
time-dependent ramps   258
tip-sample interaction   269, 270, 272, 274, 276
transient time correlation function (TTCF)   76, 85, 86
transition   7, 11–13, 34, 36, 37, 41, 47, 50, 61, 65, 114, 115, 128, 129, 140, 143, 145, 147, 151, 157–159, 162, 192, 197, 198, 208–214, 216, 218, 234, 288, 297, 319, 330, 334, 335, 380
transmission amplitude   289, 291
traveling high-field domain   330
TTCF, *see* transient time correlation function
tunneling   112–118, 143, 267, 307, 309, 313, 318, 320, 321, 325, 328, 329, 333, 334, 341–345, 347, 349–351, 353, 355, 357, 359, 361, 372, 376, 377, 379–383, 385–390, 395, 396, 400, 401, 403, 428, 440
tunneling ratchets   114, 115
two-photon photoemission (2PPE) 407, 408, 410–412, 414–423

v

van der Waals interaction   24, 29, 68, 183
VCO, *see* voltage controlled oscillator
very small arrays   255
vicinal surfaces   143, 152, 153, 155, 157, 159, 161
voltage controlled oscillator (VCO)   216
vortex clusters   440–443, 445, 447

w

work relation (WR)   76